AF352121

MULTISCALE COUPLING OF SUN-EARTH PROCESSES

MULTISCALE COUPLING OF SUN-EARTH PROCESSES

Editors

A.T.Y. Lui
The Johns Hopkins University
Applied Physics Laboratory
Laurel, MD
USA

Y. Kamide
Solar Terrestrial Env. Laboratory
Nagoya University
Aichi, Japan

G. Consolini
Ist. Fisica dello Spazio Interplanetario
Istituto Nazionale di Astrofisica
Rome, Italy

2005

ELSEVIER

**Amsterdam - Boston - Heidelberg - London - New York - Oxford - Paris
San Diego - San Francisco - Singapore - Sydney - Tokyo**

ELSEVIER B.V.
Radarweg 29
P.O. Box 521, 1000 AM
Amsterdam, The Netherlands

ELSEVIER Inc.
525 B Street, Suite 1900
San Diego, CA 92101-4495
USA

ELSEVIER Ltd.
The Boulevard, Langford Lane
Kidlington, Oxford OX5 1GB
UK

ELSEVIER Ltd.
84 Theobalds Road
London WC1X 8RR
UK

First edition 2005

Library of Congress Cataloging in Publication Data
A catalog record is available from the Library of Congress.

British Library Cataloguing in Publication Data
A catalogue record is available from the British Library.

ISBN: 0-444-51881-9

∞ The paper used in this publication meets the requirements of ANSI/NISO Z39.48-1992 (Permanence of Paper).
Printed in The Netherlands.

PREFACE

The common saying "many roads lead to Rome" applies equally to scientific research in the sense that there are many approaches to conduct scientific investigations. Research on Sun-Earth connection is no exception. In the early days of space research, dynamic processes in each space plasma region within the Sun-Earth system were explored separately with physics-based tools. With continual progress being made in our understanding of this space environment, it increasingly becomes incontrovertibly evident that tight coupling exists among various space plasma regions in the Sun-Earth system and that the dynamic processes in these regions exhibit disturbances over a wide range of scales both in time and space. This remarkable reckoning naturally leads to efforts linking the entire chain of space disturbances "from sun to mud" and results in significant academic and public interest. Emerging from this remarkable perspective is a novel approach based on concepts developed in the modern statistical mechanics to address the physical processes governing the evolution of out-of-equilibrium and complex systems. This approach fits in well with several flagship research programs such as the International Solar-Terrestrial Physics program, the National Space Weather program, the International Living With a Star program, and the Climate and Weather of the Sun-Earth System program.

These exciting developments have prompted a strong desire to hold a topical conference on Sun-Earth connection, with the goal of promoting interactions among scientists practicing the traditional physics-based approach and those utilizing modern statistical techniques. This desire was met by the Conference on Sun-Earth Connection: Multiscale Coupling in Sun-Earth Processes held on February 9-13, 2004 at Kailua-Kona, Hawaii, USA. It was supported by The Johns Hopkins University Applied Physics Laboratory, Solar-Terrestrial Environment Laboratory/Nagoya University, Committee on Space Research, International Association of Geomagnetism and Aeronomy, National Aeronautics and Space Administration, National Science Foundation, and Scientific Committee on Solar-Terrestrial Physics.

Participants came from 14 countries in Asia, North America, and Europe. A total of 79 presentations were made in this conference, covering both traditional and new innovative approaches to treat Sun-Earth connection phenomena and their multiscale coupling based on observations, computer simulations, and theoretical considerations. Presentations given in this meeting showed a diversity of space phenomena exhibiting scale free characteristics, intermittency, and non-Gaussian distributions of probability density function of fluctuations in the physical parameters of the sun, solar wind, magnetosphere, ionosphere, and thermosphere. These characteristics underscore the usefulness in cross-disciplinary exchange on theoretical work, numerical simulations, and data analysis techniques needed to unravel the underlying physical processes. They also lend to a possible unified description and prediction for space disturbances.

This Monograph is a product of this conference. Contributions are assembled into seven chapters: (1) multiscale features in complexity dynamics, (2) space storms, (3) magnetospheric substorms, (4) turbulence and magnetic reconnection, (5) modeling and coupling of space phenomena, (6) techniques for multiscale space plasma problems, and (7) present and future multiscale space missions. Each article in this Monograph has been reviewed by two referees.

Last but not the least, we gratefully acknowledge the invaluable assistance from the following individuals (in alphabetical order) in reviewing submitted manuscripts to this Monograph:
T. Abe, B.-H. Ahn, S.-I. Akasofu, G. Atkinson, D. N. Baker, W. Baumjohann, M. B. Bavassano Cattaneo, R. Bruno, L. Burlaga, V. Carbone, T. Chang, R. Clauer, S. A. Curtis, A. Egeland, M. Engbretson, C. P. Escoubet, J. Freeman, S. Fujita, P. Gary, J. Gjerloev, K.-H. Glassmeier, A. L. C. Gonzalez, M. Henderson, W. Horton, C. Huang, A. Ieda, R. L. Kessel, A. Klimas, K. Kudela, G. Le, K. Liou, W. Liu, G. Lu, W. Lyatsky, L. R. Lyons, W. M. Macek, W. Matthaeus, R. L. McPherron, Y. Miyoshi, T. Nagai, R. Nakamura, L. Rastaetter, A. Reynolds, G. Rostoker, J. Retterer, A. Saito, J. Samson, E. Sanchez, J. D. Scudder, M.A. Shea, M. I. Sitnov, L. Stenflo, S. Tam, T. Tanaka, R. A. Treumann, B. T. Tsurutani, C. Tu, D. Vassiliadis, Z. Voros, R. Walker, J. Wanliss, N. Watkins, M. Wilbur, C.-C. Wu, A. W. Yau, G. Zimbardo, Q. Zong and all the other anonymous referees. All Editors have taken part in the manuscript review as well when the need arose.

A. T. Y. Lui
The Johns Hopkins University Applied Physics Laboratory
11100 Johns Hopkins Rd., Laurel, Maryland 20723-6099, USA

Y. Kamide
Solar-Terrestrial Environment Laboratory, Nagoya University
Honohara 3-13, Toyokawa, Aichi 442-8507, Japan

G. Consolini
Istituto di Fisica dello Spazio Interplanetario, Consiglio Nazionale delle Ricerche
Via del Fosso del Cavaliere, 100, 00133 Roma, Italy.

October, 2004

Contents

Chapter 1: Multiscale Features in Complexity Dynamics

Chapter 2: Space Storms

Chapter 3: Magnetospheric Substorms

Chapter 4: Magnetic Reconnection and Turbulence

Chapter 5: Modeling and Coupling of Space Phenomena

Chapter 6: Techniques for Multiscale Space Plasma Problems

Chapter 7: Present and Future Multiscale Space Missions

CHAPTER ONE

MULTISCALE FEATURES IN COMPLEXITY DYNAMICS

Multiscale Coupling of Sun-Earth Processes
A.T.Y. Lui, Y. Kamide and G. Consolini, editors.

COMPLEXITY IN SPACE PLASMAS

Tom Chang and Sunny W.Y. Tam
*Center for Space Research, Massachusetts Institute of Technology, Cambridge, MA
02139, USA*

Cheng-chin Wu
*Department of Physics and Astronomy, University of California, Los Angeles, CA 90005
USA*

Abstract. The phenomenon of complexity related to intermittent turbulence in space plasmas is described. The ideas are based on the stochastic behavior of coherent structures that arise from plasma resonances. The concept of topological magnetic reconfiguration due to coarse-grained dissipation is also introduced.

Keywords. Complexity, Intermittent Turbulence, Magnetic Reconfiguration.

1. Introduction

Complexity has become a hot topic in nearly every field of science. Space plasmas are no exception. When the word "complexity" is mentioned, a number of questions naturally arise in one's mind. What is complexity? Does one look up the meaning of it from the Webster dictionary, or does it have a specific scientific meaning? What physical phenomena are results of complexity in plasmas? What are the available tools that one might employ to solve problems of complexity related to the physics of space plasmas? It is the purpose of this short discourse to provide some answers to such questions.

To avoid distraction from the flow of presentation of the main ideas, we will provide a narrative description of complexity in space plasmas with the insertion of only a minimum number of references. Most of the ideas described below appear in a recently published paper in Physics of Plasmas authored by *Chang et al.* [2004]. Readers are encouraged to consult this paper and its reference list for further study of this subject (web site: space.mit.edu/geocosmo).

2. Plasma Resonances and Coherent Structures

Data from in-situ space observations generally fluctuate in time (in the spacecraft frame) with varied intensities. When encountering a regime of noticeable fluctuations of such measured quantities, the standard procedure is to perform a fast Fourier transform of that section of the time series and obtain a so-called Fourier power spectrum in frequency. If one notices that there is a hump in the spectrum near certain characteristic frequency, one pronounces that the fluctuations have a strong signal within that range of frequency. Being indoctrinated by the concepts of waves in standard textbooks when discussing fluctuations, one generally goes one step further and surmises that there are strong signals of "waves" within that range of frequencies in the Fourier power spectrum. Actually, there is generally very little in-situ wave number information that is observed using the current available measuring instruments. And, the fluctuations are not really a nice superposition of plane waves. In fact if one represents the section of the fluctuations in terms of a complete set of localized, scale-dependent functions (a so-called wavelet transform), these fluctuations are generally seen localized, with different scales and amplitudes, Fig. 1. The

4

question is then: "What are these fluctuations comprised of?" We shall try to answer this question below based on a physical point of view.

Let us consider a magnetized plasma. It is known that there are generally various types of linearized waves that may propagate along the magnetic field lines. One is then tempted to express a bundle of fluctuations in terms of a sum of such waves within the observed frequency ranges. We, on the other hand, shall ask a subtler question: "Are there fluctuations that do not propagate as field-aligned plane waves?" In a set of field equations that characterizes the dynamics of the magnetized plasma, there are generally time operators, $\partial/\partial t$, and field-aligned propagation operators, $\mathbf{B} \cdot \nabla$. It is then obvious that at locations where the field-aligned propagation operator vanishes, the fluctuations cannot propagate as waves. In Fourier space, this means that the condition for non-propagation (or resonance) is when $\mathbf{k} \cdot \mathbf{B} = k_{\parallel} = 0$. In a general three-dimensional field of fluctuations, $\mathbf{k}$, and a three-dimensional magnetic field, $\mathbf{B}$, one visualizes the existence of multitudes of singular space curves that may satisfy this (resonance or non-propagation) condition.

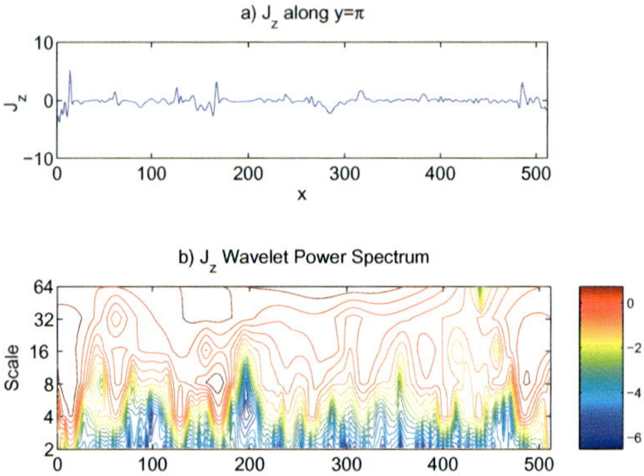

Figure 1. Example of intermittent fluctuations in a magnetized plasma. (Top) 2D MHD simulation result of current density J_z along the x-axis at a given time for homogeneous turbulence without an external magnetic field with periodic boundary conditions. The initial configuration consists of randomly distributed current filaments. (Bottom) Power spectrum of the complex Morlet wavelet transform of J_z. The x-axis and scale are in units of the grid spacing ε. We notice that the intensity of the current density is sporadic, localized, and varies nonuniformly with scale

We now ask a more physical question: "How would the fluctuations behave near these singularities?" The fluctuations will try to propagate away from the singularities as waves according to the underlying governing field equations. However, the background magnetic field and plasma medium will generally try to hold these fluctuations back close to the resonance curves. Thus, one visualizes that there are regions (resonance regions) close to the singular curves within which the fluctuations will move more or less coherently together and do not propagate as waves. Such seemingly coherent entities (to be called hereafter coherent structures) are actually bundled fluctuations of all scales that move more or less in unison (due to the nonlinear interactions). They are sporadically generated spatiotemporal structures that may meander, move, and even interact. Some of these structures may move in certain directions at uniform speeds.   These, then, could be the so-called solitary waves or they could be simply convective structures.

3. Interactions of Coherent Structures and Dynamical Complexity

In an MHD plasma embedded in a dominant background magnetic field, magnetized coherent structures are usually in the form of field-aligned flux tubes, Fig. 2.   When such coherent magnetic flux tubes with the same polarity migrate toward each other, strong local magnetic shears are created, Fig. 3.   It has been demonstrated by *Wu and Chang* [2000] that existing

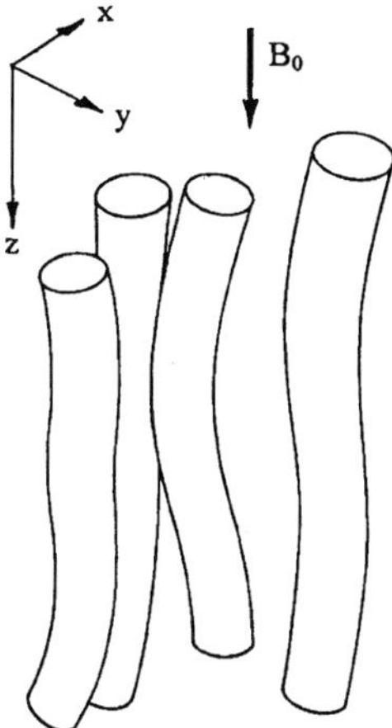

Figure 2. Field-aligned spatiotemporal coherent structures (flux tubes).

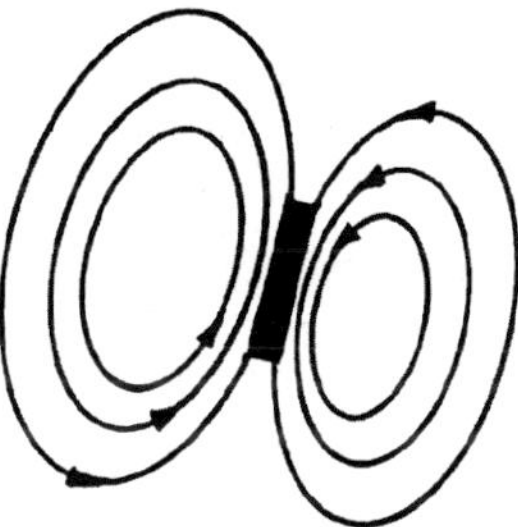

Figure 3. Cross-sectional view normal to the axes of the coherent structures (flux tubes) of the same polarity. Arrows indicate directions of the transverse magnetic field. Blackened area is an intense current sheet.

sporadic nonpropagating fluctuations will generally migrate toward the strong local shear region. Eventually the mean local energies of the coherent structures will be dissipated into these concentrated fluctuations in the coarse-grained sense and induce reconfigurations of the magnetic field geometry.

Such enhanced intermittency at the intersection regions has been observed by *Bruno et al.* [2001] in the solar wind and *Consolini et al.* [2004] in the plasma sheet. As mentioned above, the coarse-grained dissipation will then initiate "fluctuation-induced nonlinear instabilities" and, thereby reconfigure the topologies of the coherent structures of the same polarity into a combined lower local energetic state, eventually allowing the coherent structures to merge locally. And, this merging process may repeat over and over again among the coherent structures. Such phenomenon is analogous to the classical avalanching process of sand piles. On the other hand, when coherent structures of opposite polarities approach each other due to the forcing of the surrounding plasma, they might repel each other, scatter, or induce magnetically quiescent localized regions. Under any of the conditions of the above interaction scenarios, new fluctuations will be generated. And, these new fluctuations can provide new resonance sites; thereby nucleating new coherent structures of varied sizes, a phenomenon that is distinctly different from the classical concept of the avalanching process of sand piles.

All such interactions can occur at any location of a flux tube along its field-aligned direction, and the phenomenon is fully three-dimensional. The result is the generation of multitudes of coherent structures of all sizes (each being much larger than the particle sizes of the plasma medium). The nonlinear stochastic behavior of the spreading and interactions of the multitudes of the coherent structures is vastly different from that of the laminar motion of the plasma medium.

3.1 Dynamical Complexity

By definition, "dynamical complexity" is a phenomenon exhibited by a nonlinearly interacting dynamical system within which multitudes of different sizes of large scale coherent structures are formed, resulting in a global nonlinear stochastic behavior that is vastly different from that could be surmised from the original elemental dynamical equations. Thus, The dynamical behavior of the interactions of the plasma coherent structures described above is a phenomenon of "complexity". The fluctuations that are induced by such processes are sporadic and localized. Because the coherent structures are numerous and outsized, we expect the fluctuations within the interaction regions of these structures (resonance overlap regions) are generally large and can occur relatively often than those would have been expected from a medium of uniform sized plasma particles. Thus, the distribution of such fluctuations is generally non-Gaussian. As the fluctuations are generated, some can become commensurate with the plasma dispersion relation and propagate as plasma waves. Eventually a dynamic topology of coexisting propagating and nonpropagating magnetic fluctuations is created. Such is the phenomenon of intermittent turbulence in magnetized plasmas.

4. Summary and Comments

A physical description of the phenomenon of complexity in space plasmas is suggested. The description is based on the stochastic interactions of the multitude of coherent structures of varied sizes that arise naturally from plasma resonances due to nonlinear interactions. The concept of topological magnetic reconfigurations due to coarse-graining dissipation-induced nonlinear instabilities is introduced. Such magnetic reconfiguration can take place even for conserved non-dissipative plasma systems and can provide a physical explanation of the so-called magnetic reconnection processes for non-dissipative plasma media.

In a separate paper presented at this conference, Wu and Chang describe the results of direct numerical simulations of two-dimensional MHD systems that seem to confirm some of the ideas described above. The resulting complexity-induced intermittent turbulence can provide a mechanism for the energization of charged particles in the plasma medium. An application to the energization process of ionospheric ions in the auroral zone due to such type of fluctuation-particle interactions is described in a paper by Tam and Chang also at this conference.

As it is seen above, for nonlinear stochastic systems exhibiting complexity, the correlations among the fluctuations of the random dynamical fields are generally extremely long-ranged and there exist many correlation scales. The dynamics of such systems are notoriously difficult to handle analytically. On the other hand, since the correlations are extremely long-ranged, it is reasonable to expect that the system will exhibit some sort of invariance under coarse-graining scale transformations. A powerful technique that utilizes this invariance property is the method of the dynamic renormalization group. More than a decade ago, based on such invariance ideas, *Chang* [1992] suggested that the apparent low-dimensional dynamics of Earth's magnetosphere, a subject of considerable interest at the time, is a manifestation of complexity; in particular, "forced and self-organized criticality". The above results are the culmination of the development of such ideas.

Acknowledgements

Some of the afore-mentioned ideas were motivated by conversations in the early nineties with S. Akasofu at a substorm conference convened in Fairbanks, Alaska and C. Kennel at an international conference on nonlinear dynamics in La Jolla, CA. The authors are indebted to many of their colleagues for comments and discussions. They are particularly grateful to G. Consolini, R. Bruno and D. Tetreault for the discussions related to the experimental and theoretical aspects of the coherent structures in space plasmas. This research is partially sponsored by NASA and NSF.

References

Bruno, R., V. Carbone, P. Veltri, E. Pietropaolo, B. Bavassano, Identifying intermittency events in the solar wind, *Planetary and Space Science*, 49, 1201, 2001.

Chang, T., Low-dimensional behavior and symmetry breaking of stochastic systems near criticality - can these effects be observed in space and in the laboratory? *IEEE Trans. on Plasma Science*, 20, 691, 1992.

Chang, T., S.W.Y. Tam, and C.C. Wu, Complexity induced anisotropic bimodal intermittent turbulence in space plasmas, *Physics of Plasmas*, 11, 1287-1199, 2004, and references contained therein.

Consolini, G., T. Chang, and A.T.Y. Lui, Complexity and topological disorder in the Earth's magnetotail dynamics, to be published in *Non-equilibrium Transition in Plasmas*, edited by A.S. Sharma and P. Kaw (Kluwer Academic Publishers, Dordrecht, The Netherlands), 2004.

Wu, C.C., and T. Chang, 2D MHD simulation of the emergence and merging of coherent structures, *Geophys. Res. Lett.*, 27, 863, 2000.

Multiscale Coupling of Sun-Earth Processes
A.T.Y. Lui, Y. Kamide and G. Consolini, editors.

INTERMITTENCY IN SPACE PLASMA TURBULENCE

R. Bruno[1], V. Carbone[2], L. Sorriso-Valvo[2], E. Pietropaolo[3] and B. Bavassano[1]

[1] *Istituto Fisica Spazio Interplanetario-CNR, Torvergata, 00133 Rome, Italy*
[2] *Dipartimento di Fisica, Università della Calabria, Istituto Nazionale di Fisica della Materia, 87036 Rende (CS), Italy*
[3] *Dipartimento di Fisica, Università di L'Aquila, 67100 Coppito (AQ), Italy*

Abstract. We present a short overview on theoretical aspects and observational findings about intermittency in solar wind MHD turbulence. This phenomenon, which became a remarkable topic in the past few years, is due to the fact that solar wind fluctuations are neither isotropic nor scale-invariant, two of the fundamental hypotheses at the basis of K41 Kolmogorov's theory. As a matter of fact, intermittency reflects in an inhomogeneous energy transfer during the turbulent cascade. In such conditions, time series alternate intervals of very high activity to intervals of quiescence. We will first introduce the problem of intermittency from a theoretical point of view, mentioning also different models apt to reproduce statistical features of turbulent intermittent behaviour. Afterwards, we will show how intermittency affects solar wind parameters, how intermittency evolves with heliocentric distance from the sun and how typical interplanetary events causing intermittency look like.

Keywords. Solar Wind, MHD turbulence, non-linear dynamics.

1. Introduction

Turbulence is a phenomenon where chaotic dynamics and power law statistics coexist, and is characterized by randomness both in space and time. Low-frequency turbulence in plasma is the result of nonlinear dynamics of incompressible MHD equations which can be written as

$$\frac{\partial \mathbf{Z}^\sigma}{\partial t} + (\mathbf{Z}^{-\sigma} \cdot \nabla)\mathbf{Z}^\sigma = -\nabla(P/\rho) + \nu \nabla^2 \mathbf{Z}^\sigma \qquad (1.1)$$

where P, ρ and ν are the total pressure, the constant mass density and the kinematic viscosity (assumed to be equal to resistivity), respectively. We used the Elsässer variables $\mathbf{Z}^\sigma = \mathbf{u} + \sigma \mathbf{B}(4\pi\rho)^{-1/2}$ (here $\sigma = \pm 1$) which satisfy the incompressibility condition $\nabla \cdot \mathbf{Z}^\sigma = 0$. Following *Kolmogorov* [1942; *see also Frisch*, 1995], quantities of interest to characterize the statistics of turbulent flows are the increments of longitudinal fields between two points separated by a distance r. They represent characteristic fluctuations across eddies at scale r, namely $\delta \mathbf{Z}^\sigma_r = [\mathbf{Z}^\sigma(x+r) - \mathbf{Z}^\sigma(x)] \cdot \mathbf{e}$, where $\mathbf{e}$ represents the unit vector in the direction of r. These quantities are interesting first of all because they are statistically stationary, and also because, under suitable hypotheses [*see Frisch*, 1995, for the fluid flows analog] an exact relation can be obtained from (1.1), namely [*Politano and Pouquet*, 1998]

$$\left\langle \left(\delta Z_r^\sigma\right)^2 \delta Z_r^{-\sigma} \right\rangle = -\frac{4}{3}\varepsilon^\sigma r \qquad (1.2)$$

10

where ε^σ are the transfer rates, in the stationary state (brackets represent ensemble averages). Relation (1.2) can be obtained in the limit where dissipation is zero, so that it is valid in the inertial range of turbulence. Actually, since this is an exact relation, it can be used to define the extent of the inertial range in MHD turbulence [*Politano, Pouquet and Carbone,* 1998]. Relation (1.2) is the MHD analog of the famous 4/5-law obtained by *Kolmogorov* [1942] for ordinary fluid flows. As pointed out by *Carbone* [1993b, see also *Biskamp,* 1993], equations (1.1) in the limit of zero viscosity, are invariant providing lengths and velocities are scaled, through a parameter $\lambda>0$, as $r\rightarrow\lambda r$ and $\mathbf{Z}^\sigma\rightarrow\mathbf{Z}^\sigma\lambda^h$, respectively, for each value of h. Then we expect that, in the inertial range of scales where (1.2) is verified, a scaling law exists where both $\delta Z^\sigma_r\sim r^h$, the values of h being fixed by some phenomenological considerations. In the fluid-like case, by introducing the pseudo-energies dissipation rates $\varepsilon^\sigma\sim(\delta Z^\sigma_r)^2/ T^\sigma_r$, and using the eddy-turnover time $T^\sigma_r\sim t^\sigma_{NL}$ defined as $t^\sigma_{NL}\sim r/\delta Z^\sigma_r$ as the characteristic time during which the cascade for e^σ is realized, the scaling relation $(\delta Z^\sigma_r)^2\,\delta Z^\sigma_r\sim e^\sigma r$ holds and, under the hypothesis that both e^σ are constant, leads to the Kolmogorov scaling $h=1/3$. When the charged fluid is magnetically dominated, the Alfvén decorrelation time $\tau_A\sim r/C_A$ (C_A being the Alfvén speed related to the large-scale magnetic field) may become smaller that the eddy-turnover time, so that the nonlinear interactions between opposite traveling eddies are lowered. This means that the energy cascade is realized stochastically in a time $T^\sigma_r\sim t^\sigma_{NL}(t^\sigma_{NL}/\tau_A)$ [*Dobrowolny, Mangeney and Veltri,* 1980], and the scaling relation becomes $(\delta Z^\sigma_r)^2(\delta Z^\sigma_r)^2\sim C_A\,e^\sigma r$ [*Carbone,* 1993a]. When both e^σ are constant, this leads to the Iroshnikov-Kraichnan scaling $h=1/4$ [*Kraichnan,* 1965; *Iroshnikov,* 1963].

Homogeneous and isotropic turbulence has been described by the pseudo-energies density wave vector spectra $E^\sigma(k)$ which, as an order of magnitude estimate, are related to the 2nd order moment of fluctuations through $k\,E^\sigma(k)\sim\langle(\delta Z^\sigma_r)^2\rangle$. Using the scaling $h=1/3$ we obtain the usual Kolmogorov spectrum $E^\sigma(k)\sim k^{-5/3}$ while in the magnetically dominated case ($h=1/4$) we obtain the Kraichnan spectrum [*Kraichnan,* 1965] $E^\sigma(k)\sim k^{-3/2}$. To get some insight of turbulence we have to look at higher order moments of fluctuations, as we will do in the following sections. In fact, since δZ^σ_r are stochastic variables, the Probability Density Functions (pdf) $P(\delta Z^\sigma_r)$ are uniquely determined when the infinite set of moments of fluctuations are known. For a Gaussian process the 2nd order moment suffices to fully determines pdfs and then to characterize the statistics of turbulence. In the fluid-like case we have $\delta Z^\sigma_r\sim(\varepsilon^\sigma r)^{1/3}$, so that the pth order moment $S_r^{(p)}=\langle(\delta Z^\sigma_r)^p\rangle\sim(\varepsilon^\sigma r)^{p/3}$. In the magnetically dominated case $\delta Z^\sigma_r\sim(C_A\varepsilon^\sigma r)^{1/4}$, so that $S_r^{(p)}\sim(C_A\varepsilon^\sigma r)^{p/4}$. In both cases the scaling exponents ξ_p, defined through $S_r^{(p)}\sim r^{\xi_p}$, are linear in p, say $\xi_p=hp$.

The presence of scaling laws for fluctuations is a signature of the presence of self-similarity in the phenomenon. In fact, a given observable $u(r)$, which depends on a scaling variable r, is invariant with respect to the scaling relation $r\rightarrow\lambda r$ when there exists a parameter $\mu(\lambda)$ such that $u(r)=\mu(\lambda)u(\lambda r)$. The solution of this last relation is a power law $u(r)=C\,r^h$ where the scaling exponent is $h=-log_\lambda\mu$. Since as we have just seen turbulence is concerned with scaling laws, this must be the signature of self-similarity for fluctuations. In other words, if we consider fluctuations at two different scales, namely δZ^σ_r and $\delta Z^\sigma_{\lambda r}$, their ratio $\delta Z^\sigma_{\lambda r}/\delta Z^\sigma_r\sim\lambda^h$ depends only on the value of h, and this implies that fluctuations are self-similar. This means that pdfs are related through $P(\delta Z^\sigma_{\lambda r})=P(\lambda^h\delta Z^\sigma_r)$. Let us consider the standardized variables $y^\sigma_r=\delta Z^\sigma_r/\langle(\delta Z^\sigma_r)^2\rangle^{1/2}$. It can be easily shown that when h is unique, or in a pure self-similar situation, pdfs are related through $P(y^\sigma_r)=P(y^\sigma_{\lambda r})$, say by changing scale pdfs coincide. Generally, in real world pdfs do not rescale, that is, the tails of these distributions become more and more stretched at smaller and smaller scales. This means that the wings of the distributions become fatter and fatter with respect to a Gaussian, implying that, at smaller scales, extreme events become statistically more probable than if they were normally distributed. This phenomenon reflects in an anomalous scaling of the exponents ξ_p which becomes a nonlinear function of p as will be appropriately discussed in section 3. More simply, from a phenomenological point of view, a direct consequence on the data is that a given time-series alternates between bursts of activity and quiescence. This behavior is commonly defined intermittent. Intermittency (from the latin verb *Intermittere* which means to interrupt) is a well established feature of fluctuations characterizing many complex physical systems in Nature for which phases of quiescence are interrupted by activity phases.

2. Scaling Laws and Pdfs in the Solar Wind Turbulence

In situ satellite observations of solar wind turbulence reveal the presence of low-frequency high-amplitude fluctuations on all scales. Here we are interested in understanding what we can learn from solar wind turbulence about the basic features of scaling laws of the fluctuations. We will refer to velocity and magnetic field time series, and we will investigate the scaling behavior of differences separated by a (time) distance r. However, before that, it is worthwhile to remark that scaling laws, and in particular the exact relation (1.2) which defines the inertial range, is valid for longitudinal (streamwise) fluctuations.

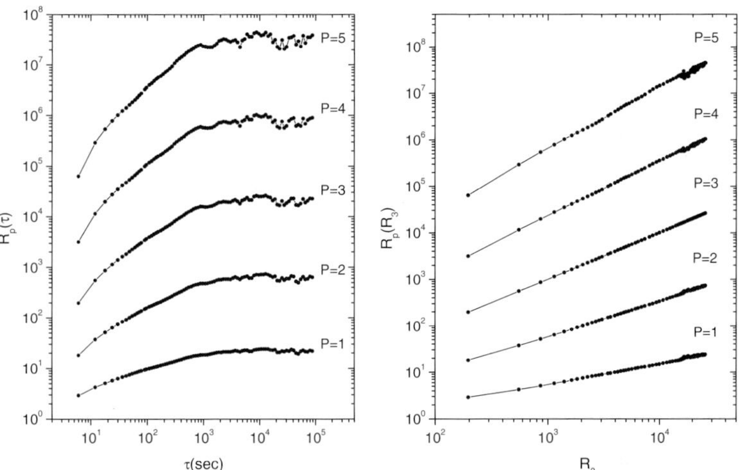

Figure 1. Left hand side panel: first 5 order SFs built with 6-sec averages of magnetic field measurements recorded by Helios 2 during day 99 of 1976, when the s/c was at 0.3 AU from the sun. Right hand side panel: first 5 order SFs plotted versus the 3rd SF. This technique, called ESS, allows to recover the scaling exponents in a much easier way than the standard procedure.

In usual fluid flows the Kolmogorov scaling is compared to the longitudinal velocity differences. In the same way for the solar wind turbulence we can investigate the scaling behavior of $\delta u_\tau = u(t+\tau)-u(t)$ where $u(t)$ represents the component of the velocity field along the radial direction. As far as the magnetic differences are concerned $\delta b_\tau = b(t+\tau)-b(t)$ we can make different choices. We can use the reference system where $B(t)$ represents the magnetic field projected along the radial direction, or the system where $B(t)$ represents the magnetic field along the local background magnetic field, or $B(t)$ represents the field along the minimum variance direction. As

12

a different case we can interpret $B(t)$ as the magnetic intensity, even if all cases are in principle equivalent. Since the solar wind moves at supersonic speed V_{SW}, the usual Taylor's hypothesis is verified, and we can get information on scaling laws on distances r by using time differences $\tau = r/V_{SW}$. As an example, let us consider the pth moment of interplanetary magnetic field fluctuations $R_p(\tau) = \langle \delta b_\tau^p \rangle$ also called p^{th} order structure function in literature (brackets being time average).

On the left hand side panel of Figure 1 we report the structure functions for some values of p, for magnetic field fluctuations recorded by Helios 2 within slow wind at ~0.3AU on the ecliptic. Starting from low values at small scales, structure functions increase towards a region where $R_p \rightarrow$ constant at the largest scales. This means that at these scales the field fluctuations are uncorrelated. A kind of inertial range, that is a region of intermediate scales r (or τ) where a power law can be recognized is more or less visible. In this range correlations exists, and we can obtain the scaling exponents ξ_p through a simple linear fit. However, does exist another technique which allows us to recover this scaling in a much easier way. This technique, called Extended Self-Similarity (ESS), was introduced some time ago by *Benzi et al.* [1993], and is used here as a tool to determine relative scaling exponents. In fact, in the fluid-like case, the 3rd order structure function can be regarded as a generalized scaling through equation (1.2). Then we can plot the pth order structure function vs. the 3rd order, to recover at least relative scaling exponents ξ_p/ξ_3. Quite surprisingly, (see right hand side panel of Figure 1) we find that the range where a power law can be recovered extends now well beyond the inertial range, covering almost all the experimental range.

In the fluid case the scaling exponents which can be obtained through ESS at low or moderate Reynolds numbers, coincide with the scaling exponents obtained for high Reynolds, where the inertial range is very well defined [*see Benzi et al.*, 1993]. This is due to the fact that by definition $\xi_3 = 1$ in the inertial range [*Frisch*, 1995], whatever their extension. In our case scaling exponents obtained through ESS can be used as a surrogate, since we cannot be sure that an inertial range exists. Relative scaling exponents for solar wind velocity ζ_p/ζ_3 and interplanetary magnetic field ξ_p/ξ_3 are reported in Table 1. There are two main considerations to be done. 1) There is a significant departure from the Kolmogorov linear scaling, that is real scaling exponents are anomalous, and look to be nonlinear functions of p, say $\zeta_p/\zeta_3 > p/3$ for $p<3$, while $\zeta_p/\zeta_3 < p/3$ for $p>3$. The same behavior can be observed for ξ_p/ξ_3 relative to magnetic field. In Table 1 we also report the scaling exponents obtained in usual fluid flows for velocity and temperature, which in that case is a passive scalar. Scaling exponents for the velocity field are similar to scaling exponents obtained in turbulent flows on earth, thus showing a kind of universality in the anomaly.

Table 1. Scaling exponents for velocity (ζ_p) and magnetic (ξ_p) variables for slow wind calculated through ESS. Errors represent the standard deviations of the linear fitting. As a reference we reported the scaling exponents of structure functions for velocity and temperature, as calculated in a wind tunnel.

p	ζ_p	ξ_p	$u(t)$ (fluid)	$T(t)$ (fluid)
1	0.37±0.06	0.56±0.06	0.37	0.61
2	0.70±0.05	0.83±0.05	0.70	0.85
3	1.00	1.00	1.00	1.00
4	1.28±0.02	1.14±0.02	1.28	1.12
5	1.54±0.03	1.25±0.03	1.54	1.21
6	1.78±0.05	1.35±0.05	1.78	1.28

This effect is commonly attributed to the phenomenon of intermittency in fully developed turbulence [see for example the book by *Frisch, 1995*]. We say that turbulence in the solar wind is intermittent, just like its fluid counterpart. 2) The degree of intermittency is measured through the distance between the curve ζ_p/ζ_3 and the linear scaling $p/3$. It can be seen that magnetic field is more intermittent than velocity field. Looking at Table 1, we can see that the same difference is observed between the velocity field and a passive scalar (in this case the temperature) in ordinary fluid flows. That is, the magnetic field, as long as intermittency properties are concerned, behaves like a passive field.

A different question concerns the maximum value of p which can be determined taking into account the finite number of points of our data set. In fact, as the value of p increases, we need an increasing number of points for an optimal determination of the structure function. Anomalous scaling laws are generated by rare and intense events which generates singularities in the gradient. The stronger their intensity the more rare are these events. Of course when the data set has a finite extent, we cannot find singularities stronger than a certain value. In that case scaling exponents ζ_p of order higher than a certain value seem to be a linear function of p. Actually the structure function depends on the pdf through

$$S_p(\tau) = \int \delta u_\tau^p P(\delta u_\tau) d\delta u_\tau \qquad (2.1)$$

and the function S_p is determined only when the integral converges. A thumb rule for the maximum value of the order p_m which we can roughly determined with a given number N of points in the data set is $p_m \sim logN$. Structure functions of order p greater than p_m cannot be determined accurately.

In Figure 2 we report the pdfs for both standardized velocity $\Delta u_\tau = \delta u_\tau / \langle (\delta u_\tau)^2 \rangle^{1/2}$ and magnetic fluctuations $\Delta b_\tau = \delta b_\tau / \langle (\delta b_\tau)^2 \rangle^{1/2}$ at three different scales τ. It appears evident that the global self-similarity in real turbulence is broken. Pdfs do not coincide, rather their shape seems to be dependent on the scale τ. In particular at large scales pdfs are almost Gaussian, becoming more and more stretched as τ decreases. At the smallest scales, pdfs are similar to stretched exponentials. This scaling dependence of the pdfs is a different way to say that scaling exponents of fluctuations are anomalous, or it is a different definition of intermittency as already said in the previous section. Namely turbulence is intermittent when the scaling laws are not globally self-similar. Note that the wings of the pdfs are higher than for a Gaussian function. This implies that the probability of occurrence of intense fluctuations is greater than if they were normally distributed. Said differently, intense stochastic fluctuations are less rare than we should expect from the point of view of a Gaussian approach to the statistics of turbulence. These fluctuations play a key role in the statistics of turbulence.

3. What is Intermittent in the Solar Wind Turbulence?

In Figure 3 we report the time behavior of δb_τ and δu_τ for three different scales τ. It appears evident that, as τ becomes small, intense fluctuations becomes more and more important, and they dominate the statistics. Fluctuations at large scales appear to be smooth, while as the scale becomes smaller, intense fluctuations becomes visible. These dominating fluctuations represent the relatively rare events we mentioned above. Actually, at the smallest scales, the time behavior of both δu_τ and δb_τ is dominated by regions where fluctuations are low interrupted by regions where fluctuations are intense and turbulent activity is very high. As a consequence, this behav-

16

The p-model [*Meneveau and Sreenivasan*, 1987; *Carbone*, 1993b] consists in an eddies fragmentation process described by a two-scale Cantor set with equal partition intervals. An eddy at the scale r with an energy derived from the transfer rate ε_r breaks down into two eddies at the scale $r/2$ with energies $\mu\varepsilon_r$ and $(1-\mu)\varepsilon_r$. The parameter $0.5 \leq \mu \leq 1$ is not defined by the model, but is fixed from the data. The model gives

$$\zeta_p = 1 - \log_2\left[\mu^{p/m} + (1-\mu)^{p/m}\right] \qquad (3.4)$$

In the model by She and Leveque [*She and Lêveque*, 1994; *Politano and Pouquet*, 1995] one assumes an infinite hierarchy for the moments of the energy transfer rates, leading to $\varepsilon_r^{(p+1)} \sim [\varepsilon_r^{(p)}]^\beta [\varepsilon_r^{(\infty)}]^{1-\beta}$ and a divergent scaling law for the infinite-order moment $\varepsilon_r^{(\infty)} \sim r^{-x}$ which describes the most singular structures within the flow. The model reads

$$\zeta_p = \frac{p}{m}(1-x) + C\left[1 - \left(1 - \frac{x}{C}\right)^{p/m}\right] \qquad (3.5)$$

The parameter $C = x/(1-\beta)$ is identified as the co-dimension of the most singular structures. In the case of fluid flows $C=2$ and the most intermittent structures are identified as filaments.

Finally, a different point of view can be introduced. That is a model that tries to characterize the behavior of the pdfs through the scaling laws of the parameters which describes how the shape of the pdfs changes going towards small scales. In its simplest form the model can be introduced by saying that pdfs of increments δZ^σ_r at a given scale is build up by a convolution of Gaussian distributions of width $\sigma = \langle (\delta Z^\sigma_r)^2 \rangle^{1/2}$ whose distribution is given by $G_\lambda(\sigma)$, namely

$$P(\delta Z_r^\pm) = \frac{1}{2\pi} \int_0^\infty G_\lambda(\sigma) \exp\left(-\frac{(\delta Z_r^\pm)^2}{2\sigma^2}\right) \frac{d\sigma}{\sigma} \qquad (3.6)$$

In a purely self-similar situation, where the energy cascade generates only a trivial variation of σ with scales, the width of the distribution $G_\lambda(\sigma)$ is zero, and invariably we recover a Gaussian distribution for $P(\delta Z_r^+)$. On the contrary when the cascade is not strictly self-similar, the width of $G_\lambda(\sigma)$ is different from zero, and the scaling behavior of the width λ^2 of $G_\lambda(\sigma)$ can be used to characterize intermittency.

4. Data Analysis: Solar Wind Turbulence and Intermittency Properties

In this section we present a reasoned look at the main aspects of what has been reported in literature about intermittency observations in solar wind turbulence.

4.1 Structure functions

Apart from the oldest investigations of fractal structure of magnetic field as observed in the interplanetary space [*Burlaga and Klein*, 1986], the starting point of the study of intermittency in solar wind dates back to 1991 when *Burlaga* [1991a] started to investigate fluctuations of the bulk velocity field at 8.5 AU using data coming from Voyager 2 satellite. Author found that anomalous scaling laws for structure functions can be recovered in the range $0.85 \leq \tau \leq 13.6$ hours. This range of scales has been arbitrarily identified as a kind of "inertial range", say a region were a linear scaling exists between log $S_\tau^{(p)}$ vs. *log* τ, and the scaling exponents have been calculated as the slope of these curves. Moreover, although the data set consisted of only about *4500* data points and these authors computed structure functions of order $p \leq 20$, the scaling was found to be quite in agreement with that found in ordinary fluid flows. Even if the data can be in agreement with the *random-β model*, from a theoretical point of view *Carbone* [1993b, 1994] showed that normalized scaling exponents ζ_p/ζ_4 calculated from Burlaga's results [*Burlaga*, 1991a] can be better fitted using a *p-model* derived from the Kraichnan phenomenology [*Kraichnan*, 1965; *Carbone*, 1993a] which provides a value for the parameter $\mu \sim 0.77$. The same author [*Burlaga*, 1991b] investigated the multifractal structure of the interplanetary magnetic field near 25 AU, and analyzed positive defined fields as magnetic field strength, temperature and density using the multifractal machinery of dissipation fields [*Paladin and Vulpiani*, 1987; *Meneveau and Sreenivasan*, 1991]. *Burlaga* [1992] showed that intermittent events observed in corotating streams at *1 AU* should be described by a multifractal geometry. Even in that cases the number of points used was very low to assure the reliability of high-order moments. Later, *Marsch and Liu* [1993] investigated the structure of intermittency of the turbulence observed in the inner heliosphere by using Helios data. This analysis, made on both bulk velocity and Alfvén speed, calculated structure functions in the whole range *40.5 sec* (the plasma instrument resolution) up to *24 hours*, and in this region p^{th} order scaling exponents were obtained. Note that also in this work the number of used data points in was too small to assure a reliability for order *p=20* structure functions reached by these authors. Using an analysis analogous to the one introduced by *Burlaga* [1991a], these authors found that anomalous scaling laws were present. A comparison between fast and slow streams at two heliocentric distances, namely *0.3 AU* and *1 AU*, allowed *Marsch and Liu* to conjecture a scenario for high speed streams where Alfvénic turbulence, originally self-similar (or poorly intermittent) near the Sun, "... loses its self-similarity and becomes more multifractal in nature" [*Marsch and Liu*, 1993], which means that intermittent corrections slightly increases from *0.3 AU* to *1 AU*. No such behavior seems to occur in the slow solar wind. From a phenomenological point of view, *Marsch and Liu* [1993] found that data can be fitted with a piecewise linear function for the scaling exponents ζ_p, namely a *β-model* $\zeta_p=3-D+p(D-2)/3$, where $D \sim 3$ for $p \leq 6$ and $D \leq 2.6$ for $p>6$. Authors say that "We believe that we see similar indications in the data by Burlaga, who still prefers to fit his whole ζ_p data set with a single fit according to the non-linear random *β-model*". However, there are two main points which should be considered related to these analyses: 1) too poor data set have been used in order to have high-order statistics, and 2) the range of scales were scaling laws have been recovered is arbitrary. To overcome these difficulties *Carbone et al.* [1996a] investigated the behavior of the normalized ratios exponents ζ_p/ζ_3 through the *ESS* procedure described above, using data coming from Helios 2 satellite. Adopting the *ESS* the whole range covered by the measurements becomes almost linear, and scaling exponent ratios can be reliably calculated. Moreover, to have a data set with an high number of points, authors mixed in the same statistics data coming from different heliocentric distances (from 0.3 AU up to 1 AU). This is not correct as far as fast wind fluctuations are taken into account, because, as found by *Marsch and Liu* [1993] and, more recently, by *Bruno et al.* [2003], there is a radial evolution of intermittency. Results showed that intermittency is a real characteristic of turbulence in the solar wind, and that the curve ζ_p/ζ_3 is a nonlinear function of p, as soon as values of $p \leq 6$ are considered. Looking at literature, it can be often realized that scaling exponents ζ_p, as observed mainly in high-speed streams of the inner solar wind, cannot be properly explained by any cascade model for turbulence. This feature has been attributed to the fact that this kind of turbulence is not in a fully-developed state with a well defined spectral index. Analyses of scaling exponents of pth order structure functions have been performed using different data sets of Ulysses. *Horbury et al.* [1995] investigated the structure functions of magnetic field fluctuations for distances between *1.7* and *4 AU* and heliographic latitude from *40°* to *80°*. By investigating the spectral index of the *2nd* order structure function, they found a decrease with heliocentric distance attributed to the radial evolution of fluctuations. Further investigations [*Ruzmaikin et al.*, 1995]

18

were obtained using structure functions to study the Ulysses magnetic field data in the range of scales $1 \le r \le 32 \ min$. These authors showed that intermittency is at work, and developed a bi-fractal model to describe Alfvénic turbulence. They found that intermittency may change the spectral index of the 2nd order structure function and this [*Carbone*, 1993a] modifies the calculation of the spectral index. *Ruzmaikin et al.* [1995] found that polar Alfvénic turbulence should be described by a Kraichnan phenomenology [*Kraichnan*, 1965] although, *Tu et al.*, [1996] showed that the same data can be fitted also with a fluid-like scaling law. These last authors, using the same idea at the basis of the models developed by *Tu et al.* [1984] and *Tu*, [1988], which were successful in describing the evolution of the power spectra as observed in space, investigated the behavior of an extended cascade model developed on the basis of the p-model [*Meneveau and Sreenivasan*, 1987; *Carbone*, 1993b]. Authors conjectured that: i) the scaling laws for fluctuations are still valid in the form $\delta Z^{\sigma}_r \sim r^h$ even when turbulence is not fully developed; ii) the energy cascade rate is not constant, rather its moments depend not only on the generalized dimensions D_p but also on the spectral index α of the power spectrum, say $<\varepsilon_r^p> \sim \varepsilon^p(r,\alpha) r^{(p-1)Dp}$, where the averaged energy transfer rate is assumed to be

$$\varepsilon(r,\alpha) = r^{-(m/2+1)} P_r^{\alpha/2} \tag{4.1.1}$$

being $P_r \sim r^{\alpha}$ the usual energy spectrum ($r \sim 1/k$). The model gives

$$\zeta_p = 1 + \left(\frac{p}{m} - 1\right) D_{p/m} + \left[\alpha \frac{m}{2} - \left(1 + \frac{m}{2}\right)\right] \frac{p}{m} \tag{4.1.2}$$

where the generalized dimensions are recovered from the usual *p-model*

$$D_{p/m} = \frac{\log_2 \left[\mu^{p/m} + (1-\mu)^{p/m}\right]}{(1 - p/m)} \tag{4.1.3}$$

In the limit of fully developed turbulence, say when the spectral slope is $\alpha = 2/m+1$ the usual expression (3.4) is recovered. Helios 2 data are consistent with this model and provide $\mu \sim 0.77$ and $\alpha \sim 1.45$ [*Tu et al.*, 1996]. Recently, *Horbury et al.* [1996, 1997] studied magnetic field fluctuations of the polar high-speed turbulence from Ulysses measurements and showed that, for structure function of order $p \le 6$, in the scaling range $20 \le r \le 300 \ sec.$, the model by *Tu et al.* [1996] seems to give a better description of the evolving turbulence observed in Ulysses data than the inertial range intermittent turbulence model of *Meneveau and Sreenivasan* [1987] with an arbitrary cascade parameter p.

In a further paper, *Carbone et al.*, [1995] investigated possible evidence for differences in the ESS scaling laws between ordinary fluid flows and solar wind turbulence. Through the analysis of different data sets collected in the solar wind and in ordinary fluid flows, it was shown that normalized scaling exponents ζ_p/ζ_3 are the same as far as $p \le 8$ are considered. This indicates a kind of universality in the scaling exponents for the velocity structure functions. Differences between scaling exponents calculated in ordinary fluid flows and solar wind turbulence are confined to high-order moments. Nevertheless the differences found in the data sets have been related to different kind of singular structures in the model [*She and Lêveque*, 1994]. In fact solar wind data can be fitted by that model as soon as the most intermittent structures are assumed to be planar sheets $C=1$ and $m=4$, that is a Kraichnan scaling is used. On the contrary ordinary fluid flows can be fitted only when $C=2$ and $m=3$, that is structures are filaments and the Kolmogorov scaling have been used. However it is worthwhile to remark that differences have been found for high-order structure functions, just where measurements are unreliable.

4.2 Probability Distribution Functions

As said in Section 2, the statistics of turbulent flows can be characterized by the pdf of field differences over varying scales. At large scales pdfs are Gaussian, while tails become higher than Gaussian (actually pdfs decay as $exp(-\delta Z^{\sigma}_r)$ at smaller scales). *Marsch and Tu* [1994] started to investigate the behavior of pdfs of fluctuations against scales and they found that pdfs are rather spiky at small scales and quite Gaussian at large scales. The same behavior has been obtained by *Sorriso-Valvo et al.* [1999] who investigated Helios 2 data for both velocity and magnetic field. These authors started from a model introduced by *Castaing et al.,* [1990] to fit the distributions and obtained extremely good results, as shown in Figure 2 by the thin line that fits experimental pdfs. This model characterizes the behavior of the pdfs through the scaling laws of the parameters which describe how the shape of the pdfs changes going towards small scales. Following this model, the pdf of increments δZ_r at a given scale (for sake of simplicity we forget for the moment about the sign of δZ_r) is build up by a convolution of Gaussian distributions of width $\sigma = <(\delta Z_r)^2>^{1/2}$ whose distribution is given by $G_\lambda(\sigma)$, namely:

$$P(\delta Z_\tau) = \frac{1}{2\pi} \int_0^\infty G_\lambda(\sigma) \exp\left(-\frac{\delta Z_\tau^2}{2\sigma^2} \right) \frac{d\sigma}{\sigma} \tag{4.2.1}$$

where:

$$G_\lambda(\sigma) = \frac{1}{\sqrt{2\pi}\lambda} \exp\left(-\frac{\ln^2 \sigma/\sigma_0}{2\lambda^2} \right) \tag{4.2.2}$$

The width of the log-normal distribution of σ is given by $\lambda^2(\tau) = \langle(\Delta\sigma)^2\rangle^{1/2}$, while σ_0 is the most probable value of σ. The expression (4.2.1) has been fitted on the experimental pdfs of both velocity and magnetic intensity, and the corresponding value for the parameter λ have been recovered. Fits drawn in Figure 2 show that the scaling behavior of pdfs, in all cases, is very well described by (4.2.1). At every scale τ, we get a single value for the width $\lambda^2(\tau)$ which can be approximated by a power law $\lambda^2(\tau) = \mu\tau^\gamma$ for $\tau < 1$ hour (not shown here). The values of parameters μ and γ obtained in the fit are $\mu = 0.75 \pm 0.03$ and $\gamma = 0.18 \pm 0.03$ for the magnetic field ($\sigma_0 = 0.90 \pm 0.05$), while $\mu = 0.38 \pm 0.02$ and $\gamma = 0.20 \pm 0.04$ for the velocity field ($\sigma_0 = 0.95 \pm 0.05$), in the range $\tau \leq 0.72$ hours, for slow-speed streams. For high-speed streams we found respectively $\mu = 0.90 \pm 0.03$ and $\gamma = 0.19 \pm 0.02$ for the magnetic field ($\sigma_0 = 0.85 \pm 0.05$), while $\mu = 0.54 \pm 0.03$ and $\gamma = 0.44 \pm 0.05$ for the velocity field ($\sigma_0 = 0.90 \pm 0.05$). This means that magnetic field is more intermittent than the velocity field.

4.3 Radial Evolution of Intermittency on the Ecliptic and at High Heliographic Latitude

As already stressed in the previous sections, intermittency can be looked at as a lack of self-similarity of the fluctuations. Pdfs loose their scaling properties evolving from Gaussian-like distributions at large scales towards more and more peaked distributions at smaller and smaller scales, characterized by fatter tails. Not being Gaussian implies that the first and second moments of the distribution are not longer sufficient to characterize the distribution itself and higher moments are needed. In particular, enhanced tails of the distribution suggest that extreme events have a probability to happen larger than they would have if they were normally distributed. One

20

of the higher moments, particularly apt to give an estimate of how far apart are the tails of the pdf of a random function $v(t)$ from a normal distribution, is the 4th moment, or flatness F

$$F = \frac{\langle (\delta v(t))^4 \rangle}{\langle (\delta v(t))^2 \rangle^2}$$
(4.3.1)

While for a Gaussian $F=3$, $F>3$ for a distribution characterized by a fatter tail. Following the concept of intermittency as given by *Frisch* [1995], *Bruno et al.* [2003] defined a given time series to be intermittent if F continually grows at smaller scales, the degree of intermittency being estimated by the growth rate of F. Moreover, a constant value of F, even larger than 3, across different scales would simply indicate a character of self-similarity although those scales would not be characterized by a Gaussian statistics. Starting from these definitions, these authors studied the radial dependence of solar wind intermittency looking at magnetic field and velocity fluctuations within the inner heliosphere, between *0.3* and *1* AU. Differently from previous studies [*Marsch and Liu*, 1993], this study was focused on the different radial behavior experienced by the flatness factor F of compressive and vector fluctuations of velocity and magnetic field within fast and slow wind. Given a generic field $\mathbf{A}$(t), wh*ile* compressive fluctuations are simply due to intensity variations $\delta |A(t)| = |A(t+\tau)| - |A(t)|$ recorded in the field at scale, say τ, vector (or directional) fluctuations $\delta A(t) = \{\sum_{i=x,y,z} [A_i(t+\tau) - A_i(t)]^2\}^{1/2}$ encompasses also directional changes due to fluctuations in the field orientation. Thus, while $\delta |A(t)|$ would be solely sensitive to compressive phenomena, mainly PBSs, shocks and TDs, $\delta A(t)$ would also be sensitive to the uncompressive influence of Alfvénic fluctuations. Since both compressive and Alfvénic character of the fluctuations strongly depend on the type of wind and heliocentric distance, we expect to see major differences. As a matter of fact the results showed that a) compressive fluctuations are more intermittent than directional fluctuations; b) magnetic fluctuations are more intermittent than velocity fluctuations; c) large scale fluctuations are always rather Gaussian, regardless of type of wind or heliocentric distance; d) slow wind is more intermittent than fast wind but it does not depend on heliocentric distance; e) intermittency for both compressive and directional fluctuations, within fast wind, increases with distance. Moreover, a study devoted to magnetic field components [*Bruno et al.*, 2003] showed that the component along the average background magnetic field is always more intermittent than the two transverse component which, in addition, behave rather similarly to each other. This last result corroborated the suspect that Alfvénic fluctuations, which mainly act on the transverse components, play a relevant role and helped to understand why directional fluctuations are always less intermittent than compressive fluctuations and why only fast wind, rich of Alfvénic fluctuations, shows radial evolution. These authors concluded that the two major ingredients of interplanetary MHD fluctuations are compressive fluctuations due to a sort of underlying, coherent structure convected by the wind and stochastic Alfvénic fluctuations propagating in the wind, recalling an idea already proposed in literature [*Bruno and Bavassano*, 1991, *Klein et al.*, 1993, *Tu and Marsch*, 1993, *Marsch and Liu*, 1993]. The coherent nature of the convected structures increases intermittency while the stochastic nature of Alfvénic modes makes the pdfs of the fluctuations more Gaussian, decreasing intermittency. In particular, directional fluctuations will have their intermittency more or less reduced depending on the amplitude of Alfvénic modes with respect to that of compressive fluctuations. As a consequence, the strong radial evolution of propagating Alfvénic fluctuations governed by non-linear interactions [see review by *Tu and Marsch*, 1995], compared to the much slower evolution of convected structures, would also explain the different radial dependence of intermittency within fast and slow wind.

Pagel and Balogh [2003] studied the radial evolution of intermittency within fast coronal wind plasma using Ulysses observations at high heliographic latitude within the framework of Castaing distribution. They found that pdfs of magnetic field fluctuations at a given scale become less and less Gaussian as the radial distance from the sun increases and ascribed this phenomenon

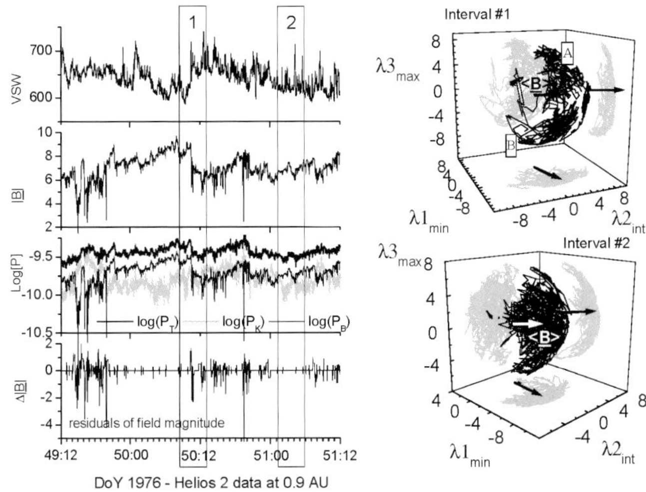

Figure 4 Left hand side panel: from top to bottom, velocity wind profile in *km/sec*, magnetic field intensity in *nT,* the logarithmic value of thermal (thick gray line), magnetic (black thin line) and total pressure (black thick line) in dyne/cm^2 and, finally, field intensity residuals in *nT* singled out by the LIM technique (see section 4.4). The two vertical boxes indicate the two time intervals chosen for comparison. At odds with interval #2, the first interval shows a strong magnetic intermittent event. Right hand side panel: from top to bottom, path followed by the tip of the magnetic vector during the two time intervals shown in the left panel. The three magnetic field components have been rotated into the minimum variance reference system, being λ_1, λ_2 and λ_3 minimum, intermediate and maximum variance directions of the relative eigenvectors, respectively. (from *Bruno et al.*, 2001, copyright of Elsevier Science Ltd., 2001)

to the growth of the inertial range towards larger scales. As a matter of fact, *Horbury et al.* [1996, 1997] and *Horbury and Balogh* [2001] showed how the inertial range of Ulysses became broader and broader as the turbulence aged or, equivalently, as the radial distance from the sun increased. Using the radial trend of the transition scale separating the *1/f* spectral region from the $f^{-5/3}$ *region* found by *Horbury et al.* [1997] and the scaling law $\lambda^2 \sim (\tau/T)^{-\beta}$, where τ is the scale and T represents the top of the inertial range, *Pagel and Balogh* [2003] demonstrated that Castaing's parameters σ^2 and λ^2 at the integral scale T did not depend on the radial distance any longer. Moreover, the same authors showed that, within the usual RTN reference system commonly used in interplanetary data, the transverse components T and N were more intermittent than the radial component R but they were unable to explain this result. In reality, the RTN reference system is not the most appropriate reference system where we should study properties of magnetic field fluctuations since the large scale direction of the background Parker's spiral breaks the spatial symmetry of interplanetary magnetic field introducing a preferential direction which generally is not radial. Thus, this kind of study would be better performed after magnetic field data were rotated into the background magnetic field reference system as showed by *Bruno et al.* [2003] and already mentioned.

22

4.4 Focusing on the Nature of Intermittent Events

The nonlinear energy cascade towards smaller scales accumulates fluctuations only in relatively small regions of space, where gradients become singular. As a rather different point of view [*Farge*, 1992] these regions can be viewed as localized zones of fluid where phase correlation exists, in some sense coherent structures. These structures, which dominate the statistic of small scales, occur as isolated events with a typical lifetime which is greater than that of stochastic fluctuations surrounding them. Structures continuously appear and disappear apparently in a random fashion, at some random location of fluid, and they carry the great quantity of energy of flows. In this framework intermittency can be considered as the result of the occurrence of coherent (non-Gaussian) structures at all scales, within the sea of stochastic Gaussian fluctuations. *Farge* [1992] suggested a numerical method, based on the wavelet decomposition of the signal, apt to separate coherent structure from the Gaussian background, namely the Local Intermittency Measure (*lim*).

The possibility of removing intermittent events helped to understand a long lasting problem concerning the anisotropy character of the power associated to the fluctuations of the components of interplanetary magnetic field as observed in the inner heliosphere. As a matter of fact, the power perpendicular to the minimum variance direction is much larger than that along it [*Bavassano et al.*, 1982] for both velocity and magnetic field fluctuations but, this anisotropy has a different radial behavior for these two fields. While velocity anisotropy doesn't evolve with distance, magnetic field anisotropy clearly increases with increasing heliocentric distance [*Bavassano et al.*, 1982]. *Bruno et al.*, [1999] showed that magnetic field intermittency, mainly due to the dynamical stream-stream interaction as the wind expands, plays a relevant role in the radial trend of magnetic anisotropy. As a matter of fact, using the *lim* technique, these authors showed that removing intermittency events from the original magnetic field would completely eliminate the observed radial trend and magnetic field anisotropy would then closely resemble velocity anisotropy. These results finally ended the long withstanding problem of reconciling the radial increase of anisotropy observed by *Bavassano et al.*, [1982] with the observed velocity and field fluctuations decoupling which would suggest an increase of isotropy [*Klein et al.*, 1993]

Using a method similar to *lim*, based on the wavelet technique, *Veltri and Mangeney* [1999], for the first time in solar wind context [see also *Salem et al.*, 2001, *Mangeney et al.*, 2001], analyzed a time series of thirteen months of velocity and magnetic data from ISEE s/c. They found that intermittent events occur on time scale of the order of few minutes and they look like one-dimensional current sheets (in agreement with *Carbone et al.*, [1995]) or shocks. Stimulated by the first results of *Veltri and Mangeney* [1999], *Bruno et al.*, [2001] performed a detailed single-case study on the characteristics of an interplanetary intermittent events which was singled out using the method introduced by *Farge* [1992]. In particular, these authors compared the behavior of solar wind and magnetic field parameters within a time interval affected by intermittency (see interval 1 in Figure 4) and another interval with no intermittent events in it (see interval 2 in Figure 4). They found that, moving across the intermittent event, the magnetic field was characterized by a large coherent rotation of its direction and, what's more, magnetic field intensity and total plasma pressure were rather discontinuous. Moreover, a minimum variance analysis performed across the field rotation, suggested that this event was a 2-D structure co-moving with the wind with no mass-flux across it. The rather good Alfvénic correlation on both sides of the discontinuity contributed to formulate the hypothesis that this coherent structure represented the border between adjacent flux-tubes within which Alfvénic, stochastic fluctuations were propagating. In this way, an observer moving across these flux-tubes would sample more coherent structures that if he was moving along them, with the consequence that the recorded sample would result more intermittent.

Further insights could be obtained studying the path followed by the tip of the vector during intermittent and non-intermittent time intervals. It was found that the interval characterized by intermittency showed a rather patchy spatial distribution while the one with no intermittency showed a more uniform distribution as shown in the right-hand-side panel of Figure 4 where intermittent and non-intermittent samples are displayed at the top panel and at the bottom panel, respectively. During the intermittent sample (top panel) the vector tends to be aligned along particular directions around which its directional fluctuations tend to cluster, at least three major spots can be easily localized. On the contrary, the bottom panel, not characterized by strong intermittent events, doesn't show similar features and the localization of analogous spots is rather difficult. *Bruno et al.* [2001] found that the passage from one spot to the next one is due to quick

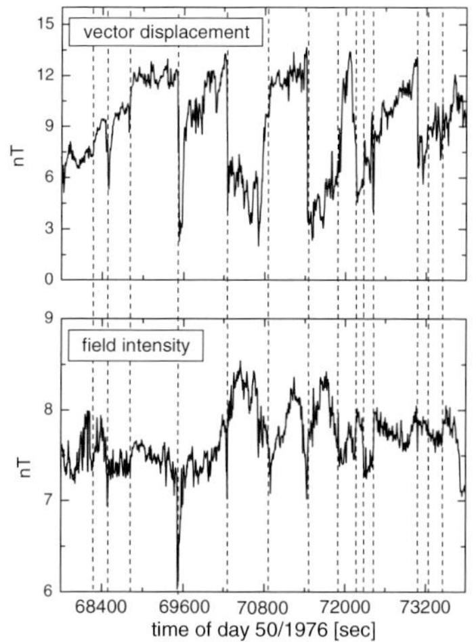

Figure 5 Top panel: intensity of vector differences between the magnetic field vector *B(t)* and the first vector of the same time series versus time. Bottom panel: intensity of magnetic field versus time.

directional jumps lasting much less than the time spent by the vector within one of those dark spots shown in the Figure. This particular behavior reminds some sort of Lévy-flight process which highlights the presence of long range correlations.

Bruno et al. [2004] tested interplanetary data adopting a Truncated Lévy Flight distribution as been proposed by *Mantegna and Stanley* [1994] in place of a standard Lévy since the second order of this distribution is infinite, at odds with real data which always have a finite variance. In particular, a TLF is characterized by a smooth cutoff which assures the finiteness of the associated variance. These authors found that both magnetic field and velocity fluctuations can reasonably be fitted by a TLF. Moreover, they showed that the pdf of these fluctuations evolved from Gaussian like, close to the sun, to possible TLF, at the earth's orbit, only within fast wind. In other words, these fluctuations, during their radial evolution, tended to loose the stochastic character proper of Alfvénic fluctuations to assume a more coherent nature. These coherent directional jumps can be better highlighted if we plot the modulus of vector differences between each

24

element of the time series and an arbitrary fixed vector which we can choose to be, for instance, the first vector of the same time series. These differences are reported in the top panel of Figure 5. This particular profile shows a series of small fluctuations interleaved by much larger jumps suggesting that the tip of the magnetic vector tends to fluctuate preferably around particular directions rather than others. Moreover, the bottom panel shows that the largest jumps noticed in the top panel often coincide with large compressive events which seem to mark the boundary between distinguishable plasma regions, possibly the same flux tubes invoked earlier in this section.

4.5 Fitting the Pdf of Interplanetary Magnetic Vector Fluctuations

As already discussed, the pdf of directional fluctuations become less and less Gaussian as we look at smaller and smaller scales. However, a detailed study of the shape of the relative pdf assumed at the shortest scales available (within the MHD range of fluctuations) has never been performed if we exclude the work by *Bruno et al.* [2004] who looked at larger scales of field and velocity fluctuations. Thus, we performed such a detailed analysis using 6 sec magnetic field averages recorded by Helios 2 between 0.3 and 1 AU. Several previous works, which dealt with a statistical approach to this same problem, considered different aspects connected to directional fluctuations focusing on the associated power, the radial evolution, the anisotropy, the nature of the fluctuations and the generation mechanisms, and so on but, none of them, to our knowledge, has ever studied how and why the orientation of these fluctuations changes with time. The first, unpublished results of this work are illustrated in Figure 6 where, on the left-hand-side panel, we show that the pdfs of interplanetary magnetic field vector differences (or directional fluctuations) within high velocity streams, normalized to their respective standard deviation, can be extremely well fitted by a double log-normal distribution. On the contrary, on the right-hand-side panel, we show that a single lognormal is sufficient to do the job within slow wind. Moreover, we found that during the radial expansion of the wind one of the two lognormals characterizing fast wind fluctuations clearly decreases leaving the second one almost unaltered. Within slow wind no such a behavior could be recorded since the relative pdf did not suffer major changes between 0.3 and 1 AU.

These results suggest that vector differences, which, by definition, are due to two distinct contributions, namely directional uncompressive fluctuations and purely compressive fluctuations, can be separated in two distinct pdfs. Given that this applies only to fast wind, notoriously more Alfvénic than slow wind, and given that within the same fast wind one of the two lognormals experiences a much stronger depletion during the radial expansion of the wind, we attribute a different nature to these two components as illustrated in Figure 6. We suggest that one of the two lognormals is mainly due to the Alfvénic component of interplanetary turbulence while the second one is mainly due to coherent structures convected by the wind. However, the interplanetary observations we have do not allow to understand whether these structures come directly from the Sun or are locally generated by some mechanism.

Recent theoretical results by *Primavera et al.,* [2003] showed that coherent structures might be locally created by parametric decay of Alfvén waves. These authors, showed that during the turbulent evolution, coherent structures like shocklets and/or current sheets were continuously created when the instability was active. The same authors showed that a similar mechanism might account for the observed behavior of the vector displacements and their statistics providing a qualitative good agreement between simulations and solar wind observations. Unfortunately, a direct quantitative comparison between the simulations and the data is still rather difficult due to the limitations of the present model.

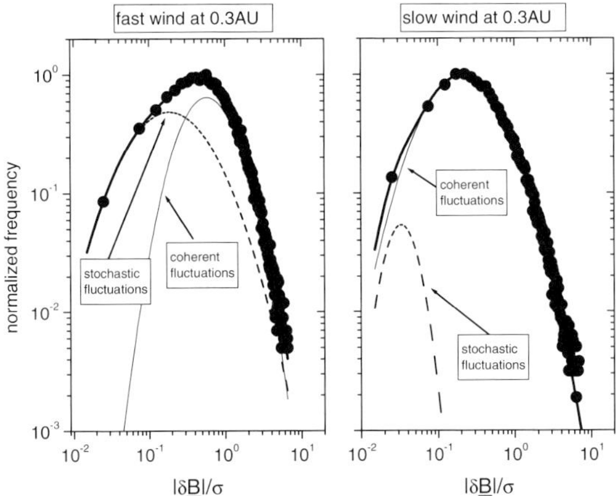

Figure 6 Left panel: pdf of normalized vector displacements $|\delta\boldsymbol{B}|/\sigma$ for fast wind at 0.3 *AU*. Thick line represents the best fit to the distribution obtained using a double log-normal function. The two lognormals concurring to the fit are represented by the dashed curve and the thin curve. The physical phenomena producing these two curves are indicated by the labels. Right panel: pdf of normalized vector displacements $|\delta\boldsymbol{B}|/\sigma$ for slow wind at *0.3 AU* in the same format as of the left panel.

Another recent and promising theoretical effort by *Chang et al.,* [2004 and references therein] models MHD turbulence in a way that recalls the interpretation of the interplanetary observations we presented in this paper. These authors stress the fact that propagating modes and coherent, convected structures share a common origin within the general view described by the physics of complexity. Propagating modes experience resonances which generate coherent structures, possibly flux tubes, which, in turn, will migrate, interact and eventually generate new modes. These theoretical efforts, which favor the local generation of coherent structures in the solar wind, fully complement the possible solar origin of the convected component of interplanetary MHD turbulence.

3. Conclusions

Observations of intermittency have been largely studied since the first results obtained by *Burlaga* [1991a] and *Marsch and Liu* [1993] more that a decade ago and our knowledge of this phenomenon, within solar wind context, is rather complete if we limit this consideration to the phenomenological behavior. As a matter of fact, we know how intermittency behaves as a function of scale, within different types of solar wind, as a function of heliocentric distance and latitude and for different solar wind parameters. Moreover, thanks to new techniques introduced in data analysis, we were able to isolate and study single intermittent events which resulted to be a sort of wall separating regions of plasma characterized by different physical conditions. These coherent structures result to be imbedded in the solar wind and convected into the interplanetary space. However, further studies on vector displacements within fast and slow wind and as a functions of the heliocentric distance, allowed to formulate the idea that the level of intermittency is due to a competing action between Alfvénic fluctuations and convected structures. As a matter of fact, data samples characterized by a strong Alfvénicity are much less intermittent than others.

Benzi and G. Parisi. North-Holland, Amsterdam, 1985.

Politano, H., A. Pouquet, Model of intermittency in magnetohydrodynamic turbulence, *Phys. Rev. E*, 52, 636-641, 1995.

Politano, H., A. Pouquet, and V. Carbone, Determination of anomalous exponents of structure functions in two-dimensional magnetohydrodynamic turbulence, *Europhys. Lett.*, 43 516-521, 1998.

Politano, H., and A. Pouquet, Dynamical length scales for turbulent magnetized flows, *Geophys. Res. Lett.*, 25, 273-276, 1998.

Primavera, L., F. Malara, and P. Veltri, Parametric instability in the solar wind: numerical study of the nonlinear evolution, in *Solar Wind 10 Conference*, Eds. M. Velli, R. Bruno, F. Malara, AIP Conference Proc., 679, 505-508, 2003.

Ruzmaikin, A., Feynman, J., Goldstein, B., and Balogh, A., Intermittent turbulence in solar wind from the south polar hole, *J. Geophys. Res.*, 100, 3395-3404, 1995.

She, Z.-S. and E. Leveque, Universal scaling laws in fully developed turbulence, *Phys. Rev. Lett.*,72, 336-339, 1994.

Salem, C., Mangeney, A., Veltri, P., Lin, R.P., Lepping, R.P., and Bougeret, J., AGU Spring Meeting Abstracts, 42, 2001

Sorriso--Valvo, L. , Carbone, V., Veltri, P., Consolini, and G., Bruno, R., Intermittency in the solar wind turbulence through probability distribution functions of fluctuations, *Geophys. Res. Lett.*, 26, 1801-1804, 1999.

Tu, C. -Y., Pu, Z.-Y, and Wei, F.-S, The power spectrum of interplanetary Alfvénic fluctuations: Derivation of its governing equation and its solution, *J. Geophys. Res.*, 89, 9695-9702, 1984.

Tu, C. -Y., The damping of interplanetary Alfvénic fluctuations and the heating of the solar wind, *J. Geophys. Res.*, 93, 7-20, 1988.

Tu, C.-Y, Marsch, E., Rosenbauer, H., An extended structure function model and its aplication to the analysis of solar wind intermittency properties, *Ann. Geophys.*, 14, 270-285, 1996.

Tu, C.-Y and Marsch, E., A model of solar wind fluctuations with two components: Alfvén waves and convective structures, *J. Geophys. Res.*, 98, 1257-1276, 1993.

Tu, C.-Y and Marsch, E., MHD structures, waves and turbulence in the solar wind: observations and theories, *Space Sci. Rev.*, 73, 1-210, 1995.

Veltri, P., and A. Mangeney, Scaling laws and intermittent structures in solar wind MHD turbulence. In *Cospar Colloquia Series:Solar Wind Nine*, Vol 471, ed. by S.R. Habbal, J.V. Hollweg, P.A. Isenberg, 543-546, 1999.

Multiscale Coupling of Sun-Earth Processes
A.T.Y. Lui, Y. Kamide and G. Consolini, editors.

SCALE-DEPENDENT ANISOTROPY OF MAGNETIC FLUCTUATIONS IN THE EARTH'S PLASMA SHEET

Z. Vörös, W. Baumjohann, R. Nakamura, A. Runov, M. Volwerk
*Space Research Institute, Austrian Academy of Sciences, Schmiedlstrasse 6,
A – 8042 Graz, Austria*

A. Balogh
Imperial College, London, UK

H. Rème
CESR/CNRS, Toulouse, France

Abstract. We use high resolution magnetic field data from the Cluster spacecraft to analyse the occurrence of specific anisotropy characteristics of magnetic fluctuations at two adjacent scales near the dissipation scale, during non-flow and bursty bulk flow (BBF) associated periods. The obtained anisotropy patterns show signatures of scale dependent anisotropy and can be explained by different physical processes. During non-flow periods the main source of the magnetic anisotropy are anisotropic ion populations within the plasma sheet boundary layer (PSBL). BBF-associated periods are characterized mainly by strong interaction of the plasma flow with the magnetic field. In both cases the local mean background magnetic field has a decisive role in the development of anisotropy in magnetic fluctuations.

Keywords. Magnetic turbulence, anisotropy, bursty bulk flows.

1. Introduction

Previous studies have confirmed that the flows in the Earth's plasma sheet are bursty and transient [*Baumjohann et al.*, 1990; *Angelopoulos et al.*, 1993, 1994]. Most of the time (80-90% of all measurements) rapid plasma flows are absent. Nevertheless, the rarely occurring bursty bulk flows (BBFs) represent mesoscale carriers [*Nakamura et al.*, 2004] of decisive amounts of mass, momentum and energy [*Schödel et al.*, 2001], energetically influencing even the near-Earth auroral regions [*Nakamura et al.*, 2001]. The analysis of associated flow fluctuations and magnetic field fluctuations [*Baumjohann et al.*, 1990; *Angelopoulos et al.*, 1993, 1994; *Hoshino et al.*, 1994; *Bauer et al.*, 1995; *Chang*, 1999; *Consolini and Lui*, 1999; *Neagu et al.*, 2002; *Lui*, 2002; *Volwerk et al.*,2003; *Vörös et al.*, 2003] led to the conjecture that the observed strong intermittent and multiscale variations in both temporal and spatial domains can be attributed to turbulence [*Borovsky et al.*, 1997]. Eddy turbulence rather than Alfvénic turbulence seems to prevail and the most important dissipation mechanisms include a multiscale cascade of energy to non-magnetohydrodynamic (non-MHD) scales and an electrical coupling of the turbulent flows to the ionosphere [*Borovsky and Funsten*, 2003].

There are several difficulties, however, which make the experimental analysis of the plasma sheet turbulence iffy. First of all, a multiscale study of spatial variability in turbulence would require as many satellite pairs as there are scales. The lack of such information is usually handled through additional hypotheses which allow us to convert temporal fluctuations to spatial ones

30

during rapid plasma flows [*Horbury*, 2000] or remnant flow intervals [*Borovsky et al.*, 1997]. Then, the range of available MHD scales in the plasma sheet spans less than two decades [*Borovsky and Funsten*, 2003]. Also, non-steady physical conditions such as the motion of the plasma sheet boundary layer (PSBL) and the time evolution of driving and/or dissipation mechanisms strongly influence the estimation of the turbulence characteristics. All these effects, together with the shortness of quasi-steady data sets and the related restricted availability of statistical moments, can lead to spurious estimations of the scaling characteristics in turbulence [*Vörös et al.*, 2004a].

In this paper we analyse anisotropy properties of 67 Hz resolution magnetic field data from the Cluster fluxgate magnetometer (FGM) [*Balogh et al.*, 2001]. Spin-resolution *(4s)* velocity data from the Cluster ion spectrometry (CIS/CODIF) experiment [*Rème et al.*, 2001] will also be used for comparison and interval selection. The high resolution magnetic field data allows an immersion into the scale or frequency ranges which are not available in velocity measurements. To avoid, at least partly, the difficulties mentioned, we restrict our analysis to second order statistics estimating the magnetic spectral power over small time-scales using the wavelet method of *Abry et al.* [2000]. Certainly the small-scale spectral power cannot characterize fully the observed fluctuations, but its estimation is less influenced by finite size effects, non-availability of statistical moments or non-stationarity of large-scale driving mechanisms or boundaries. It can still provide, however, important physical insight into the specific conditions of the evolution of BBF and non-BBF associated small-scale magnetic anisotropies.

2. The Wavelet Estimator for the Spectral Power

Abry et al. [2000] proposed a semi-parametric wavelet technique based on fast pyramidal filter bank algorithm for the estimation of scaling parameters c and α in the relation $P(f) \sim c f^{-\alpha}$, where c is a nonzero constant. The algorithm consists of several steps. First, a discrete wavelet transform of the data is performed over a dyadic grid *(scale, time)* $= (2^{j}, 2^{j}t)$. Then, at each octave $j = log_2 2^{j}$, the variance μ_j of the discrete wavelet coefficients $d_x(j,t)$ is computed through:

$$\mu_j = \frac{1}{n_j} \sum_{t=1}^{n_j} d_x^2(j,t) \approx 2^{j\alpha} c, \tag{2.1}$$

where n_j is the number of coefficients at octave j. From Equation (2.1) α and c can be estimated by constructing a plot of $y_j = log_2 \mu_j$ versus j (a so-called logscale diagram) and by using a weighted linear regression over the region (j_{min}, j_{max}) where y_j is a straight line. In this paper we use the Daubechies wavelets for which finite data size effects are minimized and the number of vanishing moments can be changed [e.g. *Mallat*, 1999]. The latter allows cancelling or decreasing the effects of linear or polynomial trends and ensures that the wavelet details are well defined.

When the parameters α and c are estimated from high resolution magnetic data over a frequency range ≥ 1 Hz, the scaling parameter α is significantly underestimated because of the influence of the magnetometer noise on the linear regression in the log-scale diagram. The estimation of the power c is less affected [*Vörös et al.*, 2004b]. For this reason we will estimate only the power of fluctuations in this paper. Furthermore, to take care of transitory character of the plasma sheet fluctuations, we perform the estimation of the parameter c within sliding overlapping windows of width 30 s with a time shift of 4 s. We also make c dimensionless, by dividing it by the time-averaged level of magnetometer noise power $<c_n>$. It has been shown that $<c_n>$ is independent of the Cluster magnetometers and spacecraft positions. The dimensionless power c shows large deviations from the noise level during rapid perpendicular plasma flows [*Vörös et al.*, 2004b].

3. Scale Dependent Power of Magnetic Fluctuations

Our goal is to investigate scale-dependent anisotropy features of magnetic fluctuations associated with BBFs and non-BBF intervals. Magnetic field fluctuations and bursty perpendicular flows are intimately associated. Figure 1 demonstrates this close relationship during the interval 17:50 - 20:00 UT on September 13, 2002, when the Cluster spacecraft were at [-7.5, 2.5, 2.5] R_E in Geocentric Solar Magnetosperic (GSM) position (if not specified, throughout the paper we will use this coordinate system). Figure 1a shows the time series of B_X magnetic component from *s/c* 1, 3 spacecraft, while the corresponding bursty perpendicular plasma flows are depicted in Figure 1b. Large perpendicular tailward and Earthward flows occur between 18:08 and 19:05 UT. Simple visual examination of B_X reveals, that high frequency magnetic fluctuations are well correlated with the period of rapid plasma flows. Before 18:08 UT and after 19:05 UT, high frequency magnetic fluctuations are absent and longer period fluctuations dominate.

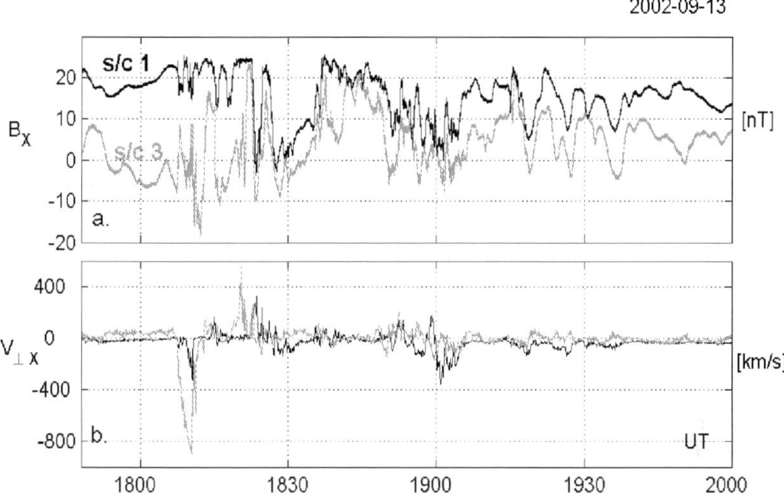

Figure 1. Event on 2002-09-13; a.) B_X component of the magnetic field from *s/c* 1,3 spacecraft; b.) proton velocity $V_{\perp X}$ perpendicular to the magnetic field from *s/c* 1,3 spacecraft.

Both magnetic field and velocity fluctuations have 4s time resolution in Figure 1. From this one can get a feeling that the smallest time scales in magnetic field fluctuations, affected by rapid flows, are of the order of seconds. Indeed, this time scale can be related to the ion-gyroperiod time scale in the plasma sheet (~seconds), where strong dissipation of MHD turbulence structures stops the inertial range cascade of energy into smaller scales. One can envision either strong damping or even destruction of turbulence at these scales [*Borovsky and Funsten*, 2003]. It was shown, however, that BBF-associated magnetic turbulence also occurs on time scales less than 1 s [*Vörös et al.*, 2003, 2004a,b]. Certainly, these fluctuations can already be affected by kinetic effects, but we prefer to approach the problem of small scales in plasma sheet magnetic turbulence experimentally. There exist several reasons supporting that approach. But the main reason is related to the problem of the cascade picture, inertial range or the scales in turbulence. An energy cascade in turbulent flows arises when a flow field is described in terms of exchange of energy, momentum, etc. between scales. The exchange of a certain quantity between scales can be described in terms of different representations, e.g. in Fourier space, using a wavelet representation or large eddy simulations, etc. The energy transfer between scales should be representation independent, but it is not [*Tsinober*, 2001]. It makes the definition of scales difficult. The definition of small scales is, however common in all representations: the small scales are always associated with the derivatives of suitable quantities, usually velocities. Generally, the small scales contain a great deal of essential physics of turbulent flows, which is poorly understood at present. These small scales are functionally as well as bi-directionally related to large scales [*Tsinober*, 2001; *Leubner and Vörös*, 2004]. Therefore, we will concentrate mainly on the small-scale description of fluctuations as is defined within the frame of the wavelet representation.

There are no physical reasons to restrict the smallest scale of magnetic fluctuations to the time scale equal or larger than the presumed MHD dissipation scale. In our study the small-scale (denoted by subscript s) is 0.2 s and comes out only from the available resolution of the magnetic data and from the requirement of the robust estimation of spectral power by the wavelet method. In our experimental approach, we computed first the small-scale power of magnetic fluctuations in a coordinate system with axes perpendicular and parallel to the local mean field. The local mean field was always computed within an interval ten times longer than the actual time scale. Then the estimation of the spectral power was repeated for gradually increased time scales. Our goal was to find the nearest neighbour to 0.2 s time scale which exhibits anisotropy features different from those at the smallest scale. The result is a two-scale wavelet study, which compares anisotropy features of magnetic fluctuations at time scales 0.2 s and 2 s. The latter is by definition our large scale (denoted by subscript L), though it is not the real large scale of the bursty flows. We note that both scales are near the presumed dissipation scale.

Figure 2 shows that large deviations from the level of noise power ($c\sim1$) occur during the intervals of BBF-associated high frequency magnetic fluctuations. Magnetometer noise can be identified through its characteristic scaling exponents. For the Cluster spacecraft these parameters describing the noise component are approximately time- and position-independent. However, they are slightly different for each spacecraft and magnetic component [*Vörös et al.*, 2004b]. Therefore, in statistical studies comprising magnetic field components or data from different spacecraft, the use of magnetic powers relative to the corresponding reference noise level is preferable [*Vörös et al.*, 2004b].

There are two main differences in the time evolution of the small-scale and large-scale powers: (a.) the large-scale power relative to the reference noise level is more than an order of magnitude larger than the small-scale power for the same interval (the corresponding absolute powers are ~1 nT^2 and ~0.01 nT^2, respectively); (b.) on average during the BBF-associated magnetic fluctuations $c_{\perp s}/c_{\parallel s} > 1$ and $c_{\perp L}/c_{\parallel L} \sim 1$.

The first observation indicates that spectral transfer of energy takes place from large scales to small scales, confirmed also by a more extensive study of *Vörös et al.* [2004b]. Inverse cascades of energy can occur, however, associated with current disruptions in the near-Earth regions [*Lui*, 1998]. The second observation indicates that magnetic fluctuations relative to the local mean magnetic field are scale dependent. The anisotropy at larger time scales can be the opposite of that observed at the small-scale. In fact, Figure 2c shows that the parallel power is stronger than the perpendicular power after 19:00 UT. Since the occurrence of scale dependent anisotropy features can facilitate a construction of appropriate physical models related to the plasma sheet turbulence, the remaining part of this paper will be devoted to the statistical study of scale-dependent magnetic anisotropies. Previous case-studies [*Vörös et al.*, 2004a,b] were devoted only to the analysis of specific events.

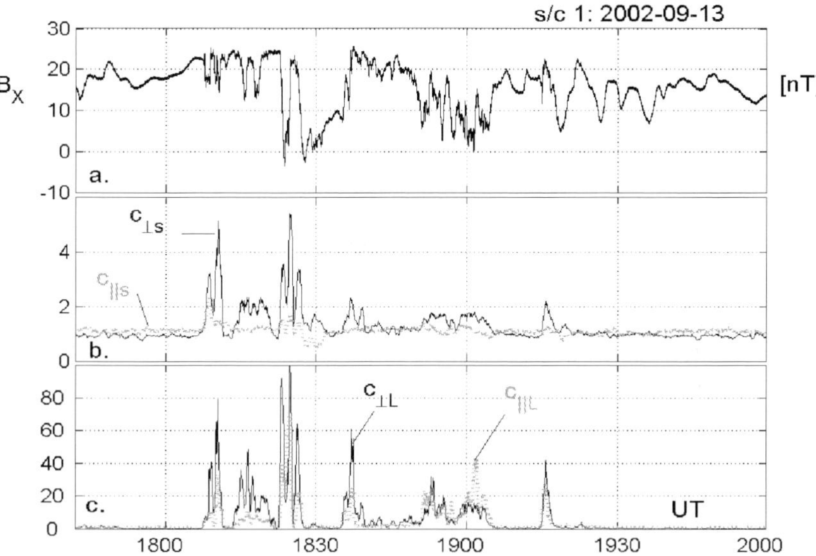

Figure 2. Event on 2002-09-13; a.) B_X component of the magnetic field from *s/c* 1; b.) small-scale perpendicular ($c_{\perp s}$) and parallel ($c_{\|s}$) relative magnetic power; c.) large-scale perpendicular ($c_{\perp L}$) and parallel ($c_{\|L}$) relative magnetic power.

We note that by choosing larger scales within the presumed inertial range, our analysis might reveal significant variations of anisotropy features of magnetic fluctuations. In such a case wider analysing windows should be chosen. Therefore, an extension of the present study into larger scales, e.g. to 20 s scale, is possible only after a careful analysis of each individual case, assuring that the estimations are statistically significant. For reliable estimation of the power of fluctuations at the scale of 20 s analysing windows of several minutes would be required. At the same time the duration of rapid flows can be shorter. Therefore, in this paper we limit our preliminary statistical analysis to the dissipative scales.

4. Scale Dependent Anisotropy of Magnetic Fluctuations

We have used the high frequency magnetic field and the spin resolution velocity data from Cluster spacecraft *s/c* 1 and 3 taken in 2001 and 2002 to select and analyse intervals with $V_{\perp x} <$ 100 km/s (non-flow or remnant flow intervals) and $V_{\perp x} > 150$ km/s (rapid flow associated intervals). The limits represent the maximum and minimum values of $V_{\perp x}$, respectively, and the condition $V_{\perp x} > 150$ km/s includes also flows with $V_{\perp x} \sim 1000$ km/s or larger. We have considered

34

6 periods of ~200 min long non-flow intervals and 6 periods of ~50 min long BBF-associated intervals. The latter intervals are shorter and can contain also a short time period of transition from Earthward to tailward flow immersed in a predominant rapid flow. To avoid any influence of the low latitude boundary layer and the Earth's magnetic field dipole effect the selected intervals fulfil the conditions $|Y_{GSM}| < 10\ R_E$ and $-20\ R_E < X_{GSM} < -14\ R_E$.

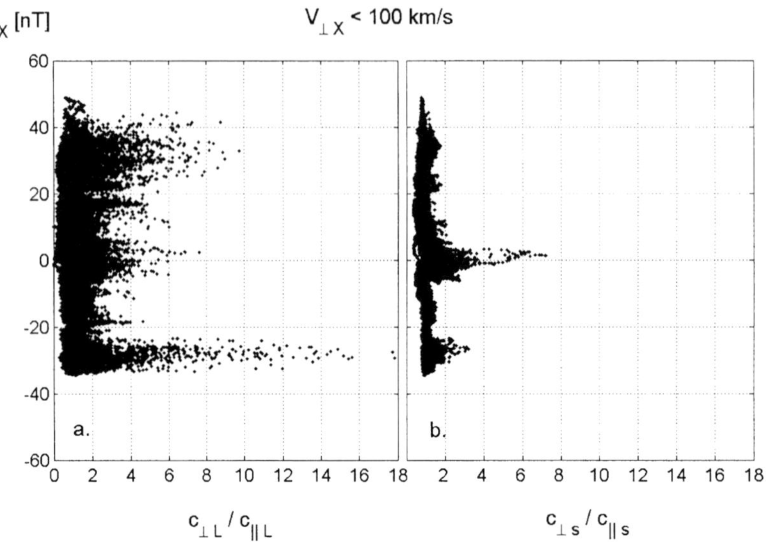

Figure 3. Non-flow associated statistics; a.) Large-scale scatter plot of magnetic anisotropy versus B_X; b.) Small-scale scatter plot of magnetic anisotropy versus B_X.

In our preliminary statistical analysis we considered scatter plots of the ratio of perpendicular to parallel power of magnetic fluctuations versus B_X. Figure 3a shows the scatter plot obtained at the large-scale and Figure 3b shows the same at the small-scale. In both cases, independent of B_X, the majority of points corresponds to isotropic fluctuations: $c_{\perp L}/c_{\parallel L} \sim c_{\perp s}/c_{\parallel s} \sim 1$. The main difference between the two scales is the appearance of significant populations of large-scale magnetic anisotropies, that is, points with $c_{\perp L}/c_{\parallel L} \gg 1$. These points seem to be organized roughly into horizontal structures of approximately constant B_X. The observed magnetic anisotropy patterns can indirectly be related to the highly anisotropic particle distributions in the PSBL. Instabilities of ion beams were shown to contribute to the spectrum of broadband electrostatic noise in kHz range within the PSBL [*Grabbe*, 1989]. Ion-beam excited ion-cyclotron modes can convert to low-frequency Alfvén, fast or slow MHD mode waves [*e.g. Treumann and Baumjohann*, 1997]. It was shown that interactions and decay of MHD waves in the presence of a mean magnetic field can produce wave turbulence with wavevectors preferentially perpendicular to the mean magnetic field [*Shebalin et al.*, 1983; *Goldreich and Sridhar*, 1995]. This can explain the increase of the perpendicular power of magnetic fluctuations and the appearance of magnetic anisotropy. Counter-streaming ion distributions occur also during quiet conditions within PSBL [*Eastman et al.*, 1984]. In our preliminary study the largest non-flow associated large-scale magnetic anisotropies occur near $|B_X|\sim30$ nT. This particular value reflects the typical

X-range of Cluster in the plasma sheet. Due to the spatial and temporal variability of the PSBL, however, the Cluster spacecraft encounter and traverse PSBL at different values of B_X in a relatively short time. Also ion beams can be transient (beamlets) or energetic conditions for the generation of electromagnetic waves would be not met. Short time enhancements of the ion-beam associated magnetic anisotropy at different values of B_X can lead to the observed horizontal structures in the $c_{\perp L}/c_{\parallel L}$ - B_X scatter plot (Figure 3a).

Much less similar horizontal structures of enhanced $c_{\perp s}/c_{\parallel s}$ are present at the small-scale (Figure 3b). When $c_{\perp s}/c_{\parallel s} \sim 1$, one finds also that $c_{\perp s} \sim c_{\parallel s} \sim 1$, which means that the majority of points belongs to the fluctuations associated with magnetometer noise in Figure 3b. Therefore there is no significant spectral transfer of energy between the two adjacent scales during non-flow periods [see also *Vörös et al.*, 2004b].

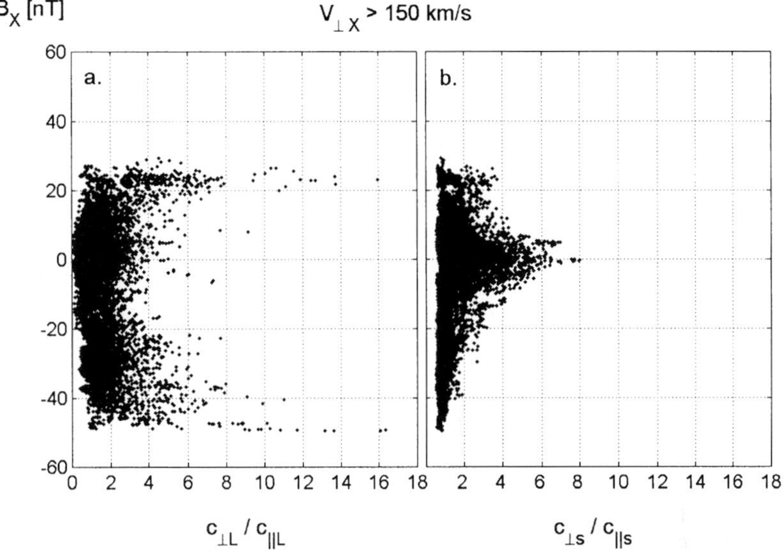

Figure 4. BBF-associated statistics; a.) Large-scale scatter plot of magnetic anisotropy versus B_X; b.) Small-scale scatter plot of magnetic anisotropy versus B_X.

Figures 4a and 4b show the two-scale anisotropy statistics for BBF-associated magnetic fluctuations. The scatter plots are non-symmetrical relative to $B_X = 0$ nT because of the smallness of our database. In comparison with non-flow statistics (Figure 3) the rapid flow associated intervals are shorter and the scatter plots contain less data points. Despite having fewer points the large scale horizontal patterns of anisotropy enhancements are less visible in Figure 4a than in Figure 3a. This indicates that besides the ion-beam instability source within the PSBL perpendicular magnetic fluctuations can be generated by other, BBF-associated mechanisms, related e.g. to large gradients of velocity, pressure or magnetic field which are nor restricted to the PSBL. Transverse magnetic structures or velocity shears can redirect the initially parallel propagating waves to highly oblique waves [*Ghosh et al.*, 1998]. In the plasma sheet BBF-associated dipolarization, an increase of B_z, can represent such transverse magnetic structures.

Figure 4b shows a peak of enhanced small-scale anisotropy around $B_X = 0$ nT. It implies that the small-scale power of perpendicular magnetic fluctuations is increasing towards the neutral sheet. It can partly be explained by occurrence of decaying magnetic turbulence in a high-beta plasma, energized by BBFs. Since it is the result of the interaction of the plasma flow with the magnetic field, we can consider the enhanced level of magnetic anisotropy as a small-scale counterpart of BBF-associated magnetic field dipolarization. At the same time, the occurrence of the narrow maximum in Figure 3b at $B_X = 0$ nT indicates that at least a portion of the observed anisotropies is independent of the plasma flow, being a pure current sheet effect.

Near $B_X = 0$ nT and for $|B_X| \geq 20$ nT the anisotropy is scale dependent. In the first case the small-scale anisotropy (Figure 4b) is larger than the large scale anisotropy (Figure 4a). In the second case the opposite is true. In absence of rapid flows the occurrence of small-scale fluctuations near $B_X = 0$ nT is strongly reduced because of the lack of spectral transfer of energy to small scales (Figure 3b).

5. Conclusions

We used a wavelet method for the estimation of the perpendicular and parallel power of magnetic fluctuations relative to the local mean magnetic field. The magnetic power is estimated at two adjacent scales. Both scales were chosen close to the ion gyroperiod, where strong dissipation of MHD structures presumably stops the inertial range cascade of energy into smaller scales. Our goal was to find two neighbouring scales with different anisotropy characteristics of magnetic fluctuations. The two-scale study allowed us to identify characteristic anisotropy patterns associated with non-flow and rapid flow (BBF) periods. Physical mechanisms are proposed which could explain the observed anisotropy patterns and their sources. Non-flow associated magnetic anisotropies appear mainly within the PSBL due to the occurrence of anisotropic ion beams. BBF-associated magnetic anisotropies can be driven by the interaction of the rapid plasma flows with the background magnetic field and essentially are not restricted to the PSBL. BBFs can energize decaying magnetic fluctuations exhibiting scale dependent anisotropy features near the neutral sheet and close to the PSBL. These dissipative scale anisotropy patterns can be regarded as a counterpart of the BBF-associated dipolarizations of the magnetic field. During non-flow intervals magnetic fluctuations appear rarely near the neutral sheet and seem to be related to a pure current sheet effect.

Acknowledgements

We thank H.-U. Eichelberger for help with FGM data. The work by MV was financially supported by the German Bundesministerium für Bildung und Forschung and the Zentrum für Luft und Raumfahrt under contract 50 OC 0104.

References

Abry, P., P. Flandrin, M.S. Taqqu, and D. Veitch, Wavelets for the analysis, estimation and synthesis of scaling data, in *Self-similar network traffic and performance evaluation*, edited by K. Park, and W. Willinger, Wiley Interscience, New York, 39, 2000.
Angelopoulos, V., et al., Characteristics of ion flow in the quiet state of the inner plasma sheet, *Geophys. Res. Lett.*, 20, 1711-1714, 1993.

Angelopoulos, V., C.F. Kennel, F.V. Coroniti, R. Pellat, M. G. Kivelson, R.J. Walker, C.T. Russel, W. Baumjohann, W.C. Feldman, and J.T. Gosling, Statistical characteristics of bursty bulk flow events, *J. Geophys. Res., 99,* 21257-21280, 1994.

Balogh, A., et al., The Cluster magnetic field investigation: overview of in-flight performance and initial results, *Ann. Geophys., 19,* 1207-1217, 2001.

Baumjohann, W., G. Paschmann, and H. Lühr, Characteristics of high-speed ion flows in the plasma sheet, *J. Geophys. Res., 95,* 3801-3809, 1990.

Bauer, T.M., W. Baumjohann, R. Treumann, N. Sckopke, and H. Lühr, Low-frequency waves in the near-Earth plasma sheet, *J. Geophys. Res., 100,* 9605-9618, 1995.

Borovsky, J. E. , R.C. Elphic, H.O. Funsten, and M.F. Thomsen, The Earth's plasma sheet as a laboratory for flow turbulence in high-β MHD, *J. Plasma Phys., 57,* 1-34, 1997.

Borovsky J.E., and H.O. Funsten, MHD turbulence in the Earth's plasma sheet: Dynamics, dissipation and driving, *J. Geophys. Res., 108,* 1284, doi:10.1029/2002JA009625, 2003.

Chang, T., Self-organized criticality, multi-fractal spectra, sporadic localized reconnections and intermittent turbulence in the magnetotail, *Phys. Plasmas, 6,* 4137-4145, 1999.

Consolini, G., and A.T.Y. Lui, Sign-singularity analysis of current disruption, *Geophys. Res. Lett., 26,* 1673, 1999.

Eastman, T.E., L.A. Frank, W.K. Peterson, and W. Lennartsson, The plasma sheet boundary layer, *J. Geophys. Res., 89,* 1553-1572, 1984.

Goldreich, P., and S. Sridhar, Toward a theory of interstellar turbulence. II. Strong Alfvénic turbulence, *Astrophys. J., 438,* 763-775, 1995.

Ghosh, S., W. H. Matthaeus, D.A. Roberts, and M.L. Goldstein, The evolution of slab fluctuations in the presence of pressure-balanced magnetic structures and velocity shears, *J. Geophys. Res., 103,* 23691, 1998.

Grabbe, C., Wave propagation effects of broadband electrostatic noise in the magnetotail, *J. Geophys. Res., 94,* 17299-17304, 1989.

Hoshino, M., A. Nishida, T. Yamamoto, and S. Kokubun, Turbulent magnetic field in the distant magnetotail: Bottom-up processes of plasmoid formation?, *Geophys. Res. Lett., 21,* 2935-2938, 1994.

Horbury, T.S., Cluster II analysis of turbulence using correlation functions, in *Proc. Cluster II Workshop, ESA SP-449,* 89-97, 2000.

Leubner, M.P., and Z. Vörös, A nonextensive entropy approach to astrophysical plasma turbulence tested on interplanetary probability distributions, submitted to *Astrophys. J.*

Lui, A.T.Y, Multiscale and intermittent nature of current disruption in magnetotail, *Phys. Space Plasmas, 15,* 233-238, 1998.

Lui, A.T.Y., Multiscale phenomena in the near-Earth magnetosphere, *J. Atmosph. Sol. Terr. Phys., 64,* 125-143, 2002.

Nakamura, R., W. Baumjohann, R. Schödel, M. Brittnacher, V.A. Sergeev, M. Kubyshkina, T. Mukai, and K. Liou, Earthward flow bursts, auroral streamers, and small expansions, *J. Geophys.Res., 106,* 10791-10802, 2001.

Nakamura, R., et al., Spatial scale of high-speed flows in the plasma sheet observed by Cluster, *Geophys. Res. Lett., 31,* L09804, doi:10.1029/2004GL019558, 2004.

Neagu, E., J.E. Borovsky, M.F. Thomsen, and S. P. Gary, Statistical survey of magnetic field and ion velocity fluctuations in the near_Earth plasma sheet: Active Magnetospheric Particle Trace Explorers/Ion Release Module (AMPTE/IRM) measurements, *J. Geophys. Res., 107,* doi:10.1029/2001JA000318, 2002.

Mallat, S., A wavelet tour of signal processing, Academic Press, London, 1999.

Réme, H., et al., First multispacecraft ion measurements in and near the Earth's magnetosphere with the identical Cluster ion spectrometry (CIS) experiment, *Ann. Geophys., 19,* 1303-1354, 2001.

sional wave flux and IMF direction. This function correlated with the energy consumption rate of the magnetosphere that they defined as depending on D_{st} and AE indices.

Earth's magnetosphere's is made up of magnetic fields from different sources and involves a complicated set of processes: storms, substorms, current flows, and power redistribution. It is generally accepted that storms are driven by the solar wind. Much attention has focused on Coronal Mass Ejections (CME's) especially during the recent solar maximum when CME's were prevalent and caused large solar storms. Tsurutani and Gonzalez [1994] discussed the causes of geomagnetic storms during solar maximum, primarily focusing on CME's. Li et al. [2003] claimed that the declining phase of the solar cycle is the most geo-effective phase. They equated geo-effectiveness to strong increases in two parameters: outer radiation belt flux or so-called "killer" electrons (SAMPEX 2-6 MeV electrons) and the D_{st} index of the strength of magnetic storms. Their Figure 1 illustrates the strength of their claim for the last solar cycle with the strongest negative D_{st} and electron fluxes occurring during the declining phase. The declining phase of the solar cycle is characterized by long-lasting recurrent high speed solar wind streams, many leading to strong magnetic storms. Li et al. [2003] suggest that CME's don't occur as often or last as long as high speed streams and thus are not as geo-effective.

Many papers have shown a linkage between high speed solar wind streams, "killer" electrons and ULF waves [e.g., Rostoker et al., 1998, Baker et al., 1998, Mathie and Mann, 2000]. Kessel et al. [2003] showed that the high speed streams carry enhanced Pc5 power as compressional fluctuations (1-8 mHz). They also showed a clear correlation between the enhanced Pc5 power in the solar wind streams and enhanced Pc5 power in the magnetosphere as determined from the Kilpisjarvi ground station. Their figures 2 and 3 showed the correlation for power in the magnetic field parallel and transverse components, respectively. These fluctuations are not generated in the foreshock because they are initially observed at Wind, 200 R_E upstream from Earth. The Pc5 power may be more geoeffective than that in the Pc3/4 range. Lower frequency peaks (from a few 10^{-7} Hz to mHz) also exist in high speed streams as is typical of the solar wind spectrum for both magnetic field fluctuations [e.g., Goldstein et al. (this issue)] and speed fluctuations [e.g., Burlaga and Forman, 2002].

It is commonly agreed that toroidal Pc5 pulsations are caused by an external source in the solar wind. The most frequently cited source of Pc5 pulsations in the magnetosphere is the Kelvin-Helmholtz instability at the magnetopause [e.g., Dungey, 1955; Miura, 1992; Anderson, 1994; Engebretson et al., 1998]. Mann et al. [1999] have recently shown that for very large flow speeds at the magnetopause flanks, the Kelvin-Helmholtz instability can energize body type waveguide modes. Other possible sources of Pc5 pulsations have been proposed such as upstream shock-related pressure oscillations that drive magnetopause surface waves with periods in the Pc5 range [Sibeck et al., 1989]. Fairfield et al. [1990] suggested that upstream pressure variations may be linked to magnetospheric compressions. Engebretson et al. [1998 and sources therein] suggested that if the compression regions at the leading edges of high speed streams contain waves in the Pc5 range they could provide a source of wave energy to the magnetosphere or that the waves could act as seed perturbations to drive boundary displacements that are amplified by the Kelvin-Helmholtz instability. By contrast, Kepko et al. [2002] showed observations of pressure fluctuations at the same discrete frequencies in the solar wind as in the magnetosphere and suggested that the solar wind may be a direct source for discrete Pc5 pulsations. Wright and Rickard

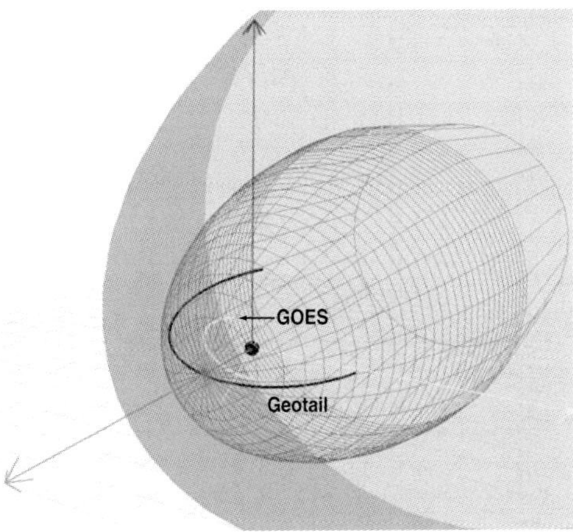

Figure 1: View from Wind satellite with Geotail skimming the magnetopause and GOES at geostationary orbit.

[1995] showed that broadband fluctuation power in the solar wind can lead to enhanced excitation of the magnetospheric cavity or waveguide modes, even if the spectral content of the upstream and magnetospheric waves are different.

In the next section we briefly describe our data and methodology. We then show observations of fluctuations in the solar wind and near the magnetopause, and pulsations at geostationary orbit for one high speed stream in 2002. We suggest here that the solar wind input for this high speed stream is first dependent on the Pc5 power in dynamic pressure fluctuations, ρV^2, and to a lesser degree on the Pc5 power in compressional magnetic field fluctuations, B^2. These inputs are sustained for the entire high speed stream as are Pc5 pulsations observed in the magnetosphere at geostationary orbit for the duration of the high speed stream passage. We discuss the local time effects at the magnetopause and inner magnetosphere which highlight the magnetospheric impacts and energy transfer mechanisms.

2. Orbits, Instruments and Methods

For this study we used data from the Wind, Geotail, and GOES satellites for a high speed stream starting 29 March 2002. Wind was upstream between 70 and 80 R_E. Geotail was skimming the magnetopause, crossing between the magnetosphere, boundary layer, and magnetosheath as shown in Figure 1, and GOES was at geostationary orbit. We used data from the Magnetic Field Investigation (MFI) [Lepping et al., 1995] and the Solar Wind Experiment (SWE) [Ogilvie et al., 1995] on Wind. We used magnetometer data from the Magnetic Field Measurement (MGF) [Kokubun et al., 1994] and the Low Energy Particle (LEP) [Mukai et al., 1992] instrument on Geotail. We used the magnetic field instrument on both GOES 8 and GOES 10 [Singer et al., 1996].

42

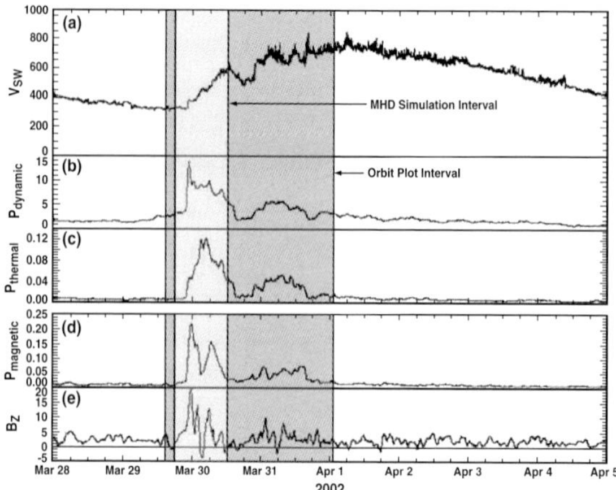

Figure 2: High speed solar wind stream (Wind data). Two nested intervals are shown (arrows indicate end points). The longer interval corresponds to the orbit shown in Figure 1 and the shorter (lighter in color) interval corresponds to the MHD simulation.

For the solar wind spectral analysis we used both dynamic pressure (nV_{sw}^2) and magnetic field magnitude (compressional waves). The pressure was converted to the same units as the magnetic field (nT) in order to compare quantitatively as well as qualitatively. Near the magnetopause (Geotail) and in the magnetosphere (GOES) we used magnetic field magnitude (compressional fluctuations) to compare with the solar wind fluctuations. We used 1 minute resolution data to provide detail on the Pc5 portion of the spectrum (1-8.3 mHz). Each data set was detrended by subtracting the background field using the IDL smooth function with a window of 61 points (one hour). Other windows from 20 minutes to 4 hours were examined for comparison; the spectral analyses were not significantly different. The resulting fluctuations were tapered using a Parzen window. The data were processed using a fast Fourier transform algorithm with a sliding 128 point (2 hour) window to create power spectral densities (PSD) and then the PSDs were summed from 1 to 8.3 mHz to get total power in the Pc5 range. FFT methods were determined from Ramirez [1985] and Press et al. [1986].

We used the Lyon-Fedder-Mobarry (LFM) global MHD code [Fedder et al., 1995] to examine plasma flows and vortices. The code couples a 3D MHD magnetospheric model with a 2D electrostatic ionospheric model and was then driven by solar wind input provided by satellite data.

3. Observations

The top panel of Figure 2 begins with a nominal solar wind followed by a high speed stream at the end of March 29, 2002. The high speed stream is identified by an initial increase in speed with the speed remaining high for 2 days before gradually trailing off over 3 days. At the leading edge of the stream there is a compression region as seen by the increases in dynamic pressure, thermal pressure and magnetic pressure in panels (b) - (d). This compression region occurs at the onset of the increase in speed and is evident 3/4 of a day. There is a second increase in pressure of smaller magnitude also evident for about 3/4 of a day. Then the pressures return to a nominal value where they remain for the remainder of the high speed stream. We indicate the interval corresponding to the orbit of Geotail shown in Figure 1. This interval encompasses the two compression regions just discussed. We also indicate the interval corresponding to an MHD simulation run using the LFM model that corresponds to the first compression region. We will discuss both data from the MHD simulation and from satellite observations later in this section.

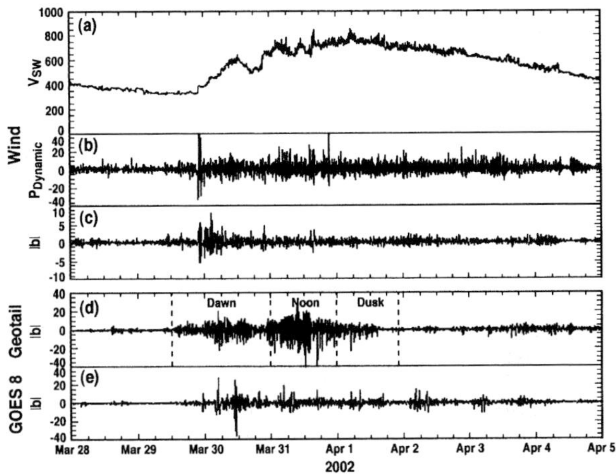

Figure 3: Compressional ULF fluctuations during high speed solar wind stream 28 March - 5 April 2002.

We next look at the fluctuations associated with this high speed stream in Figure 3. The top panel shows the speed again to orient the viewer. For the dynamic pressure (b) and magnetic field magnitudes (c - e), we detrend the data by stripping out the background. The resulting fluctuations intrinsic to the solar wind are shown in panels (b) and (c) of Figure 3, and those in the magnetosphere in (d) and (e). The pressure fluctuations have been converted to the same units (nT) as the magnetic field fluctuations so that they can be compared quantitatively as well as qualitatively. In the solar wind, the fluctuations in the dynamic pressure are larger than those in the magnetic field and are sustained at a higher level throughout the entire high speed stream interval. The fluctuations in the magnetic field magnitude are largest in the compression region at the onset of the high speed stream and are maintained at a lower level for the remainder of the high speed stream.

We compare the solar wind fluctuations with those near the magnetopause and in the

44

magnetosphere in the bottom two panels of Figure 3. Panel (d) shows fluctuations in the magnetic field magnitude at Geotail which is skimming the magnetopause during the first part of this interval as seen in Figure 1. The fluctuations are highest near the nose of the magnetopause. The fluctuations on the dawn flank are less than those at the nose but more than those at dusk. These regions are indicated in Figure 3(d). Geotail is crossing between the magnetosheath and boundary layer during this interval as will be discussed later. Figure 3(e) shows fluctuations at GOES 8 inside the magnetosphere at geostationary orbit. These fluctuations are higher during the passage of the high speed stream compared to during times of nominal solar wind (March 28-29), and highest during the first compression region ending March 30 ~1500.

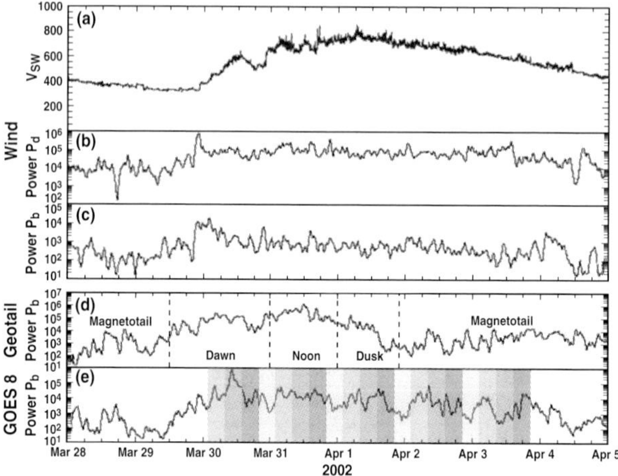

Figure 4: Top panel shows high speed solar wind stream on 28 March - 5 April 2002. (b) shows Pc5 power in solar wind dynamic pressure fluctuations. The lower 3 panels show Pc5 power in compressional magnetic field fluctuations (c) in solar wind, (d) near magnetopause (Geotail), and (e) at geostationary orbit in 6 hour strips at midnight, dawn, noon, and dusk.

Figure 4 shows the power in these various fluctuations along with the solar wind speed in the top panel to orient the viewer. The power was determined by performing a FFT on the fluctuations shown in Figure 3. The power in all cases was enhanced in the Pc5 frequency range (1-8.3 mHz) during high speed streams. Power was also evident at lower frequencies as in a typical IMF spectrum, but we focus on the Pc5 frequency range as it appears to be more geo-effective. We first examine the power intrinsic to the high speed stream in panels (b) and (c). The power in the dynamic pressure (b) is larger by at least an order of magnitude compared to the power in the magnetic field compressional component (c). The power in the dynamic pressure is also sustained for the entire interval of the high speed stream compared to the power in the magnetic field compressional component which is highest at the beginning of the stream, and then reduced. Near the magnetopause and inside the magnetosphere the power is enhanced during the passage of the high speed stream

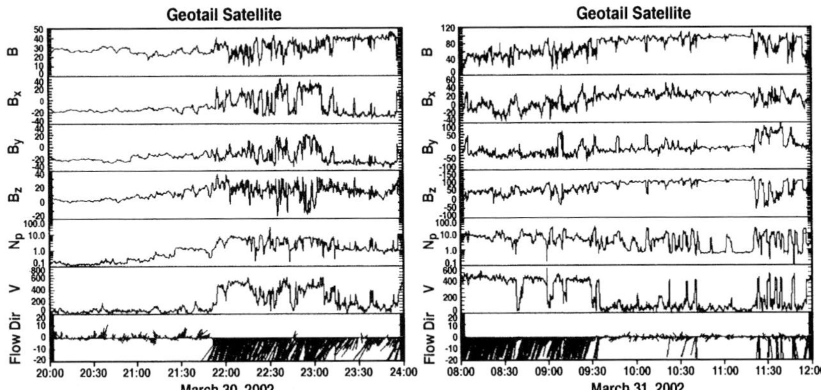

Figure 5: Magnetic field and plasma characteristics near the magnetopause (a) on the dawn flank (4-5 LT) and (b) at the nose (10-12 LT).

as shown in panels (d) and (e), respectively, of Figure 4.

As discussed in reference to Figure 3 the power at Geotail shown in Figure 4(d) is enhanced most significantly near the nose of the magnetopause (noon), followed by the dawn and dusk flanks, respectively. The power in the magnetotail region may be slightly enhanced during the high speed stream compared to in the solar wind, though no analysis of this region was undertaken in this study. Figure 4d shows some peaks and valleys in the power as Geotail was skimming the magnetopause on March 30 and 31, and April 1. Geotail was crossing back and forth between the magnetosheath and the boundary layer; generally the power in the magnetosheath was higher than in the adjacent boundary layer. The depressions within each region occur in the boundary layer. We show Geotail boundary crossings below.

At GOE8 (Figure 4e), the power during the passage of the high speed stream is enhanced compared to the interval of nominal solar wind. We further subset the data in Figure 4e into adjacent 6 hour strips to look for local time effects. The midnight sector is identified as a series of light gray strips, the dawn sector is shown as slightly darker strips, the noon sector is darkest strips, and the dusk sector is white. The geostationary power just preceding the onset of the high speed stream shows equal peaks pre-midnight, at dawn and at the nose, with significantly less power elsewhere. During the passage of the high speed stream, the largest enhancement is in the first dawn strip on March 30. Subsequent dawn strips show no enhancement compared to adjacent regions and even a decrease mid-stream and on. The midnight region is at about the same level throughout the passage, but slightly weakening as time goes on. The noon region (darkest strips), like the midnight region, is enhanced throughout though it often has a double peak and power is slightly less than at midnight. Full characterization of these pulsations is outside the scope of this paper, but we discuss possible source regions in the next section.

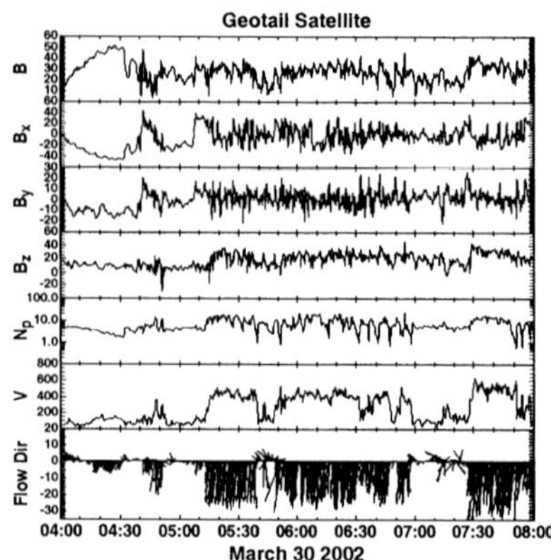

Figure 6: Geotail data on the dawn side flank of the magnetopause.

We next examine, in more detail, Geotail data in the vicinity of the magnetopause on the dawn flank and at the nose. During March 30 and 31 Geotail was crossing back and forth between the magnetosheath and the boundary layer. Figure 5 shows two time intervals during these two days: (a) shows data from the dawn flank between 04 and 05 LT (x from -5 to -10R_E on the nightside) and (b) shows data near noon (10 to 12 LT). Both (a) and (b) show the magnetic field magnitude and cartesian GSE coordinates (top four panels) followed by the ion density, speed and flow direction from the LEP instrument. The flow direction (bottom panels) is in the $+(-)x$ direction if pointing up (down) and in the $+(-)y$ direction if pointing right (left). The magnetosheath is identified by turbulent magnetic fields, high density and speed, and flows (on the flanks) primarily in the $-x$ direction. The boundary layer is distinguished by more orderly magnetospheric fields, low densities and speeds, and random flow directions. Figure 5a shows the last four hours of March 30. Geotail was in the boundary layer at 2000 and crossed into the magnetosheath around 2200 followed by multiple traversals between the two regions. Figure 5b shows four hours of March 31. Geotail was in the magnetosheath at 0800 with a couple of short forays into the boundary layer before moving into the boundary layer at about 0930. Multiple traversals back into the magnetosheath are evident later in this interval. The magnetic field fluctuations in the boundary layer between 2000 and 2200 were less than in the magnetosheath and later boundary layer regions in both (a) and (b). It may be that there are more fluctuations and power in the boundary layer near the nose than on the flank. We discuss this in the following section.

We also show an interval of Geotail data from earlier on March 30, 2002 in Figure 6 when Geotail is at about (-20, -15)R_E on the dawn flank. The magnetic field and plasma parameters from 0400 to 0800 are in the same format as Figure 5 and likewise show multiple traversals between the magnetosheath and boundary layer. The solar wind pressure is very high at this time (Figure 2) so that the magnetopause is contracted. The magnetopause contraction also explains why Geotail remained near the dawn flank much longer than it was

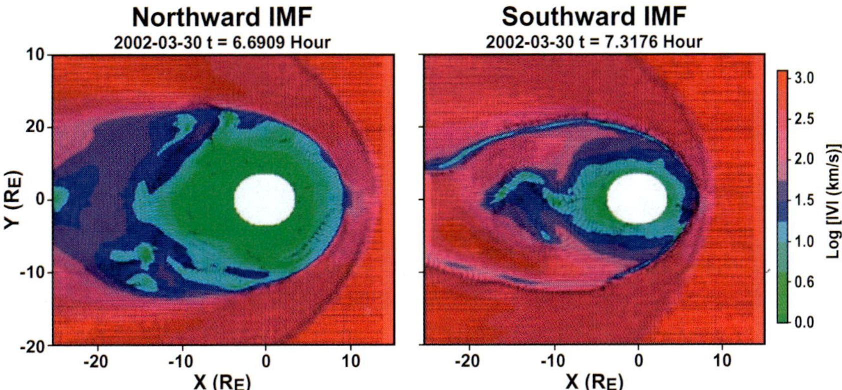

Figure 7: Typical MHD flows under (left) northward and (right) southward IMF.

on the dusk flank when the magnetosphere had relaxed back to its more nominal position. The IMF was strongly northward and the solar wind speed was increasing from about 400 to 600 km/s during this interval. The speed as observed by Geotail varied between less than 50 km/s and 600 km/s. The density was highly variable especially from 0510 to 0700, including during intervals that would be labeled as magnetosheath by other features (high speed with fairly constant flow direction and turbulent fields). The regions of highly variable density may not be easily classified as either magnetosheath or boundary layer, but may represent mixing regions. It is common to find turbulent regions behind the quasi-parallel bow shock and makes study of this region problematic. It is clear that the boundary is in motion; the boundary may also be wave-like as Geotail is skimming through.

Figure 2 showed an interval for which an MHD simulation was run. During this interval the solar wind speed started at a nominal value around 300 km/s and increased to about 600 km/s. The IMF was northward for much of this period and the dynamic, thermal and magnetic pressure all increased dramatically. We focused on plasma flows produced from the MHD simulation, and found a distinct difference between flows during times of northward and southward IMF. Figure 7 shows typical MHD flows under (left) northward and (right) southward IMF. The background is the colormap of the log of the velocity magnitude (speed). Black arrows indicate the flow direction of unit length in the equatorial plane. Under northward IMF large scale vortices developed and were swept downtail with the solar wind flow, and stagnant regions formed (Figure 7 left). At the top and bottom of this MHD snapshot, vortical regions can be seen as darker green regions of small V surrounded by regions of light blue and then darker blue (increasing V). Black arrows in these regions indicate circular flow; these can be seen more clearly in Figure 8 which is a 3D depiction of one vortex region.

Under southward IMF a thin boundary layer region was evident (Figure 7 right) and the magnetospheric region was compressed compared to times of Northward IMF. In this MHD

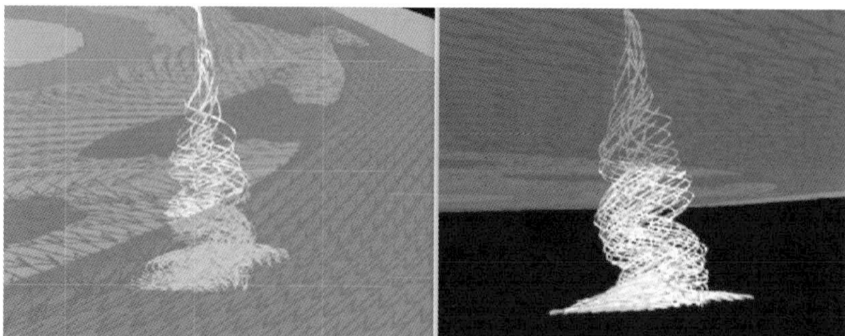

Figure 8: 3-dimensional vortex produced during northward IMF pictured from (a) above and (b) below the ecliptic plane.

snapshot, the thin boundary layer at the top (dusk) can be seen clearly, but the boundary layer at the bottom (dawn) appears broken. At other times of southward IMF, the MHD snapshots show a clear dawn side boundary layer. Referring back to the B_z component of IMF shown in Figure 2, the simulation results looked like those in Figure 7 (left) for about the first third of the run (northward IMF). This was followed by a few hours of southward IMF where the results looked like Figure 7 (right). The remainder of the MHD run consisted of a few more hours of northward IMF and a few short southward intervals. The vortices shown in Figure 7a could be associated with a Kelvin-Helmholtz type instability. The Geotail data on the dawn side around this time is highly turbulent and does not allow easy identification of K-H features, but also does not preclude K-H. Most observations of K-H instabilities have been made on the less turbulent dusk flank [e.g., Fairfield et al., 2000; Fairfield et al., 2003; Hasegawa et al., 2004]. The turbulence and plasma mixing are kinetic effects, not fluid effects and so do not show up in MHD simulation.

Finally, we introduce a new analysis that may be useful in understanding energy transfer across the magnetopause. We show one vortex from our MHD simulation as a 3D object in Figure 8 from (a) above and (b) below the ecliptic plane. Vortices examined from the interval of this MHD simulation are located just inside the magnetopause surface and extend above and below the ecliptic plane a few Earth radii. The one shown in Figure 8 flows in a counterclockwise direction but others flow clockwise. This vortex appears to form below the ecliptic plane and then dissipate into a small region above the ecliptic plane. At this time there were four vortices evident on the dawn flank. Vortices also formed on the dusk flank though generally in smaller number. We are left with many questions from this initial sample. For example, do counterclockwise vortices always form below the ecliptic plane and clockwise vortices above the plane? Is the axis of the vortex always in the z-direction? This tantalizing view of a 3D vortex structure opens up new avenues to pursue. We hope that further exploration in this area will result in a better understanding of the structure of the magnetopause and the energy transfer into the magnetosphere.

4. Summary and Conclusions

Previous studies have shown that Earth's magnetosphere is particularly geo-effective during the declining phase of the solar cycle due to recurrent high speed solar wind streams [e.g., Li et al, 2003]. Many high speed streams are associated with large magnetic storms and D_{st} of -120 to -200 [e.g., Posch et al., 2003], triggered by a southward turning of the IMF usually preceded by a long interval of northward IMF. Some high speed streams do not have large southward IMF or large D_{st} and yet have significant Pc5 pulsations and energetic electrons [Kessel et al., 2004]. The high speed stream in this paper was one of the latter cases, D_{st} was modest, only reaching about -40 (see, e.g., OMNIWeb.gsfc.nasa.gov). One solar wind driver that is consistent across all high speed streams is ULF waves or fluctuations. Akasofu [1981] identified two types of solar energy flux, essentially compressional magnetic field and dynamic pressure fluctuations, that carry energy in the ULF band. He was primarily focused on magnetic storms and correlated compressional magnetic field fluctuations and southward IMF with D_{st} and AE. Kessel et al. [2003] showed 5 months of solar wind high speed streams in 1995 that carried enhanced Pc5 power in compressional magnetic field fluctuations; the solar wind power was correlated with Pc5 power at the Kilpisjarvi ground-based station. In the current paper we have also examined the power in solar wind dynamic pressure fluctuations for one high speed stream in 2002. We are interested here to see how solar energy flux in both compressional magnetic field and dynamic pressure fluctuations affects Earth's magnetosphere for high speed streams without large magnetic storms. A later paper, Kessel et al. [2004] will make comparisons between high speed streams with and without large magnetic storms.

In this paper we have shown power at Pc5 frequencies in the solar wind in both compressional magnetic field and dynamic pressure fluctuations. We first discuss the dynamic pressure fluctuations because they appear to be dominant for high speed streams without large magnetic storms. Power in solar wind dynamic pressure fluctuations remained high for the entire high speed stream shown here and was an order of magnitude higher than power in the largest compressional magnetic field fluctuations (Figure 4). Likewise, Pc5 pulsations observed by GOES 8 overall have higher, sustained power for the entire interval of the high speed stream. But when we look at local time effects with GOES 8 at geostationary orbit, we find that some local time regions have fairly constant sustained power and other regions have more variable power. GOES 8 observations of power in the noon and midnight sectors remained at about the same levels throughout the passage of the high speed stream, with the midnight sector being slighty larger than the noon sector until near the end of the stream interval. Dawn and dusk sectors were variable. The dawn sector showed significant power at the beginning of the stream compared to all other sectors, but then quickly weakened and was less than the power at noon and midnight for the rest of the high speed stream interval. The dusk sector was stronger in the first half of the high speed stream interval and weaker in the second half, but weaker than all other local time sectors at all times during the stream.

At the magnetopause with Geotail data, so far we have examined only compressional magnetic field fluctuations. Without simultaneous measurements in the magnetosheath at dawn, noon, and dusk we cannot tell whether or not compressional magnetic field fluctuations follow the same pattern as in the solar wind, i.e., are strongest at the onset of the high speed stream and then weaken. We expect that dynamic pressure fluctuations will also be strong in the magnetosheath region, and probably sustained throughout the high speed stream. We have observed that the compressional magnetic field fluctuations vary with local time as Geotail skims the magnetopause; they are highest in the noon sector, followed by dawn, dusk, and then weakest in the magnetotail (Figure 4). The magnetosheath is well known as a turbulent region with most fluctuations originating in the solar wind or associated with the quasi-parallel bow shock [e.g., Crooker et al., 1981; Sibeck et al., 2000]. At the nose of the magnetopause, particles and fields pile up so it is not surprising to find the highest power in this region.

We also used Geotail data to identify magnetosheath and boundary layer regions through both magnetic and plasma characteristics (Figure 5). What we found and show in Figure 4 is elevated power in the boundary layer as well as in the magnetosheath. The power in the boundary layer was less than that in the magnetosheath so only some of the energy crossed the magnetopause. In addition, we found that the power at Pc5 frequencies in the noon boundary layer was higher than that in the dawn or dusk boundary layer. How is this energy transferred to the magnetosphere? Several authors have suggested that pressure pulses drive magnetospheric pulsations [e.g., Sibeck et al., 1989; Fairfield et al., 1990; Engebretson et al., 1998]. The dynamic pressure fluctuations observed by Wind and Geotail were not a pressure 'pulse', but were sustained over many days. Likewise, the magnetospheric pulsations in the noon and midnight sectors were sustained over many days. It is possible that these regions are driven by the fluctuations in dynamic pressure or magnetic pressure and that the energy may enter the magnetosphere in the noon region. This compressional power may decay evanescently farther into the magnetosphere, but be sufficient to reach geostationary orbit. Kepko et al. [2002] suggested that the solar wind may be a direct source for discrete Pc5 pulsations. Alternately, the solar wind/magnetosheath fluctuations at Pc5 frequencies may excite field line resonances (FLR) that are observed at geostationary orbits. Pc5 pulsations are common at latitudes between about 60° and 70° (Hughes, 1994 and sources therein) that map to L-shells that include geostationary orbit. The noon sector GOES 8 observations are more probably explained by one of these scenarios. The midnight sector might also be tied to particle injections from the magnetotail [e.g., Posch et al, 2003 and sources therein]. However, particle injections tend to occur during storm periods and our interval includes only a modest storm as mentioned above.

The dawn sector GOES 8 observations might have a different or additional source than dynamic pressure fluctuations that are sustained at approximately the same level throughout the high speed stream. GOES 8 observations (Figure 4e) showed the strongest dawn pulsations on March 30 from 0800 - 1400 UT; subsequent dawn pulsations observed by GOES 8 were significantly weakened. In this paper we showed solar wind compressional magnetic field fluctuations that were strongest at the onset of the high speed stream (March 29 2000 - March 30 0800), and then decreased by about a factor of 10 for the remainder of the stream (Figure 4c). GOES 8 dawn sector observations were strongest immediately following the strongest solar wind compressional magnetic field fluctuations. This could be a coincidence or it could be that solar wind compressional magnetic field fluctuations drive Pc5 pulsations on the dawn side.

The IMF was strongly northward during most of this interval (March 29 2000 - March 30 0200, and March 30 0400 - 0700), coinciding with vortex formation along the flanks as seen in MHD simulations. The vortices disappeared when the IMF turned southward

(Figure 7). The boundary was in motion as was evidenced by multiple crossings of the magnetopause by Geotail. Engebretson et al. [1998 and sources therein] suggested that solar wind or magnetosheath input waves/fluctuations could act as seed perturbations to drive magnetopause boundary displacements that were amplified by the Kelvin-Helmholtz instability. These can drive geomagnetic pulsations, most likely on the dawn side. A Kelvin-Helmholtz instability can be driven by velocity shear alone at the magnetopause flanks [e.g., Miura, 1984; Otto and Fairfield, 2000]. However, for this high speed stream interval the speed increased and then remained high *after* the strongest dawn pulsations, but the dawn pulsations weakened like the compressional magnetic field fluctuations. It is possible that the solar wind compressional fluctuations control the level of magnetospheric pulsations. A K-H instability could have enhanced the initially driven boundary displacements, and then power was transferred to magnetospheric dawn pulsations. This would account for the various observations, but may not be the only explanation.

In this paper we have shown one high speed stream in 2002 during the declining stage of the current solar cycle. Although more cases must be examined before drawing definitive conclusions we suspect that both types of solar energy flux shown here will be found to be common in high speed streams and will serve as effective drivers of magnetospheric Pc5 pulsations, though they may contribute differently to energizing Earth's magnetosphere. Based on this study we make the following statements. Wind observations showed significant, sustained power at Pc5 frequencies intrinsic to a high speed stream starting March 29, 2002. The enhanced power was evident in dynamic pressure fluctuations and to a lesser extent in compressional magnetic field fluctuations. Geotail observations near the magnetopause suggest that more energy may be transferred across the nose than at the flanks during high speed streams. This energy may be responsible for the sustained geomagnetic pulsations seen at GOES 8 in the noon and midnight sectors, although the midnight sector might have additional sources. GOES 8 observations of dawn pulsations follow the trend of the compressional magnetic field fluctuations, that is higher at first and then reduced, and may control the level of Pc5 pulsations on the dawn side. Northward IMF along with Geotail observations of magnetopause motion and MHD simulations of vortices along the flanks suggest that a Kelvin-Helmholtz (K-H) instability may contribute to transport of the compressional energy into the magnetosphere.

Acknowledgements

We thank R. Boller for his development of the 3D vortex anaylsis. We obtained most of the data for this study from CDAWeb.

References

Akasofu, S.-I., Energy coupling between the solar wind and the magnetosphere, Space Sci. Rev., 28, 121, 1981.

Akasofu, S.-I. and S. Chapman, *J. Geophys. Res.*,68, 125, 1963.

Anderson, B.J, An overview of spacecraft observations of 10s to 600s period magnetic pulsations in the Earth's magnetosphere, in *Solar Wind Sources of Magnetospheric Ultra-Low-Frequency Waves,* edited by M.J. Engebretson, K. Takahashi, M. Scholer, AGU, Wash., D.C., 1994.

Baker, D. N., et al., Coronal mass ejections, magnetic clouds and relativistic magnetospheric electron events: ISTP *J. Geophys. Res.*, 103, 17279-17291, 1998.

Burlaga, L.F., and M. Forman, Large-scale speed fluctuations at 1 AU on scales from 1 hour to 1 year: 1999 and 1995, *J. Geophys. Res.*, 107 (A11), 1403, 2002.

Crooker, N.U., T.E. Eastman, L.A. Frank, E.J. Smith, and C.T. Russell, Energetic magnetosheath ions and the interplanetary magnetic field orientation, *J. Geophys. Res.*, 86, 4455, 1981.

Dungey, J.W., Electrodynamics of the outer atmosphere, in *Proceedings of the Ionosphere Conference*, p. 225, The Physical Society of London, 1955.

Engebretson, M.J., K.-H. Glassmeier, and M. Stellmacher, The dependence of high-latitude Pc5 power on solar wind velocity and phase of high-speed solar wind streams, *J. Geophys. Res.*, 103, 26271, 1998.

Fairfield, D.H. et al., Upstream pressure variations associated with the bow shock and their effects on the magnetosphere, *J. Geophys. Res.*, 95, 3773, 1990.

Fairfield, D.H. et al., Geotail observations of the Kelvin-Helmholtz instability at the equatorial magnetotail boundary for parallel northward fields, *J. Geophys. Res.*, 105, 21159, 2000.

Fairfield, D.H. et al., Motion of the dusk flank boundary layer caused by solar wind pressure changes and the Kelvin-Helmholtz instability, *J. Geophys. Res.*, 108, 20-1, 2003.

Fedder, J. A., J. G. Lyon, S. P. Slinker, and C. M. Mobarry, Topological structure of the magnetotail as function of interplanetary magnetic field and with magnetic shear, *J. Geophys. Res.,* 100, 3613, 1995.

Goldstein, M.L, D.A. Roberts, A.V. Usmanov, Numerical simulation of solar wind turbulence, this issue.

Hasagewa, H. et al., Transport of solar wind into Earth's magnetosphere through rolled-up Kelvin-Helmholtz vortices, *Nature*, 430, 755, 2004.

Hughes, W.J., Magnetospheric ULF waves: a tutorial with a historical perspective, in *Solar Wind Sources of Magnetospheric Ultra-Low-Frequency Waves,* edited by M.J. Engebretson,
K. Takahashi, M. Scholer, AGU, Wash., D.C., 1994.

Kepko, L., H.E. Spence, and H.J. Singer, ULF waves in the solar wind as direct drivers of magnetospheric pulsations, *J. Geophys. Res.*, 29, 39-1, 2002.

Kessel, R.L., I.R. Mann, S.F. Fung, D. Milling, and N. Oconnell, Correlation of Pc5 wave power inside and outside the magnetosphere during high speed streams, *Ann. Geophys.*, 21, 1-13, 2003.

Kessel, R.L., et al, Comparison of high speed streams with and without large magnetic storms, manuscript in preparation, 2004.

Kokubun, S., T. Yamamote, M.H. Acuna, K. Hayashi, K. Shiokawa, and H. Kawano, The GEOTAIL magnetic field experiment, *J. Geomag. Geoelectr.*, 46, 7-21, 1994.

Lepping, R.P., M.H. Acuna, L.F. Burlaga, W.M. Farrell, et al., The Wind Magnetic Field Investigation, in The Global Geospace Mission, ed. C.T. Russell, **Kluwer** *Academic Publishers*, 207-229. 1995.

Li, X., D.N. Baker, D. Larson, M. Temerin, G. Reeves, S.G. Kanekal, The predictability of the magnetosphere and space weather, EOS, 361, 2003.

Mann, I.R., A.N. Wright, K.J. Mills, and V.M. Nakariakov, Excitation of magnetospheric waveguide modes by magnetosheath flows, *J. Geophys. Res.*, 104, 333-353, 1999.

Mathie, R.A., and I.R. Mann, A correlation between extended intervals of ULF wave power and storm-time geosynchronous relativistic electron flux enhancements,

Geophys. Res. Lett., **27**, 3261, 2000.

Miura, A, Anomalous Transport by Magnetohydrodynamic Kelvin-Helmholtz instabilities in the solar wind magnetosphere interaction, *J. Geophys. Res.*, 89, 801, 1984.

Miura, A, Kelvin-Helmholtz instability at the magnetospheric boundary: Dependence on the magnetosheath sonic Mach number, *J. Geophys. Res.*, 97, 10665, 1992.

Mukai, T. et al., Low Energy Particle Experiment (LEP), in p. 97, Geotail Prelaunch Report, ISAS, 1992.

Ogilvie, K.W., D.J. Chornay, R.J. Fritzenreiter, F. Hunsaker, et al., SWE, Comprehensive Plasma Instrument for Wind Spacecraft, in Global Geospace Mission, ed. C.T. Russell, **Kluwer Academic Pub.**, 55, 1995.

Otto, A., and D.H. Fairfield, Kelvin-Helmholtz instability at the magnetotail boundary: MHD simulation and comparison with Geotail observations, *J. Geophys. Res.*, 105, 21175, 2000.

Perreault, P and S.-I. Akasofu, Geophys. J. Roy. Astron. Soc., 54, 547, 1978.

Posch, J.L. et al., Characterizing the long-period ULF response to magnetic storms, *J. Geophys. Res.*, 108, 18-1, 2003.

Press, W.H., B.P. Flannery, S.A. Teukolsky, and W.T. Vetterling, Numerical Recipes, the art of scientific computing, Cambridge University Press, Cambridge, 1986.

Ramirez, Robert W., The FFT, Fundamentals and Concepts, Prentice-Hall, Inc., Englewood Cliffs, N.J., 1985.

Rostoker Gordon, et al., On the origin of relativistic electrons in the magnetosphere assocaited with geomagnetic storms, *Geophys. Res. Lett.*, 25, No. 19, 3701, 1998.

Sibeck, D.G., W. Baumjohann, R.C. Elphic, D.H. Fairfield, et al., The magnetospheric response to 8-minute period strong-amplitude upstream pressure variations, *J. Geophys. Res.*, 94, 2505, 1989.

Sibeck, D.G., T.-D. Phan, R.P. Lin, R.P. Lepping, T. Mukai, and S. Kokubun, A survey of MHD waves in the magnetosheath: International Solar Terrestial Program observations, *J. Geophys. Res.*, 105, 129, 2000.

Singer, H.J, L. Matheson, R.Grubb, A.Newman, and S.D.Bouwer, Monitoring Space Weather with GOES Magnetometers, SPIE Proceedings, Volume 2812, 1996.

Tsurutani, B.T. and W.D. Gonzalez, The causes of geomagnetic storms during solar maximum, EOS, 49, 1994.

Wright, A.N. and G.J. Rickard, A numerical study of resonant absorption in a magnetohydrodynamic cavity driven by a broadband spectrum, Astrophys. J., 444, 458-470, 1995.

ON THE IDENTIFICATION OF SOC DYNAMICS IN THE SUN-EARTH SYSTEM

R. Woodard[*]
D. E. Newman
University of Alaska Fairbanks
Fairbanks, Alaska 99775-5920, USA

R. Sánchez
Universidad Carlos III de Madrid
28911 Leganés, Madrid, Spain

B. A. Carreras
Oak Ridge National Laboratory
Oak Ridge, Tennessee 37831-8070, USA

Abstract. We use a self-organized criticality (SOC) model to show that 1) such a system can have different characteristic signatures depending on the level of external forcing while still having the same underlying dynamics and that 2) current time series of Sun–Earth processes are too short to compare with all dynamical regions of a SOC model. SOC is a concept that has been applied to various aspects of the Sun–Earth system, such as rearrangement of magnetic flux loops on the Sun, AE indices and substorm statistics. The basic tenet of SOC is that simple local interactions produce complex global signatures that are not simply predicted by the low level physics. Simple SOC models, such as the sandpile, are used to compare their signatures with those observed in a physical system.

1. Introduction

Self-organized criticality (SOC) *Bak et al.* [1987, 1988] has been suggested as a model for the dynamics of various aspects of the Sun–Earth system. The basic tenet of SOC is that simple local interactions produce complex global signatures that are not simply predicted by the low level physics. The Sun–Earth system and SOC models share similar dynamical and statistical signatures, such as power law scaling of event size distributions *Uritsky et al.* [2002] and of the power spectra of some characteristic time series, discussed below. Another shared signature is a Hurst exponent that indicates correlated dynamics over some time scales longer than an autocorrelation time *Price and Newman* [2001]. These very brief statements clearly do not do justice to all that is implied about the Sun-Earth system when viewed from the perspective of SOC. Some examples of more thorough treatments of the subject include *Chang* [1992], *Uritsky and Pudovkin* [1998], *Angelopoulos et al.* [1999], *Chang* [1999], *Klimas et al.* [2000], *Lui et al.* [2000], *Chapman and Watkins* [2001] and *Chang et al.* [2003].

When deciding whether or not a system is SOC, a usual practice is to compare specific signatures of a defined SOC model and the physical system under study. Rather than saying: "If these signatures are similar, then the system must be SOC," the more cautious approach is to say: "If these signatures are similar, then this system is consistent with SOC dynamics." Fundamentally, it is the characteristics of the dynamics that are of interest. Here, we investigate two basic issues involved in whether or not even this second statement can be safely made. It is important to note

---

[*]Corresponding author: ryan@timehaven.org

56

that one cannot prove that a system is SOC unless it has been constructed as such. But it is the shared dynamics among systems that matters most, not the name.

Before presenting those two issues, we briefly mention the measures that we use in this study: the power spectrum and rescaled range (R/S) analysis. We will define and discuss them further below, but for now we simply say that they are two measures that quantify a time series in the frequency domain and time domain, respectively. Both measures can produce distinct straight line regions when plotted on doubly logarithmic axes, indicating power law scaling. Two important features of a SOC time series in this regard are that it 1) produces multiple power law regions and 2) the number and behavior of these power law regions are very different from any other known model, including random uniform or Gaussian noise, fractional Gaussian noise, fractional Brownian motion and random superposition of pulses. We elaborate on the significance of power laws below.

For our results, we first show that the spectral and R/S signatures of a SOC system can change drastically when the external forcing of the system changes. Because of this, there is no single reference signature that a system must match in order to be considered SOC. An application of this idea is in considering the fluctuations in the solar wind as the external driver of the magnetosphere, for which SOC has been suggested as a model. Since the solar wind can range from very strong and steady to practically nonexistent, the reaction to this changing forcing can easily produce different spectra over different time scales.

This result, changing spectrum with changing forcing, can be seen in two ways: by changing the strength of the external drive and by changing the level of correlations in the external drive. Strength of drive is the intuitive notion of how much of some quantity is being deposited in the system per unit time. In terms of the solar wind, for example, very weak drive was seen during the period 10-12 May 1999, when it almost disappeared *Le et al.* [2000b, a]. (In general, though, SOC is concerned with dynamics on much longer time scales.) The level of correlations deals with the issue of whether the external drive is completely random or not. For instance, the solar wind has been seen to be a correlated source so that the subsequent external driving of the magnetosphere is not completely random. We will only discuss strength of driving here. Refer to *Sánchez et al.* [2002] for studies of the effect of correlations in the drive in a SOC system.

Second, we show that the longest available time series of a space climate process is likely far too short to show all of the power law regions in the spectra and R/S measures associated with SOC. 'Too short' here implies a needed time scale on the order of a century. That is, many more generations of scientists will pass before long enough time series are acquired and this issue can be settled.

The rest of this paper is organized as follows. We first review power laws, the power spectrum and rescaled range analysis in Section 2 so that we can refer to them in Section 3, which is a brief overview of self-organized criticality, our model and some recent work in space physics from the perspective of SOC. We will present and discuss our results of the strength of forcing and the effects of the length of a time series in Sections 4 and 5. Conclusions are drawn in Section 6.

2. Power Laws, the Power Spectrum and R/S Analysis

A characteristic associated with SOC since its introduction have been power laws. A power law is any function of the form $y = cx^a$. On doubly logarithmic axes, this function appears as a straight line with slope a; a is also referred to as the scaling exponent. Power laws appear in many measures of many systems, such as probability distributions, power spectra and R/S analysis. We will only discuss the spectral and R/S analyses. For a thorough discussion of power laws in probability distributions refer to *Sornette* [2000] and references therein.

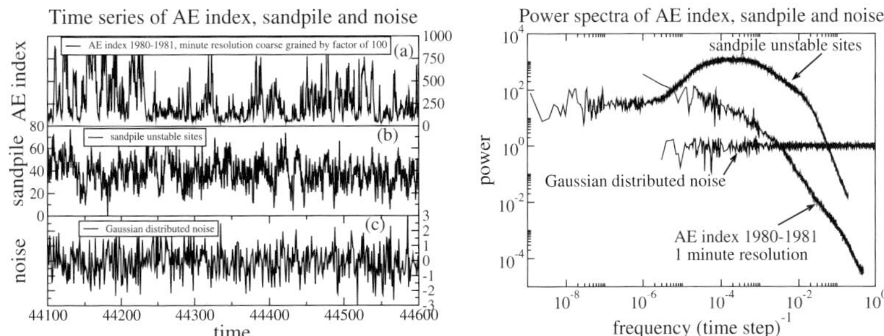

Figure 1. Time series and power spectra of AE index, sandpile and Gaussian noise.

When discussing power laws, a usual unstated assumption is that the power law is an appropriate fit of the data over a reasonably wide domain. 'Reasonably wide' is vague so this criteria must be separately established for each case. A common critical, yet reasonable, observation made in regard to these studies is that any function—power law, exponential, sine curve, etc.—can be fit very well to any data if one zooms in to a small enough scale on a plot. Moreover, when two neighboring power law regions are claimed, with different scaling exponents, the additional issue of whether or not there is a distinct breakpoint between the two regions is raised. Needless to say, these are critical issues and the scientist must be aware of them. We are, and elsewhere we have investigated them thoroughly for the data that we present here *Woodard et al.* [2004a]. In general, we find that a decade (power of 10) is a reasonable minimum for establishing a power law as a good fit.

Having established that a power law is a good fit to a region and that the limits of the region are identified by breakpoints that separate neighboring regions that may or may not also be fit by a power law, the most important task is to identify the process or processes in the system that produce such a signature. This is where understanding of the particular measure is needed. In practice we find that using more than one measure is invaluable in attempts to understand such systems and signatures. This allows us to use multiple measures to distinguish and quantify separate regions. Hence we complement the power spectrum with the lesser known R/S analysis, with which we estimate the Hurst exponent H.

The power spectrum of a discrete time series $f(\tau)$ allows one to examine the data in frequency space. It is defined as $S(f) = |F(f)|^2$, where $F(f)$ is the Fourier transform of $f(\tau)$; power is plotted versus frequency. The spectrum of random noise (with Gaussian or uniform distribution) is flat ($\sim f^0$) so that the power at all frequencies is the same.

The task of any spectral analysis is to understand why a system has a spectrum that differs from the flat one of a completely random series. A simple example is the spectrum of a sine curve; it is the same flat spectrum as that of the random one except for a spike at the characteristic frequency, indicating where most of the power is concentrated. (Analytically, the spectrum is a delta function; we will deal mainly with finite discrete time series.) A spectrum that has more than one such peak indicates multiple periodic processes in the system.

Besides distinct peaks, another feature of some power spectra is a power law dependence on frequency, where the spectrum scales as $f^{-\beta}$ with $\beta \neq 0$; β can be positive or negative. The power law extends over a finite frequency band (for a finite time series) and there may be more than one such scaling region. Again, the task is to understand why this spectrum is different from that of a random one. Specifically, one must understand why each region has a particular value of β and why the breakpoints of the region occur where they do.

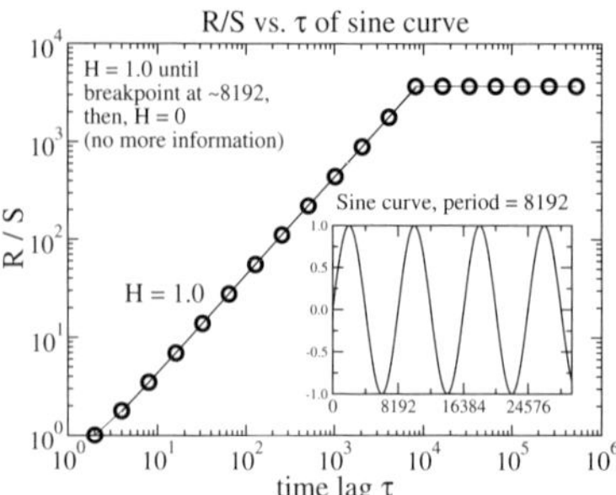

Figure 2. Time series and R/S analysis of sine curve.

Some spectral power laws are not due to dynamics in a system and can be analytically derived. We refer to this type of power law as one due to pulse shape, which is the highest frequency component of a series. The smoother a function is, the faster its spectrum falls off with increasing frequency. This is quantified by taking derivatives of the function. The spectrum falls off as f^{-2k} if the kth derivative becomes impulsive. For instance, a sawtooth wave (a triangular pulse) of width w has an impulsive second derivative so its spectrum scales as f^{-4} as $f \to \infty$. For frequencies below w^{-1}, the spectrum is flat. One can extend this and think of the sine wave as a function that does not have any impulsive derivatives; its spectrum, then, the delta function, falls off infinitely fast with increasing frequency *Bracewell* [2000]. Another analytically derived power law spectrum is that of a random superposition of square pulses, which falls off as f^{-2} for frequencies above the inverse of the widest pulse *Jensen et al.* [1989]; *Kertész and Kiss* [1990].

Part of the interest in physical systems that have power law scaling regions in their power spectra is because the time series themselves are far from the simple examples above. Most time series of 'real' physical processes, such as the AE index or fluctuations in the solar wind velocity, are not simple shapes or superpositions of simple shapes. They look noisy, almost random. Yet they still have spectra that are very different from that of a completely random process (Figure 1). Spikes in the spectra can usually be explained by known periodicities in the system (rotation of the earth, 11 year solar cycle, etc.). But the observed values of β and the locations of breakpoints between scaling regions, in most cases, are not well understood and are probably important indicators of the dynamics.

Qualitatively, a spectrum that falls off as $f^{-\beta}$ with $\beta > 0$, a negative slope on log-log axes, means that lower frequencies are most important in characterizing the time series. For values of $\beta < 0$, a positive slope, the higher frequencies dominate the signal. So signals with $\beta > 0$ are smoother than those with $\beta < 0$. Because of this, the smoother signals with positive β are said to be correlated and the rougher ones with negative β are anticorrelated. Here, correlated implies that a current trend in the data continues. The value of β, then, is often used as a measure of correlations in a time series.

A different and perhaps better measure of correlations in a time series is the Hurst exponent, $H \in [0, 1]$. A value of $H = 0.5$ implies a data set that is completely random, with no correlations. Series with $H > 0.5$ are correlated and those with $H < 0.5$ are anticorrelated. The closer H is to 0 the rougher and more anticorrelated is the signal; the closer to 0.5, the more uncorrelated; and the closer to 1 the smoother and more correlated. For instance, $H = 0.5$ for a time series of fair dice being rolled and $H = 1$ for a sine curve up to its period (Figure 2). *Price and Newman* [2001] has estimated the Hurst exponent of the AE index as $H \approx 0.7$, indicating that the AE process has long time correlations.

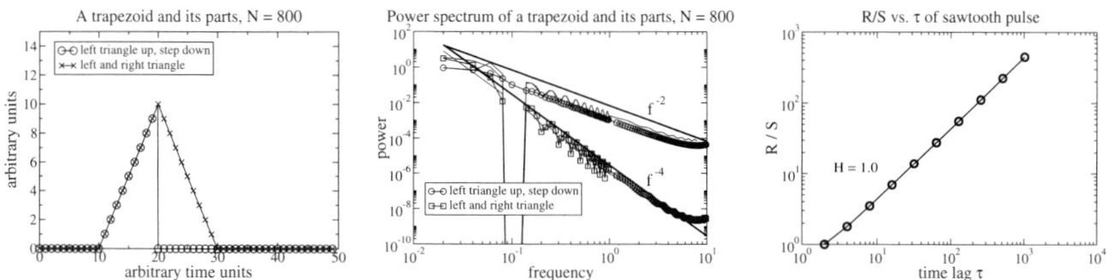

Figure 3. Time series, power spectrum and R/S analysisof a sawtooth pulse.

One technique of estimating H is through rescaled range (R/S) analysis *Hurst* [1951]; *Mandelbrot* [2002]; there are other methods *Bassingthwaighte and Raymond* [1994]; *Bassingthwaighte et al.* [1994]; *Bassingthwaighte and Raymond* [1995]. For a series of data ξ_t, the rescaled range is defined as:

$$R'(s) \equiv R(s)/S(s) \quad \text{(rescaled range)}$$

$$R(s) = \max_{1 \leq t \leq s} X(t,s) - \min_{1 \leq t \leq s} X(t,s), \quad \text{(range)}$$

$$X(t,s) = \sum_{u=1}^{t} (\xi_u - \langle\xi\rangle_s) \quad \text{(cumulative deviation)}$$

$$S(s) = \left[\frac{1}{s} \sum_{t=1}^{s} (\xi_t - \langle\xi\rangle_s)^2 \right]^{1/2} \quad \text{(standard deviation)}$$

$$\langle\xi\rangle_s = \frac{1}{s} \sum_{i=1}^{s} \xi_i \quad \text{(mean)}$$

If the rescaled range of the time series scales as $R'(s) \sim s^H$, the slope of the plot of $R'(s)$ versus the time lag s on a doubly logarithmic plot is an estimate of the Hurst exponent, H.

H can be related to the more familiar measure, variance. Consider classical Brownian motion (a random walk), the increments of which are simply a Gaussian distributed random noise. For a large ensemble of random walks, the expectation value of the variance of the motion scales linearly with time, $\sigma^2 \sim t$. For the noise series, $H = 0.5$. In general, the variance of the motion is related to the Hurst exponent of the noise by $\sigma^2 \sim t^{2H}$.

Algorithms exist *Mandelbrot* [2002] that create synthetic time series with Gaussian distributions and a given value of $0 < H < 1$. Such a series is called fractional Gaussian noise (fGn) and when integrated produces fractional Brownian motion (fBm). For these series, H is analytically related to the slope of the power spectrum of the fGn via $\beta = 2H - 1$. For discrete fGn data, though, this relation does not hold so well at all values of H. A thorough discussion of H and β in the context of fGn is given in *Malamud and Turcotte* [1999].

It is very important to note that the relation $\beta = 2H - 1$ is not always true for physical data. This is equivalent to saying that fGn is not an appropriate model for the system under study. A basic example is the time series of a single sawtooth pulse mentioned above. It has a power spectrum that falls off as f^{-4} and a Hurst exponent, calculated via R/S analysis, of $H \approx 1$ (Figure 3). Here, $\beta \neq 2H - 1$; the spectrum and the R/S analysis describe two different aspects of the same series. $\beta = 4$ is a statement about the discontinuity of the first derivative of the sawtooth pulse and $H \approx 1$ is a statement of the strong correlation that a defined shape has with itself for time scales up to its width.

Self-organized criticality is a model where $\beta \neq 2H - 1$ over different time scale regions. But the reasons are not as simple as for that of a single pulse. Here, the dynamics of a SOC system has correlations, anticorrelations and lack of correlations on distinctly different time scales and on time scales larger than that of the largest fundamental single pulse. The changing levels of correlations appear in the power spectrum and R/S analysis of a SOC system as separate and distinct power law regions.

Looking at only the spectrum or the R/S alone does not adequately explain the regions of a SOC system. But together the measures give a clear picture of SOC dynamics. We emphasize this to encourage the use of multiple measures in any system; taken alone, the power spectrum does not always explain the dynamics of a system.

3. Self-Organized Criticality and the Sun-Earth System

The underlying idea behind self-organized criticality is that, in many complex systems, simple local interactions produce complex global signatures that are not easily predicted by the local low level physics. That is, the fundamental physics of a system is often understood but some of the observed signatures are not captured by models that are built on that physics. The problem is that, inevitably, all models must leave something out, must make some approximations. An example is the power spectrum of the AE index seen in Figure 1.

The main approximation that a SOC model makes is to reduce all of the local physics in a system to a simple physical rule: if the local gradient of some quantity between two nearest neighbors exceeds some critical gradient, then reduce the gradient by transporting some of the quantity from one neighbor to the other. A 'neighbor' is purposefully vague and can be different for each system. It represents the notion that gradients exist on macroscopic scales and the source of the gradients is not as important as the fact that they grow, shrink and interact. Crucial to the SOC dynamics are that the time scales of the driving and relaxing processes are very different: the gradients are reduced much faster than they are produced. Within these bounds, a plethora of models can be constructed but all adhere to this one rule; we describe our model, the sandpile, below. The interesting dynamics appears because the transport from one neighbor to the next may make the next local gradient exceed the critical gradient, causing a new transport event. In this way, disturbances can propagate throughout the model.

In the jargon of SOC, the disturbances are often referred to as avalanches. Avalanches can range in size from the smallest possible (one transport event) up to the size of the system. Over much of this range of scales, SOC avalanches are distributed as a power law, indicating no preferred spatial scale within that range. This is the criticality part of SOC. It refers to statistical mechanics, where disturbances in a material at a phase transition can propagate throughout the entire sample. In such a case, a control parameter—the temperature—must be tuned to reach this critical state. In contrast, a SOC system arrives at criticality with no external tuning of a parameter [†]. The system self-organizes to the critical and steady state.

Time series of the avalanches can be constructed and analyzed. Regions of the power spectra of these time series scale as power laws, indicating correlations and anticorrelations on different time scales. These temporal power laws together with the spatial power laws of the avalanche size distributions are important signatures of SOC systems that are similar to signatures of many observed physical systems. Because of these similarities and of the underlying physics, the Sun-Earth system has been studied as possibly SOC.

[†]This is a matter of discussion in the world of SOC. See, for example, *Jensen* [1998] and references therein for a discussion.

The model that we use is the one dimensional running sandpile, studied extensively in *Hwa and Kardar* [1992]; *Woodard et al.* [2004a]. The sandpile consists of L cells labeled by an index $n \in [1, L]$. Each cell stores an amount of sand h_n and the local gradient between two cells is defined as $Z_n = h_n - h_{n+1}$. U_0 grains of sand are dropped randomly on every cell at each iteration with probability P_0. The external drive per cell is thus $S_0 = U_0 P_0$ grains per time step. SOC dynamics appears because of the existence of a critical slope Z_{crit} that, when locally overcome, triggers the removal of N_f grains of sand to the next cell in the downhill direction (increasing n). The sandpile is initialized with $h_n = (Z_{\text{crit}} - 1)(L - n)$ and run for $T_{\max}$ time steps. We study the sandpile in steady state so that transient time steps before T_T are ignored in the following analysis.[‡] We used $U_0 = 1$ (therefore $S_0 = P_0$), $Z_{\text{crit}} = 8$ and $N_f = 3$, the same parameters used in a study of confined plasmas *Newman et al.* [1996]. In this study, we use sandpile sizes of L = 200 and 1000 and driving rates of $P_0 L = 0.2$, 0.1 and 0.001. We study a more complete range of five orders of magnitude of driving rate and three orders of magnitude of system size in *Woodard et al.* [2004a].

Effective driving rate is given by $P_0 L^2$ *Woodard et al.* [2004a]. The idea of effective driving rate is that a fixed driving rate P_0, which is in units of grains per cell per time step, is effectively higher for a small sandpile than for a larger one. Since a larger sandpile has a greater capacity, P_0 will take longer to fill it up when compared with the same rate of sand falling onto a smaller system. In the past, $P_0 L$ has been used as the measure of driving rate because its units are in grains per time step for the entire system. But since the average avalanche size is larger in a larger sandpile, two sandpiles of different sizes but with the same $P_0 L$ can be in very different drive regimes because the quiet times will be shorter in the larger system. We find that a better measure, effective driving rate, should be used. Systems with different size and/or P_0 but with the same value of $P_0 L^2$ have power spectra that are related via a rescaling function. We elaborate on effective driving rate in *Woodard et al.* [2004a].

The time series that we analyze is called the flips. Consider the total number of unstable sites (where $Z \geq Z_{\text{crit}}$) at each time step in a sandpile model in steady-state. An unstable cell spills N_f grains of sand to the next cell; this action is a flip. The total flips at each time step can be thought of as the instantaneous (potential) energy dissipation in the system. The sandpile is driven by a random process but the number of flips fluctuates with time in a way that is not entirely random, as we show.

4. Effect of Strength of External Forcing and System Size

We now show that systems with very different driving rates can have very different spectral and R/S signatures. Figure 4 shows the power spectra from three different sandpile runs. Runs (a) and (b) differ in their driving rate $P_0 L$ and runs (b) and (c) differ in their driving rate and system size, L. Run (a) is the smallest system with the highest driving rate; run (b) is the same size but with a lower driving rate; run (c) has a lower driving rate and larger system size than (b). The effective driving rate $P_0 L^2$, then, decreases from run (a) through (c), with (c) being the lowest effective driving rate. The spectra can be seen to change systematically.

The spectra in Figure 4 show different power law scaling regions for all driving rates. Each spectrum actually has 4 or 5 separate regions, depending on effective driving rate *Woodard et al.* [2004a]; we have sketched lines on each indicating only three of these regions. The highest frequency region with $\beta \approx 3.5$ indicates correlations at short time scales, where avalanche pulse shapes correlate with themselves. On longer time scales, the middle frequency region has a slope $0 < \beta \leq 1$ that changes with driving rate. These positive values of β indicate correlations among separate and overlapping avalanches. The time scales for this region are greater than the maximum duration of

---

[‡]The initialization saves computer time; the same results hold when the sandpile is started from any initial condition. The transient time T_T must be adjusted accordingly.

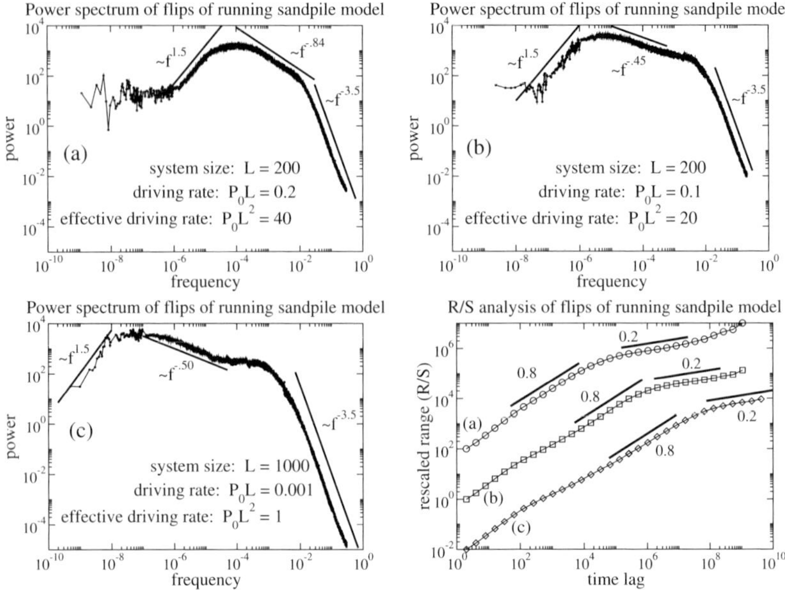

Figure 4. Power spectra and R/S analysis of flips of running sandpile model for three different values of effective drive, $P_0 L^2$. Spectra and rescaled range have been shifted along the y-axes so that all plots are over the same range. Estimates of the Hurst exponent are shown for the straight line segments in the R/S plot. These are not fits of the data but lines of the indicated slope to guide the eye.

a single avalanche. The lowest frequency region with $\beta \approx -1.5$ indicates anticorrelations on the longest time scales. These are the time scales of global discharge events, where one such event is unlikely to be followed by another similar event.

The R/S analysis of sandpile flips also shows multiple regions. We have only indicated two in Figure 4. These correspond to the middle and long time scales of the power spectra. The estimated values of the Hurst exponent can be compared with the values of β in the spectra. For both cases, $\beta \neq 2H - 1$. Most importantly, the middle time scale region has a constant value of $H \approx 0.8$ while β changes with driving rate. This is because H measures correlations in the sandpile that are produced by the memory in the system due to the system rules. The rules do not change as driving rate changes, hence H stays relatively constant. Because of the changing values of β in this region, it is not clear what aspect of the correlations is measured by the power spectrum.

The goal of these results is to show that the SOC running sandpile model will produce very different power spectra when the level of external forcing changes. This is why comparing the values of the slope or the number of regions of the spectrum of a physical system like the Sun-Earth system with those of a single run of a SOC model can be misleading. If the values of the slope do not match or if the number of regions are different, this does not necessarily mean that the system is not SOC. Instead, it could be that the system is in a different drive regime than that of the model being used.

An important aspect of the running sandpile model is that over a very wide range of driving rate it is still in a state of SOC. That is, it still exhibits all of the qualitative signatures that collectively are called SOC. These include power law distributions of event sizes, power law regions in the power spectra, correlated dynamics indicated by Hurst exponents $H \neq 0.5$ and regions and event sizes that scale with system size. More importantly, the physics in the models is identical *regardless of driving rate*. Sand is still added at a constant probability that triggers avalanches that may or may not spread throughout the entire system. The memory of the avalanches is retained in the local gradients of each cell and this is the source of long time correlations. The only difference is the rate at which the avalanches and the time scales over which the correlations occur.

This idea can be applied to data that is already in the literature. The first power spectrum of the AE index showing the broken power law behavior was seen in *Tsurutani et al.* [1990] for data from three separate periods, 1967-1970, 1971-1974 and 1978-1980. Similar spectra were shown subsequently by *Consolini et al.* [1996], *Uritsky and Pudovkin* [1998], *Price and Newman* [2001] and *N. W. Watkins* [2002]. The time periods studied in all of these works, as well as the slopes of the spectra in the two regions and the location of the breakpoints are shown in Table 1. That they vary over different periods is completely consistent with an SOC model that runs for a time at one driving rate and then for another time at a different driving rate. In general, the slopes found for shorter periods of observations differ the most from the average of all samples. In the language of the sandpile, there are two possible causes for this.

First, the driving rate can change over time so that the average input of the solar wind over the period 1 January through 19 February 1975 was very different from the period 1978-1979. These two periods were chosen as examples because they show the greatest difference in slopes for both spectral regions among the values presented. That one period is near the solar minimum and the other near the solar maximum should be noted and taken as a possible example of the level of external forcing changing the power spectrum, as in the running sandpile model.

A second and, in this case, more likely possibility that can account for changing spectra for the time periods shown is that over a long time scale, say the roughly 40 year period for which we have AE index data, the input to the system (the solar wind) is constant in the same way that the mean of a series of random numbers is constant. But when one looks at a small subset of the random series, the mean may be very different and there will certainly be fluctuations far from the overall mean. In other words, data is scattered within the errors bars of a sampling. All of the values of β in Table 1 are within 2σ of the mean.

Both of these possibilities are intuitively appealing because the Sun, the driver in the Sun-Earth system, is known to not have a constant output from year to year. But whether that change is due to a fundamental change in the drive regime of the Sun or to intrinsic fluctuations within a steady state is not clear.

As a quick test of this idea, we compared power spectra for three different three day periods, 10-12 May 1999, 26-28 January 1999 and 12-14 November 1999. The solar wind essentially vanished during the first period *Le et al.* [2000b, a], falling more than 98% for a period of approximately 30 hours. The other two periods were closer to the yearly mean. There is no significant difference in the spectra among the three cases. We attribute this to the very short observation time, as the spectra show much ringing at high frequencies, making determination of any sort of fit statistically irrelevant. We must wait for longer periods of no solar wind to pursue this idea further.

In addition to those of measured AE indices, the literature also holds power spectra of a continuum model that "provides a link between the sandpile model studies ... and a realistic plasma physical study of SOC dynamics in the plasma sheet *Klimas et al.* [2000]." That study refers to the Lu model of *Lu* [1995], which "can be viewed as an idealized one-dimensional resistive field reversal model in which the resistivity is generated self-consistently." This is a model where a scalar field is evolved in time while coupled to a variable diffusion coefficient and source term that are space

Table 1. Time period, resolution, slopes of first two spectral regions and breakpoint of AE index data found in *Tsurutani et al.* [1990], *Consolini et al.* [1996], *Uritsky and Pudovkin* [1998], *Price and Newman* [2001] and *N. W. Watkins* [2002]. Breakpoint for *Consolini et al.* [1996] taken between labeled second and third regions. Breakpoint for *Price and Newman* [2001] 1978-1979 estimated from plot at intersection of two power law fits. Slope for *N. W. Watkins* [2002] taken as best fit with a straight edge, slope estimated from axes.

Study	Period	Res.	β_A	β_B	Break (mHz)
Tsurutani et al. [1990]	1967–1970	5 min	2.42	1.02	0.059 (4.7 hr)
Tsurutani et al. [1990]	1971–1974	1 hr	2.2	0.98	0.050 (5.5 hr)
Uritsky and Pudovkin [1998]	1973–1974	1 hr	2.10	0.95	0.056 (5.0 hr)
Consolini et al. [1996]	1/1–19/2 1975	1 min	2.65	1.14	0.073 (3.8 hr)
N. W. Watkins [2002]	1978	5 min	2.1	1.1	0.056 (5.0 hr)
Price and Newman [2001]	1978–1979	1 min	1.85	0.82	0.033 (8.4 hr)
Tsurutani et al. [1990]	1978–1980	1 hr	2.2	1.00	0.056 (5.0 hr)
Price and Newman [2001]	March 1979	1 min	1.89	n/a	n/a
Mean $\pm\ \sigma$			2.4 ± 0.26	1.0 ± 0.10	

and time dependent. In terms of the sandpile model, the scalar field of the Lu model represents the height or gradient at each cell, the variable diffusion represents the avalanche rule and the source term represents the rain of sand. [§]

Power spectra of this model are shown in *Klimas et al.* [2000] for varying levels of the source term and the diffusion operator. As in the spectra of the running sandpile for varying driving rates, these spectra exhibit a wide range of behavior. Also, as in the sandpile model, the same physics from the same system produces these different spectra. This says that a system in a state of SOC can show very different signatures depending upon the drive regime in which it operates.

Concerning the anticorrelated region of the spectra, reference *Klimas et al.* [2000] states that the comparison of that model with the running sandpile appears to fail in this region because no system-wide discharges where all grid points are simultaneously unstable are seen in the continuous model. Instead, the largest events observed show wave-like behavior in the hydrodynamic regime. This behavior is, in fact, consistent with the sandpile model because of the following.

We have performed additional sandpile runs *Woodard et al.* [2004a] that append the results of *Hwa and Kardar* [1992] and show that all sites are rarely, if ever, simultaneously unstable. System-wide discharges do not refer to a single time step where all cells are unstable. Instead, these large events occur over a short period of time. Recall that sand only exits the sandpile through the bottom cell, which has a small and fixed amount of sand, N_f, that it can transport at a single time step. The nature of the sandpile is such that, in what is called the $N_f/2$ limit, a cell in the middle of a spatially extended avalanche alternates between stable and unstable until the avalanche has either washed passed it or died. The drive regimes studied by *Hwa and Kardar* [1992] and *Woodard et al.* [2004a] are well below the $N_f/2$ limit. Furthermore, animated visualizations *Woodard et al.* [2004b] of the sandpile show that these system-wide discharge events are really a series of many avalanches in a short period of time and that they are very wave-like in nature. So then the observed behavior of the Lu model discussed in *Klimas et al.* [2000] is consistent with that of the running sandpile. This is another motivation for looking for similar SOC regions in physical data, as we next discuss.

[§] In the interest of cross-field communication, we mention that a very similar model to that of *Lu* [1995] has been applied to the study of the running sandpile model in the context of plasma transport in confined plasma devices *Garcia et al.* [2002]. Citations within that work refer to other studies of sandpile dynamics in confined plasmas.

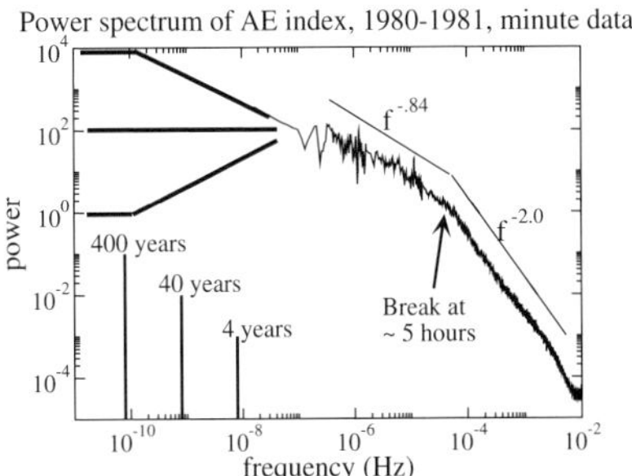

Figure 5. Power spectrum of AE index 1980-1981 and cartoon of time scales of additional spectrum.

5. Effect of Length of Time Series

The comparison of regions between the AE index and the sandpile model end at region B because of the limited time series for the AE index. The question arises: How long must the AE index time series be in order to see a new region at lower frequencies? This question assumes that the long-term process that drives the Sun-Earth system and that some properties of the Sun, magnetosphere and the space in between remain somewhat constant.

This question also assumes that the lower region seen in the spectrum of the AE index *will* end. Of course, the lifetime of a star ensures this on the longest time scales but this ending falls outside the bounds of the first assumption. We claim that this region will end for dynamical reasons. The positive slope $\beta > 0$ of a spectral region and Hurst exponent $H > 0.5$ imply long time correlations. But the correlations must end on some time scale because a system of finite size driven randomly will reflect the random drive at the longest time scales *Woodard et al.* [2004a].

Given a steady state Sun-Earth system and a randomly driven yet deterministic universe, then the correlations must end at some time scale. Anticorrelations may arise in this system in the same way that they do in the running sandpile: all gradients in a finite system eventually grow close to critical and then trigger a system-wide discharge. This speaks to the finite capacity of the system. Perhaps an energy and mass balance calculation between the fluctuations of the input from the solar wind and output through the magnetotail, magnetopause and other parts of the system can give an estimate of a time scale on which the magnetosphere reaches capacity.

Figure 5 shows the same spectrum of the AE index as that in Figure 1 but now we have drawn lines showing possible behavior of the spectrum at longer time scales based on the behavior seen in sandpile spectra. We adamantly state here: the breakpoints and slopes of these lines are not predictions of breakpoints and slopes that would be seen given arbitrarily long AE time series. Instead, they are drawn to demonstrate how long the time series would have to be to see this behavior. At best, we now have roughly 40 years of AE index data. Since we look at scaling behavior on log-log axes, the full 40 years of data would only slightly extend the plot to the point labeled. An order of magnitude greater time series, 400 years, would extend the plot as shown. Observation of the spectra of sandpile models, AE index and other physical systems shows that nature does not cut her spectra off abruptly at low frequency. This is, of course, a rough and qualitative argument. But we do expect to see the spectrum of the AE index roll over to, possibly, an anticorrelated region with $\beta < 0$ and/or, definitely, a flat, uncorrelated region. Based on the slope of the best current spectra and longest available time series, we will not be around to see that new region.

6. Conclusions

We have shown that a rich diversity of spectral signatures can be produced by a self-organized critical system when the effective driving rate is changed. This has implications in the ongoing investigations of studying SOC as a model of various physical systems, including the Sun-Earth system. Our results show that failure of the power spectrum of a physical system to match that of the running sandpile model with a certain driving rate does not at all exclude SOC as an appropriate model for the system. A system can be in a wide range of drive regimes yet still be considered SOC even though the spectral signatures differ among the different regimes. This is a statement on the lack of tuning that is required for a SOC system to be critical; it is the self-organized portion of the name.

The system remains critical because the physics does not change when the driving rate changes. In the sandpile, changing the driving rate merely changes how fast or how slowly sand is added to the pile. The rules of the system have not changed. A memory of past events is stored in the local gradients of the system regardless of how fast or slowly sand is added. The analogy to a physical system is this: the underlying physics of the transport of a system, such as the Sun-Earth system, does not change when it is driven more strongly or weakly. Gradients still grow and shrink by the fundamental physical processes. But the rate at which this happens changes and this change is reflected in the different spectra of a process over different periods of time.

We also discussed that the current longest available time series of the AE index only shows two regions in the power spectrum. Inevitably, the region at lower frequency with $\beta \approx 1$ must either flatten or turn down to an anticorrelated region with $\beta < 0$ before flattening at the lowest frequencies. The flat f^0 spectrum is a signature of a fundamental random process that is driving the system. An anticorrelated region reflects a system of finite size and capacity that non-periodically relaxes in a series of long bursts over a short period of time.

The running sandpile model has a well-defined clock which makes it diffferent from other SOC models. More importantly, this feature, along with its dynamics that is similar to many physical processes, make it an appropriate model to use in the study of physical systems that are suspected to be SOC, such as that of the Sun and Earth.

Acknowledgments

R.W. thanks N. W. Watkins for noticing, H. J. Fletcher for comments and support and W. C. Ferenbaugh for assistance with unit conversion. We also thank the DOE for grants DE-FG03-99ER54551 and DE-FG03-00ER54599

References

Angelopoulos, V., T. Mukai, and S. Kokubun, Evidence for intermittency in earth's plasma sheet and implications for self-organized criticality, *Phys. Plasmas*, *6*(11), 4161–4168, 1999.

Bak, P., C. Tang, and K. Wiesenfeld, Self-organized criticality: an explanation of $1/f$ noise, *Phys. Rev. Let.*, *59*, 381, 1987.

Bak, P., C. Tang, and K. Wiesenfeld, Self-organized criticality, *Phys. Rev. A*, *38*(1), 364–374, 1988.

Bassingthwaighte, J. B., and G. M. Raymond, Evaluating rescaled range analysis for time series, *Annals of Biomedical Engineering*, *22*, 432, 1994.

Bassingthwaighte, J. B., and G. M. Raymond, Evaluation of the dispersional analysis method for fractal time series, *Annals of Biomedical Engineering*, *23*, 491, 1995.

Bassingthwaighte, J. B., L. S. Liebovitch, and B. J. West, *Fractal Physiology*, Oxford University Press, 1994.

Bracewell, R. N., *The Fourier Transform and Its Applications*, 3rd ed., McGraw-Hill, 2000.

Chang, T., S. W. Y. Tam, C.-C. Wu, and G. Consolini, Complexity, forced and/or self-organized criticality, and topological phase transitions in space plasmas, *Space Sci. Rev.*, *107*, 425–445, 2003.

Chang, T. S., Low-dimensional behavior and symmetry breaking of stochastic systems near criticality—can these effects be observed in space and in the laboratory?, *IEEE Trans. on Plasma Science*, *20*, 691–695, 1992.

Chang, T. S., Self-organized criticality, multi-fractal spectra, sporadic localized reconnections and intermittent turbulence in the magnetotail, *Phys. Plasmas*, *6*, 4137–4145, 1999.

Chapman, S. C., and N. W. Watkins, Avalanching and self organised criticality: a paradigm for magnetospheric dynamics?, *Space Sci. Rev.*, *95*, 2001.

Consolini, G., M. F. Marcucci, and M. Candidi, Multifractal structure of auroral electrojet index data, *Phys. Rev. Let.*, *76*(21), 4082–4085, 1996.

Garcia, L., B. A. Carreras, and D. E. Newman, A self-organized critical transport model based on critical-gradient fluctuation dynamics, *Phys. Plasmas*, *9*(3), 841–848, 2002.

Hurst, H. E., Long-term storage capacity of reservoirs, *Trans. Am. Soc. Civ. Eng.*, *116*, 770, 1951.

Hwa, T., and M. Kardar, Avalanches, hydrodynamics, and discharge events in models of sandpiles, *Phys. Rev. A*, *45*, 7002, 1992.

Jensen, H. J., *Self-Organized Criticality: Emergent Complex Behaviour In Physical And Biological Systems*, Cambridge University Press, Cambridge, 1998.

Jensen, H. J., K. Christensen, and H. C. Fogedby, $1/f$ noise, distribution of lifetimes, and a pile of sand, *Phys. Rev. B*, *40*(10), 7425–7427, 1989.

Kertész, J., and L. B. Kiss, The noise spectrum in the model of self-organised criticality, *J. Phys. A*, *23*, L433–L440, 1990.

Klimas, A. J., J. A. Valdivia, D. Vassiliadis, D. N. Baker, M. Hesse, and J. Takalo, Self-organized criticality in the substorm phenomenon and its relation to localized reconnection in the magnetospheric plasma sheet, *J. Geophys. Res.*, *105*(A8), 18,765 – 18,780, 2000.

Le, G., P. Chi, W. Goedecke, C. Rusell, A. Szabo, S. Petrinec, V. Angelopoulos, G. Reeves, and F. Chun, Magnetosphere on may 11, 1999, the day the solar wind almost disappeared: Ii. magnetic pulsations in space and on the ground, *Geophys. Res. Let.*, *27*(14), 2165–2168, 2000a.

Le, G., C. Rusell, and S. Petrinec, The magnetosphere on may 11, 1999, the day the solar wind almost disappeared: I. current systems, *Geophys. Res. Let.*, *27*(13), 1827–1830, 2000b.

Lu, E. T., Avalanches in continuum driven dissipative systems, *Phys. Rev. Let.*, *74*(13), 2511–2514, 1995.

Lui, A. T. Y., S. C. Chapman, K. Liou, P. T. Newell, C. I. Meng, M. Brittnacher, and G. K. Parks, Is the dynamic magnetosphere an avalanching system?, *Geophys. Res. Let.*, *27*(7), 911–914, 2000.

Malamud, B. D., and D. L. Turcotte, Self-affine time series: I. generation and analyses, *Adv. Geophys.*, *40*, 1–87, 1999.

Mandelbrot, B. B., *Gaussian self-affinity and fractals*, Springer-Verlag, 2002.

N. W. Watkins, Scaling in the space climatology of the auroral indices: is SOC the only possible description?, *Nonlinear Processes in Geophysics*, *9*, 389–397, 2002.

Newman, D. E., B. A. Carreras, P. H. Diamond, and T. S. Hahm, The dynamics of marginality and self-organized criticality as a paradigm for turbulent transport, *Phys. Plasmas*, *3*, 1858, 1996.

Price, C. P., and D. E. Newman, r/s analysis of the ae index, *J. Atmos. Sol. Terr. Phys.*, *63*, 1387–1397, 2001.

Sánchez, R., D. E. Newman, and B. A. Carreras, Waiting-time statistics of self-organized criticality systems, *Phys. Rev. Let.*, *88*(6), 068,302, 2002.

Sornette, D., *Critical Phenomena in Natural Sciences—Chaos, Fractals, Selforganization and Disorder: Concepts and Tools*, Springer, 2000.

Tsurutani, B. T., M. Sgiura, T. Iyemori, B. E. Goldstein, W. D. Gonzalez, S. I. Akasofu, and E. J. Smith, The nonlinear response of ae to the imf b_s driver: A spectral break at 5 hours, *Geophys. Res. Let.*, *17*(3), 279 282, 1990.

Uritsky, V. M., and M. I. Pudovkin, Low frequency $1/f$-like fluctuations of the ae-index as a possible manifestation of self-organized criticality in the magnetosphere, *Annales Geophysicae*, *16*, 1580–1588, 1998.

Uritsky, V. M., A. J. Klimas, D. Vassiliadis, D. Chua, and G. Parks, Scale-free statistics of spatiotemporal auroral emissions as depicted by polar uvi images: Dynamic magnetosphere is an avalanching system, *J. Geophys. Res.*, *107*(A12), 7-1 – 7-11, 2002.

Woodard, R., D. E. Newman, R. Sánchez, and B. A. Carreras, Building blocks of self-organized criticality, part i: the very low drive case, submitted to PRE, 2004a.

Woodard, R., D. E. Newman, R. Sánchez, and B. A. Carreras, Building blocks of self-organized criticality, part ii: transition from low to high drive, submitted to PRE, 2004b.

CHAPTER TWO

SPACE STORMS

LONG-STANDING UNSOLVED PROBLEMS
IN SOLAR PHYSICS AND MAGNETOSPHERIC PHYSICS

S.-I. Akasofu

International Arctic Research Center, University of Alaska Fairbanks

Abstract. For many decades, there have existed a number of long-standing, unsolved problems in solar physics and magnetospheric physics. It is suggested that some of them remain unsolved because the guiding concepts, or paradigms, have no sound foundation. Here several paradigms are chosen for examination, in light of the related observations. They are the ionization rate of the solar wind, sunspots, solar flares/CMES, magnetic clouds, the concept of magnetic flux transfer in the magnetosphere, the diversion of the cross-tail current, and the dipolarization and substorm onset.

It is obviously not the intent of this paper to provide answers to those difficult problems. Rather, here are posed basic questions about the sources of the established paradigms, and a suggestion that the established paradigms are not necessarily the final answers.

Keywords. Solar flares/CMEs, magnetospheric substorms.

1. Introduction

In spite of considerable progress in the discipline of solar-terrestrial physics in recent years, a number of long-standing fundamental problems have remained unsolved for many decades. It may well be that some of these problems remain unsolved because no doubt has been cast on the guiding concept behind the prevailing ideas, not necessarily because we are presently incapable of solving them.

The guiding concept behind a prevailing idea is called a *paradigm*. This term was coined by Thomas Kuhn, a scientific philosopher. His definition of a paradigm:

"In the history of science, there are periods during which there is a high degree of agreement, both on theoretical assumptions and on the problems to be solved within the framework provided by those assumptions. The resulting coherent tradition of scientific research is called a paradigm. Scientists whose research is based on shared paradigms are committed to the same rules (including established viewpoints) for scientific practice."; see Akasofu (2002).

2. Solar Physics

In solar physics, existing paradigms include:

1. The solar corona is so hot that it can blow out against the solar gravitational field.
2. The solar wind is fully ionized plasma.
3. Sunspots appear when a magnetic flux tube breaks through the photosphere.
4. Solar flares/CMEs result from a sudden annihilation of magnetic energy (magnetic reconnection).

For this paper, I chose subjects 2, 3, and 4.

2.1 The Neutral Component of Solar Wind Particles

It is generally understood that the onset of a geomagnetic storm is characterized by a step-function-like change of the earth's magnetic field, the Storm Sudden Commencement (SSC). However, in the early 1960s, Sydney Chapman and I faced the problem of understanding the so-called 'gradually commencing storm' which has no SSC (Figure 1). Since the SSC is supposed to signal the arrival of the interplanetary shockwave and plasma 'clouds', compressing the dayside of the magnetosphere and causing the SSC, it was difficult to understand why gradually commencing storms can even occur. Since the Chapman-Ferraro theory of the SSC was well accepted then (and is also now), I thought that the only way to explain gradually commencing storms was to assume a cloud of neutral hydrogen atoms, which can penetrate into the magnetosphere without disturbing the dayside magnetopause (without causing the SSC) and can become energetic ring current protons after exchanging charges with terrestrial protons (Akasofu, 1964).

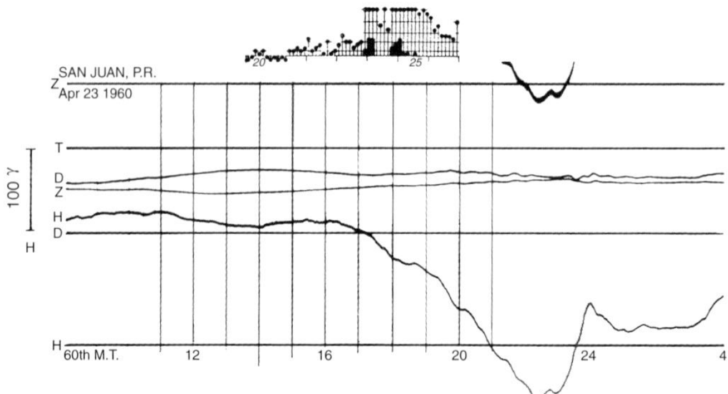

Figure 1: An example of a gradually commencing storm. An intense main phase can develop without a sign of the storm sudden commencement (SSC).

In order to examine such a possibility, I examined exploding prominences that must contain neutral hydrogen atoms; we know this because prominences are observed by hydrogen emission lines (neutral hydrogen atoms can produce emissions, but not ionized hydrogen atoms [protons]). I found that exploding prominences are visible for many hours in the corona and thus concluded, after some estimates, that they can escape from the sun, in spite of exposure to the sun's ionizing radiation. At that time, it was thought foolish to consider neutral hydrogen atoms in the solar wind, because they would be ionized by the solar radiation before escaping from the sun. However, Collier et al. (2001) recently detected neutral hydrogen atoms in the solar wind, although they may not be of solar origin.

2.2 Sunspots as a Result of a Magnetic Tube of Force Breaking through the Photosphere

All of us learned during our student days that the differential rotation of the sun winds up solar magnetic fields azimuthally and produces magnetic flux tubes of force just below the photosphere. When a tube of force breaks through the photosphere as magnetic buoyancy becomes large enough, the two intersecting areas of the tube can be identified as a pair of sunspots. Certainly, it is well known that sunspots form a pair, one with the magnetic field pointing toward the earth (positive) and the other one pointing away from the earth (negative).

I have pondered a naive question for a long time, one which even an elementary school child could pose: why does a simple spot (called the p-spot) form first before the following spot (called the f-spot) appears? This is hard to explain if a flux tube of force forms a sunspot pair. Solar physicists have not yet given me any reasonable answer to this simple question. I want to emphasize that there is no conclusive observation of magnetic flux tubes below the photospheric surface, so that the magnetic flux tubes are still nothing but a hypothesis.

In studying various shapes of sunspots, I came across a beautiful sunspot group that looked very much like a hurricane (Figure 2). It may be very difficult to explain such a spot group in terms of emerging flux; in fact, so far as I am aware, I am the only one who wrote a paper on this particular sunspot group, although suggestions similar to mine were made by several pioneering solar physicists. I believe that such a shape of sunspots suggests strongly that there is a powerful converging flow. It is well established that there is a diverging flow from the top of sunspots. It is also well known that weak magnetic fields are distributed on the photosphere. Indeed, it appears that a p-spot tends to form in a weak positive field (which is a remnant field of an old f-spot) and a f-spot in the weak negative period (which is also a remnant field of an old f-spot). A pair of spots can be formed across the weak positive and negative fields; when a converging flow occurs in a weak positive (or negative) field, the convergence can be induced in the conjugate area of the weak negative (or positive) field. Therefore, there is no impediment to the generation of sunspot pairs inherent in this idea (Akasofu, 1984).

74

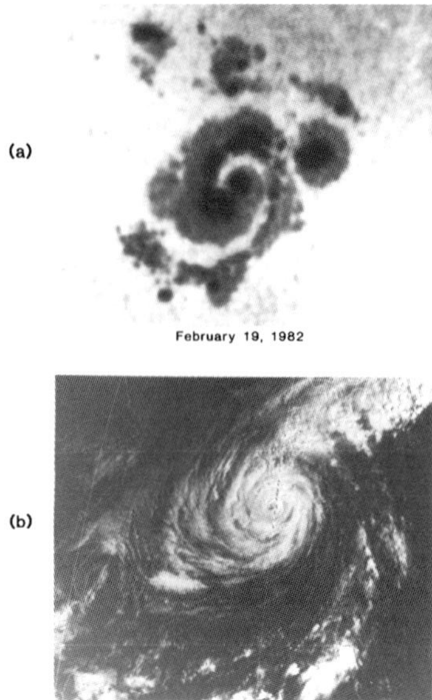

Figure 2:   Upper: An observed sunspot group in the northern hemisphere (Kitt Peak Solar Observatory). Lower: A photograph of a hurricane near Japan (Japan Meteorological Agency).

2.3   Magnetic Reconnection as the Cause of Solar Flares

In a simple pair of sunspots, the magnetic field may be approximated by the potential field. However, there are complicated flows around a pair of sunspots, and as a result the field becomes considerably distorted, resulting in what is called 'magnetic shear.' It has generally been believed that magnetic energy in the sheared magnetic field becomes available for the formation of solar flares. If the source of the energy for solar flares is contained in the sheared field, the shear energy should *decrease* at flare onset. Therefore, I asked Hal Zirin and his colleagues at Cal Tech to examine this particular point, using their excellent data obtained at the Big Bear Solar Observatory. They examined a number of X class flares and found that, without exception, the magnetic shear increases at flare onset (Wang et al., 1994), rather than decreases (as magnetic reconnection predicts); see Figure 3.

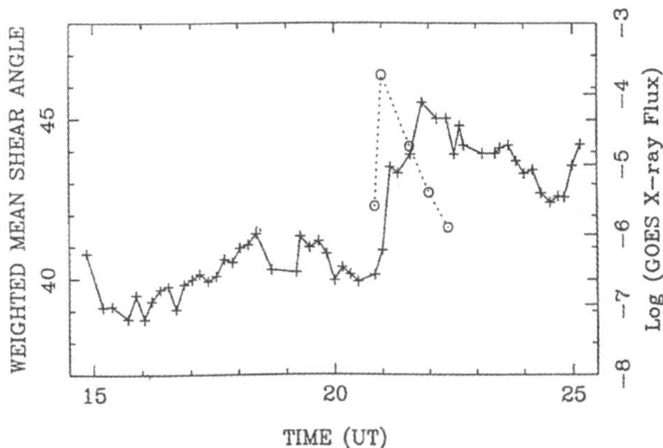

Figure 3:   An example of large increase of the magnetic shear at flare onset indicated by a sudden increase of the x-ray flux (dot-circle). (Wang et al., 1994)

I asked the Cal Tech people to examine this point because I believe that a flare occurs when the power of the dynamo process in the photosphere exceeds a certain value, rather than when the stored magnetic energy is expended (as practically all solar physicists believe). It is clear from the Zirin group's work and their following works that the solar flares occur when the power of a dynamo process exceeds a certain limit.

Solar physicists use the following MHD equation to understand solar flares, which are an electromagnetic energy dissipation process:

$$\frac{\partial B}{\partial t} = \nabla \times (V \times B)$$

$$V_p \equiv V_n$$

where V_p and V_n denote the velocity of plasma and the neutral gas, respectively; the second term of the first equation $\nabla^2 B/4\pi\sigma$ is ignored.. I believe that these equations are not useful for studying solar flare phenomena, which are manifestations of electromagnetic energy dissipation processes, while the two equations above can be used only for non-dissipative process. We tend to forget that the photosphere is a very weakly ionized gas, the ionization rate being 10^{-6} - 10^{-7}, so that the neutral component tends to control the ionized component. The dissipation process will reduce V_p, and $(V_n - V_p)$ supplies the necessary energy for solar flares (Kan et al., 1982). Thus, when using the above equations to describe solar flares, one throws away the solution one seeks at the outset. It is doubtful that the dissipation term can be ignored simply because one is dealing with a large-scale phenomenon. It is likely that some processes could be understood by assuming that the photosphere is an infinitely conductive plasma (rather than a substance with conductivity similar to that of sea water), but such an assumption prohibits the understanding of solar flares.

Solar physicists consider that the magnetic energy for solar flares is contained in force-free fields, namely $J \times B = (\nabla \times B) \times B = 0$. This obviously means that the field deviating from a potential field is caused by magnetic field-aligned current $J_{\parallel}$, which must be generated by a dynamo process. By definition, a dynamo generates $J_{\perp}$, so that divergence of $J_{\perp}$, namely $\nabla \bullet J_{\perp}$, produces $J_{\parallel}$. Like auroral substorms, it may be possible that a solar flare can occur when $J_{\parallel}$ increases above a certain value. This is the reason why I asked Zirin and his colleagues to examine changes of the magnetic shear at flare onset. Their results are given in Figure 3. The weakly ionized photosphere is likely to be the layer where the dynamo process operates, because it is the layer where $J_{\perp}$ and $J_{\parallel}$ can be generated.

I suspect that solar physicists may be trying to avoid facing the problem of the dynamo process by hypothesizing magnetic flux tubes.

Variations of the interplanetary magnetic field (IMF) associated with magnetic storms have been studied in terms of passage of magnetic clouds. Burlaga (1988) found that, at least in some cases, the magnetic structure can be approximated by a cylindrical force-free field. It is not difficult to infer the intensity of the field-aligned current in such a magnetic cloud. It is estimated to be about 10^9 amperes. Again, one must consider how such a current can be generated in the sun and flow along magnetic field lines as a result of solar flares.

3. Magnetospheric Physics

We also have a number of paradigms in magnetospheric physics. They include:

1. Macroscopic magnetospheric processes must be explained in terms of magnetic flux transfer, transfer/convection and magnetic reconnection.
2. The westward electrojet results from diversion of the cross-tail current.
3. Magnetic storms and auroral activities are most intense during the period of sunspot maximum years.

In this paper, we deal only with subjects 1 and 2.

3.1 Magnetic Flux Transfer

Since Dungey proposed it in 1961, a prevailing concept in magnetospheric physics has been the concept of magnetic flux transfer. This MHD-based concept assumes that interplanetary magnetic field lines and magnetospheric field lines 'reconnect' or connect on the dayside magnetopause and are transferred to the magnetotail. Subsequently, the transferred field lines 'reconnect' in the magnetotail and are transferred back to the dayside. The transfer process is associated with the convection of plasma in the magnetosphere, namely, the so-called 'frozen-in-field' concept. Further, it has been believed that the reconnection process in the magnetotail converts magnetic energy explosively, causing a magnetospheric substorm. This concept has been the basis of studying magnetospheric processes and phenomena for four decades. Indeed, it has worked as one of the guiding principles in explaining a number of magnetospheric phenomena.

However, we may consider magnetospheric processes (including the magnetotail which consists of two solenoidal currents) in a different way. An increase of the ram pressure of the solar wind can be reproduced by a combination of the image dipole at the front of the magnetosphere (as the Chapman-Ferraro theory demonstrated) and two solenoidal currents in the magnetotail. It is reasonable to assume that the intensity of the solenoidal currents depends on the power of the solar wind-magnetosphere dynamo.

Such a consideration may be less sophisticated than the MHD treatment (with computer simulation). However, it is possible to understand the physical processes involved in this process better than in the MHD simulation. It may well be that one can use our simple approach because the magnetospheric currents tend to flow mostly in thin layers, rather than all over in the main body.

The MHD concept encounters difficulty when it attempts to explain an important observation in the magnetotail. The flux transfer concept has no choice but to return magnetic flux in the magnetotail to the dayside by reconnecting it at substorm onset. If the occurrence of substorms happens as a result of magnetic reconnection, the magnetic flux in the magnetotail should *decrease* at substorm onset (Figure 4). However, this is not necessarily the case. The magnetic flux in the magnetotail often *increases* after substorm onset. In fact, the size of the polar cap and the AE index are well correlated; note that the magnetic field lines originating in the polar cap occupy the lobes, north and south, of the magnetotail. This observation is difficult to explain in terms of the magnetic flux transfer concept.

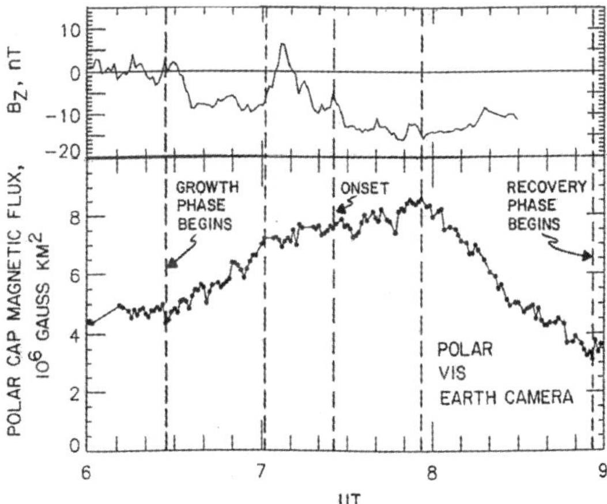

Figure 4: An example of increasing magnetic flux in the lobe of the magnetotail after substorm onset. (Frank et al., 1998)

The source of the firm belief that magnetic energy in the magnetotail decreases is the oft-quoted paper by Caan et al (1975), which demonstrates that the magnetic field in the magnetotail decreases during substorms. One of their figures, which is reproduced here as Figure 5, shows changes of both the lobe-field magnetic energy density and the IMF Bz. Note that the IMF Bz sharply turned northward when the magnetic energy density in the lobe began to decrease, indicating that the power of the dynamo ε decreased, so that the intensity of the solenoid current must also have decreased. Therefore, the decrease of the lobe field need not be a result of magnetic reconnection and the subsequent transfer back of the tail flux as generally believed. Indeed, it makes perfect sense to consider that the decrease occurred because the power ε and subsequently the intensity of the solenoidal currents began to decrease. Considering the fact that the lobe field can increase or decrease during substorms, it is reasonable to consider that the lobe field increases when the power is greater than the total loss rate L in the magnetosphere, while the lobe field decreases when the loss L is greater than ε.

Based on the above observations, instead of the transfer of magnetic flux from the dayside of the magnetosphere to the tail lobe and the transfer back from the tail lobe to the dayside, it is possible to interpret the observations in terms of increase and decrease of the solenoidal current and the cross-tail current as ε increases and decreases.

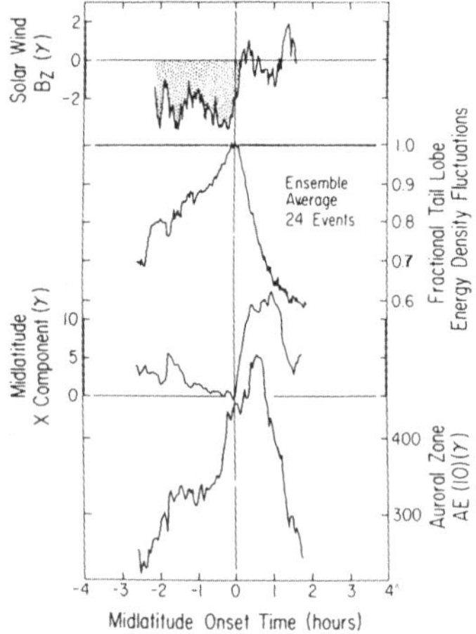

Figure 5: A statistical result of magnetospheric changes, the energy density in the lobe (the mid-latitude H component, and the AE index), together with the IMF Bz change. (Caan et al., 1975)

On the other hand, one should not overlook an important hint hidden in these results. It is significant that substorm onset at the time of a sudden northward turning of the IMF occurs *after* the IMF Bz southward turning (Lyons 1996). This suggests that *substorm onset occurs when the cross-tail current is suddenly reduced after it was first enhanced during the growth phase*. The northward turning should thus reduce ε and the cross-tail current.

There may be other causes that can reduce the cross-tail current. Another possible cause of a sudden reduction of the cross-tail current is plasma instability, which has been extensively examined by Lui (2001) and others. A reduction of the cross-tail current *after an increase of the cross-tail current* obviously causes the so-called 'dipolarization'. In the past, however, it has been believed that such instability in the plasma sheet diverts the cross-tail current to the ionosphere, driving the westward electrojet. I was one of the first to propose the diversion of the cross-tail current (Akasofu, 1972).

3.2   Westward Electrojet

In this section, we deal only with the westward current that occurs in association with substorm onset, mainly the so-called "DP-1" current, not the "DP-2" current. If the westward electrojet results from the diversion of the cross-tail current, part of the potential drop (~50-100kV) across the magnetotail should appear between both ends of the electrojet. Therefore, a westward electric field of 20kV/m (=100kV/50km) must drive the westward electrojet. However, incoherent scatter radar observations show clearly that the westward electrojet is driven by a southward electric field (Banks and Doupniket al., 1975); see Figure 6. Therefore, *the westward electrojet must result mainly from the Hall current*. Indeed, I showed in 1960 that this is in fact the case; I demonstrated that the southward electric field is of the order of 20V/km (Akasofu, 1960), which is in agreement with radar observations conducted more than 10 years later, but this fact has been forgotten.

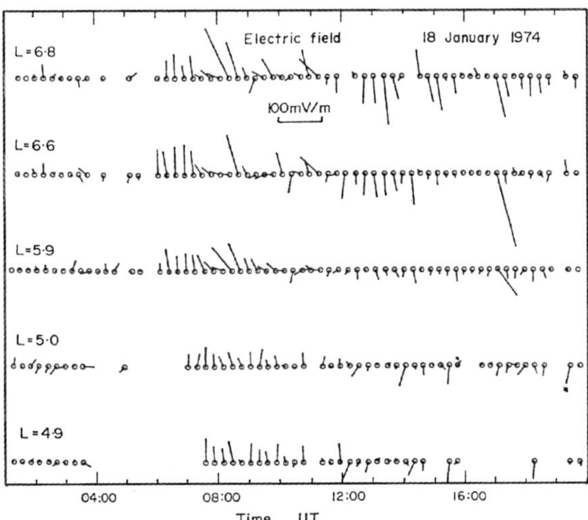

Figure 6:   Electric fields observed during substorms. Northward fields were observed in the evening, while southward fields were observed when the westward electrojet was flowing overhead. (Banks and Doupnik, 1975)

Since the polar ionosphere does not possess sufficient conductivity to allow the return current from the westward electrojet to flow there, the electrojet has to drive field-aligned currents from the westward and eastern ends, closing the circuit by driving an eastward current on the equatorial plane in the magnetotail, against the cross-tail current. This equatorial return current can reduce or overwhelm the cross-tail current, depending on the intensity of the electrojet. Figure 7 shows the complete westward electrojet circuit proposed by Boström (1964). Note that the electric field indicated by the (+, -) sign also drives the meridional current; its upward current, carried by the accelerated electrons, causes the aurora. The origin of this electric field will be discussed shortly.

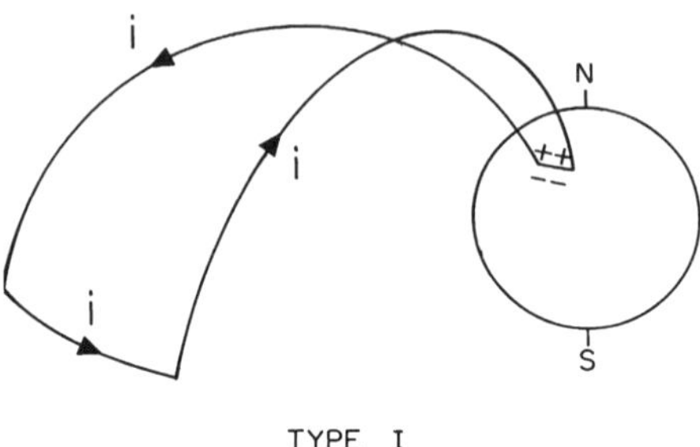

Figure 7: The complete westward electrojet circuit proposed by Boström (1964). Note that the electric field indicated by the (+, -) sign also drives the meridional current; its upward current, carried by the accelerated electrons, causes the aurora.

The presence of the equatorial return current can be demonstrated as follows. If the dipolarization is due simply to the diversion of the cross-tail current, the complete diversion returns the stretched earth's dipole to the dipole field. At the geosynchronous distance, the magnitude of the dipole field is about 100~110nT. When the magnitude of the field at the location of the geosynchronous satellite is examined, it sometimes exceeds the dipole value (Akasofu et al., 1974), in spite of the fact that plasma pressure ($\sim10^{-8}$ erg) is comparable to magnetic pressure ($B^2/8\pi$). If the so-called 'dipolarization' is simply the tendency for the stretched dipole field to return to the original dipole field, the increase should not exceed the dipole value, but it can happen.

This may be a good example of a case in which reexamination of two paradigms can spur important progress towards understanding a complicated problem, such as substorm onset. Substorm onset appears to be triggered by a sudden reduction of the cross-tail current *after an enhancement by the southward turning of the IMF vector.*

The sudden reduction of the cross-tail current automatically causes the dipolarization. As described elsewhere (Akasofu, 2002), the dipolarization appears to activate the Boström current system, which can explain the substorm onset. The activation is a result of the growth of a radially inward directed electric field. Akasofu (2002) suggested that the electric field associated with $\partial Bz/\partial t$ may be responsible. On the other hand, Lui and amide (2003) suggested charge separation during the dipolarization, because electrons in the plasma sheet move with the field lines toward the earth during the dipolarization, while positive ions are left behind, because their gyro-radius is larger compared with the thickness of the plasma sheet. This is a result of the breakdown of the MHD frozen-in-field concept. Thus, by abandoning the paradigm of the cross-tail current diversion and of the MHD formalism, I feel that we are finally one step closer to understanding substorm onset Figure 8.

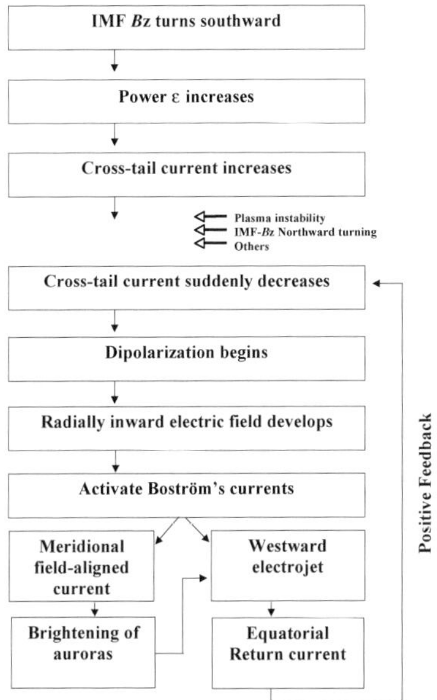

Figure 8: A possible chain of processes leading to substorm onset after the 'southward turning' of the IMF.

It should be noted that Boström's current establishes a positive feedback system. This is because the auroral ionization increases the Hall conductivity for the westward electrojet. The enhanced electrojet further reduces the cross-tail current and enhances the dipolarization and so on. The mechanism for substorm onset requires such a positive feedback.

4. Concluding Remarks

It is obviously not the intent of this paper to provide answers to these long-standing unsolved problems. The purpose of the paper is to suggest that some of the unsolved problems will never be solved, unless we reexamine carefully the foundations of the present paradigms. The last subjects (Sections 4) are good examples of how such a reexamination could possibly provide a new approach to the long-standing problem of understanding substorm onset. We have many paradigms and should examine each of them carefully from different points of view based on observations, such as those shown in Figures 2, 3, 4, 5, and 6. We should pay much more attention to the observations. Too many important observations are 'put under the rug' of the present paradigms.

Flexible minds and open-mindedness are needed in accomplishing this task. One route may be to discuss the unsolved problems with new graduate students. Do not overeducate them. If we do, they will long remain mired in the same approaches we have used. Always, graduate students and post-docs should be encouraged to ask 'dumb' questions.

References

Akasofu, S.-I., Large scale auroral motions and polar magnetic disturbances-I: A polar disturbance at about 1100 hours on 23 September 1957, *J. Atmos. Terr.*, 19, 10, 1960.

Akasofu, S.-I., The neutral hydrogen flux in the solar plasma flow-I, *Planet. Space Sci.*, 12, 905, 1964.

Akasofu, S.-I, *Magnetospheric substorms: A model, Solar-Terrestrial Physics - 1970; Part III*, E.R. Dyer and J.R. Roederer (eds.), p. 131, D. Reidel Pub. Co., 1972.

Akasofu, S.-I., An essay on sunspots and solar flares, *Planet. Space Sci.*, 32, 1469, 1984.

Akasofu, S.-I., Vortical distribution of sunspots, *Planet. Space Sci.*, 33, 275, 1985.

Akasofu, S.-I., *Exploring the Secrets of the Aurora*, Kluwer Academic Publishing, October 2002.

Akasofu, S.-I., S. Deforest, and C. McIlwain, Auroral displays near the 'foot' of the field line of the ATS-5 satellite, *Planet Space Sci.*, 22, 25, 1974.

Banks, P.M. and J.R. Doupnik, A review of auroral zone electrodynamics deduced from incoherent scatter radar observations, *J. Atmos. Terr. Phys.*, 37, 951, 1975.

Boström, R., A model of the auroral electrojets, *J. Geophys. Res.*, 69, 4983, 1964.

Burlaga, L.F., Magnetic clouds and force-free fields with constant alpha, *J. Geophys. Res.*, 93, 7217, 1988.

Caan, M.N., R.L. McPherron, and C.T. Russell, Substorm and interplanetary magnetic field effects on the geomagnetic tail lobes, *J. Geophys. Res.*, 80, 191, 1975.

Collier, M.R., T.E. Moore, K.W. Ogilvie, D. Chornay, J.W. Keller, S. Boardsen, J. Burch, B. El Marji, M.-C. Fok, S.A. Fuselier, A.G. Ghielmetti, B.L. Giles, D. C. Hamilton, B.L. Peko, J.M. Quinn, E.C. Roelof, T.M. Stephen, G.R. Wilson, and P. Wurz, Observations of neutral atoms from the solar wind, *J. of Geophys. Res.*, 106, No. A11, 24,893, 2001.

Frank, L.A., J.B. Sigwarth, and W.R. Paterson, *High-resolution global images of earth's auroras, Substorms-4*, ed. by S. Kokubun and Y. Kamide, Kluwer Academic Pub. Co., Dordrecht, Holland, 1998.

Kan, J.R., S.-I. Akasofu, and L.C. Lee, A dynamo theory of solar flares, *Solar Phys.*, 84, 153, 1982.

Lui, A.T.Y., Current controversies in magnetospheric physics, *Review of Geophysics*, 39, 535, 2001.

Lui, A.T.Y., and Y. Kamide, A fresh perspective of substorm current system and its dynamo, *Geophys. Res. Lett.*, 30, No. 18., SSC12, 2003.

Lyons, L.R., Substorms: Fundamental observational features, distinction from other disturbances, and external triggering, *J. Geophys. Res.*, 101, 13011, 1996.

Wang, H., M.W. Ewell, Jr., and H. Zirin, Vector magnetic field changes associated with X-Class flares, *Ap. J.*, 424, 436, 1994.

Multiscale Coupling of Sun-Earth Processes   83
A.T.Y. Lui, Y. Kamide and G. Consolini, editors.

SPACE WEATHER EFFECTS ON SOHO AND ITS LEADING ROLE AS A SPACE WEATHER WATCHDOG

P. Brekke, B. Fleck, S. V. Haugan, T. van Overbeek, and H. Schweitzer,
European Space Agency, GSFC Greenbelt,MD 20771, USA
and
B. Simonin
EADS Astrium, GSFC Greenbelt,MD 20771, USA

Abstract. Since its launch on 2 December 1995, the Solar and Heliospheric Observatory (SOHO) has provided an unparalleled breadth and depth of information about the Sun, from its interior, through the hot and dynamic atmosphere, and out to the solar wind. In addition SOHO has several times demonstrated its leading role in the early-warning system for space weather. SOHO is in a halo orbit around L1 Lagrangian point where it views the Sun 24 hours a day. Thus, it is situated outside the Earth's protective magnetosphere, which shields other satellites from high-energy particles and the solar wind. We present a summary of the observed effects on the instruments and electronics on SOHO throughout the mission. In particular we focus on a number of large particle events during the recent years while the Sun was near maximum activity, and how they affected both the scientific data as well as hardware components.

1. The SOHO Spacecraft

The SOHO mission is a major element of the International Solar Terrestrial Programme (ISTP), and, together with Cluster, forms the Solar Terrestrial Science Programme (STSP), the first cornerstone in ESA's long-term science programme 'Horizons 2000'. ESA was responsible for the spacecraft's procurement, integration and testing. It was built in Europe by an industry team lead by Matra Marconi Space (now called EADS Astrium). Weighing in at 1,850 kg, the SOHO spacecraft measures about 9.5 m across with its solar panels extended and is 4.3 m high. Figure 1 provides a schematic view of the SOHO spacecraft. NASA provided the launcher, launch services and ground-segment system and is responsible for in-flight operations. Mission operations are conducted from NASA/Goddard Space Flight Center (GSFC).

SOHO was launched by an Atlas II-AS from Cape Canaveral on 2 December 1995 and was inserted into its halo orbit around the L1 Lagrangian point on 14 February 1996, six weeks ahead of schedule. Commissioning of the spacecraft and the scientific payload was completed by the end of March 1996. The launch was so accurate and the orbital maneuvers were so efficient that enough fuel remains on board to maintain the halo orbit for several decades, many times the lifetime originally foreseen. An extension of the SOHO mission for a period of five years beyond its nominal mission duration (2 years), i.e. until March 2003, was approved in 1997 by ESA's Science Programme Committee (SPC). A second extension of another four years, i.e. until March 2007, was granted by the SPC in 2002. This will allow SOHO to cover a complete 11-year solar cycle.

SOHO has a unique mode of operations, with a "live" display of data on the scientists' workstations at the SOHO Experimenters' Operations Facility (EOF) at NASA/Goddard Space Flight Center, where the scientists can command their instruments in real-time, directly from their workstations. SOHO enjoys a remarkable "market share" in the worldwide solar physics community: over 1500 papers in refereed journals and over 1500 papers in conference proceedings and other publications, representing the work of over 1500 scientists. At the same time, SOHO's easily accessible, spectacular data and basic science results have captured the imagination of the space science community and the general public alike. In September 2003 the SOHO team was awarded the Laurels for Team Achievement Award of the International Academy of Astronautics (IAA).

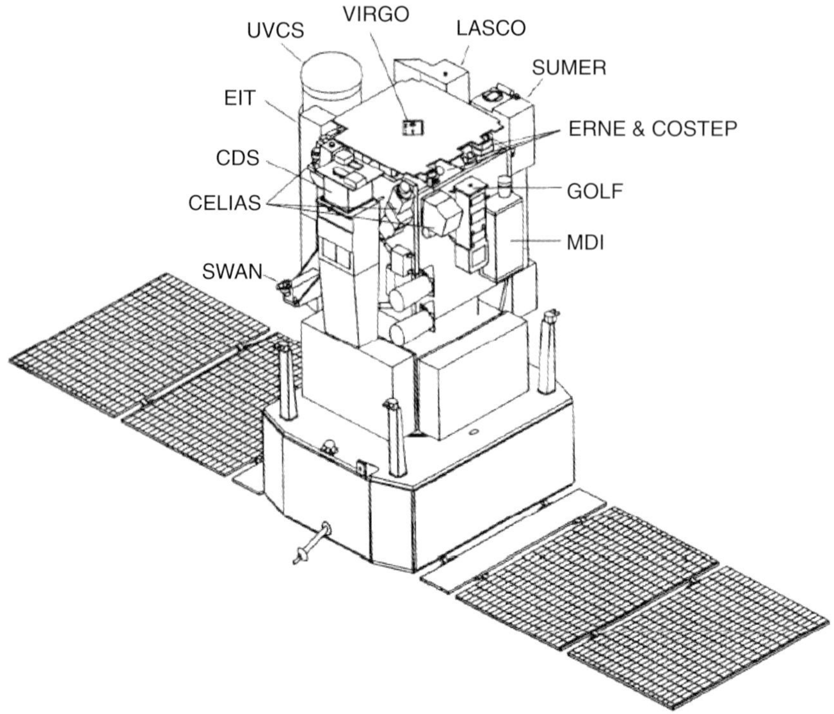

Figure 1. SOHO spacecraft schematic view.

2. The Loss and Recovery of SOHO

On 25 June 1998 at 04:43 UT our worst nightmare was beginning to unfold - contact with SOHO was lost, not to be re-established for more than six weeks. Through a series of unfortunate events, the spacecraft had lost its sun-pointing attitude, which ultimately resulted in loss of power, telemetry and thermal control.

In one of the most dramatic recovery efforts in space, the SOHO recovery team - of more than 160 people drawn from ESA, Matra Marconi Space, NASA and AlliedSignal - succeeded in bringing the spacecraft back to normal sun-pointing attitude on 16 September 1998, after thawing the frozen hydrazine in the fuel tank, pipes and thrusters.

After re-commissioning all spacecraft subsystems, SOHO was brought back in normal mode nine days later. The re-commissioning of the scientific payload was successfully completed on 5 November. The fact that the spacecraft and the 12 instruments came through this ordeal almost entirely unscathed is quite remarkable and constitutes a great tribute to the skill, dedication and professionalism of the scientists and engineers who designed and built them.

Two of the three gyros on SOHO were damaged during this ordeal. When the last onboard gyro failed on 21 December 1998, SOHO went into Emergency Sun Reacquisition (ESR) mode. In a race against time - the ESR thruster firings consumed an average of about 7 kg of hydrazine per week - engineers at ESTEC and Matra Marconi Space developed software to exit ESR mode without a gyro and allow gyroless operation of the spacecraft.

On 1 February 1999 the first gyroless reaction wheel management and station-keeping manoeuvre was performed, making SOHO the first three-axis-stabilised spacecraft to be operated without a gyro. A new Coarse Roll Pointing (CRP) mode, which uses the reaction wheel speed measurements to both monitor and compensate for roll rate changes, was successfully commissioned in September 1999. The CRP mode is almost two orders of magnitude more stable than using gyros. It also acts as an additional safety net between the normal mode and ESR mode, making SOHO perhaps more robust than ever. Conservative estimates of spacecraft fuel usage yields a remaining lifetime of more than 20 years.

In early May 2003, the East-West pointing mechanism of SOHO's High Gain Antenna (HGA) started missing steps; by late June it appeared stuck. Using both primary and redundant motor windings simultaneously, the mechanism was parked in a position that maximizes the time it can be used throughout a 6-month halo orbit, with the spacecraft rotated by 180 degrees for half of each orbit and with "keyhole periods" twice per orbit (see Figure 2). During the keyholes the Low Gain Antenna can be used with larger DSN stations to receive science telemetry, but data losses of varying magnitude occur depending on the competition for these resources.

3. SOHO - a Space Weather Watchdog

Observations of the solar corona with the Large Angle Spectrometric Coronagraph (LASCO) and the Extreme ultraviolet Imaging Telescope (EIT) instruments on SOHO provide an unprecedented opportunity for continuous real-time monitoring of solar eruptions that affect space weather. Coronal mass ejections (CMEs) are one of the most energetic and important solar phenomena. They propels plasma at speeds up to 2500 km/s into the heliosphere, causing space weather effects here on Earth. These events were first discovered in 1972 by the OSO-7 spacecraft [Tousey, 1973] and later observed with the ATM coronagraph on Skylab, the Solwind coronagraph on the P78-1 satellite and by the Coronagraph/Polarimeter on Solar Maximum Mission (SMM). CMEs have also been observed from ground-based coronagraphs. The most comprehensive monitoring of CMEs is from the LASCO instrument on SOHO.

3.1. LASCO and EIT

LASCO takes images of the solar corona by blocking the light coming directly from the Sun itself with an occulter disk, creating an artificial eclipse within the instrument. CMEs moving outward from the Sun along the Sun-Earth line can, in principle, be detected when they have expanded to a size that exceeds the diameter of the coronagraphs occulting disk. CMEs directed toward or away from the Earth should appear as expanding halo-like brightenings surrounding the occulter, so-called halo CMEs. LASCO best observes limb CMEs, but its extreme sensitivity even allows unprecedented detection of halo CMEs. EIT provides images of the solar atmosphere at four extreme ultraviolet wavelengths and reveals flares, dimmings and other associated events in the atmosphere. EIT can usually determine whether CMEs seen by LASCO originated on the near or far side of the Sun, based on the presence or absence of corresponding events on the near side.

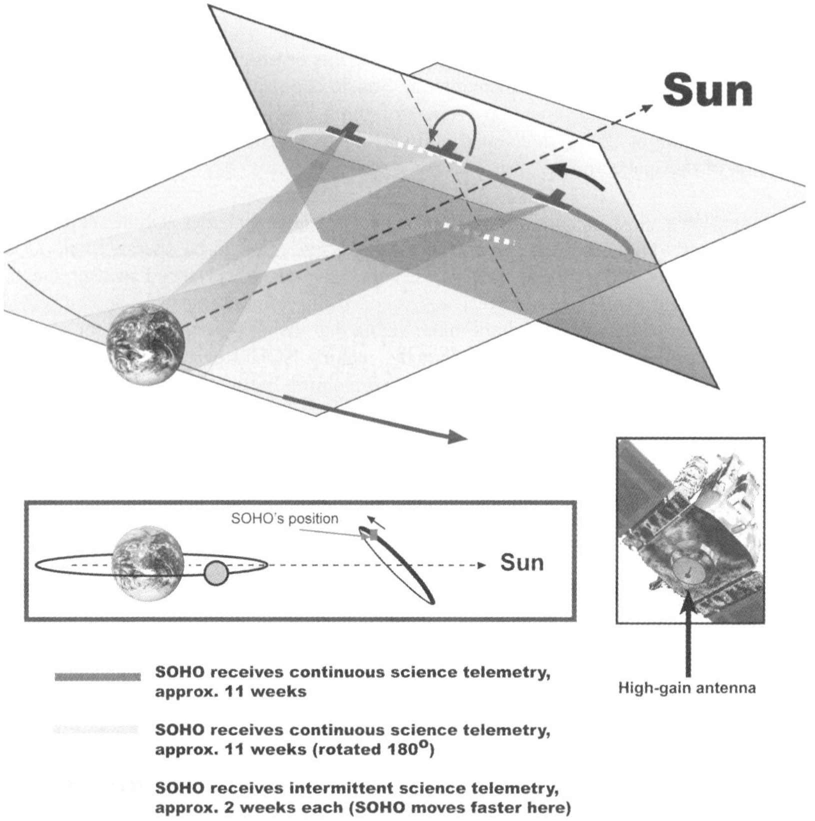

Figure 2. Schematic illustration of SOHO's orbit, indicating the geometry of the now parked High Gain Antenna

LASCO has been collecting an extensive database for establishing the best statistics ever on CMEs and their geomagnetic effects. Nearly 8000 CMEs have been detected over a period of 8 years (1996-2003)*. An example of a halo-CME is shown in Figure 3 as recorded by the LASCO C3 detector on 6 June 2000.

The properties of 841 CMEs observed by the LASCO C2 and C3 white-light coronagraphs from January 1996 through the SOHO mission interruption in June 1998 were studied and compared with those of previous observations by other instruments [St. Cyr et al 2000]. The CME rate for solar minimum conditions was slightly higher than had been reported for previous solar cycles, but both the rate and the distribution of apparent locations of CMEs varied during this period as expected.

---

* A complete list of all detected CMEs with LASCO can be found at: http://lasco-www.nrl.navy.mil/cmelist.html

A more recent study by Yashiro et al. [2004] provides a summary of the statistical properties of all CMEs between 1996 and 2001. Just for the period from January 1996 to June 1998 they identified over 200 additional CMEs compared to the St Cyr et al [2000] catalog. It was found that the average speed of the CMEs increased towards the solar maximum from 300 km/s to 500 km/s, and slightly decreased to 474 km/s in 2001. The increase in average speeds is thought to be caused by the increase of the number of flare-associated CMEs. There were 11 CMEs with speeds more than 2000 km/s, and the fastest CME with a speed of 2606 km/s occurred on 2000 May 12. During solar minimum most CMEs were ejected approximately around the equator region while during solar maximum CMEs originated from all latitudes. Gopalswamy et al. [2003] found that the CME rate increased by an order of magnitude from solar minimum (0.5/day) to solar maximum (6/day). The rate during solar maximum is almost twice the rate estimated for previous cycles. Furthermore, the peak of the CME rate occurred in 2002, roughly 2 years after the peak in the sunspot number.

The statistics of halo CMEs have also been studied in detail [St. Cyr et al 2000]. Using full disk EIT images they found that 40 out of 92 of these events might have been directed toward the Earth. A comparison of the timing of those events with the Kp geomagnetic storm index in the days following the CME found that 15 out of 21 (71%) of the Kp > 6 storms could be accounted for as SOHO LASCO/EIT frontside halo (FSH) CMEs.

More recently it was found that 22 out of 27 (81%) major geomagnetic storms occurring between 1996-2000 were identified with FSH CMEs [Zhang et al. 2003]. Of these 16 (59%) were associated with unique FSH CMEs while 6 (22%) where related to multiple FSH CMEs. They also find that while these geoeffective CMEs are either full-halo CMEs (67%) or partial-halo CMEs (30%), there is no preference for them to be fast CMEs or to be associated with major flares and erupting filaments. Again, this illustrates that SOHO has been providing new valuable information to better understand CMEs as well as being the only monitoring system for Earth directed CMEs until more ideal missions are launched (e.g. STEREO)

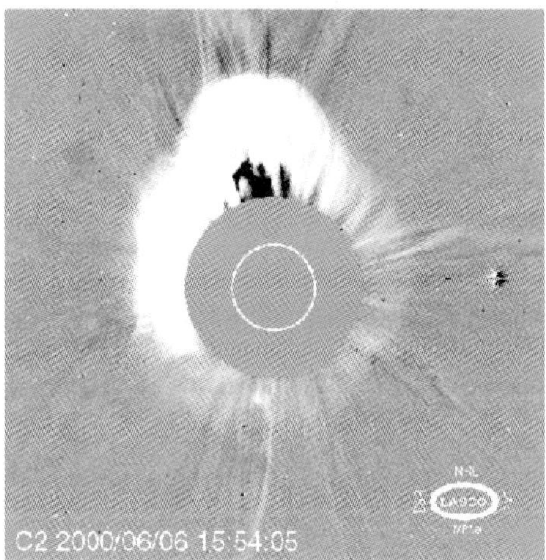

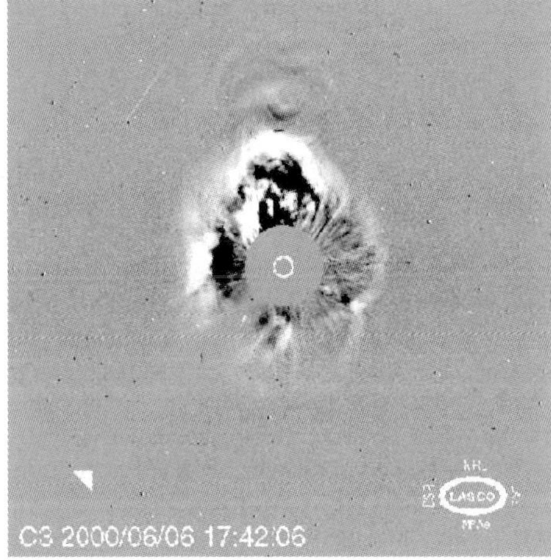

Figure 3. Example of a full halo CME observed by LASCO C2 (left panel) and C3 (right panel) coronagraphs. The field of view of the images is 2-6 and 3.5-30 solar radii.

88

3.2 SOHO's role at the Space Environment Center

The Space Weather Operations Center at the Space Environment Center (SEC) in Boulder uses SOHO images daily. The forecasting operations have come to rely on SOHO on a routine basis as a key input. LASCO provides the only direct observation of coronal mass ejections. Prior to LASCO they had to rely on activity they knew to be well associated with CMEs, but none of these associations are reliable. They use direction, size and velocity information from LASCO images to help determine the arrival time and effectiveness of the disturbance.

EIT also plays an important role at SEC to pin down the source of any eruption. In addition EIT is a very good source for identifying erupting prominences and to identify coronal hole locations. Coronal holes have become an increasingly important part of the geomagnetic forecasting process. In fact at this point in the solar cycle, coronal hole activity has become the predominant driver of geomagnetic activity.

Finally, forecasters use the MDI data on SOHO in order to track sunspot growth and decay and the magnetograms are used to track magnetic field strengths and complexity, a valuable input for flare forecasting.

3.3 Automated Detection of CMEs

The visual detection of CMEs in the flood of incoming LASCO data is a labour-intensive task. Until today it is essentially the human eye that detects a CME occurrence and a scientist that collects all the CME parameters. An automated detection system called "Computer Aided CME Tracking" (CACTus) has been developed for the LASCO images [Berghmans et al. 2002]. The software detects bright ridges in [height, time] maps using the Hough transform and creates a list of events with principal angle, angular width and velocity estimation for each CME. In contrast to lists assembled by human operators, these CME detections by software can be faster and possibly also more objective. The first version CACTus has been evaluated and it obtained a success rate of about 75%. This number is expected to improve in later versions. The software also detected some CMEs that were not reported in the official human created catalogs.

3.4 Solar Wind Shockspotter

The CELIAS/MTOF/PM instrument on SOHO measures the solar wind speed, density and temperature. A group at the University of Maryland recently implemented a "Shockspotter" program to identify interplanetary shocks in near-real time using proton monitor data. The program is based on semi-empirical algorithms using only solar wind proton data (since no magnetometer data is available on SOHO). Shock candidates are classified into 4 distinct zones, with confidence levels ranging from about 40% to 99%. Results have been used to study the frequency distribution of interplanetary shocks over the solar cycle. The Shockspotter program is now part of the proton monitor real time data page at http://umtof.umd.edu/pm. The program can alert users (via email, upon request) whenever a shock front passes the SOHO spacecraft approximately 30-60 minutes prior to the arrival at Earth. A catalog of interplanetary shocks is also maintained at http://umtof.umd.edu/pm/figs.html. The Maryland CELIAS group has also developed Web pages that show the solar energetic particle flux deduced from proton monitor background levels (http://umtof.umd.edu/pm/flare) and the solar soft X-ray flux from SEM measurements (http://umtof.umd.edu/sem/).

4. SOHO – Promoting Space Weather to the Public

Late October 2003, SOHO appeared to be in everyone's focus as the Sun turned from an almost spotless orb into an ominously scarred source of mighty fireworks in just a few days. Around 17 October 2003, solar activity was very low. The face of the sun was nearly blank with only a few tiny sunspots. This was not unexpected: Solar maximum was in 2001, and solar activity had been declining in line with expectations since then.

What happened next may go into the history books as one of the most exciting times for scientists observing the Sun with modern space-based instruments. First, sunspot region 10484 rotated into view, growing fast. It soon caught the attention of sungazers around the world. The sunspots in the region covered more than 1700 millionths of the visible solar surface, or 10 times the entire surface of the Earth!

SOHO scientists already suspected, however, that more trouble was coming around the corner. Region 10486 was rotating onto the solar disk, showing even more signs of activity. And this particular region had caught the attention of solar physicists while it was still on the far side of the Sun! Using a technique known as helioseismic holography, the MDI instrument is routinely used to construct maps of magnetic activity on the far side of the Sun. The far side images had shown considerable strengthening of this region over a short period of time. And it did not disappoint while traveling across the visible disk: By 31 October 2003, its sunspot area had grown to over 2600 millionths, or 15 times the entire surface of the Earth. It was now officially the largest sunspot region of this solar cycle (see Figure 4). In the mean time, the new region 10488 had grown from nothing into 1750 millionths in just a few days, right in front of the eyes of the world.

Ten X-class flares were registered from the three regions on their journey to the limb. Some flares grabbed worldwide attention when Earth's regional power distribution systems went on high alert for the impact of associated coronal mass ejections.

Just as solar scientists were ready to start breathing normally again, with the last of the three regions (10486) disappearing behind the solar limb, yet another mega-flare erupted. This one saturated the X-ray detectors on the NOAA's GOES satellites; the jury was therefore out for a while on the definitive classification of the flare. The GOES X-ray monitor saturates at X17.3, but after a day of analyzing the data, NOAA's Space Environment Center estimated the event to be an X28 flare. In other words, the strongest X-ray flare ever recorded since X-ray observations were initiated from satellites in the mid 70's. The three giant sunspots unleashed eleven X-class flares in only fourteen days—equaling the total number observed during the previous twelve months. Two of the other flares went into the top 20 strongest flares. Two of the events also had CMEs moving with speeds exceeding 8 million kilometers per hour, reaching Earth in less than 20 hours. This is almost 5 times faster than typical, and makes them some of the fastest CMEs recorded, close to the record super storm in 1859 [Tsurutani et al. 2003].

One of the magnetic clouds slammed into Earth's magnetosphere on 29 October 2003, creating a G5 geomagnetic storm, the strongest category. Some of these storms created a beautiful aurora as far south as Spain and Florida. Satellites, power grids, radio communication and navigation systems were significantly affected. Airline passengers reported bright auroras on most night flights. A Japanese satellite was lost completely while a number of others experienced problems. Several thousand people lost power in the Southern part of Sweden. Air traffic control moved transatlantic flights further south to avoid loss of radio communication. Climbers in the Himalayas experienced problems with their satellite phones. These are just a few of the effects on our society from these storms [Webb and Allen, 2004].

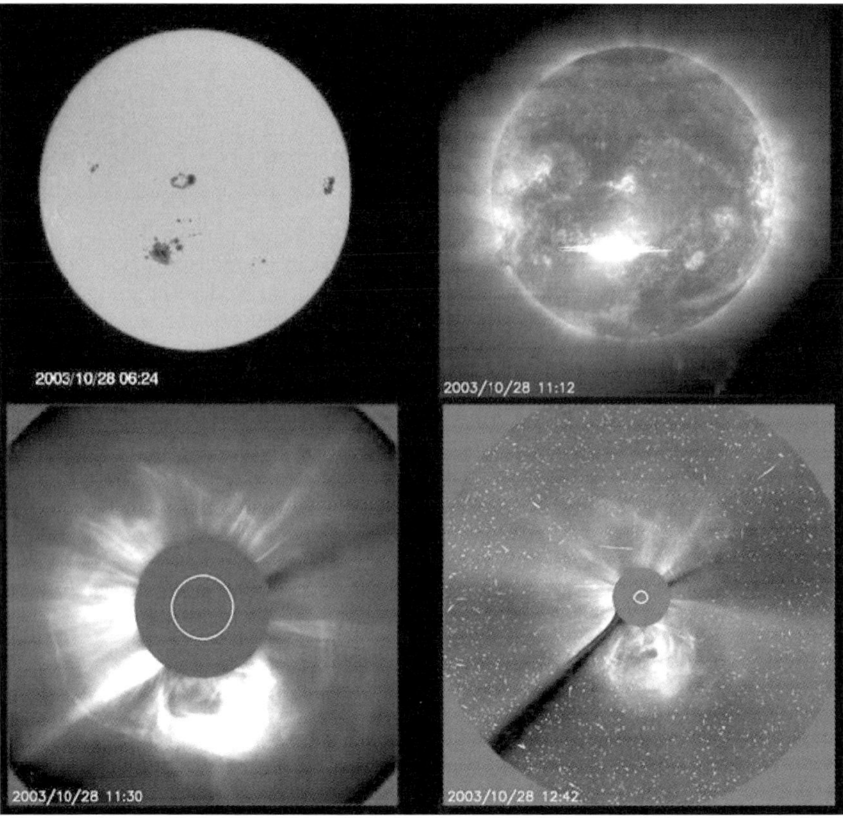

Figure 4. Region 10486 unleashed a spectacular show on 28 October 2003. An X 17.2 flare, a fast-moving CME and a strong solar energetic particle event. From top left: Giant sunspot regions 10484, 10486 and 10488 seen by MDI in white light. Flare as seen by EIT in 195 Å emission. The fast-moving CME in the LASCO C2 coronagraph, then in the LASCO C3 coronagraph (with particle shower becoming visible as 'snow' in the image).

Six distinct proton events were detected during this stormy period. The largest was a 29,500 Particle Flux Units (1 pfu = 1 p cm^{-2} sr^{-1} s^{-1}), greater than-10-MeV proton event. This severe storm was the second largest proton event in this cycle, and ranks fourth in the all-time list dating back to 1976. The Bastille Day proton event of July 14, 2001 reached 24,000 pfu [SEC, 2004].

The events caused unprecedented attention from the media and the public. Images from SOHO and quotes from SOHO scientists appeared in nearly every major news outlet. Stories including SOHO images were featured on the front page in most newspapers around the world. NASA TV estimated that the story reached "all" newspapers and 2000 US TV channels. This made a significant impact on the awareness of space weather effects in our society. The attention wiped out all existing SOHO web traffic records (requests/data volume): Monthly (31 million/4.3 TB), weekly (16 million/2.6 TB), daily (4.8 million/0.7 TB), and hourly (0.4 million/33 GB). The daily and hourly record volumes were bandwidth limited.

5. Space Weather Effects on the Service Module

SOHO is designed to withstand the effects of the varying flux of high-energy particles encountered in its L1 Halo orbit. This section discusses the effects on the service module and solar panels. The following section discusses the effects on the different scientific instruments. Brief summaries of efforts to prevent interruptions to the operation of the spacecraft and instruments are also presented.

5.1. Permanent Degradation of Subsystems

During their lifetime, spacecraft components receive an integrated radiation dose that degrade their performance and can cause the failures. The only permanent effects so far are the degradation of the solar arrays and the Fine Pointing Sun Sensor (FPSS) due to energetic particles from solar eruptions and extragalactic sources. This degradation is due to "displacement damage": Energetic particles interact with the solar cell lattice or the FPSS sensor substrate, producing defects which enhance electron and hole recombination thus reducing the solar cell's output voltage and current, and decreasing the FPSS sensitivity.

The actual degradation of the solar array is given in Figure 5. The degradation due to proton events is evident with significant drops during the July 14, 2000, November 4 & 23, 2001 events and the October/November 2003 events. After 100 months in space, the degradation is 15.95%. This is an annual average degradation of 1.91%, well within the 4% per year requirement. SOHO can operate down to 70% sensitivity without taking any energy saving action. The remaining 14% margin should last at least another 7 years at the average rate of degradation. Extending the mission even further would require switching off selected instruments without turning on corresponding substitution heaters.

The present performance of the FPSS is still sufficient, but in the long run it might eventually require a new calibration of the output level. This is a simple on-board parameter change.

5.2. Radiation Induced Background

Radiation impinging on detectors or associated electronics can produce an increase of the background noise. The Star Sensor Unit consists of an optical system with thermal sensors for calibration of the focal length of the optics and a CCD detector (377 x 283 pixels), mounted on a Peltier cooler with thermal control for the CCD temperature (– 40°C) and for the electronics of the detector drivers and data pre-processing. The background noise of the Star Sensor Unit has been very stable since the beginning of the mission.

5.3. Single Event Upsets (SEUs)

Cosmic rays or heavy ion impacts can provoke single event upsets, disrupting the operation of sensitive electronics in a number of subsystems.

5.3.1 Electronic units self switch-off

A fair number of self switch-off events occurred, which are attributed to Single Event Upsets (SEUs). Three of them caused transitions to the spacecraft safe mode (Emergency Sun Reacquisition – ESR), causing major disruptions of science operations. Five times, the battery discharge regulators switched themselves off and there were about 7 occurrences where instrument boxes were switched off or required rebooting. Many of the self switch-offs are probably caused by false triggering of internal protection circuits, which are designed to protect against over-voltage or over-current. In all cases, no permanent damage occurred and the systems could be re-activated successfully.

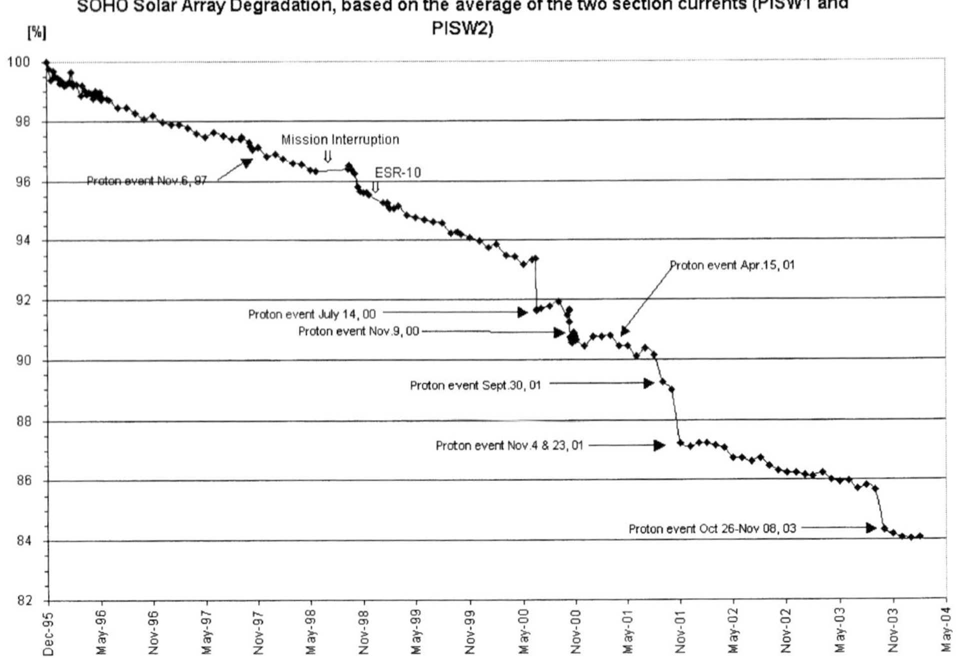

Figure 5. SOHO solar array degradation since the launch in 1995. The drop in sensitivity due to individual proton events is evident.

5.3.2. Solid State Recorder

A major temporary radiation effect is the SEUs in the Solid State Recorder (SSR), resulting in bit flips in the memory. The EDAC (Error Detection And Correction) detects and corrects these single errors (in the same word). Double errors are detected but not corrected.

Single errors are very common for the SSR 2Gbit memory:

- at solar min: 1 SEUs/minute
- at solar max: 0.5 SEUs /minute
- during proton events: up to 76 SEUs /minute (July 14, 2000 event)

Except during proton events, the majority of SEUs are caused by galactic cosmic rays. During solar maximum, magnetic fields are drawn out from the Sun by the large number of coronal mass ejections, providing a shield from galactic cosmic rays for the inner heliosphere.

So far there has been only 1 double error since launch, which was corrected when the affected memory location was overwritten with new data. A plot of the SEU's/ minute/2GB over the entire mission is given in Figure 6.

5.3.3. Star Sensor Unit

Another temporary radiation effect is observed in the Star Sensor Unit (SSU). When particles hit the CCD (Charge Coupled Device) of the SSU, they generate electrons, charging the pixels just like the regular photons, and thereby producing bright star-like signatures. The SOHO star tracker tracks five stars in small tracking windows. If a particle hits the tracking window, it can result in a wrong assessment of the tracked star's barycenter and/or magnitude.

The SSU interprets this as a movement of the star it was tracking thus providing wrong information to the attitude and orbit control (AOCS) software, resulting in turn in wrongful attitude correction orders to the wheels. Furthermore, the Star Tracker itself is moving its tracking window to the new wrong barycenter, and sometimes loses the true star in doing so.

The star tracker lost the guide star 54 times the first 3 years. Most of these resulted in loss of nominal attitude (fall-back into Roll Maneuver Wheels mode/gyro mode). This in turn caused a reducion in science gathering during the special operation to recover to nominal configuration.

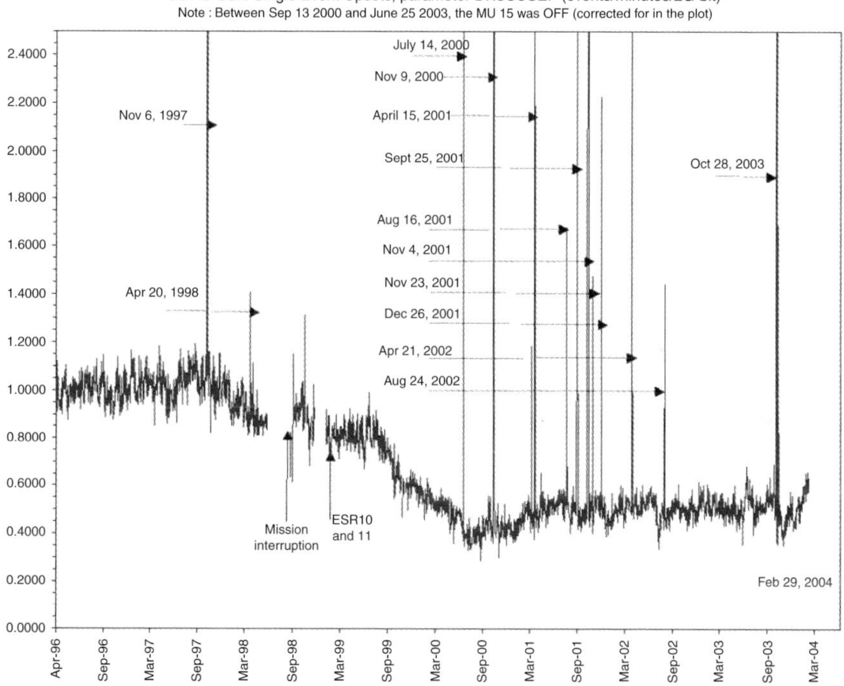

Figure 6. Number of SEU's per minute per 2GB over the entire mission. The solar cycle effect can be clearly seen with more SEU's during solar minimum (weaker solar magnetic field). The individual proton events during the SOHO mission can easily bee seen.

94

5.4. Improvements to Onboard Software

The onboard software was designed to be upgraded, and a series of improvements have taken place throughout the mission. Two improvements to increase the robustness against SEUs have been implemented:

The star tracker's internal software was modified:

- to filter out transient changes in the star barycenter (Position Jump Filter).
- to filter out transient changes in the star's magnitude

The result is that for both types of transients, no false event report is sent to the AOCS software. The AOCS overall task is to provide the spacecraft with the requisite pointing performance during the various spacecraft activities. The AOCS software was modified with a staircase filter to delay the effect of false event reports of the SSU to the attitude control computer. This filter was first implemented as a patch, but was later included in the gyroless software.

The other improvement added an automatic "star swap" capability to the attitude control computer gyroless software. Before 1998, the control mode was automatically changed from Normal Mode to Roll Maneuver Wheels mode when the guide star was lost or flagged invalid due to an SEU. In Roll Maneuver Wheels mode the roll control is transferred from the star tracker to the gyros, when the guide star was lost or simply flagged invalid due to a SEU.

SOHO no longer has gyros for backup, but the new software can now automatically use one of the 4 other stars that it is tracking as new guide star. Thanks to this, SOHO was able to remain in normal mode during the Bastille Day (July 14, 2000; 3 star swaps), the November 9, 2000 event (also 3 star swaps) and the October 28 2003 event (1 star swap). There have been 12 star swaps in all since the gyroless software was commissioned in October 1999.

Since the new gyroless software with the star swap feature was uploaded, there has been no loss of nominal attitude due to SEUs. SEUs can still cause the stars to be flagged "invalid" for a while, but they have always returned to valid on their own. With these new upgrades SOHO is now extremely stable.

6. Effects on the Scientific Instruments and Operation

As with spacecraft electronics and detectors, several instruments are also subject to effects from energetic particle events. For some in situ sensor instruments, the particles are the subject matter; for some, the particles are mostly a nuisance causing image degradation, but some instruments have health and safety concerns, due to e.g. high voltages on their detectors, the potential for arcing and permanent damage.

6.1. Image Degradation

As can be seen in Figures 7 and 8, the image degradation experienced during energetic particle events can be quite severe. Not only does the (relatively short-lived) degradation render images nearly useless for scientific analysis and space weather purposes – they also cause them to be much less compressible by the on-board software, in the case of EIT and LASCO. With a limited amount of telemetry and on-board storage, this results in the instrument getting "backed up", with a shifting of scheduled observations to a later time. While not necessarily critical under regular circumstances, certain joint observing programs rely on a closely coordinated timeline between a number of instruments both on board SOHO and on other spacecrafts, as well as ground based observations. This can only be corrected by intervention from the ground, using near-real-time (NRT) commanding to flush queues, skip observations, or upload new plans.

Figure 7. Images taken by the LASCO C3 coronagraph during the July 2000 solar energetic particle event showing severe effects on the detector from radiation background. Note that even though the images appear to be totally swamped during a proton storm, we are scaling the images to show the subtle coronal changes so that the particles are enhanced. They really don't saturate the detectors.

Figure 8. The Extreme Ultraviolet Imaging Telescope (EIT) observing during a proton event.

6.2. Health and Safety Effects

With several types of instruments operating detectors that have high voltage "image intensifiers" of different types, energetic particle showers are not purely an inconvenience. Although no incident has yet damaged any of the SOHO instruments, precautions have been put in place to ensure that the likelihood of damage is being kept as low as possible. Since, in general, the image-intensified instruments' data during particle events are not very useful anyhow, the loss of science data is not of concern; health and safety takes priority for those that feel a "better safe than sorry" approach is appropriate. The instruments that do take precautions of various kinds are: Coronal Diagnostic Spectrometer - CDS (continuous detector readouts to prevent charge build-up), Ultraviolet Coronagraph Spectrometer - UVCS and Solar Ultraviolet Measurements of Emitted Radiation - SUMER (high voltages turned down).

6.3. Operational Implications

The main operational "warning system" is the spacecraft solid state recorder. Since the SSR SEU counter is being monitored on the ground while the spacecraft is in contact (to prevent the SEU counter from overflowing), the impacts to normal operations are minimal when there is no particle event. If the SEU counter needs to be reset more often than once per hour, the Science Operations Coordinators are contacted, alerting instrument teams about the situation according to their own criteria.

For times when the spacecraft is not in contact, the warning system is based on NOAA GOES data from the web. Of course, with no spacecraft contact, nothing can be done about the instruments, so the status is only checked some time in advance of station passes with commanding ability. In addition, a 24/7 system based on automatic paging of the SOCs is in place, using NOAA GOES data from the web.

6.4. Long-term Effects

No serious long-term adverse effects have yet been noted, although the high energy particle environment does contribute to the gradual degradation of instruments. In particular, contaminant "doping" of refractive optics changes the absorption coefficients (impacting the optics temperatures), and the indices of refraction (focus changes). In addition, parts of the gradual sensitivity losses experience by many instruments can be attributed to contamination of detector electronics.

7. Mission Status and Future Plans

SOHO has several times demonstrated its leading role in the early-warning system for space weather. It provides a continuous real-time monitoring of solar eruptions that affect the Earths environment and space weather forecast operations have become to rely on SOHO on a routine basis.

Although long past the design lifetime of 2 years, SOHO is doing remarkably well. Fuel reserves of 120 kg should last 15 more years according to conservative estimates, and the solar array degradation is at only 16%, with a remaining margin of 14% before conservation measures must be applied. The gradual degradation of instruments and multi-layer insulation due to EUV exposure and high-energy particles is as expected, and not a cause of concern. Barring unexpected events, there seems to be no technical reason why SOHO and its instruments should not be able to complete observations of a full solar cycle or more.

Acknowledgements

We would like to thank Christopher Balch (SEC) and Fred Ipavich (Univ. of Maryland) for very useful input and comments to this paper.

References

Berghmans, D., Foing, B. H., and Fleck, B. Automated detection of CMEs in LASCO data, *ESA* SP-508, P. 437, 2002.

Gopalswamy, N. Lara, A., Yashiro, S., and Howard, R. A. Coronal mass ejections and solar polarity reversal, *ApJ,* 598, L63, 2003.

SEC User Notes, Issue 43, p3, 2004.

St.Cyr, C. et al. Properties of coronal mass ejections: SOHO LASCO observations from January 1996 to June 1998, *JGR.,* 105, 185.169, 2000.

Tousey, R., The Solar Corona, *Space Res.,* 13, 713, 1973.

Tsurutani, B. T., Gonzales, W. D., Lakhina, G. S., Alex, S. The extreme magnetic storm of 1-2 September 1859, *JGR,* 108, pp. SSH1-1, 2003.

Webb, D. F., and Allen, J. H. Spacecraft and ground anomalies related to the October-November 2003 solar activity, *Space Weather,* Vol 2, No. 3, S03008, 10.1029, 2004.

Yashiro, S. et al., A Catalog of white light coronal mass ejections observed by the SOHO spacecraft, *JGR,* in press, 2004.

Zhang, J., Dere, K. P., Howard, R. A., Bothmer, V. Identification of solar sources of major geomagnetic storms between 1996 and 2000, *ApJ,,* 582, 520, 2003.

STATISTICAL PRECURSORS TO SPACE STORM ONSET

J. A. Wanliss

*Department of Physical Sciences, Embry-Riddle Aeronautical University, 600 S. Clyde
Morris Blvd, Daytona Beach, FL 32114-3900, USA*

Abstract. Fractal fluctuation analysis is applied to ground-based data during
quiet times and during magnetic storm times between 1981-2002. For the
purposes of this study, quiet times were required to low activity at high latitudes,
as measured by Kp, for at least two days. Active times typically corresponded to
intense space storms. The data analyzed was from the H_{sym} time series for the
selected periods. The new technique computes nonlinear statistics from these
ground-based data, and monitors the variation of the statistics with time.
Variations in nonlinear statistics indicates a difference in statistical variability for
quiet times and storm times, and suggests an interesting new technique to predict
space storm onset.

1. Introduction

In recent years it has become evident that many space physics problems can be profitably
studied with the tools of statistical physics. For instance, the behavior of the solar wind driver is
commonly described in terms of geometric Brownian motion, and demonstrates a scaling
behavior [*Burlaga and Klein*, 1986]. It is probable that to some degree, the solar wind scaling
behavior propagates into the magnetosphere, although somewhat different scaling laws are
observed in ground-based data, for example geomagnetic indices [*Chapman et al.*, 2004 this
volume] and magnetometer data [*Wanliss and Reynolds*, 2003], and magnetospheric space-based
data [*Ohtani*, 1995].

Space storms, commonly referred to as magnetic storms, characterize the most dynamic
magnetospheric behavior. They frequently begin with a sudden worldwide increase in the ground-
based horizontal magnetic field by tens of nanoTesla, that lasts several minutes to several hours,
known as the initial phase. Following the initial phase comes the main phase, which typically
lasts about a day, and features large perturbations in the horizontal component on the order of
hundreds of nanoTesla. Subsequent to the main phase, the worldwide horizontal magnetic field
slowly returns to pre-storm values during a recovery phase that lasts several days. Space storms
include a rich variety of complex plasma electromagnetic processes extending from the surface of
the earth into the magnetosphere, with the primary locus of activity being in the near-earth
geospace environment [*Baker et al.*, 1997; *Li et al.*, 1997; *Reeves*, 1998]. These include energetic
particle injection and precipitation [*Reeves and Henderson*, 2001; *Horne*, 2003], acceleration of
relativistic electrons [*Li et al.*, 2001; *Summers et al.*, 2002; *Meredith et al.*, 2003], ring current
enhancement, decay, and composition changes [*Daglis et al.*, 1999; *Liemohn et al.*, 2001; *Kozyra
et al.*, 2002]. Recent studies on the causes of magnetic storms have found that coronal mass
ejections and extreme values of the southward interplanetary magnetic fields appear to be the key
factors in storm development [*Tsurutani et al.*, 1992; *Gonzalez et al.*, 1994; *Kamide et al.*, 1998;
Richardson et al., 2001]. Coupling and feedback between the ionosphere and magnetosphere also
plays an important role in the initiation and development of space storms, and interaction between
these two spheres is highly nonlinear [*Lui*, 2002; *Daglis et al.*, 2003 and the references therein].

Space storms thus form a complex system of nonlinear phenomena that include components of solar and terrestrial origin [*Benkevitch et al.*, 2002; *Daglis et al.*, 2003]. Intense storms have been identified as the cause of extensive damage to many ground and space-based systems [*Joselyn*, 1995; *Odenwald*, 2002] and, as such, understanding their dynamics is crucial to space weather studies.

The most widely used statistical descriptor of magnetic storm activity is the D_{st} index. This index is considered to reflect variations in the intensity of the symmetric part of the ring current that circles Earth at altitudes ranging from about 3-8 earth radii (R_E), and is proportional to the total energy in the drifting particles that form the ring current. It is calculated as an hourly index from the horizontal magnetic field component at four observatories, namely, Hermanus (33.3° south, 80.3° in magnetic dipole latitude and longitude), Kakioka (26.0° north, 206.0°), Honolulu (21.0° north, 266.4°), and San Juan (29.9° north, 3.2°). These four observatories were chosen because they are close to the magnetic equator and thus are not strongly influenced by auroral current systems. Convolution of their magnetic variations forms the D_{st} index, measured in nanoTesla, which is considered to provide a reasonable global estimate of the variation of the horizontal field near the equator. It is calculated once every hour.

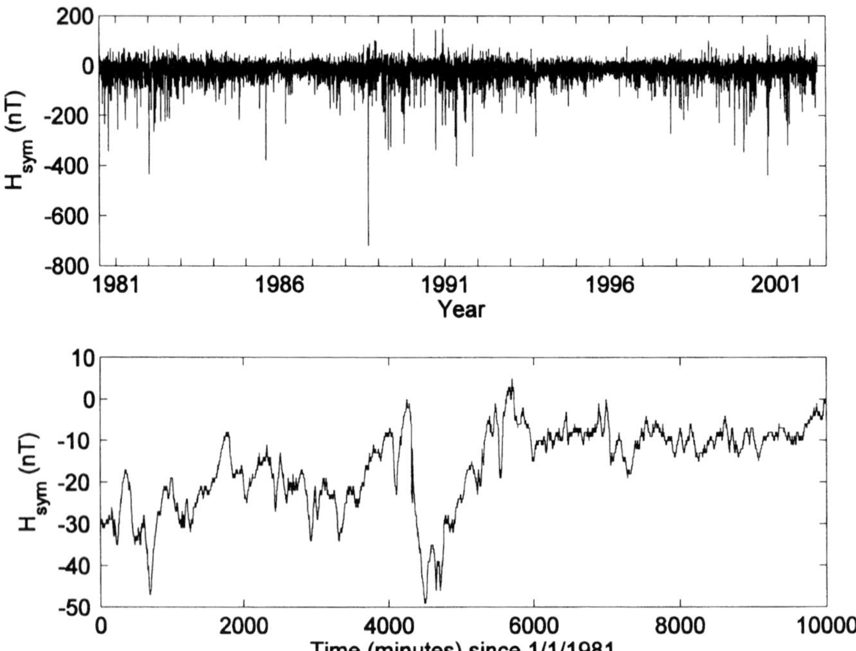

Figure 1. (Top) The entire H_{sym} time series from 1981 through 2002 appears to fluctuate around a zero mean with large intermittent negative perturbations. (Bottom) The first 10,000 minutes of H_{sym} data for 1981 appear to demonstrate long-range dependence.

In this work I examine the statistical nature of the nonlinear scaling properties in the ground-based SYM-H magnetic index [*Iyemori et al.,* 1999], hereafter referred to as H_{sym}. This index was developed as part of an effort to describe geomagnetic disturbance fields in mid-latitudes with high-time resolution. It is essentially the same as the D_{st} index, although it uses one-minute values from a different set of stations and a slightly different coordinate system. As such, this index also provides an excellent measure of the large-scale behavior of the ring current and magnetic storm dynamics.

The statistics of the space storm fluctuations are related to relaxation, and energization processes, and to transport phenomena in space plasmas. It is clear from Figure 1 that the H_{sym} time series displays evidence of both long-range dependence and intermittency. The large negative spikes in H_{sym} correspond to intervals of intense space storms. In this paper, I will examine H_{sym} to determine the characteristic nonlinear statistical differences, if any, between quiet (Q) and active (A) magnetospheric dynamics. The idea being pursued is that statistically stable, but dissimilar, nonlinear processes are involved in Q and A periods.

The H_{sym} time-series can be tested for nonlinear correlations in numerous ways. A general methodology is to estimate how a fluctuation measure, denoted here by F, scales with the size n of the time window considered. Specific methods, such as Hurst's rescaled range analysis [*Hurst*, 1951], power spectral analysis, structure function analysis [*Abramenko et al.*, 2002], or detrended fluctuation analysis [*Peng et al.*, 1995], all essentially calculate such a fluctuation measure, although the measure is different for each technique. Typically, $F \propto n^{\alpha}$, where α is the scaling exponent. For a time series that follows fractional Brownian motion (fBm) the relationships between the scaling exponents of the various methods are simple.

I employ a detrended fluctuation analysis (DFA) to the H_{sym} data. For this analysis technique, an uncorrelated time series gives $\alpha = 1/2$, as for standard Brownian motion. If $\alpha > 1/2$ ($\alpha < 1/2$) the data demonstrate correlation (anticorrelation). The analysis is employed on the H_{sym} time-series covering 22 years of data, from 1981 through 2002. The detrended fluctuation analysis was applied to Q and A periods selected for each year. The scaling exponent for Q periods was generally close to 1/2, but for A periods was usually greater than 1/2.

The rest of this paper is organized in the following fashion. In Section 2 I describe the data selection methodology. Section 3 gives background information about fBm and a description of the DFA technique. The results of the DFA of H_{sym} are shown in Section 4, and Section 5 discusses the results and their relevance to space storm prediction.

2. Data selection

To ensure the data were selected that were representative of a magnetospheric quiet or active state, I relied not only on H_{sym}, but also on the Kp index. Generally, when H_{sym} indicates significant activity, there is usually significant Kp activity also. Since H_{sym} is calculated exclusively from low- to mid-latitude magnetometer stations, and Kp includes higher-latitude stations, quiet interval data selection based on Kp ensures that data are selected for which the entire magnetosphere is as close as possible to a ground state.

On the other hand, during active times such as space storms, Kp is large and H_{sym} reaches large negative values. However, sometimes H_{sym} shows only small activity even when Kp is large, demonstrating dynamic activity (for example, magnetospheric substorms) at higher-latitude regions of the magnetosphere. Thus use of Kp to select events ensures that data are selected for which the magnetosphere is truly Q or A over a wide range of latitudes.

For each year, 10,000 consecutive minutes (i.e. 10,000 data points) that have the largest/smallest mean Kp were selected as representative of A/Q states. For each of the 22 years, the active times corresponded to magnetic storms. The average mean values of Kp for each year are shown in Figure 2. The upper curve (stars) shows Kp for the most active intervals, and the lower curve (circles) shows the corresponding average values for quiet intervals.

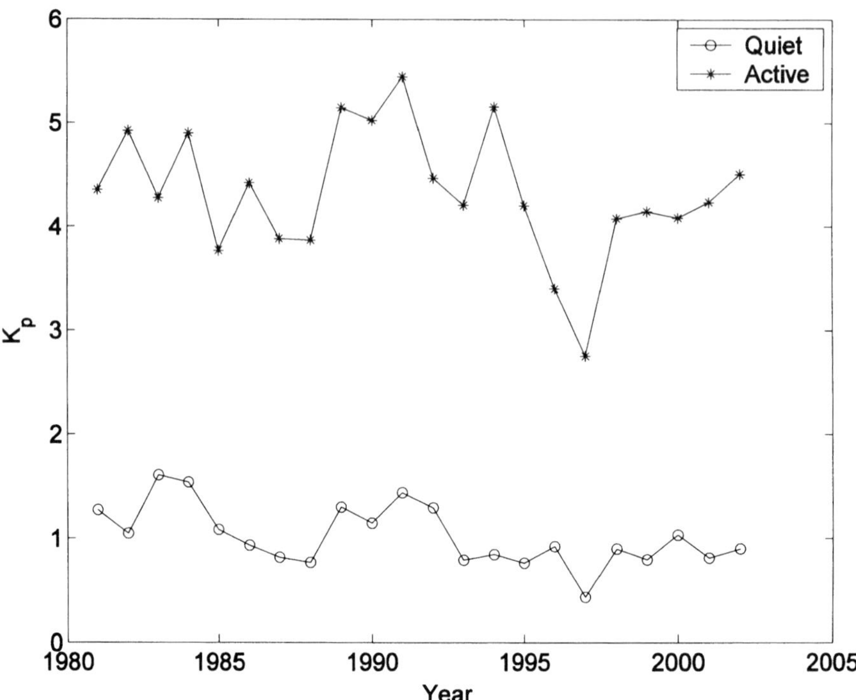

Figure 2. Smallest (circles) and largest (stars) mean values of Kp for 10,000 consecutive minutes in each year from 1981 through 2002.

The solar cycle influence on the data selection criterion is weakly evident as both Q/A events have smallest values near solar minimum (solar minimum around 1986 and 1996). Figure 3 shows the minimum H_{sym} values from each of the yearly 10,000 samples. Intervals of low activity, representing Q periods, generally have $H_{sym} > -40$ nT (100% of events) and A periods have $H_{sym} < -80$ nT (82% of events). For the A events, most of the data correspond to intense space storm events [*Gonzalez et al.*, 1994]. The mean and standard deviations are indicated as the dashed lines, with appropriate quiet and active symbols, to the right of the figure.

In Figure 4 the averages of H_{sym} for all 22 years for both Q and A events are shown. This figure indicates that the A periods generally encompass extreme space storms, and that they typically begin with the main phase of a space storm (featuring large negative H_{sym} values), and include a large proportion of the dynamic recovery phase. Q periods tend to trend upwards with mean values between -10 and +10 nT.

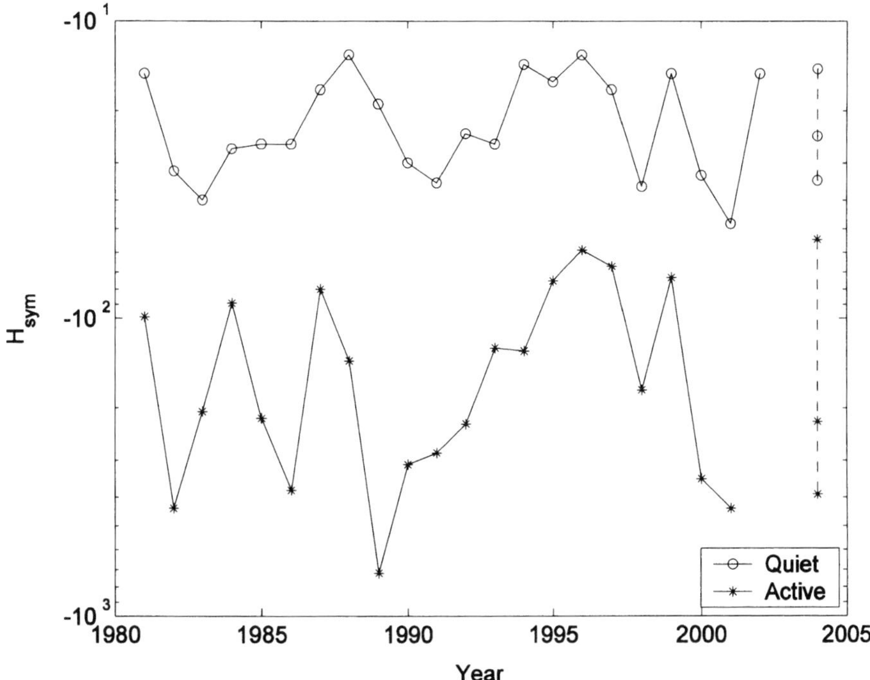

Figure 3. Smallest value of H_{sym} for each years' dataset for Q (circles) and A (stars) intervals.

3. Analysis technique

Many naturally occurring physical phenomena can be effectively modeled using fractal processes. Among the simplest models that display long-range dependence, one can consider the fractional Brownian motion. Fractional Brownian motion (fBm) is the process defined as the fractional integration of a Gaussian pure white noise. Typically, fBm is nonstationary, and thus detection of the presence of memory is a delicate task. It has been observed in the variety of the and scientific fields, including hydrology [*Neuman and Federico*, 2003], geophysics [*Frisch*, 1997], biology [*Collins and De Luca*, 1994], telecommunication networks [*Taqqu et al.*, 1997], and other fields.

Fractional Brownian motion has been observed in numerous space physics data, although they may also be described in other ways. The scaling properties of space physics data during dynamic magnetospheric activity were investigated by *Ohtani et al.* [1995]. They found that magnetic fluctuations in the magnetotail were well described as self-affine data with a power law spectrum. Studies of geomagnetic indices have served as particularly fruitful examples of fractional Brownian motion in space physics [*Sharma*, 1995; *Takalo et al.*, 1999; *Price and*

Newman, 2001; *Wanliss and Reynolds*, 2003]. It will therefore not come as a complete surprise to discover that the H_{sym} series also presents a space physics example data that may be fBm.

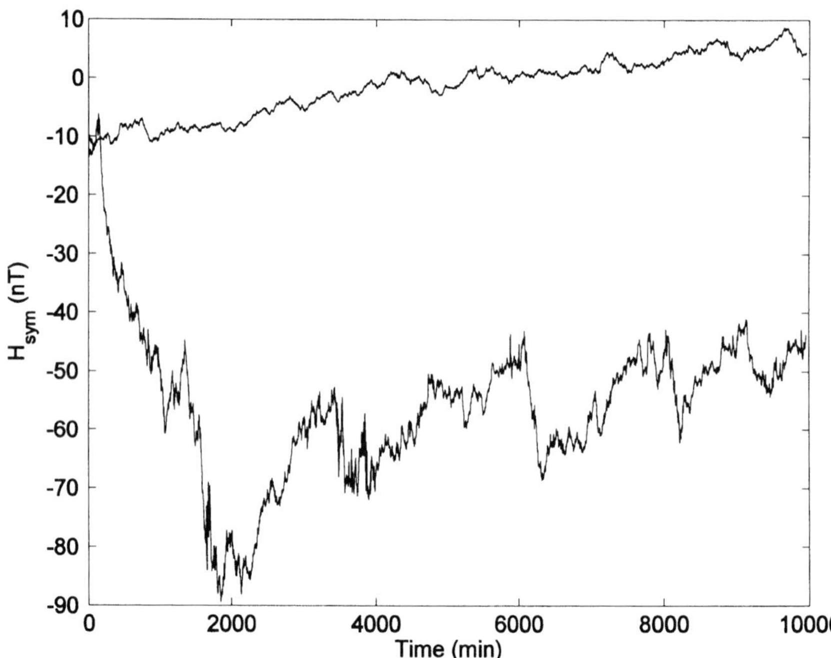

Figure 4. Average value of H_{sym} for the 22 years' data. The upper curve is for Q intervals, and the lower curve is for A events.

In this paper, I present a relatively new method of nonlinear analysis that has only recently seen use in space science research [*Ivanova et al.*, 2003]. Novel ideas from statistical physics led to the development of the detrended fluctuation analysis (DFA) [*Peng et al.*, 1995]. The method is a modified root mean squared analysis of a random walk designed specifically to be able to deal with nonstationarities in nonlinear data, and is among the most robust of statistical techniques designed to detect long-range correlations in time series [*Taqqu et al.*, 1996; *Cannon et al.*, 1997; *Blok*, 2000]. DFA has been shown to be robust to the presence of trends [*Hu et al.*, 2001] and nonstationary time series [*Kantelhardt et al.*, 2002; *Chen et al.*, 2002].

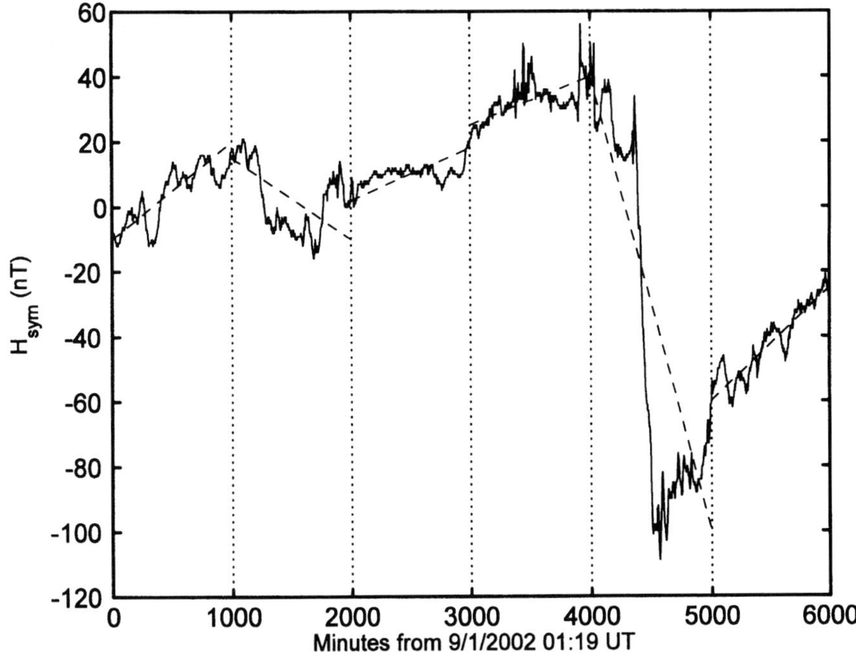

Figure 5. 6000 minutes of H$_{sym}$ data are shown in this figure. The vertical dotted lines indicate box of size n=1000, and the dashed straight line segments represent the trend estimated in each box by a least-squares fit.

The technique begins by the division of the time series into boxes of varying length *n* (Figure 5). In this example, *n=1000*. After this, a least-squares linear fit to the data signal is performed for each box; this linear fit represents the local trend in each box. Next, for each box the root mean squared deviations $F(n)$ of the signal from the local trend is determined. This procedure is repeated for different box sizes. Finally, the log-log plot of the deviation, $F(n)$, versus box size is used to calculate the slope which gives the scaling exponent, α.

4. Results

Figure 6 shows the Q/A Hsym data selected for epoch 1985. As was evident in the yearly means, the Q period trends up and the A period includes an extreme space storm near the beginning of the data interval. The Q period is from June 13, 1985 at 3 UT, and the A period is from April 19, 1985 at 12 UT.

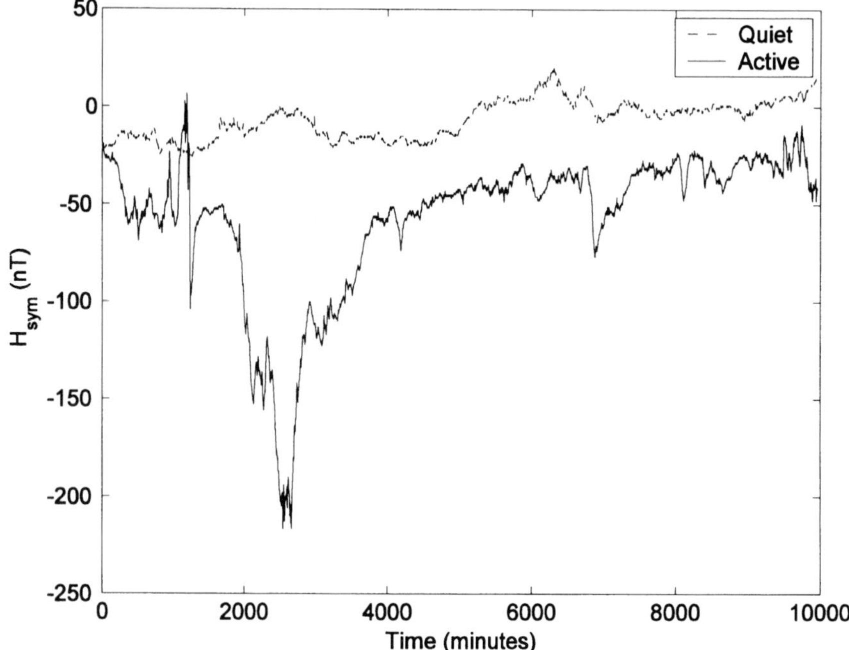

Figure 6. Q and A periods selected for 1985.

The log-log plot of fluctuation versus box size for epoch 1985, calculated from DFA, is given in Figure 7. The upper curve (stars) shows the fluctuations for the A interval, and the lower curve (circles) shows the fluctuations for the Q interval. Both curves are well fit by a straight-line, the upper yielding a scaling exponent $\alpha = 0.60$, quite different from the dashed reference curve, which has $\alpha = 0.50$, corresponding to a random walk. This means that the A data are correlated, or persistent. The nonlinear statistical behavior of the Q data is quite different. They fit $\alpha = 0.50$, consistent with a random walk.

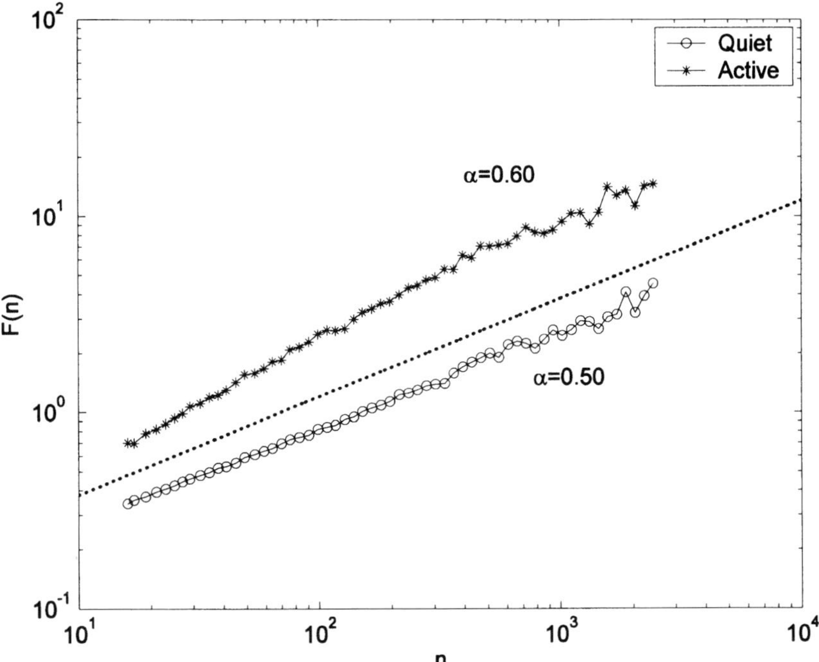

Figure 7. Fluctuation versus box size for Q and A periods for 1985.

In Figure 8 I show the calculated scaling exponents for each of the 22 years in the dataset. The mean and standard deviations are indicated as the dashed lines, with appropriate quiet and active symbols, to the right of the figure. Around solar minimum (near 1996) the scaling exponents for Q/A intervals are quite similar. This is probably an indication that during solar minimum the Kp selection criterion fails to properly separate dynamical behaviors. During the solar minimum the selection criterion gave Kp only slightly larger than 3 for the A interval (Figure 1). All other A intervals typically had Kp~4, usually larger. Even so, for all but two years $\alpha_A > \alpha_Q$.

The variability appears most sensitive for the Q intervals -- the scaling parameters for Q/A intervals, are similar when Kp values are close. For all events, the averages are $\alpha_Q = 0.498 \pm 0.039$ and $\alpha_A = 0.548 \pm 0.044$. To determine whether these averages are significantly different from the null hypothesis - that the difference is due purely to randomness - I apply the students-t test, and find $t = 3.880$, $p = 3.726 \times 10^{-4}$. These results imply that the difference between the statistics computed for Q and A intervals are significant; the likelihood that the means are different due to random processes is only a small fraction of 1%.

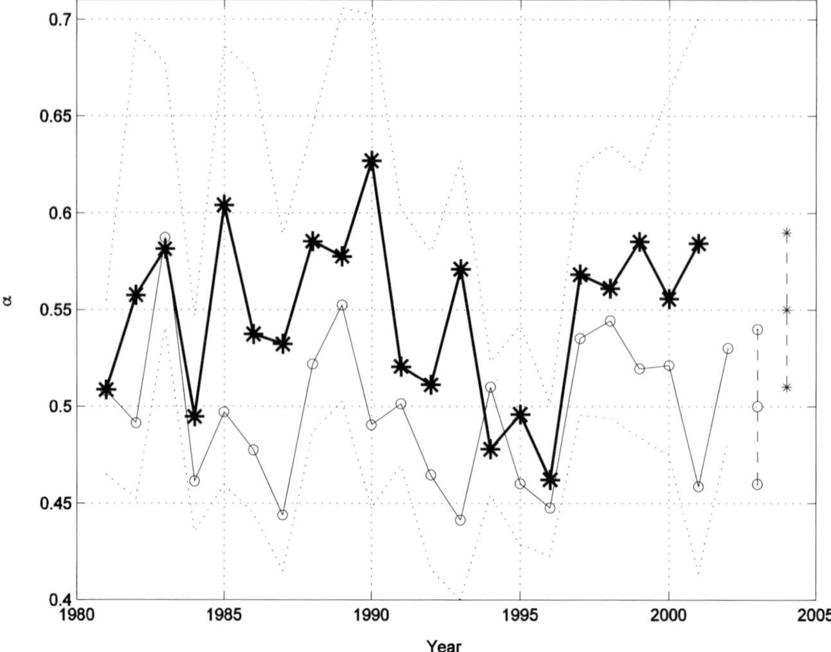

Figure 8. Scaling exponents for both Q and A periods for 1981-2002. One standard deviation is shown for the A periods (upper dashed curve) and Q periods (lower dashed curve).

5. Discussion

The goal of this research has been to characterize the differences in the scaling behavior between magnetospheric quiet and active dynamics over two solar cycles. To this end, I have applied the DFA method to quantify long-range correlations embedded in the nonstationary H_{sym} time series. The DFA method was selected because it copes very well with nonlinearity and nonstationarity. The method is particularly well-suited to analysis of the nonstationary H_{sym} data because it avoids spurious detection of correlations that are artifacts of nonstationarity. In particular, I have analyzed the most quiet and active periods for each of the 22 years between 1981 and 2002. For statistical robustness, each Q and A data subset selected for analysis comprised 10,000 consecutive H_{sym} values, at one-minute sampling interval. Through selection of only one representative Q and A interval for each year, it is possible to characterize the variation of the scaling exponents over two solar cycles.

The nonlinear statistical properties between Q and A events were noticeably different. In general, however, for each year, Q and A subsets generally clustered around their own unique values. A statistically consistent difference was found between the nonlinear scaling exponents for Q and A the events; $\alpha_Q = 0.498 \pm 0.039$ and $\alpha_A = 0.548 \pm 0.044$.

When the calculated scaling exponents are compared to the minimum H_{sym} for each selected interval, an interesting trend is apparent, as shown in Figure 9; less negative H_{sym} corresponds to smaller scaling exponents. Similarly, more negative H_{sym}, most common for the active intervals (stars in Figure 9) tends to result in larger scaling exponents. This trend will be investigated in future research with a larger subset of the H_{sym} series.

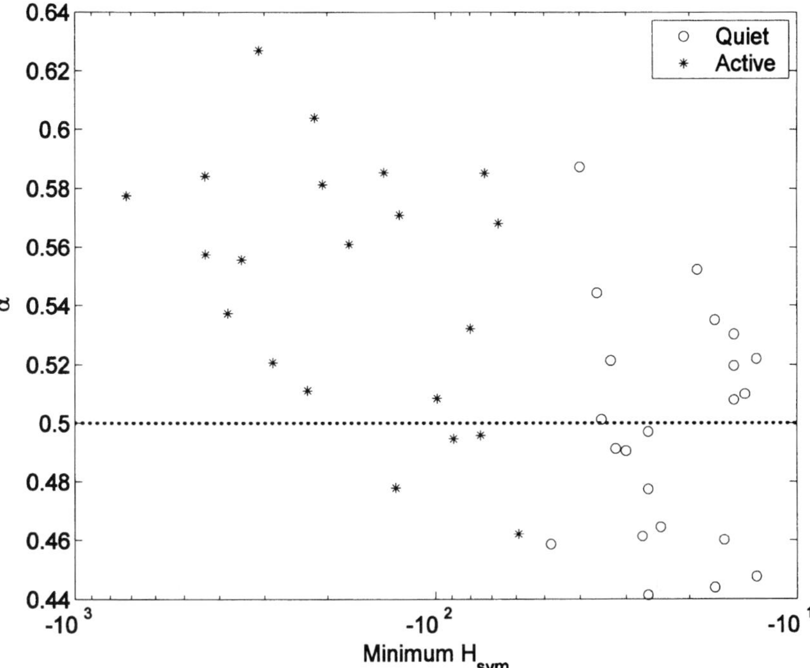

Figure 9. Comparison of the computed scaling exponent versus the minimum H_{sym} for the chosen interval.

In conclusion, these results demonstrate that the nonlinear statistical behavior of the magnetosphere, as derived from H_{sym}, is markedly different during Q and A intervals. The transition from Q to A phases during a space storm represent a period of potentially increased statistical variability, because it represents a transition from a random walk ($\alpha \sim 0.5$) to weakly correlated ($\alpha \sim 0.6$) regulation of the magnetospheric dynamics. The active intervals selected in this study included space storms in 100% of the cases. Since space storms frequently are preceded by quiet intervals, and are themselves characterized by global dynamic activity, the results presented here raise interesting questions that will be investigated further. For example, does the transition to storm times feature a repeatable change in nonlinear statistical behavior? Preliminary results indicate that the onset of storm times (Q to A intervals) are presaged by alterations in the scale-invariant properties of H_{sym}. By monitoring the variations of these scale-invariant properties one may be able to develop a method to predict the coming storm onset (to be presented in another paper).

References

Abramenko, V. I., V. B. Yurchyshyn, H. Wang, T. J. Spirock, and P. R. Goode, Scaling behaviour of structure functions of the longitudinal magnetic field in active regions on the sun, *ApJ*, 577: 487-495, 2002.

Baker, D. N., X. Li, N. Turner, J. H. Allen, and L. F. Bargatze, et al., Recurrent geomagnetic storms and relativistic electron enhancements in the outer magnetosphere: ISTP coordinated measurements, *J. Geophys. Res.*, 102, 14,141, 1997.

Benkevitch, L. V., W. B. Lyatsky, A. V. Koustov, G. J. Sofko, and A. M. Hamza, Substorm onset times as derived from geomagnetic indices, *Geophys. Res. Lett.*, *29*(10), 1496, doi:10.1029/2001GL014386, 2002.

Blok, H. J., On the nature of the stock market: Simulations and experiments, PhD thesis, University of British Columbia, Canada, 2000.

Burlaga, L. F. and L. W. Klein. Fractal structure of the interplanetary magnetic field., *J. Geophys. Res.*, *J. Geophys. Res.*, 91(A1), 347-350, 1986.

Cannon, M. J., D. B. Percival, D. C. Caccia, G. M. Raymond and J. B. Bassingthwaighte, the Evaluating Scaled Windowed Variance Methods for Estimating the Hurst Coefficient of Time Series, *Physica A,* **241**, 606-626, 1997.

Chen, Z., P. Ch. Ivanov, K. Hu, and H. E. Stanley, Effect of nonstationarities on detrended fluctuation analysis, *Phys. Rev. E*, 65, 041107, 2002.

Chapman, S.C., B. Hnat, G. Rowlands, N. W. Watkins, and M. P. Freeman, Scaling in long term data sets of geomagnetic indices and solar wind ϵ as seen by WIND spacecraft, *Conf. Sun-Earth Connections*, Kona, HI, Feb. 2004.

Collins, J. J. and C. J. De Luca, Upright, correlated random walks: A statistical-biomechanics approach to human postural control system, *Chaos*, 5 (1), 57-63, 1994.

Daglis, I. A., R. M. Thorne, W. Baumjohann, and S. Orsini, The terrestrial ring current: Origin, formation, and decay, *Reviews of Geophysics*, *37*, 407-438, 1999.

Daglis, I. A., J. U. Kozyra, Y. Kamide, D. Vassiliadis, A. S. Sharma, M. W. Liemohn, W. D. Gonzalez, B. T. Tsurutani, and G. Lu, Intense space storms: Critical issues and open disputes, *J. Geophys. Res.*, 108 (A5), 1208, doi:10.1029/2002JA009722, 2003.

Frisch, U., *Turbulence*, Cambridge University Press, 1997.

Gonzalez, W. D., J. A. Joselyn, Y. Kamide, H. W. Kroehl, G. Rostoker, B. T. Tsurutani, and V. N. Vasyliunas, What is a geomagnetic storm?, *J. Geophys. Res.*, 99, 5771-5792, 1994.

Horne, R. B., and R. M. Thorne, Relativistic electron acceleration and precipitation during resonant interactions with whistler-mode chorus, *Geophys. Res. Lett.*, *30*(10), 1527, doi:10.1029/2003GL016973, 2003.

Hu, K., P. Ch. Ivanov, Z. Chen, P. Carpena, and H. E. Stanley, Effect of trends on detrended fluctuation analysis, *Phys. Rev. E*, 64, 011114, 2001.

Hurst, H. E.: The long term storage capacity of reservoirs, *Trans. Am. Soc. Civil. Eng.*, 116, 770-799, 1951.

Ivanova, K., T. P. Ackerman, E. E. Clothiaux, P. C. Ivanov, H. E. Stanley, and M. Ausloos, Time Correlations and 1/f Behavior in Backscattering Radar Reflectivity Measurements from Cirrus Cloud Ice Fluctuations, *J. Geophys. Res.*, 108, doi:10.1029/2002JD003000, 2003.

Iyemori, T., T. Araki, T. Kamei, and M. Takeda, Mid-latitude Geomagnetic Indices "ASY" and "SYM" for 1999 (Provisional), online at http://swdcwww.kugi.kyoto-u.ac.jp/aeasy/asy.pdf, 1999.

Joselyn, J. A., Geomagnetic activity forecasting: The state of the art, *Rev. Geophys.*, 33, 3, 1995.

Kamide, Y., W. Baumjohann, I. A. Daglis, W. D. Gonzalez, M. Grande, J. A. Joselyn, R. L. McPherron, J. L. Phillips, E. G. D. Reeves, G. Rostoker, A. S. Sharma, H. J. Singer, B. T.

Tsurutani, and V. M. Vasyliunas, Current understanding of magnetic storms: Storm-substorm relationships, *J. Geophys. Res.*, 103, 17,705-17,728, 1998.

Kantelhardt, J. W., S. A. Zschiegner, E. Koscielny-Bunde, S. Havlin, A. Bunde, and H. E. Stanley, Multifractal detrended fluctuation analysis of nonstationary time series, *Physica A*, 316, 87-114, 2002.

Kozyra, J. U., M. W. Liemohn, C. R. Clauer, A. J. Ridley, M. F. Thomsen, J. E. Borovsky, J. L. Roeder, V. K. Jordanova, and W. D. Gonzalez, Multistep development and ring current composition changes during the 4-6 June 1991 magnetic storm, *J. Geophys. Res.*, 107(A8), 1224, doi:10.1029/2001JA000023, 2002.

Li, X., D. N. Baker, M. A. Temerin, T. E. Cayton, G. D. Reeves, R. A. Christensen, J. B. Blake, M. D. Looper, R. Nakamura, and S. G. Kanekal, Multi-satellite observations of the outer zone electron variation during the November 3-4, 1993, magnetic storm, *J. Geophys. Res.*, 102, 14,123, 1997.

Li, X., D. N. Baker, S. G. Kanekal, M. Looper, and M. Temerin, SAMPEX Long Term Observations of MeV Electrons, *Geophys. Res. Lett.*, 28, 3827, 2001.

Liemohn, M. W., Kozyra, J. U., M. F. Thomsen, J. L. Roeder, G. Lu, J. E. Borovsky, and T. E. Cayton, Dominant role of the asymmetric ring current in producing the storm time Dst*, *J. Geophys. Res.*, 106, 10,883-10,904, 2001.

Lui, A. T. Y., Multiscale phenomena in the near-Earth magnetosphere, *J. Atmos. Sol.-Terr. Phys.*, 64, 125-143, 2002.

Meredith, N. P., R. M. Thorne, R. B. Horne, D. Summers, B. J. Fraser, and R. R. Anderson, Statistical analysis of relativistic electron energies for cyclotron resonance with EMIC waves observed on CRRES, *J. Geophys. Res.*, 108(A6), 1250, doi:10.1029/2002JA009700, 2003.

Neuman, S. P. and V. D. Federico, Multifaceted Nature of Hydrogeologic Scaling and Its Interpretation, *Rev. Geophys.*, 41, 3, doi:10.1029./2003RG000130, 2003.

Odenwald, S., The 23rd Cycle: Learning to Live with a Stormy Star, Columbia University Press, 2002.

Ohtani, S., Higuchi, T., Lui, A.T.Y., and Takahashi, K.: Magnetic fluctuations associated with tail current disruption: Fractal analysis, *J. Geophys. Res.*, 100, 19135-19145, 1995.

Peng, C.-K., S. Havlin, H. E. Stanley, A. L. Goldberger, Quantification of Scaling Exponents and Crossover Phenomena in Nonstationary Heartbeat Timeseries, *Chaos*, 5 (1), 82-87, 1995.

Price, C. P. and Newman, D. E., Using the R/S statistic to analyze AE data, *J. Atmos. Solar-Terr. Phys.*, 63, 1387-1397, 2001.

Reeves, G. D., Relativistic electrons and magnetic storms: 1992-1995, *Geophys. Res. Lett.*, 25, 1817-1820, 1998.

Reeves, G. D. and M. G. Henderson, The storm-substorm relationship: Ion injections in geosynchronous measurements and composite energetic neutral atom images, *J. Geophys. Res.*, 106, 5833-5844, 2001.

Richardson, I. G., E. W. Cliver, H. V. Cane, Sources of geomagnetic storms for solar minimum and maximum conditions during 1972-2000, *Geophys. Res. Lett.*, 28, 2569-2572, 2001.

Sharma, A. S., Assessing the magnetosphere's nonlinear behaviour: Its dimension is low, its predictability is high, *Rev. Geophys.*, 35, 645, 1995.

Summers, D., C. Ma, N. P. Meredith, R. B. Horne, R. M. Thorne, D. Heynderickx, and R. R. Anderson, Model of the energization of outer-zone electrons by whistler-mode chorus during the October 9, 1990 geomagnetic storm, *Geophys. Res. Lett.*, 29, 2174, doi:10.1029/2002GL016039, 2002.

Takalo, J., Timonen, J., Klimas, A., Valdivia, J., Vassiliadis, D., Nonlinear energy dissipation in a cellular automaton magnetotail field model, *Geophys. Res. Lett.*, 26, 1813-1816, 1999.

Taqqu, M. S., V. Teverovsky, and W. Willinger, Estimators for long-range dependence: An empirical study, *Fractals*, 3, 185, 1996.

112

Taqqu, M., V. Teverovsky, and W. Willinger, Is network traffic self-similar or multifractal?, *Fractals*, 5, 63-73, 1997.

Tsurutani, B. T., W. D. Gonzalez, F. Tang, and Y. T. Lee, Great Magnetic Storms, *Geophys. Res. Lett.*, 19, 73-76, 1992.

Wanliss, J. A., and M. A. Reynolds, Measurement of the stochasticity of low-latitude geomagnetic temporal variations, *Ann. Geophys.*, 21, 1-6, 2003.

SOLAR WIND DRIVERS FOR STEADY MAGNETOSPHERIC CONVECTION

Robert L. McPherron[1], T. Paul O'Brien[2], Scott Thompson[1]

[1]*Institute of Geophysics and Planetary Physics, University of California, Los Angeles, CA 90095-1567, USA*

[2]*Space Sciences Department, Aerospace Corporation, M2-260, P.O. Box 92957, El Segundo, CA 900009-92957, paul.obrien@aero.org*

Abstract. Steady magnetospheric convection (SMC) also known as convection bays, is a particular mode of response of the magnetosphere to solar wind coupling. It is characterized by convection lasting for times longer than a typical substorm recovery during which no substorms expansions can be identified. It is generally believed that the solar wind must be unusually steady for the magnetosphere to enter this state. However, most previous studies have assumed this is true and have used such conditions to identify events. In a preliminary investigation using only the AE and AL indices to select events we have shown that these expectations are generally correct. SMC events seem to be associated with slow speed solar wind and moderate, stable IMF Bz. In this report we extend our previous study including additional parameters and the time variations in various statistical quantities. For the intervals identified as SMCs we perform a detailed statistical analysis of the properties of different solar wind variables. We compare these statistics to those determined from all data, and from intervals in which substorms but not SMCs are present. We also consider the question of whether substorms are required to initiate and terminate an SMC. We conclude that the intervals we have identified as SMC are likely to be examples of the original Dungey concept of balanced reconnection at a pair of x-lines on the day and night side of the Earth.

Keywords. Convection bay, steady magnetospheric convection, solar wind coupling.

1. Introduction Heading

Dungey [1961] introduced the idea that magnetic reconnection is the driver of auroral activity. According to his concept activity begins when the solar wind magnetic field merges with the Earth's magnetic field at an x-line located at the sub-solar point on the dayside magnetopause. The solar wind transports the interconnected field lines over the polar caps creating a long magnetic tail. The open field lines reconnect at a nightside x-line and then move Sunward around the Earth to return to the dayside. The process creates a 2-celled convection pattern in the ionosphere that corresponds to the observed motion of ionospheric plasma. Akasofu [1964] showed that auroral and magnetic activity occurs sporadically near midnight in an organized sequence of events called the substorm. The concept of a substorm was later modified becoming the near Earth neutral line model [Hones et al., 1973; McPherron et al., 1973; Nishida and Nagayama, 1973]. This model postulated that the expansion phase of a substorm is caused by the formation of a second nightside x-line close to the Earth. An essential feature of this model is unbalanced flux transfer. Because the distant x-line in the Dungey model is so far downstream it can not immediately respond to the onset of reconnection on the dayside. As a consequence

114

magnetic flux from the dayside accumulates in the tail lobes causing Earthward motion and thinning of the tail current. The thinning leads to instability and an x-line forms close to the Earth. This line reconnects the lobe flux and allows it to return to the dayside more rapidly than reconnection at the distant x-line.

The original Dungey model was revitalized when Pytte et al. [1978] discovered magnetic disturbances with a 2-celled convection pattern, but with no evidence of a substorm expansion. These disturbances were named convection bays. It was speculated that they were a manifestation of balanced reconnection of the type envisioned by Dungey. For this to occur the nightside x-line must be relatively close to the Earth so that it can balance dayside reconnection. Pytte et al. [1978] speculated that this happens when a substorm forms a near-Earth x-line that does not move far away from the Earth. They suggested that steady solar wind driving might be the cause of these events. Sergeev and coworkers have carried out many studies of similar events establishing that they represent a unique mode of response of the magnetosphere to the solar wind [Sergeev et al., 1996]. The name 'convection bay has been replaced by the new term 'steady magnetospheric convection (SMC)' emphasizing the steady nature of these events.

In most studies of SMCs it is generally assumed that conditions in the solar wind are steady when such events occur. In fact this condition has been used as one of the selection criteria in choosing events. It is therefore not surprising that the solar wind is steady during steady magnetospheric convection. O'Brien et al. [2002] pointed out that this is a circular argument and that it is necessary to define events using only magnetospheric data and then show that the properties of the solar wind during these events are steadier than in the typical solar wind. O'Brien et al. [2002] accomplished this by using only the AE and AL indices to define events. They found that the median solar wind velocity during SMC (~380 km/s) is significantly lower than characteristic of all disturbed times (450 km/s). In contrast, the IMF Bz is slightly more negative (-3.0 nT) than is typical of disturbed times (-2 nT). Also they showed that the fluctuations in velocity and Bz are smaller during SMC than during disturbed times. Thus, the SMC events identified in this study correspond to rather weak but steady driving of the magnetosphere by the solar wind. The purpose of this paper is to extend the work of O'Brien et al. [2002] showing some additional properties of the intervals selected as SMCs.

2. Data and Event Selection

The data used in this study are the IMP8 plasma and magnetic field and AE indices for the interval January 1, 1978 - June 30, 1988. Because the circular orbit of IMP8 is inside the magnetosphere for about half of each orbit we have solar wind data for only about 40% of the AE samples. The original IMP-8 data were interpolated to one-minute resolution and then propagated to the Earth using simple ballistic propagation.

Steady magnetospheric convection events were defined by the following three conditions:

1. AE > 200 nT   Geomagnetically disturbed times
2. d(AL)/dt > -25 nT/m   No substorm expansions
3. The above conditions persist for more than 90 minutes to guarantee that substorm recovery phases have ended.

There is little question that AE>200 nT is a disturbed condition. The most probable value of AE in our study interval was ~50 nT and the median was 150 nT. Neither is there argument that most recovery phases are about 90 minutes long. The second criteria, however, needs some justification. A moderate substorm expansion of AL = -500 nT developing in 20 minutes corresponds to -25 nT/min. Alternatively assume that there are typically three substorm expansions per day and that each expansion is 20 minutes long. Then in a day there is a probability of expansion equal to 0.0417 (3*20/1440). We have calculated the cumulative probability distribution for the rate of change of AL in our study interval and identified the rate of change corresponding to this probability. We find a value of -26 nT/min. We assume all rates less than this threshold correspond to a substorm expansion.

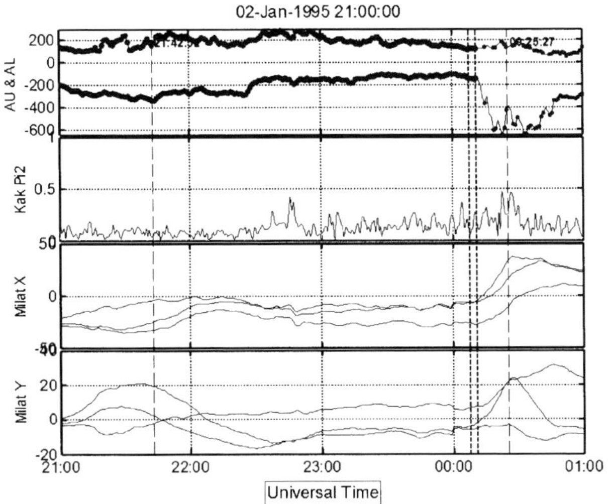

Figure 1. An example of an interval selected as a steady magnetospheric convection event is shown in top panel by heavy markers. Isolated dots show points satisfying first two selection criteria, but not the third (see text). Lower panels show Kakioka Pi 2 power and a selection of X-Y traces from midlatitude stations near midnight.

Every time point in the AE and AL time series was tested by the first two criteria. When at least 90 consecutive points satisfied these conditions the interval was labeled an SMC event. For each interval the beginning of the >90-minute interval was taken as the start time of the event. The total number of points satisfying these criteria in any interval was recorded as the duration of the event. A total of 2365 events were identified of which 600 had duration longer than 150 minutes and 47 were longer than 300 minutes. The longest event had duration of 670 minutes.

3. Results

An example of an SMC event selected by the preceding criteria is presented in Figure 1. The top panel displays the AU and AL indices for a 4-hour interval at the end of January 2, 1995. Heavy lines denote an SMC interval while isolated dots show points that satisfy the first two criteria, but because there are intervening points that do not, they fail the third. It is evident that this SMC interval was terminated by a moderate substorm. Lower panels show Pi 2 power at Kakioka near the dawn meridian and the X and Y magnetometer traces from midlatitude stations near midnight. Kakioka is too far from midnight to be a reliable indicator of substorm onset, but the magnetometer traces clearly show the onset of the expansion phase about 0010 UT.

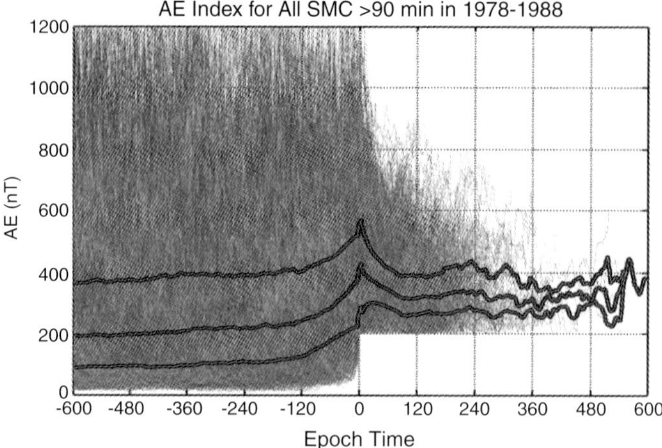

Figure 2. The ensemble of AE segments for all 2365 SMC events are shown by thin lines. Heavy lines are the quartiles of the probability distribution. Each trace is terminated at the end of the event so subsequent data does not distort the ensemble statistics.

O'Brien et al. [2002] used the list of SMC events to create the cumulative distribution for SMC duration. They showed that the probability that duration exceeds a value given by the abscissa is fit by a subexponential. They interpreted this result as an indication that there is a positive correlation between successive time points in an event. This implies there is a tendency for the magnetosphere to stay in the SMC state for some time once it enters this state.

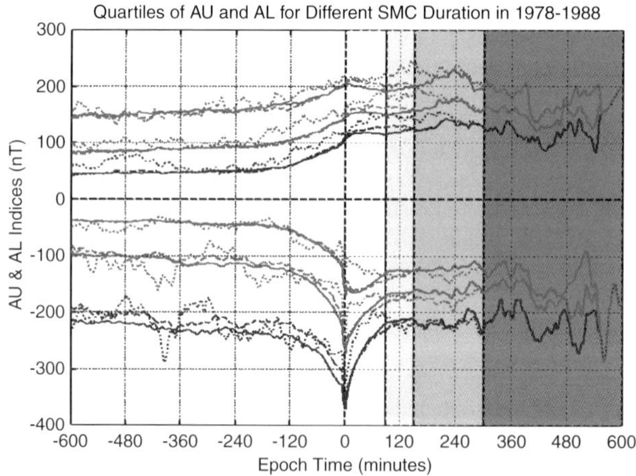

Figure 3. Quartiles of AU and AL indices for all SMC (solid), for all SMC dur>150 min (dashed), and dur>300 min (dotted). Gray bars show ranges of duration used in statistics.

The ensemble of all 2,365 SMC events as seen in the AE index is shown in Figure 2. Every event is plotted as a thin line. Quartiles of the ensemble at each instant of time are represented by thick black lines. No data beyond the end of an event were used in the display or in the statistics so that the quartiles only represent properties of the SMC events. The effect of the first criterion (AE > 200 nT) is evident in the plot from the sharp lower edge of the ensemble of traces. Note that the AE index occupies a much more limited range during SMC than it does during general activity (there are few large values of AE). Also, SMC begin with 90-minute decay from previously elevated activity (maximum quartiles at epoch zero). This suggests that a preceding substorm expansion is required to start these events.

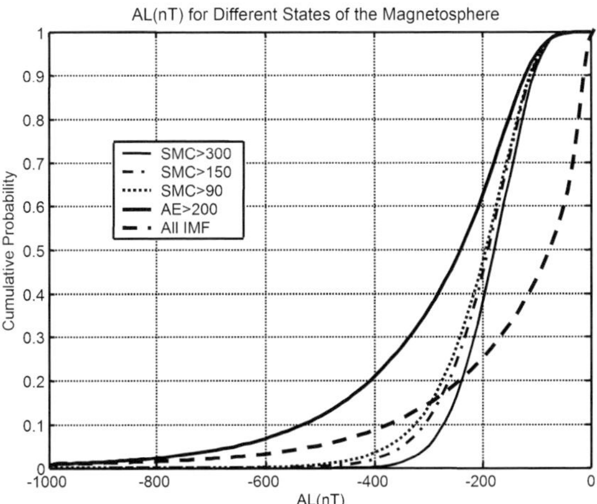

Figure 4. The cumulative distribution functions of the AL index for different subsets of the data shown by legend. The cdfs for SMC events of different duration (3 inner traces) are very similar with medians of order -200 nT

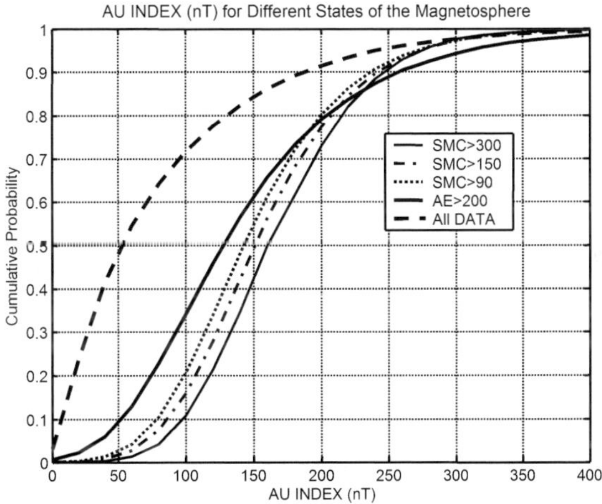

Figure 5. The cdfs for the AU index. Median cdf for all SMC events is about +160 nT and higher than typical in disturbed times (130 nT).

The temporal development of the quartiles of the AU and AL ensembles are presented in Figure 3. The median AL is most disturbed at epoch zero. Subsequently it decays to a constant value of about -150 nT where it remains for the remainder of the event. In contrast, the AU index is somewhat larger during the SMC event than it is at epoch zero or any time before. The equilibrium value for the median AU is nearly equal to that of the AL index in contrast to the more typical situation where the ratio AU/AL ~ 0.5.

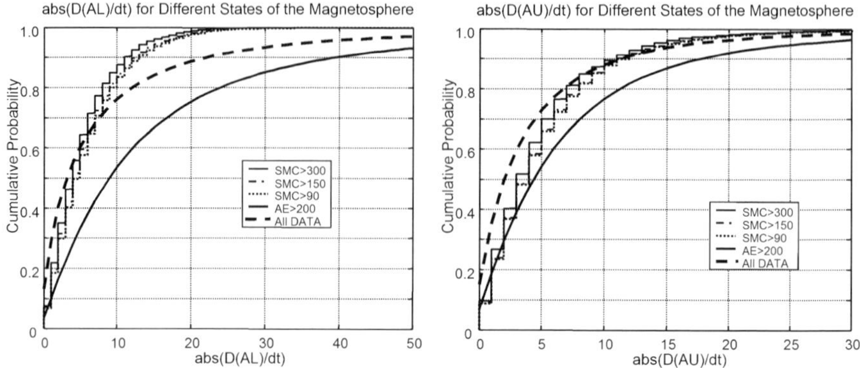

Figure 6. The cdfs of the magnitude of fluctuations in the AL index (left panel) and the AU index (right panel) for different subsets of the data shown by legend. The cdfs for SMC of different duration are virtually identical (stair step traces) and are characterized by much weaker fluctuations than general activity

The cumulative probability distribution for AL averaged over different subsets of the data is shown in Figure 4. The left-most line shows the cdf for all disturbed times. The right-most line is for all data. The median AL for disturbed conditions (AE>200 nT) is ~-300 nT. The cdfs for SMC of three different durations is shown by the cluster of three, overlapping traces all having a median value of only ~-170 nT. The dynamic range of AL during SMC events is quite limited [-300 to -100 nT] as compared to its range in generally disturbed times. A similar plot for the AU index is presented in Figure 5. The left-most line is for all data and the next line is for all disturbed times. For SMC the median AU values are clustered around -160 nT as compared to general disturbance when it is about -130 nT. Also for SMC events the dynamic range of AU is somewhat larger than that of AL.

The distributions of the magnitude of fluctuations in AL and AU during different conditions are shown by cdfs in Figure 6. The distributions for SMC events are shown by stair-step curves in each graph. For both indices SMC events exhibit less variability than is the case for general activity (curves at right side of graphs). This tendency is more pronounced for the AL index. There is no dependence of the size of fluctuations on the duration of the events.

We have calculated cdfs for a variety of solar wind variables. The results for solar wind velocity and IMF Bz are displayed in Figure 7. For the velocity in the left panel the cdfs are well separated indicating a strong dependence of SMC occurrence and duration on the value of solar wind velocity. The right-most curve in the left panel is the cdf for general activity (AE>200 nT) and the next curve is for all activity. The three left-most curves are for progressively longer duration SMC events. The graphs indicate that SMC events are associated with lower solar wind velocity, and also that duration increases as velocity decreases. The velocity cdfs during SMC events also have more limited dynamic range with far fewer high values of velocity than is characteristic of general activity.

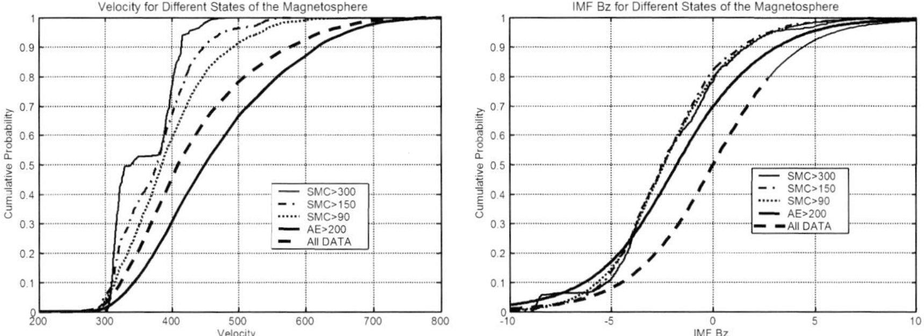

Figure 7. The cdfs for the solar wind velocity and IMF Bz are presented in the left and right panels respectively. The three left-most curves are for progressively longer SMC events. The right-most curve is for general activity.

Results for the IMF GSM Bz component are presented in the right panel of Figure 7. The cdfs of Bz for different duration SMC are identical and clustered at the left side of the graph. These distributions are significantly different from the cdf of all solar wind data shown by the right-most curve with a median value of 0.0 nT. The distributions for SMC also have a more negative median (-3 nT) that does the cdf for general activity (-2 nT). However, the dynamic range of Bz during SMC is much more limited to values close to the median than is the case during general activity.

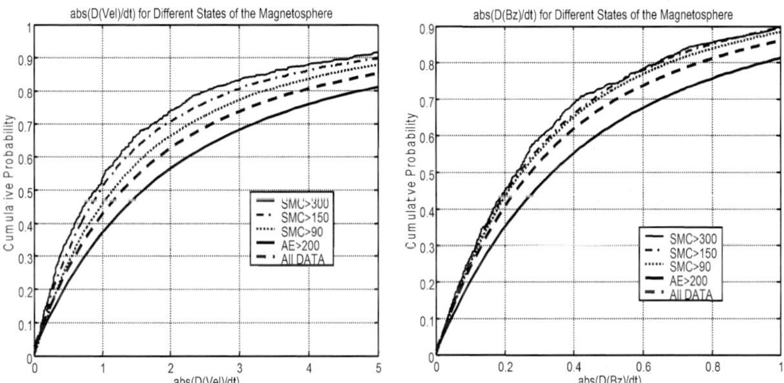

Figure 8. The cdfs of fluctuations in solar wind velocity and IMF Bz are shown in the left and right panels respectively. The three left-most curves in each panel are for progressively longer SMC events. The right-most curves are for all data.

The left panel of Figure 8 shows the cdfs for the magnitude of fluctuations in solar wind velocity. These curves are ordered in the same manner as are the velocity cdfs. Thus SMC events

120

occur during times of weaker fluctuations in solar wind velocity, and long duration SMC events are associated with the lowest level of fluctuations in solar wind velocity. Analysis of the thermal velocity of protons produces identical results. SMC events are associated with lower temperatures and weaker fluctuations in temperature than is the case for general activity. Solar wind density for SMC events is slightly higher than during general activity, but there is no dependence of duration on density.

The cdf for the magnitude of fluctuation in Bz is shown in the right panel of Figure 8. There is no dependence of the cdfs on SMC duration (clustered curves on left), but the cdfs are significantly different from the cdf for general activity (right-most curve). The results demonstrate that intervals identified as SMC are associated with weaker fluctuations in IMF Bz than occur during general activity.

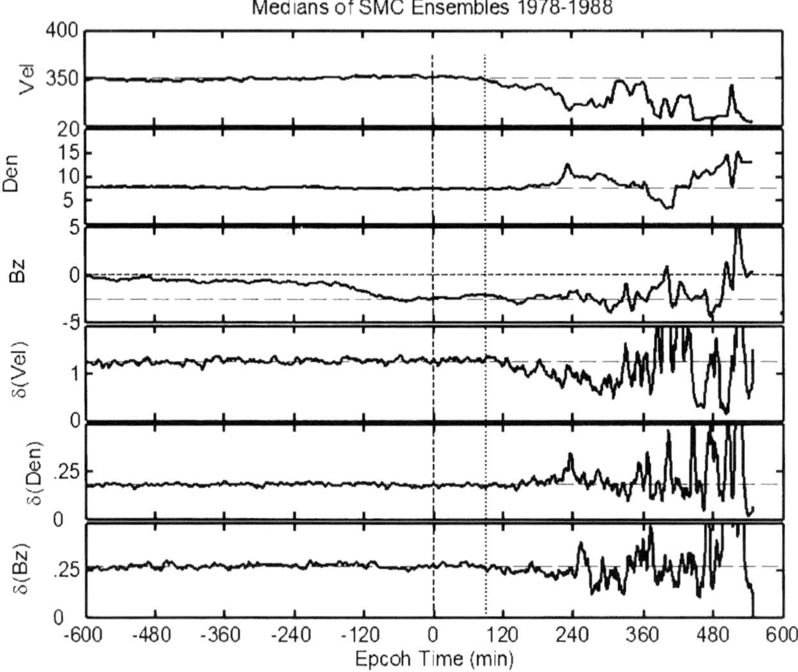

Figure 9. The time development of the medians of solar wind velocity, density, IMF Bz, and their fluctuations during SMC.

The temporal behavior of the medians of the solar wind variables during SMC events provides considerable insight into the possible cause of SMC events. The results are presented in Figure 9. The top three panels of this figure show the medians of velocity, density, and IMF Bz for 10 hours before and after an SMC begins. The bottom three panels show their fluctuations. The median velocity is a low, steady 350 km/s before and 90 minutes into the SMC. Thereafter (see vertical line at 90 minutes) it appears to steadily decrease by about 40 km/s in an interval of about 8 hours. The magnitude of fluctuations also decreases in concert with the velocity. As the SMC duration becomes longer the results become less clear. There are so few long-duration events that the statistics become suspect. The behavior of density is the inverse of velocity. As the velocity decreases density and its fluctuations at first increase, and then begin to fluctuate wildly. Median IMF Bz is negative throughout the entire 20 hours of the plot. About 2.5-3 hours before the SMC starts median Bz turns more southward to -2.5 nT where it remains throughout the event. However, fluctuations in Bz behave similar to those in velocity beginning to decrease after about 90 minutes. The data suggest they stabilize at a lower level after about four hours.

4. Discussion

Steady magnetospheric convection (SMC) is characterized by a well developed 2-celled convection pattern and the absence of any indication of substorm expansion phases [Pytte et al., 1978; Sergeev et al., 1996]. Ideally one would select events based on ionospheric measurements of convection and a comprehensive list of substorm onsets. Neither of these datasets exists at the present time. In the absence of such information we have chosen to use proxies for these quantities. The best available data set containing information about convection is the AE index. AE is calculated as the difference between the AU and AL indices. AU is a measure of the strength of the eastward electrojet, and AL is a measure of the strength of the westward electrojet. In the absence of substorms these two indices or their combination, AE, characterize the DP-2 current system created by ionospheric convection. We have selected a value of AE = 200 nT well above quiet-time values as a threshold for geomagnetic activity and the presence of enhanced convection. Substorms are usually identified by the development of the polar magnetic substorm in association with the expansion phase of the auroral substorm. The polar magnetic substorm is a manifestation of the DP-1 current system. This current is the ionospheric closure of the substorm current wedge. It flows downward along field lines in the early morning hours, across the auroral bulge, and outward along field lines from the westward traveling surge. This current is in the same direction as the DP-2 westward electrojet, but is more concentrated around midnight. During the substorm expansion phase AL rapidly decreases due to an increase in the strength of DP-1. To eliminate such intervals from our analysis we have taken the conservative criterion that the rate of decrease of AL at any instant should be > -25 nT/min. As discussed in Section 2 this value eliminates most substorm expansions and probably a number of other instances as well. Any instant of time when AE and the derivative of AL satisfy these two criteria is labeled as a point in a potential SMC event. The final criterion is that the disturbance is not just the recovery phase of a substorm, and is part of a different state of magnetospheric convection. The typical duration of a substorm recovery is 90 minutes. Therefore we require that both conditions must be continuously satisfied for at least this length of time. The start of the SMC event is taken as the beginning of the continuous interval satisfying the first two criteria. As shown above these intervals appear to begin in the recovery phase of a substorm.

The results presented above are based on the analysis of 11 years of AE index data from 1978 through 1988. During this time IMP-8 was used as the solar wind monitor providing roughly 40% coverage of the solar wind. We have used these data propagated to the subsolar bow shock and interpolated to one minute resolution to characterize the solar wind during the intervals identified as SMC events.

Several colleagues have expressed concern about the use of the AE indices to identify intervals of steady magnetospheric convection. For example, Kamide (personal communication) points out that the AE index has no physical meaning, and that both AU and AL indices have systematic errors associated with universal time and level of activity. To investigate possible effects of these errors in the indices we have calculated the probability of observing SMC as a function of universal time, day of year, and phase of the solar cycle. We find no dependence of SMC occurrence on universal time. However there is a strong dependence of occurrence on both

122

season and phase of the solar cycle. These results are shown in Figure 10. The grayscale map shows the joint probability of observing SMC as a function of solar cycle (abscissa) and season (ordinate). The line plot to the right shows the average seasonal dependence and the line plot at the top shows the average solar cycle dependence. It is obvious that there is a strong dependence of SMC occurrence on season with a maximum in northern summer. Also SMC occurrence peaks near solar maximum. These characteristics are those of the AU index as shown by Ahn et al. [1999]. During northern summer the conductivity of the northern ionosphere is much higher than it is in winter due to solar illumination. For a given convection electric field the eastward electrojet is stronger in summer and hence produces a larger AU index. But the AE index we used to select disturbed times is the sum of AU and AL. Thus it is more likely that a given weak disturbance will meet our selection criteria during northern summer than during winter. The solar cycle effect can be understood in a similar way. At solar maximum increased solar activity produces higher levels of ultraviolet radiation. Provided the northern hemisphere is illuminated, i.e. northern summer, then the ultraviolet radiation increases the conductivity relative to its value at solar minimum. As also shown by Ahn et al. [1999] the AL index behaves quite differently from AU. The AL index is most disturbed at equinox due to the Russell-McPherron effect [Russell and McPherron, 1973].

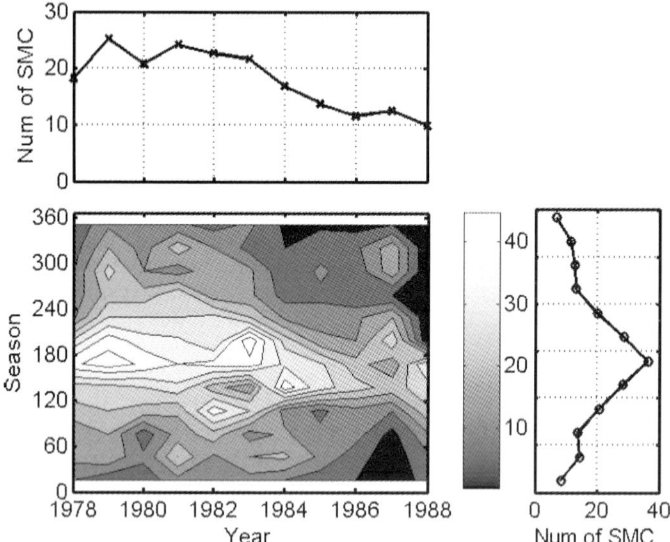

Figure 10. The joint probability distribution for observing SMC as a function of solar cycle (abscissa) and season (ordinate). Line plots show marginal distributions for seasonal (right) and solar cycle (top) dependence.

From the foregoing it is clear that our AE selection criterion does affect the statistics of SMC occurrence. However, our main objective was to find a large set of SMC events for which we could characterize the solar wind, and the seasonal and solar cycle dependence of SMC occurrence are unlikely to strongly affect these characteristics. Our results generally confirm what most researchers believe to be true regarding SMC events – they occur during intervals of unusually steady solar wind velocity and IMF Bz. A comparison of our statistical results with five well-studied SMC events described by Sergeev et al. [1986] reveals few differences. His events have average AE ranging from 190 to 360 nT, average Bz from -6.0 to -2.5 nT, and solar wind velocity between 287 and 381 km/sec. The convection bay event studied by Pytte et al. [1978] was somewhat stronger than our statistical averages or Sergeev's events with average AE ~ 750 nT and Bz ~ -5 nT. Our conclusion is that SMC are generally associated with relatively low solar wind velocity (350 km/s) and weak IMF Bz (-3 nT). Also the AL index and AU index are relatively weaker than during typical activity. Together the median values of velocity and Bz correspond to an interplanetary electric field of only 1 mV/m. Although previously studied SMC have properties close to our values it is possible that our selection criteria affect the outcome.

The most likely reason for identifying weak activity as an SMC is the criterion that the rate of change of AL must exceed -25 nT/min. During weak convection it is likely that AL fails this criterion only during a substorm expansion. However, in strong convection it is possible that ordinary fluctuations in the location or strength of the DP-2 current system occasionally cause a rapid decrease in AL. Possibly a relative criterion based on the ratio of the current slope to current value of AL would discriminate such points. Better yet would be the use of a comprehensive list of substorm onsets ruling out all points between an onset and the beginning of its associated recovery phase. Unfortunately, during very disturbed times it is difficult to identify onsets without careful examination of many different data sets. Also at such times several expansions may occur in sequence, with the later ones exceedingly difficult to detect. Such situations are not easily treated with a simple procedure using only one type of data as we have done here.

The main experimental result of this work is that the magnetosphere frequently (~240 times a year) enters a state in which there is evidence of enhanced convection (elevated AE index) but no evidence of substorm expansions (no rapid decrease in AL index). This state appears to emerge from the recovery phase of a previous substorm (peak in AL). This state is associated with an IMF Bz that remains continuously southward (~-3 nT) for the duration of the SMC. During the events the solar wind velocity decreases from ~350 km/s by about 50 km/s and the solar wind density increases. Fluctuations in both velocity and Bz decrease as those in density slightly increase. Improvements in the procedure for identifying SMC and ruling out substorms are likely to change these results. Our criterion that d(AL)/dt > -25 nT probably discriminates against strong SMC events.

A possible physical explanation for the magnetospheric state called SMC is the original Dungey hypothesis [Dungey, 1961]. Two x-lines, one on the dayside and one on the night side reconnect magnetic flux at equal rates. Together the two x-lines drive a balanced 2-cell convection system in which no flux is depleted or accumulated in any region of the magnetosphere. Consequently there is no change in magnetic configuration and hence no need to form an additional nightside x-line to form Earthward of the pre-existing x-line. Thus there is no substorm expansion. Our results suggest that this state is created when an initial substorm establishes a near-Earth x-line that reconnects magnetic flux at a rate equal to that on the dayside. Then, provided that the solar wind remains constant and the nightside reconnection rate does not change, the convection will continue without distortion of the magnetosphere. In fact, our results suggest that the dayside input must decrease with time for this state to persist. This may indicate that the nightside reconnection rate decreases with time, possibly due to retreat of the x-line.

It is interesting to ask, what terminates an SMC? Although this question was not the goal of this study our results provide some insight. An ensemble plot of AE in which we did not suppress data beyond the end of SMC events shows that nearly half of the events ended with a substantial increase in AE. This suggests that another substorm expansion was the cause of the termination of this fraction of the event. Figure 1 is an example of such an event. A similar plot of IMF Bz shows that at least 25% of the events were terminated by a northward turning of the IMF. Whether these are gradual or sudden turnings is not evident from the plots. Additional work using the end of the SMC events as epoch zero in the analysis is needed to clarify the cause of termination.

Acknowledgements

The authors would like to acknowledge support by the NSF Space Weather project through grant ATM 0208501. Additional support for the study of solar wind data at 1 AU was provided by a NASA Grant NNG04GA93G. RLM would like to thank Y. Kamide for his stimulating comments on an earlier version of this work.

References

Ahn, B.H., B.A. Emery, H.W. Kroehl, and Y. Kamide, Climatological characteristics of the auroral ionosphere in terms of electric field and ionospheric conductance, *Journal of Geophysical Research, 104*(A5), 10031-40, 1999.

Akasofu, S.-I., The development of the auroral substorm, *Planetary Space Science, 12*(4), 273-282, 1964.

Dungey, J.W., Interplanetary magnetic field and the auroral zones, *Phys. Res. Letters, 6*, 47-48, 1961.

Hones, E.W., Jr., J.R. Asbridge, S.J. Bame, and S. Singer, Substorm variations of the magnetotail plasma sheet from X_{sm} = -6 Re to X_{sm} = -60 Re, *Journal of Geophysical Research, 78*(1), 109-132, 1973.

McPherron, R.L., C.T. Russell, and M. Aubry, Satellite studies of magnetospheric substorms on August 15, 1978, 9. Phenomenological model for substorms, *Journal of Geophysical Research, 78*(16), 3131-3149, 1973.

Nishida, A., and N. Nagayama, Synoptic survey for the neutral line in the magnetotail during the substorm expansion phase, *Journal of Geophysical Research, 78*(19), 3782-3798, 1973.

O'Brien, T.P., S.M. Thompson, and R.L. McPherron, Steady magnetospheric convection: statistical signatures in the solar wind and AE, *Geophysical Research Letters, 29*(7), 10.1029/2001Gl014641, 2002.

Pytte, T., R.L. McPherron, E.W. Hones, Jr., and H.I. West, Jr., Multiple-satellite studies of magnetospheric substorms: distinction between polar magnetic substorms and convection-driven negative bays, *Journal of Geophysical Research, 83*(A2), 663-79, 1978.

Russell, C.T., and R.L. McPherron, Semiannual variation of geomagnetic activity, *Journal of Geophysical Research, 78*(1), 92-108, 1973.

Sergeev, V.A., R.J. Pellinen, and T.I. Pulkkinen, Steady magnetospheric convection: a review of recent results, *Space Science Reviews, 75*(3-4), 551-604, 1996.

Multiscale Coupling of Sun-Earth Processes
A.T.Y. Lui, Y. Kamide and G. Consolini, editors.

GEOEFFECTIVENESS OF SHOCKS IN POPULATING THE RADIATION BELTS

J. B. Blake, P. L. Slocum, J. E. Mazur, M. D. Looper, and R. S. Selesnick
Space Sciences Department, The Aerospace Corporation, Los Angeles, USA

K. Shiokawa
Solar-Terrestrial Environment Laboratory, Nagoya University, Japan

Abstract. On 24 March 1991, ions and electrons with energies of 10s of MeV were observed by CRRES to be injected into the Earth's magnetosphere. The injection created a stable new radiation belt that persisted for years. Later in the decade, several geomagnetic storms and associated energetic solar particle events led to new radiation belts, but none with comparable stability until two injections occurred in November 2001 that lasted for two years. Using data from SAMPEX, Polar, 1997-068, and the 210 Magnetic Meridian Chain, we examine observations of storms that lead to new long-lasting belts. We find that stable belts arise after a deep injection into the slot region (L ~ 2.5). Associated with these injections is an unusually rapid rise in the magnetic field impulse as seen by ground-based magnetometers.

1. Introduction

On 24 March 1991, instruments aboard CRRES observed electrons and ions with energies of 10s of MeV injected into the slot region of the Earth's magnetosphere (L = 2 to 3). These observations were a big surprise, especially the electron injection with energies well above 10 MeV. The particle lifetime was observed to be years. Unfortunately, no interplanetary data were available and, as a result, the solar wind input is unknown. This event has been studied in detail (cf. Blake *et al.*, 1992) and has led to substantial new interest in the inner magnetosphere.

Two other, obvious ion injections were observed during the (15-month) CRRES mission. The first, on 1 August 1990, was at L = 4.3, consisted of several MeV protons only, and lasted just 10 days. The center of the new belt was observed to move inward around 0.1 in L value during its life. The second proton injection was at lower L and lasted until it was obliterated by the event of 24 March.

Assiduous searching over the next several years (Lorentzen *et al.* 2002) found short-lived proton injections but no March 1991-like electron injections. Finally, on 6 November 2001, a clear, long-lasting ion injection was observed; it was followed on 24 November 2001 by a similar one.

In this paper, we will describe some representative shock-induced injections of the last thirteen years, starting with 24 March 1991, and the correlation of the injection events with ground-based magnetograms from the 210 Magnetic Meridian Chain.

We find that long-lasting events require injection well into the slot region, L ~ 2.5, and that the sudden-commencement, magnetic-field impulse associated with such injections has an unusually rapid risetime, as seen in ground magnetograms. This study shows that a key question,

126

requiring both theoretical analyses and more detailed observations, is: What properties of the solar wind and interplanetary shocks lead to strong magnetic-field impulses with a rapid rise time? An understanding does not exist at the present time.

2. Observations

Figure 1 shows the shock-associated electron injection of 24 March 1991 as observed by CRRES. The satellite was inbound, around L = 3, and in the post-midnight sector. The times of the Storm Sudden Commencement (SSC) onset as seen by the low-latitude magnetometer station WEP in Australia and the Japanese METSAT GMS4 are indicated. Within a minute or two, intense pulses of highly energetic electrons and protons were observed at CRRES. The initial electron pulse was followed by six sawtooth pulses. (This was similar for the protons, not shown.) There were four such large pulses followed by two smaller ones as CRRES moved into the inner zone, below L = 2. These multiple pulses are drift echoes, well known through many studies of dispersed substorm injections at GEO altitude. A detailed study of a substorm injection and particle echoes in GEO orbit has been published recently by Li *et al.* 2003..

In examining Figure 1 it is important to note that both electron channels are integral channels; that is, each channel measures all electrons with energies above the channel threshold. However, the nature of the detector is such that the geometric factors associated with each channel differ significantly. The most energetic electrons return first because they longitudinally drift faster

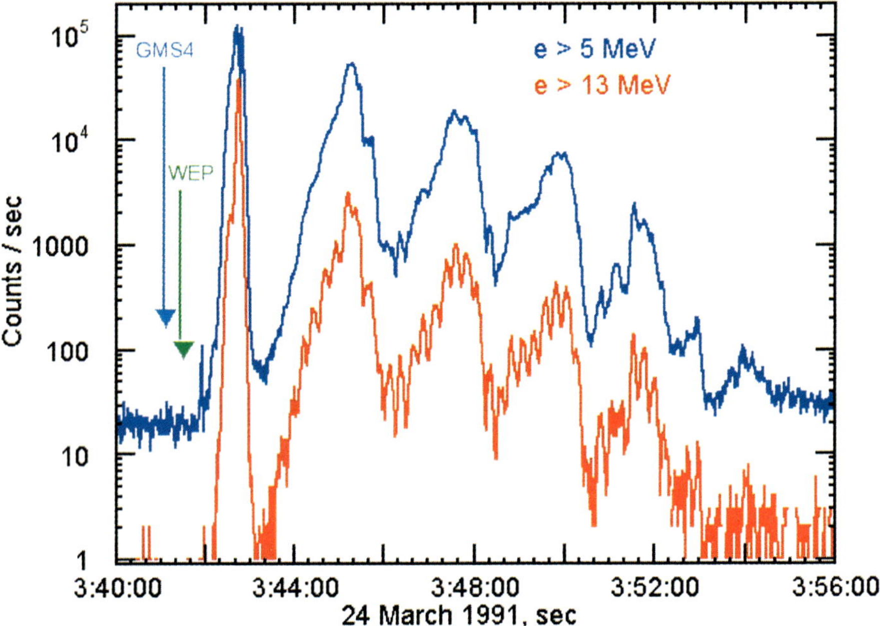

Figure 1. Electron count rate observed aboard the CRRES satellite during the 24 March 1991 shock-associated injection event. The arrows denote the arrival time of the shock at the Weipa (WEP) magnetometer station and at the Japanese satellite GMS4 in GEO orbit in the afternoon local-time sector.

than electrons of lower energy; thus, the time history of the intensity of the returning electrons gives their energy spectrum. The geomagnetic field served as a giant magnetic electron spectrometer, providing detailed information that could not have been obtained otherwise.

The onset of the first returning pulse, at 03:43:20, is due to electrons with an energy of ~50 MeV. Both channels then ramp up as electrons of increasingly lower energy and greater intensity reach the CRRES satellite after drifting around the Earth. The steep slope of the rise in electron countrate indicates that the energy spectrum is steep, $\sim E^{-6}$. At approximately 03:45:20 the count rates reach a peak in <u>both</u> channels and then <u>both</u> drop abruptly; the intensity peak corresponds to an electron energy of ~15 MeV. Thus, the number of electrons with an energy below 15 MeV is nil. The fact that both energy channels show the same temporal profile confirms this conclusion. If there were electrons in the pulse below 13 MeV, the 5-MeV channels would have continued to rise after the 13-MeV channel started to fall.

Although CRRES ceased operation in October 1991, the evolution of these relativistic electrons was followed by SAMPEX into 1994. The remnants of the electrons slowly faded away into the background; they could be seen by SAMPEX for over three years. The SAMPEX orbit is approximately circular at an altitude of only 600 km; however, the electrons' pitch-angle scattered to low altitude and were followed by the PET instrument aboard SAMPEX (Looper *et al.*, 1994).

The 24 March 1991 shock-injection event can be characterized as follows: the injection was immediate, within a minute or two, the injection was well into the slot region, the particles had energies of tens of MeV, and the lifetime of the particles was years.

SAMPEX and Polar provided coverage of ion injections from July 1992 and February 1996, respectively. In November 1997, our coverage of the magnetosphere was increased by the launch of the HEO satellite 1997-068. An advantage of a HEO satellite over Polar for monitoring studies is the fact that the argument of perigee does not advance, and thus provides identical coverage in B,L space year after year.

Figure 2 shows the time history of the radial profile of 15-MeV protons as measured by 1997-068 from 1998 through 2003. At the lowest L values, inside L = 2, the red shows the CRAND protons of the inner zone. This red region is seen to fade with time as the minimum L value reached by 1997-068 first rises from L ~ 1.8 to L ~ 2.2 and then decreases. (The perigee altitude of a HEO satellite varies over a 7-year period.) The narrow, vertical streaks are energetic solar particles. The lowest L reached by these 15-MeV protons in a given solar particle event can be seen to vary with time, illustrating the variations in the proton geomagnetic cutoff. The plot shows that the 15-MeV proton intensities in the slot region (2 < L < 3) were low and variable prior to November 1991. The two substantial injection events in November 2001 can be seen clearly, as well as the fact that they persisted, although the L of the outer edge decreased with time.

However, before discussing the two November 2001 injections, we begin the discussion of Figure 2 with the short-lived injection event of late August 1998. This injection can be seen in Figure 2 as an irregular blob at the low-L end of the solar-proton penetration around Day 240, 1998. Compared with the event of November 2001, it is easy to overlook. In marked contrast to the injection of 24 March 1991, no protons were seen in the August 1998 event for over 36 h after the sudden commencement (the gap is not visible in this time-compressed plot).

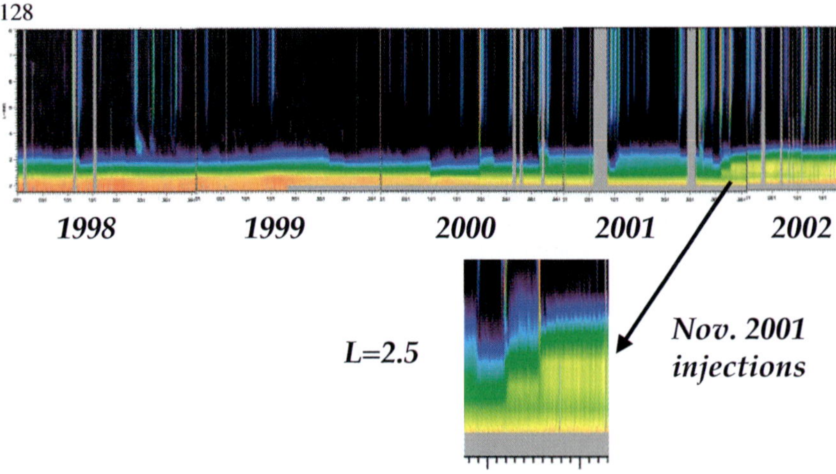

Figure 2. L vs Time profile of protons with E > 15 MeV, from the 1997-068 satellite. The top panel shows data from the time period 1998-2002. The inset details the injections of 6 and 24 November 2001.

In addition to the "late appearance" of the onset of this proton injection, once they appeared, the proton intensities increased slowly over a time period of hours. Figure 3 shows the radial profile of protons in three energy channels around the time of the first appearance, as seen by 1997-068, and as seen a day later. The more intense, inner-zone particles can be seen to be stable whereas the intensity of the new belt (in the slot) increases with time. Not only did the proton intensities increase substantially over many hours, but the more energetic protons increased more slowly. These observations suggest that the proton population in this new belt was relatively slowly energized by magnetospheric processes.

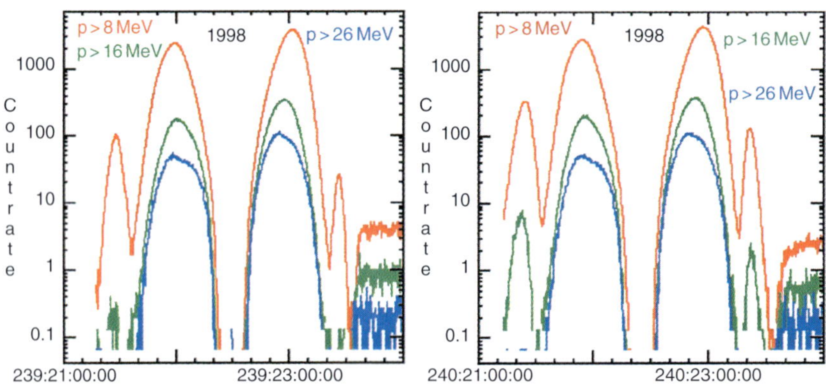

Figure 3. Count rates of protons with E > 8 MeV, E > 16 MeV, and E > 26 MeV from 1997-068. The left panel shows the proton count rates on the day they first appeared, 27 August 1998. The right panel shows the corresponding data on 28 August 1998.

Figure 4 shows the time history of the decay of these protons. It can be seen that by Day 242, the proton intensities had already decreased from those on Day 240, and that within a fortnight, the new belt had almost disappeared. We believe that the loss of the protons probably was due to a breakdown of adiabatic motion for these particles. In Figure 5, we plot the time history of the magnitude of the mirror-point magnetic field of an 8-MeV proton as numerically traced in the T96 model of the geomagnetic field. This plot is for Dst = 0, 100, and 200. For the case of Dst = 100, the mirror-point field is varying by tens of percent, and for Dst = 200, the proton soon plunges into the atmosphere. Thus, the proton lifetime in the new belts may well be determined by a breakdown of adiabatic motion.

This conclusion is supported by the observations cited above of an injection at L = 4.3 in October 1990. For more than a week after this injection, geomagnetic activity was exceptionally low, Kp = 0, 1. As soon as geomagnetic activity increased a bit, the protons vanished.

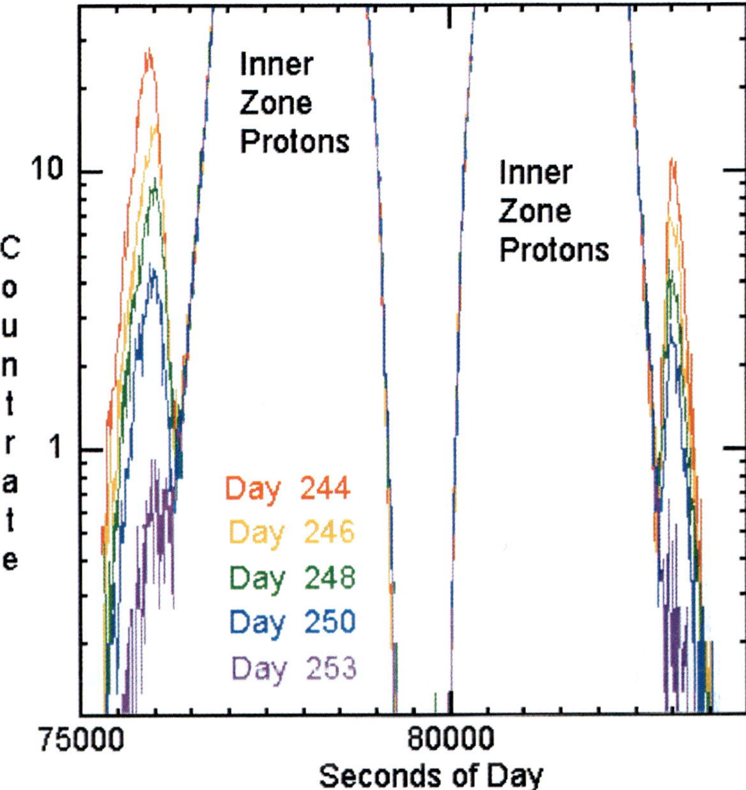

Figure 4. Count rates of protons with E > 8 MeV from 1997-068 satellite, starting on day 242 (30 August) of 1998 and continuing on at 2–3 day intervals, until the new proton belt has almost disappeared on day 253. The data after the first day (242) are time-shifted by a couple of minutes in order that the two, big, inner-zone peaks be aligned; the orbital period of 1997-068 is not quite 12 h.

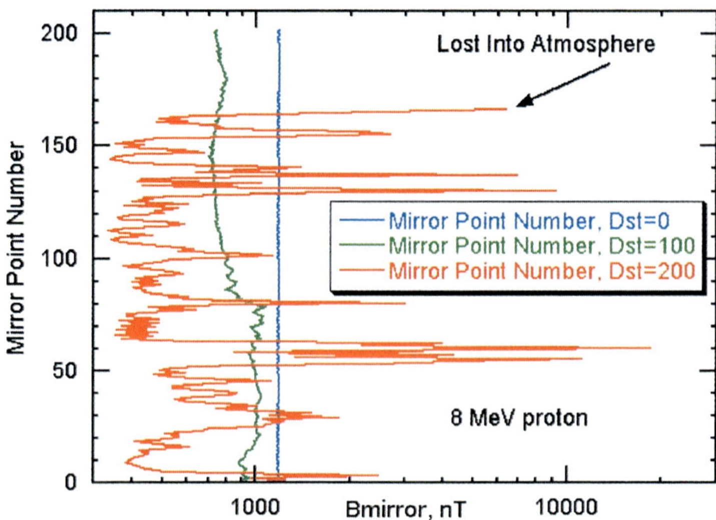

Figure 5. Results of a numerical simulation tracing the motion of a proton with E = 8 MeV in the Tsyganenko 96 (Tsyganenko and Stern, 1996) magnetic field model for several values of Dst. The plot shows the values of the magnetic field at each mirror point.

The August 1998 event can be characterized as follows: the new-belt protons appeared more than a day after the shock; the intensities increased over hours; no associated electrons were detectable (but could have been masked by the strong electron injection by what appeared to be the usual processes); the new proton belt lifetime was two weeks. Thus, the large August 1998 geomagnetic storm and solar particle event was not geoeffective in terms of creating a new, long-lived radiation belt.

However, the August 1998 storm was highly effective in terms of increasing the outer-zone relativistic electron population. Figure 6 shows the outer-zone electron content as a function of time throughout 1998 for three electron energies. At 1 MeV, many events were geoeffective. But at 5.6 MeV, only three events were geoeffective, and the August one was the largest by a factor ~ 5. Note that the decay of the these three events has an exponential character, independent of what was happening to the electron populations at lower energy. The August geomagnetic storm was very geoeffective in terms of the relativistic-electron increase. In describing a geomagnetic storm as geoeffective, one must specify what effects/changes are being considered.

Figure 2 shows that the November 2001 injections differ from the others by the low L value of the injection and the intensity of the new belt of protons. There was another dramatic difference, shown in Figure 7. Heavy ions were observed by SAMPEX to have been injected into the new belt. The L values of these two injections are the same as the location of the proton injections observed near the geomagnetic equator by 1997-068, shown in Figure 2. Note also that the trapped heavy ions do not have a solar composition, nor do the two injections have the same composition. The ions were observed for several weeks. Energy loss to the residual atmosphere at the low SAMPEX altitude of ~ 600 km limits the lifetime of these low-energy heavy ions.

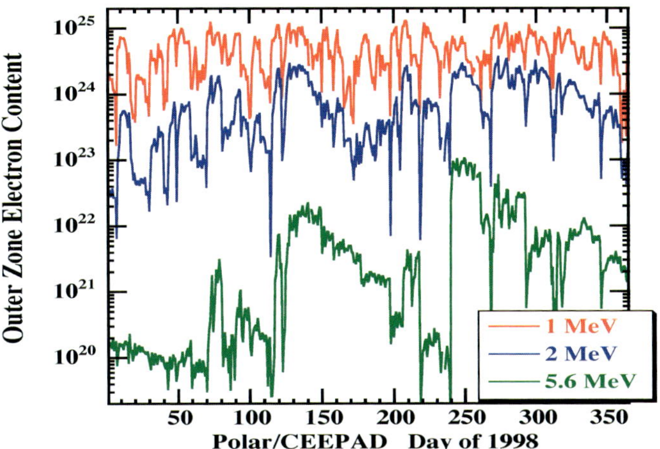

Figure 6. Outer-zone electron content based upon data from the CEEPAD instrument on Polar. Shown are the results for electrons with E > 1 MeV, E >2 MeV, and E > 5.6 MeV from the year 1998.

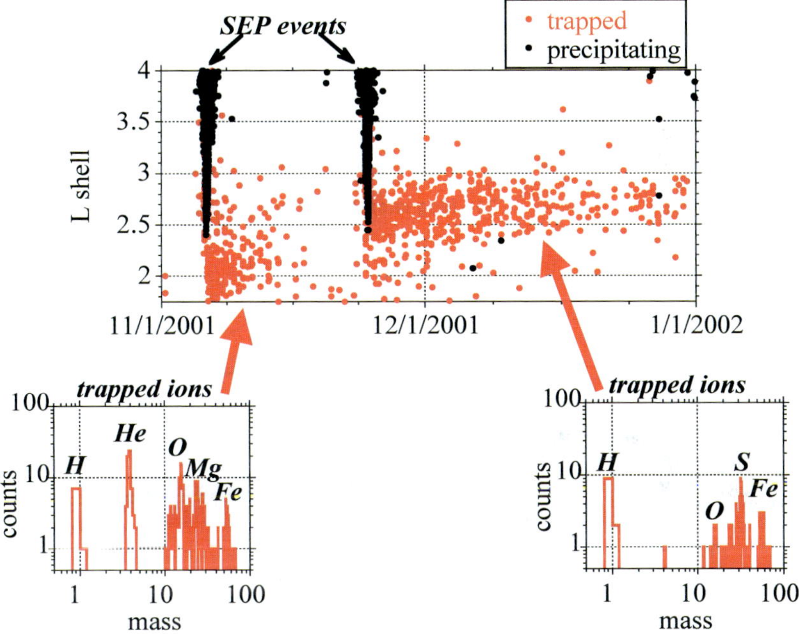

Figure 7. Heavy ions measured using the LICA detector on SAMPEX for the shock-associated events of 6 and 24 November 2001. Each data point represents one ion. The "trapped" and "precipitating" ions are differentiated according to pitch angle.

132

Protons were observed by SAMPEX also, not shown; they too disappeared relatively quickly. A heavy-ion sensor is not aboard 1997-068, and therefore we can not confirm that heavy-ions were injected with large equatorial pitch-angles. However, the symmetry of the situation argues that almost certainly heavy ions were injected along with the protons and survived for an extended period. Studies of the CRAND proton populations in the past indicate slow radial diffusion of these protons. We would expect similar radial diffusion of the long-lived, shock-injected solar ions in events such as November 2001. It seems reasonable to believe therefore that there are energetic heavy ions inside of L = 2 along with the CRAND and solar protons.

Examination of these data have led us to conclude that in order for a shock to create a long-lived radiation belt, it must inject the new ions (and electrons) well into the slot, well inside of L = 3. The outer boundary is not fixed but depends upon the level of geomagnetic activity. Inside L ~ 2.8, multi-MeV particles can be trapped for months and more. Thus, the question becomes: *What are the characteristics of a shock that lead to the injection of energetic particles deep into the slot region?*

This question led us to examine the characteristics of the shock as evidenced by observations of the sudden commencement as seen in ground-based magnetometers, in particular, the 210 Magnetic Meridian Chain. Figure 8 shows the H component of the SSC as seen by the 210 Chain; the stations are near-equatorial, either DRW or WEP. It can be seen that the rise time of the three events (910324, 011106, 011124) that injected particles deeply enough into the slot region that the lifetime was years and substantially faster than the other events. The other events had dramatic geophysical effects but did not create long-lived radiation belts. The event of 980826 resulted in an extremely large increase in relativistic electrons in the outer zone discussed above. The event of 980504 pushed the magnetopause into L = 4.5. Clearly, the effectiveness of a shock in creating long-lived new belts of particles is not the same as those required to create major geomagnetic storms.

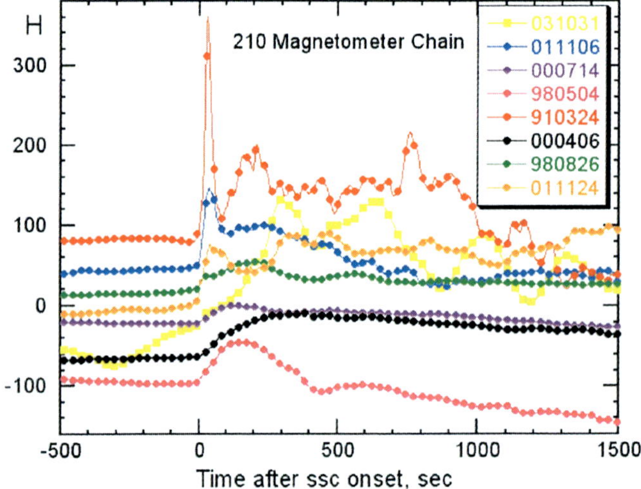

Figure 8. Horizontal (H) component of the magnetic field perturbation from two near-equatorial ground magnetometer stations (DRW and WEP). Each plot represents one geomagnetic storm, corresponding to the date shown in the key.

3. Summary

Several shock-associated injection events have been observed during the last solar cycle that exhibited a variety of differing features, including: promptness of the injection relative to the arrival of the shock at the Earth; location of the new belt in the magnetosphere (L value); particle species injected; and the relationship to "normal" relativistic electron injections. Two things have become apparent: long-lived belts require injection into L ~ 2.5; deep injections are associated with fast risetimes in ground-based magnetometer observations of the SSC. The need for an injection deep into the magnetosphere is straightforward to understand. Obviously, magnetospheric loss processes do not depend upon the injection process. So in order for an injection to be long-lived, it must be into a region where loss processes usually are weak, thus, deeply into the magnetosphere.

Further progress requires high-duty-cycle measurements in the slot region with a sensor with good differential energy resolution and energies to more than 10 MeV for electrons and several 10s of MeV for ions; ion composition information; simultaneous field measurements.

Additional modeling of shock events is of primary importance. What solar wind characteristics lead a fast SSC risetime and subsequent ULF power? What is the temporal cadence of solar wind measurements needed for modeling SSC risetimes of a minute or less?

References

Blake, J. B., W. A.Kolasinski, R. W.Fillius, and E. G.Mullen, Injection of electrons and protons with energies of tens of MeV into L > 4 on 24 March 1991, *Geophys. Res. Lett.*, **19**, 821, 1992.

Hudson, M. K. , A. D. Kotelnikov, X. Li, I. Roth, M. Temerin, J. Wygant, J. B. Blake, and M. S. Gussenhoven, Simulations of proton radiation belt formation during the March 24, 1991 SSC, *Geophys. Res. Lett.*, **22**, 291, 1995.

Hudson, M. K., S. R. Elkington, J. G. Lyon, V. A. Marchenko, I. Roth, M. Temerin, J. B. Blake, M. S. Gusenhoven, and J. R. Wygant, Simulation of radiation belt formation during storm sudden commencements, *J. Geophys. Res.*, **102**, 14087, 1997.

Li, X., I. Roth, M. Temerin, J. R. Wygant, M. K. Hudson, and J. B. Blake, Simulations of the prompt energization and transport of radiation belt particles during the March 24, 1991 SSC, *Geophys. Res. Lett.*, **20**, 2423, 1993.

Li, X., T. E. Sarris, D N. Baker, W. K. Peterson, and H. J. Singer, Simulation of energetic particle injections associated with a substorm on August 27, 2001, *Geophys. Res. Lett.*, **30**, No. 1, 10.1029/2002GL015967, 2003.

Looper, M. D., J. B. Blake, R. A. Mewaldt, J. R. Cummings, and D. N. Baker, Observations of the Remnants of the Ultrarelativistic Electrons Injected by the Strong SSC of 24 March 1991, *Geophys. Res. Lett.*, **21**, 2079, 1994.

Lorentzen, K. R., J. E. Mazur, M. E. Loper, J. F. Fennell, and J. B. Blake, Multisatellite observations of MeV ion injections during storms, *J. Geophys. Res.*, **107**, 1231, 2002.

Slocum, P. L., K. R. Lorentzen, J. B. Blake, J. F. Fennell, M. K. Hudson, M. D. Looper, G. M. Masson, and J. E. Mazur, Observations of ion injections during large solar particle events, *EOS Trans. AGU*, **83**, 2002.

Tsyganenko, N. A. and D. P. Stern, Modeling the Global Magnetic Field of the Large-Scale Birkeland Current Systems, *J. Geophys.Res.*, **101**, 27187–27198, 1996.

Multiscale Coupling of Sun-Earth Processes
A.T.Y. Lui, Y. Kamide and G. Consolini, editors.

COUPLING OF RELATIVISTIC ELECTRONS IN THE INNER MAGNETOSPHERE TO SPACE WEATHER PHENOMENA

T. Obara

Applied Research and Standards Dept., National Institute of Information and Communication Technology, 4-2-1, Nukuikita, Koganei, Tokyo 184-8795, Japan.

T. Goka and H. Matsumoto

Inst. of Space Technology and Aeronautics, JAXA, Sengen, Tsukuba, Japan.

Abstract.   We have observed satellite anomalies when the flux of highly energetic electrons is extremely high. Highly energetic electrons in the outer radiation belt disappear during the main phase of the magnetic storm, and rebuilding of highly energetic electrons is made during the recovery phase of the magnetic storm. Increase is caused by the supply of intermediate-energy electrons by the substorms and these seed electrons are accelerated internally up to the MeV energy range by the intense plasma waves. The distance of the new peak of highly energetic electron flux from the Earth is inversely proportional to the amplitude of the magnetic storm. Integration of the magnetic activity index is a good proxy for the increase of the flux of MeV electrons not only at the geostationary orbit altitude but also in the outer radiation belt, which gives us a clue for the radiation belt forecast.

1. Introduction

It has been mentioned that large enhancements in the flux of highly energetic (MeV) electrons causes satellite anomalies. We first introduce one example, which we have experienced in our country. It has been found that the Earth sensor, installed on the DRTS (Data Relay Technology Satellite), produces false signals when MeV electron flux is high. Figure 1 shows the time history of the noise count and MeV electron flux for a particular time interval in 2003. Since there seems to be a good correlation between them, we made scatter plots for this interval. The result is given in Figure 2, where the vertical axis shows MeV electron fluence integrated for one day and the horizontal axis shows noise counts per day. We have marked satellite anomalies by red circles. Anomalies actually occurred when the flux was extremely high.

To the satellite operation centers in JAXA, we are now providing current information on highly energetic electron fluxes and predicting energetic electron variation at geostationary orbit altitude. We further aim to forecast total MeV electron content in the outer radiation belt; i.e. when, where and how high MeV electron flux will increase in the outer radiation belt. The project is still going on and it is not completed. However, we think it is meaningful to report on the status of our project. In the following, we first demonstrate where the outer belt is rebuilt in terms of L-value. We next consider the maximum levels of energetic electron flux increase in the outer radiation belt. We then discuss the efforts for forecasting MeV electron content in the outer radiation belt.

136

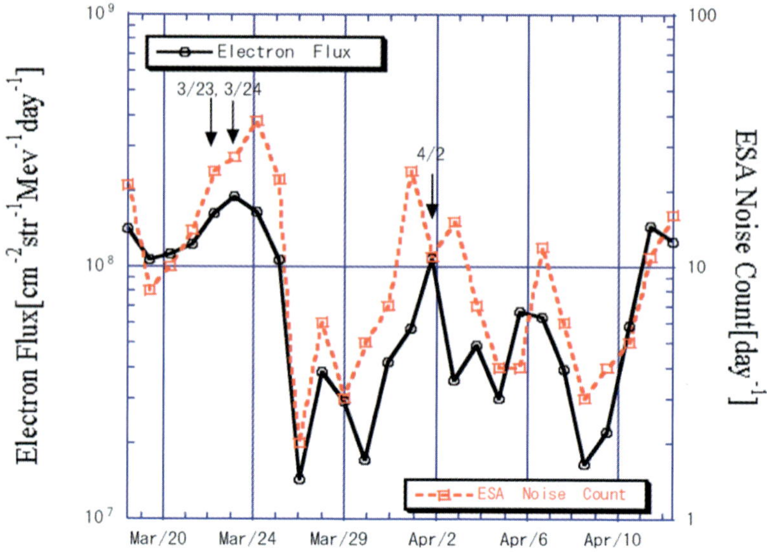

Figure 1. Noise count vs. MeV electron flux. Arrows show satellite anomalies

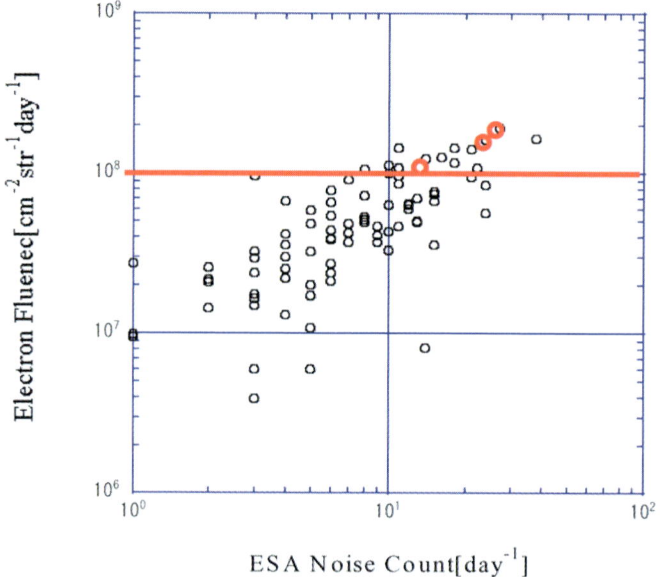

Figure 2. Earth sensor noise count vs. MeV electron fluence (integrated electron flux). We have marked satellite anomalies by red circles. Anomalies occur when the flux was extremely high.

2. MDS-1 Measurements

Mission demonstration satellite 1 (MDS-1) was launched on Feb.4, 2002 into the geostationary transfer orbit (GTO) with an inclination of 28 degree and an orbital period of 10 hours. The MDS-1 satellite was renamed TSUBASA, meaning wings after launch.

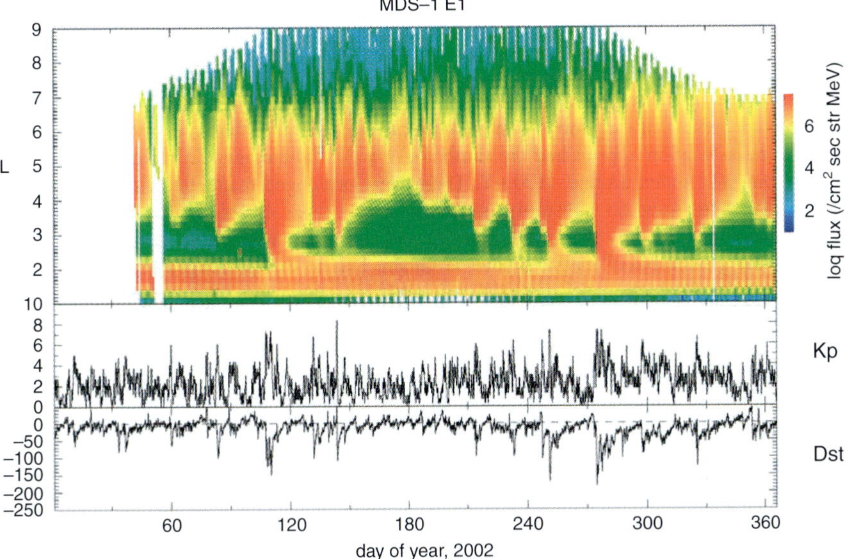

Figure 3. L-t diagram of 0.4 MeV to 0.91 MeV electrons and Dst index in 2002.

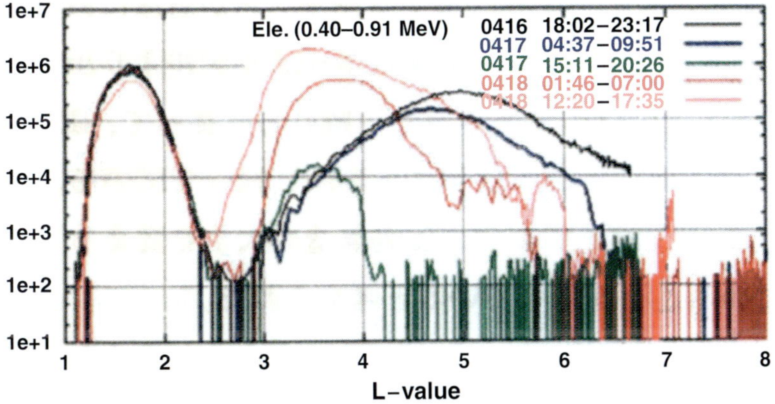

Figure 4. Variation of radiation belt profile measured by MDS-1 satellite.
Each line demonstrates flux of electrons in the unit of (cm² sec st. MeV)$^{-1}$,
covering three days from April 16 to April 18.

138

On this satellite a high energy particle detector was installed. The results are given in Figure 3. Top panel in the figure shows L - t diagram, where the vertical axis shows L - value ranging from $L = 1$ to 9 and the horizontal axis gives time covering one year with a start of April 1, 2002. Middle and bottom panel in the figure shows Kp and Dst indexes. Around noon of April 17, a magnetic storm started and the Dst reached –120 nT early of April 18. The storm developed once again on April 19 and the Dst reached its minimum (-150nT) early on April 20 and then recovered gradually. Individual measurements of MDS-1 are given in Figure 4.

Around 18UT on April 16, there was a peak of outer electron belt at $L = 5$ and it persisted till 09UT on April 17. The flux then decreased largely by 15UT on April 17; only a few hours after the commencement of the storm. The peak flux decreased more than one order of the magnitude and the peak location approached the Earth: i.e. $L = 3.5$. The magnetic storm continued on April 18, keeping Dst index around –100 nT. There were prolonged magnetic activities over April 18 producing quasi-periodic substorm injections every three hours or so. First injection was identified around 3UT by the LANL satellite and strong disturbances of the magnetic field were seen by the ground-based magnetometer at Guam. Due to this prolonged magnetic activities the flux of outer belt electrons increased significantly, exceeding pre storm level within one day. It should be mentioned that a new peak location of the outer radiation belt shifted toward the Earth; i.e. $L = 3.5$, which is consistent with Akebono simultaneous measurements and also consistent with statistical result by Obara et al.(2000).

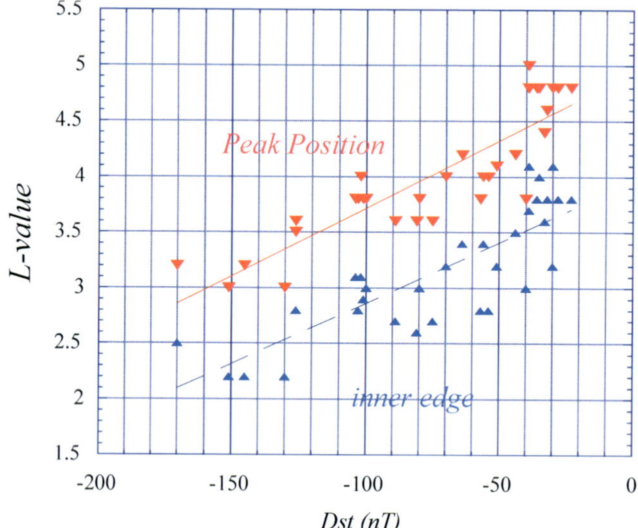

Figure 5. Peak position of newly developed outer belt as a function of Dst.
Inner edge of the newly formed belt is also plotted.

During the life-time of MDS-1 satellite; i.e. one year and half, there were almost 30 enhancements in the flux of outer belt electrons with energy of 0.4 – 0.95 MeV. The results are given in Figure 5, in which a peak position of newly formed outer belt during the storm recovery phase is plotted as a function of minimum Dst of each storm. There seems to be a correlation of the peak position and Dst value; a large storm produces a new outer electron belt closer to the Earth. This result is largely consistent with Obara et al. (2000), O'Brien et al.(2003) and references therein.

Large enhancements of highly energetic electron flux occur when the *Dst* variation is quite small; i.e. less than –30 nT. This type of enhancement is seen around April 1, June 3 and June 10 in Figure 3. *Dst* variation was actually –30 nT, -30 nT and –36 nT, respectively. Green and Kivelson (2001), Meredith et al. (2002 and 2003) have already found the energetic electron flux increase while *Dst* remained above –30 nT. An important point to note from ours as well as the past results, is that these increases took place around $L = 5$. However, as shown in Figure 1, electron flux can increase as far out as geostationary orbit location. We will discuss exact processes which cause large enhancement in the energetic electron flux in the next section.

3. Discussion

MDS-1 measurements show a clear relationship between the peak position of the energetic electron flux and *Dst* variations. This tendency is largely consistent with previous works by Tverskaya (1996), Obara et al. (2000) and O'Brian et al (2003). This is consistent with being caused by the transport of seed electrons during the main phase or early recovery phase as shown by Obara et al. (2001). Advantage of MDS-1 measurement is that the observation has been made near the magnetic equator. We can clearly identify the inner edge of newly developed outer electron belt as given in Figure 5. Location of inner edge is much more closer to the Earth than that of peak location. This means that the transport of seed electrons is quite efficient near the Earth than as expected. Detail processes are however unclear at present. We need to investigate energetic electron transport mechanisms much more thoroughly.

The MDS-1 observations also showed that large enhancements of the energetic electron flux when the *Dst* variation is quite small. The Akebono satellite measured VLF chorus emissions along the orbit. Figure 6 shows where the intense chorus emission occurred. The peak location of wave amplitude is given by the triangle in Figure 6, which largely coincides with that of energetic electrons as given in Figure 5. Kanazawa University team is now investigating the details and further explanations will be given shortly. Here we note that there are chorus emissions when *Dst* is small. For example when the *Dst* is around –30nT the chorus emission peaks around $L = 4.8$. This is largely consistent with the result as shown in Figure 5.

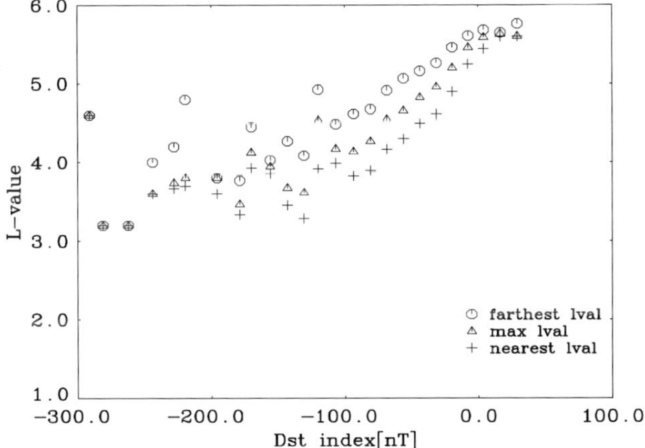

Figure 6. Akebono VLF measurements of intense whistler mode chorus emission
(courtesy of Y. Kasahara and I. Nagano from Kanazawa Univ.)

Meredith et al. (2002) insist that whistler mode chorus emission accelerates seed electrons up to MeV range. MDS-1 and Akebono measurements show the coexistence of energetic electron enhancements and the VLF chorus emissions. These observations confirm many results found using the previous satellites in the different orbit. Next step will be the quantitative study how the VLF waves produce relativistic electrons in the heart of outer radiation belt.

Figures 5 and 6 describe the energetic electron flux enhancement earthward of $L = 5.5$. As demonstrated in the introduction, enhancements of MeV electron flux also occur outside of $L = 5.5$. at geostationary orbit. However, we did not see any significant increase of VLF emissions at geostationary orbit which is consistent with the result by O'Brian et al. (2003). Instead, we have experienced ULF wave intensification during the magnetic storm. Figure 7 demonstrates the line plots of ULF wave intensity and energetic electron flux at geostationary orbit altitude, which shows a good correlation between the two.

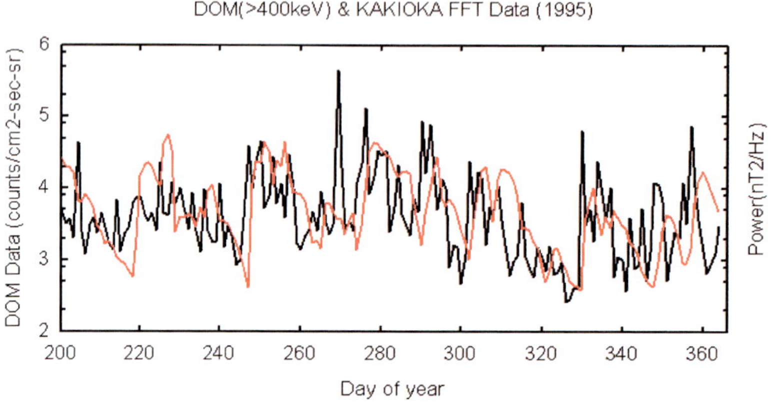

Figure 7. MeV electron flux (red) and FFT power of ULF waves with 150-600 sec period (black).Each enhancement of MeV electron is caused by the magnetic storm. There seems to be a good correlation between them. By closer inspection, an increase of ULF waves is a little bit earlier than that of electrons.

The lack of VLF wave power outside $L = 5.5$ and the correlation of MeV electron flux with ULF wave power at geostationary orbit suggests that, in this region, acceleration is dominated by the interaction with ULF waves. We need to do much more work to understand the acceleration process. Indeed, it is an urgent requirement from the operation side to know when the MeV electron flux will increase in the outer radiation belt. We therefore move to this more practical aspect in the following section.

4. Toward A Radiation Belt Forecast

In order to avoid a dangerous situation, we have been providing space radiation information to the satellite operation centers in our country. By using a liner prediction filter, variation of MeV electron at GEO is being calculated from the real-time solar wind velocity from ACE satellite and is being demonstrated on the web.
(http://sees.tksc.nasda.go.jp/Japanese/Data/Updata_ja/WeeklyDataDrtswElectron.htm)

In order to clarify the control parameter for the large increase of the outer belt electron flux, we have investigated the relationship between MeV electron flux enhancements and the magnetic activity during the recovery phase of storms. The result is given in Figure 8.

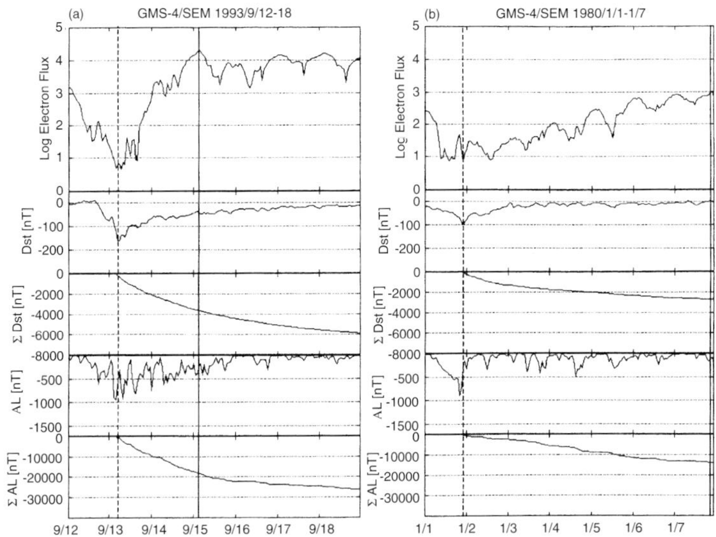

Figure 8. Increase in the flux of MeV electrons obtained by the geostationary satellite GMS for two magnetic storms. The left case demonstrates magnetically active case during the recovery phase, while the right case shows rather quiet case during the recovery phase.

Top panel in Figure 8 shows MeV electron flux observed by GMS satellite at geostationary orbit altitude. The vertical dot line in the panel shows the start time of the recovery phase of the magnetic storm. The second panel demonstrates *Dst* index. In both cases *Dst* index has a minimum value –100 nT, showing that a major storm took place. Third panel shows an integration of *Dst* index during the storm recovery phase. We can see the recovery phase was longer in the left case than that in the right case. The fourth panel demonstrates the *AL* index, showing that there were prolonged magnetic activity during the recovery phase in the left case. The bottom panel shows an integration of the *AL* index during the storm recovery phase. From the comparison of the *AL* index and the electron flux increase, a large increase in the flux can be attributed to the large magnetic activities during the storm recovery phase. In order to see this, we have examined more storms. Figure 9 shows a scatter plot of MeV electron peak flux and the integrated *AL* index. This result demonstrates that there is a relationship between them.

Since the investigation was limited to geostationary orbit altitude, we should expand the area by using geostationary transfer orbit satellite such as MDS-1. Such an effort is now being made and the close relationship of total MeV electron flux in the outer radiation belt and the integrated magnetic activities is obtained. Our result shows very good agreement with the previous work by P. Buhler during his stay in Japan. He analyzed STRV satellite data and the result is given in Figure 10, where a total electron flux integrated from $L = 4.0$ to 6.8 is plotted as a function of integrated *Ap* index.

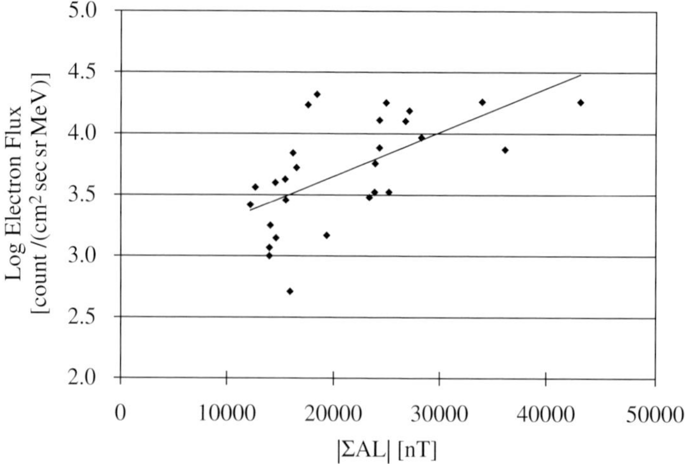

Figure 9. MeV electron flux obtained by GMS at geostationary orbit altitude
and the integrated *AL* index during the storm recovery phase.

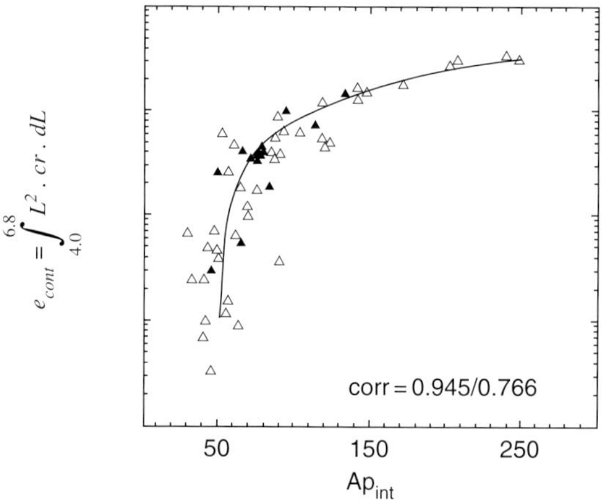

Figure 10. Total electron flux integrated from $L = 4.0$ to 6.8 and
the integrated magnetic *Ap* index.(courtesy of P. Buhler)

As a natural expansion of this work, we plan to calculate the total electron content of MeV electrons in the outer radiation belt together with a peak location in terms of *L*-value. Real-time simulation of the magnetosphere and the polar ionosphere is being made by T. Tanaka by using MHD simulation with an input of real-time solar wind data by ACE (see T. Tanaka, in this book).T. Tanaka and S. Fujita are now trying to calculate *AE* index in real-time. Using the real-time AE calculation, we would like to perform real-time prediction of the outer belt variation as shown in Figure 11. To do this, a super computer system in our institute will be used.

In order to check whether the calculated result is accurate or not, satellite measurements are very important. On the JAXA satellites, a space radiation monitoring package has been installed, providing radiation belt information in real-time. Right now, JAXA is operating DRTS satellite at geostationary orbit and JAXA will launch ALOS and ETS-8 shortly. On these satellites, particle monitors will be installed. In order to avoid dangerous situation in geospace region JAXA and NiCT would like to perform a reliable service on space radiation environment.

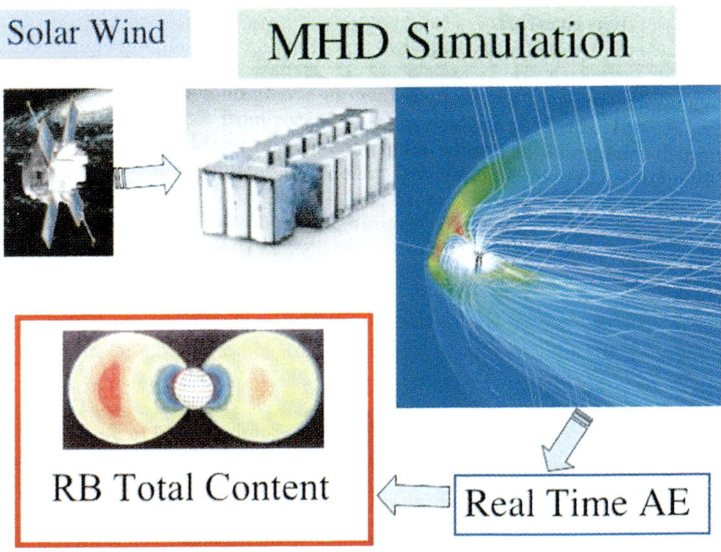

Figure 11. Outline of radiation belt prediction system at NiCT, Japan.

5. Concluding Remarks

We showed an example of DRTS satellite anomalies. When the flux of highly energetic electrons is extremely high, it might be expected that satellite anomalies take place not only at GEO but also in the heart of the outer radiation belt. MDS-1 performed experiments on how highly energetic electrons produce errors in some devises. MDS-1 confirmed the results of laboratory experiments which were done before launch. Since the supply of devices is getting difficult nowadays, the information and prediction of space radiation environment become important. Hence, we are trying to perform radiation belt forecast.

MDS-1 observations show that energetic electron flux decreases during the main phase of the magnetic storms and increases during the recovery phase. The distance of the newly produced outer radiation belt is inversely proportional to the amplitude of the storm, which is largely consistent with the locations of intense plasma waves during the magnetic storm. This result allows us to predict where the new outer belt will form based on the magnitude of the storm.

Integration of magnetic activity index is a good proxy for the increase in the flux of MeV electrons in the outer radiation belt, which gives us a clue for the radiation belt forecast. A real-time calculation of magnetic activity index AE is being made based on the MHD simulation. Getting this result and the *Dst* prediction at NiCT (Watanabe et al., 2003), a total content of outer

144

belt electron will be obtained shortly. We would like to provide more information on the space radiation to the satellite operation centers not only in Japan but also in the world.

Acknowledgements

We are grateful to Prof. I. Nagano, Dr. Kasahara, Dr. Y. Miyoshi and Dr. P. Buhler for providing valuable figures for this article. We would like to acknowledge Dr. M. Nakamura and Mr. K. Koga and Dr. H. Koshiishi for their valuable discussions and Prof. B. Shizgal for his carefully reading of the manuscript and valuable suggestions.

References

Green, J. C. and M. G. Kievelson, A tale of two theories: How the adiabatic response and ULF waves affect relativistic electrons, *J. Geophys. Res.*, **106**, 25777, 2001.

Meredith, N. P., R. B. Hone, R. H. A. Iles, R. M. Thone, D. Heyndrickx, and R. R. Anderson, Outer zone relativistic electron acceleration associated with substorm-enhanced whistler mode chorus, *J. Geophys. Res.*, **107**, A7, 1144, doi:10.1029/2001JA900146, 2002.

Meredith, N. P., M. Cain, R. B. Hone, R. M. Thone, D. Summers, and R. R. Anderson, Evidence for chorus-drven electron acceleration to relativistic energies from a survey of geomagnetically disturbed periods, *J. Geophys. Res.*, **108**, A6, 1248, doi:10.1029/2002JA009764, 2003.

Obara, T., M. Den, Y. Miyoshi, and A. Morioka, Energetic electron variation in the outer radiation zone during early May 1998 magnetic storm, *JASTP*, **65**, 1407-1412, 2000.

Obara, T., Y. Miyoshi, and A. Morioka, Large enhancement of the outer belt electrons during magnetic storms, *Earth Planets Space*, **53**, 1163-1170, 2001.

O'Brien, T. P., K. R. Lorentzen, I. R. Mann, N. P. Meridith, J. B. Blake, J. F. Fennell, M. D. Looper, D. K. Milling, and R. R. Anderson, Energization of relativistic electrons in the presence of ULF power and MeV mibcrobursts: Evidence for dual ULF and VLF acceleration, *J. Geophys. Res.*, **108**, A8, 1329, doi:10.1029/2002JA009784, 2003.

Tverskaya, L. V., The latitude position dependence of the relativistic electron maximum as a function of *Dst*, *Adv. Space Res.*, **18**, 135-140, 2002.

Watanabe, S., E. Sagawa, K. Ohtaka, and H. Shimazu, Operational model for forecasting *Dst*, *Adv. Space Res.*, **31**, 829-834, 2003.

THE LOWEST POSSIBLE LATITUDE OF THE WESTWARD ELECTROJET DURING SEVERELY DISTURBED PERIODS

B.-H. Ahn
Department of Earth Science, Kyungpook National Univerisity, Daegu, 702-701, Korea

G. X. Chen
Institute of Geology and Geophysics, Chinese Academy of Science, Beijing, 100029, China

W. Sun
Geophysical Institute, University of Alaska, Fairbanks, AK, 99775-7340, USA

J. W. Gjerloev
Applied Physics Laboratory, The Johns Hopkins University, Laurel, MD, 20723-6099, USA

Y. Kamide
Solar-Terrestrial Environment Laboratory, Nagoya University, Toyokawa, Aichi 442-8705, Japan

J. B. Sigwarth
Goddard Space Flight Center, Electrodynamics Branch, NASA, Greenbelt, MD, 20771, USA

L. A. Frank
Department of Physics and Astronomy, University of Iowa, Iowa City, IA, 52242-1479, USA

Abstract. It has been reported that the AE index cannot, at times, adequately monitor the auroral electrojets because as magnetic activity increases, it shifts equatorward from the standard AE stations, resulting in a serious underestimation of the auroral electrojet intensity. To evaluate quantitatively the equatorial shift of the westward electrojet, an extensive database obtained from CANOPUS, Alaska and IMAGE chains of magnetometers are utilized in this study. The data thus assembled enable us to determine how the westward electrojet shifts equatorward with increased magnetic activity. We are particularly interested in the latitude of the center of the westward electrojet during intense magnetic storms. The tendency of equatorward shift is confirmed from this study. However, the peak of the westward electrojet seems to shift only approximately 60° in magnetic latitude, regardless of magnetic activity levels. Therefore the current AE network, which covers as low as about 62°, does not have any serious problems in monitoring the auroral electrojet. The relative location of the westward electrojet with respect to the global auroral image taken from the Polar satellite is also examined. It is found that the center of the westward electrojet does not flow over the brightest auroral region but slightly poleward of it, with less luminous region. It indicates that the electric field is more important in intensifying the auroral electrojet than the ionospheric conductivity.

1. Introduction

The auroral electrojets shift equatorward during magnetically disturbed periods particularly during intense geomagnetic storms (e.g., Feldstein et al., 1997, Ahn et al., 2000). Due to the equatorward shift of the auroral electrojets below the latitude of the standard AE stations, it has widely been recognized that the AE index is seriously underestimated. Akasofu [1981] examined the relationship between the AE and Dst indices, finding that AE tends to saturate for large values of Dst. As one of several possible causes of the saturation, he suggested that the equatorward shift of the auroral electrojet beyond the standard AE network make it difficult for the AE stations to monitor the auroral electrojets properly.

Recently Ahn et al. [2000] also reported that the AL index cannot properly represent the auroral electrojet because it shifts equatorward below the latitude of the standard AE network particularly during a specific universal time interval. They further reported that the equatorial shift is more serious than the longitudinal AE station gaps. To estimate the equatorward shift of the westward electrojet quantitatively, an extensive ground magnetic database obtained from the CANOPUS, Alaska and IMAGE chains of magnetometers are analyzed. The relative location of the westward electrojet with respect to auroral imagery is also examined. For this purpose an event is examined based on an auroral image taken from the Polar satellite on October 22, 1999, during which a major magnetic storm was in progress. This would shed a light on the relative contribution of the ionospheric conductivity and the electric field to the westward electrojet.

2. The Latitudinal Center of the Westward Electrojet

The latitudinal center of the westward electrojet is determined from magnetic disturbance data obtained from the CANOPUS, Alaska, and IMAGE chains of magnetometers during 1998. Since we are interested in the location of the westward electrojet during disturbed periods, days are selected when at least one of the stations along each chain of magnetometers recorded ΔH less than $-1,000$ nT. The number of days that satisfied this selection criterion was 40, 38, and 72 for the CANOPUS, Alaska, and IMAGE chains, respectively. Although the IMAGE chain operates 23 magnetic stations, for the purpose of comparing with the other chains, we utilized only nine stations, from north to south, Hornsund, Bear Island, Soroya, Masi, Muonio, Pello, Lycksele, Hankasalmi, and Nurmijarvi. They are located within about $\pm 5°$ from $104°$ meridian of corrected geomagnetic longitude and cover a wide latitudinal range from $74.06°$ to $56.85°$ of corrected geomagnetic latitude. We tentatively assume that the magnetic stations belonging to one magnetic chain lie along the same magnetic meridian.

Each selected day is subdivided into epochs of five-minute intervals and the highest disturbance along the magnetic meridian chain for a given epoch is identified. Although due to the selection criterion the highest disturbance during a given data day under consideration always exceeds $-1,000$ nT, every epoch of five-minute intervals does not always record such a highly disturbed level. Previous studies (e.g., Ahn et al., 1984; 1986, Sun et al., 1993) approximated the latitudinal profile of the auroral electrojet as a Gaussian distribution. Thus the region where the maximum magnetic disturbance is recorded can be assumed as the center of the auroral electrojet in the latitudinal direction during the given epoch. The highest disturbance level thus identified during each five-minute epoch is further binned by every 200 nT level. Figures 1, 2 and 3 are constructed by utilizing the information on the latitude and magnetic local time of the station that recorded the highest magnetic disturbance along the corresponding magnetic chain of magnetometers during each five-minute interval. For example, if Gillam recorded the highest disturbance, –850 nT, among CANOPUS chain stations during a given epoch and that happened to occur at 01 MLT, we would mark a plus sign (+) at the location of 65.65° (corrected geomagnetic latitude of Gillam; all positions are expressed by corrected geomagnetic coordinate unless otherwise specified) and 01 MLT in the top panel of the right hand side of Figure 1. Since we are interested in the westward electrojet, the magnetic disturbances recorded between 2000 and 0900 MLT through the midnight sector are considered in this study. We examine every 200 nT level, from the activity level of $0 < |\Delta H| < 200$ nT up to $1400 < |\Delta H| < 1600$ nT, altogether eight activity levels. The plus signs show the latitudinal and magnetic local time distribution of the center of the westward electrojet for a given level of magnetic activity. It is a kind of probability distribution about the center of westward auroral electrojet for a given level of magnetic activity in terms of corrected geomagnetic latitude and MLT.

Figure 1 shows the distribution of the center of the westward electrojet over CANOPUS meridian chain during 1998. The eastern meridian line, Churchill line, of CANOPUS chain is utilized in this study. The seven stations from north to south are Taloyoak, Rankin Inlet, Eskimo Point, Fort Churchill, Gillam, Island Lake and Pinawa. Since the magnetic stations of CANOPUS cover a wide latitudinal region, from 79.03° to 60.60°, it is ideal in monitoring the westward electrojet for a wide range of magnetic activity. During very quiet periods, say $|\Delta H| <$ 200 nT, the probability of recording this level of activity is the same regardless of latitude and magnetic local time, as shown in the top panel of the left hand side of Figure 1. Although it does not carry any meaningful information on the location of the center of the westward electrojet at this particular activity level, the filled circle with error bars represents the average location of the westward electrojet with ± one standard deviation both of the latitudinal and MLT directions. During activity level of $200 < |\Delta H| < 400$ nT, however, one can clearly see that the probability of recording such an activity level, say, at the highest latitude station, Taloyoak (79.03°), becomes very low. The average center of the westward electrojet during this activity level is found to be at 67.26° ± 3.47° and 2.43 ± 3.33 MLT.

As magnetic activity further increases, the center of the westward electrojet migrates equatorward gradually without showing any significant local time shift. During the activity level of $800 < |\Delta H| < 1000$ nT, i.e., the top panel of the right hand side of Figure 1, the center is found to be at 64.55° ± 3.37° and 2.69 ± 2.71 MLT. During an extremely disturbed situation, $1400 < |\Delta H| < 1600$ nT, shown in the bottom panel of the right hand side of Figure 1, it expands as low as 62.80° ± 2.98°. From the distribution pattern of the plus signs with a finite latitudinal range, one cannot exclude the possibility that the center of the westward electrojet is located further equatorward of Pinawa (60.60°), the lowest station of CANOPUS chain. It is unlikely, however, that the westward electrojet keeps expanding eqautorward as magnetic activity increases. Although not shown here, even during the activity level of $1600 < |\Delta H| < 1800$ nT, several lower latitude stations of CANOPUS chain properly recorded such a level of disturbance. The distribution pattern of the center of the westward electrojet during this activity level does not show any significant difference from that of the activity level, $1400 < |\Delta H| < 1600$ nT. Thus the center of the westward electrojet does not seem to expand below 60° regardless of magnetic activity level.

Alaska chain of magnetometers consists of six stations from north to south, Fort Yukon, Poker Flat, Gakona, Talkeetna, Anchorage, and Sitka. The six rows of plus signs appearing in Figure 2 correspond to the six stations. For the magnetic activity level of $0 < |\Delta H| < 200$ nT, one can note that the six rows of the plus signs are continuous without any interruption, indicating that the probability of observing this level of magnetic activity is the same regardless of either latitude or magnetic local time. In other words, during very low activity period, such a level of magnetic activity can be observed at any latitude and any MLT sector. Thus, it is meaningless to determine the average center of the latitudinal location of the westward electrojet at such an activity level. The latitudinal center of the westward electrojet denoted by the filled circle, 62.75° ± 2.46°, corresponds simply to the average latitude of Alaska meridian chain stations. This is lower in latitude than in CANOPUS chain for the same activity level. It is simply because the Alaska chain is generally located at lower latitude region than CANOPUS chain is. On the other hand, the longitudinal center, 3.21 ± 3.78 MLT, is located at slightly later hours than 2.5 MLT, the half way point between 2000 and 0900 MLT. It is simply because there were more data points in the later time sector than 2.5 MLT.

The second panel of the left hand side of Figure 2 shows the distribution of the auroral electrojet center during the activity level of $200 < |\Delta H| < 400$ nT. Although the location of the electrojet center in terms of either latitude or MLT do not show any noticeable change compared to the lowest activity level, a couple of interesting changes are worth mentioning. First, the probability of observing this level of activity at the lower latitude stations tends to decrease. Second, the number of data points in earlier local time, premidnight sector has decreased significantly. As magnetic activity further intensifies, such a tendency becomes more apparent. It is also interesting to note that the probability of observing the centers at later local time diminishes too, for example during the activity level of $600 < |\Delta H| < 800$ nT. Even if the magnetic activity exceeds -1,000 nT, the same trend persists. The average center of the auroral electrojet during the activity level of $1000 < |\Delta H| < 1200$ nT is found to be at 63.50° ± 2.36° and 1.62 ± 2.05 MLT. The westward electrojet does not seem to expand equatorward considerably with magnetic activity, but it simply shifts to earlier local times, more than one hour and half, compared to the lowest activity level. We also examined the location of the center during very disturbed periods, $1400 < |\Delta H| < 1600$ nT, finding the similar characteristics noted at the lower activity levels.

Although no station lower than Sitka is utilized in this study, it is unlikely that the center of the westward electrojet expands below it during very disturbed periods. If the center shifts equatorward considerably during severely disturbed periods, the distribution pattern of the plus signs would also shift accordingly, thus leaving even lower latitude stations with fewer data points. In other words, more plus signs would appear at or around the latitude of Sitka. However, no such a trend is noted even during very disturbed periods; see activity level of $1400 < |\Delta H| < 1600$ nT. Although it is not shown here, the same trend persists even during the magnetic activity level of $1800 < |\Delta H| < 2000$ nT, the most active period ever recorded over the Alaska meridian chain during 1998. On the other hand, the center of the westward electrojet tends to shift to earlier local times. During severely disturbed periods, say $|\Delta H| > 1400$ nT, the center of the westward electrojet fluctuates around 00 MLT. Thus it is concluded that as magnetic activity increases, the equatorward expansion is insignificant, while the center of the westward electrojet tends to shift to earlier local times as far as the Alaska meridian chain is concerned.

The locations of the center of the westward electrojet based on IMAGE chain of magnetometers during 1998 are shown in Figure 3. The same characteristics shown in Figures 1 and 2 are also noted. The center of the westward electrojet during quiet period, say $|\Delta H| < 200$ nT, can be located at any latitude, while during disturbed period, say $800 < |\Delta H| < 1000$ nT, the top panel of the right hand side of Figure 3, it shifts to 64.88° ± 2.38° and 3.51 ± 2.91 MLT, almost the same latitudinal region   recorded from CANOPUS chain for the same activity level. During

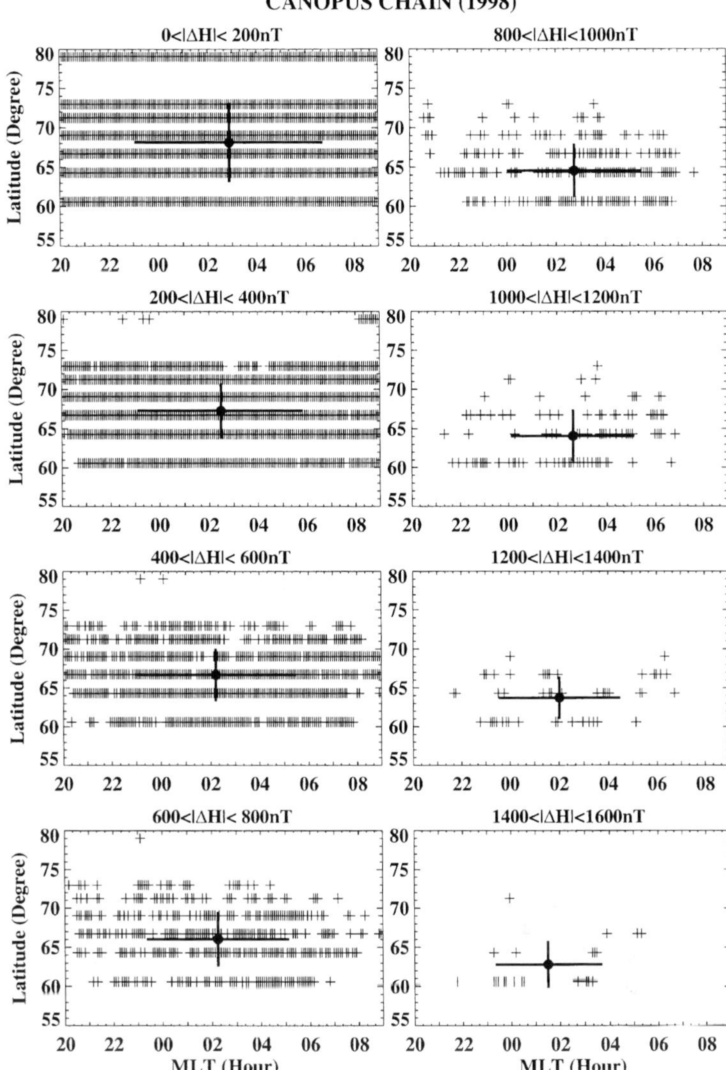

Figure 1. The distribution of the center of the westward electrojet for different levels of magnetic activity. The plus signs represent the latitude and magnetic local time of the center of the westward electrojet for a given level of magnetic activity. A total of 40 days were examined for the case of CANOPUS chain. The filled circle with error bars is used to indicate the average latitude and MLT of the center of the westward electrojet during a given level of magnetic activity. The total length of the error bars represents two standard deviations in the latitudinal and local time directions. A total of eight activity levels are examined.

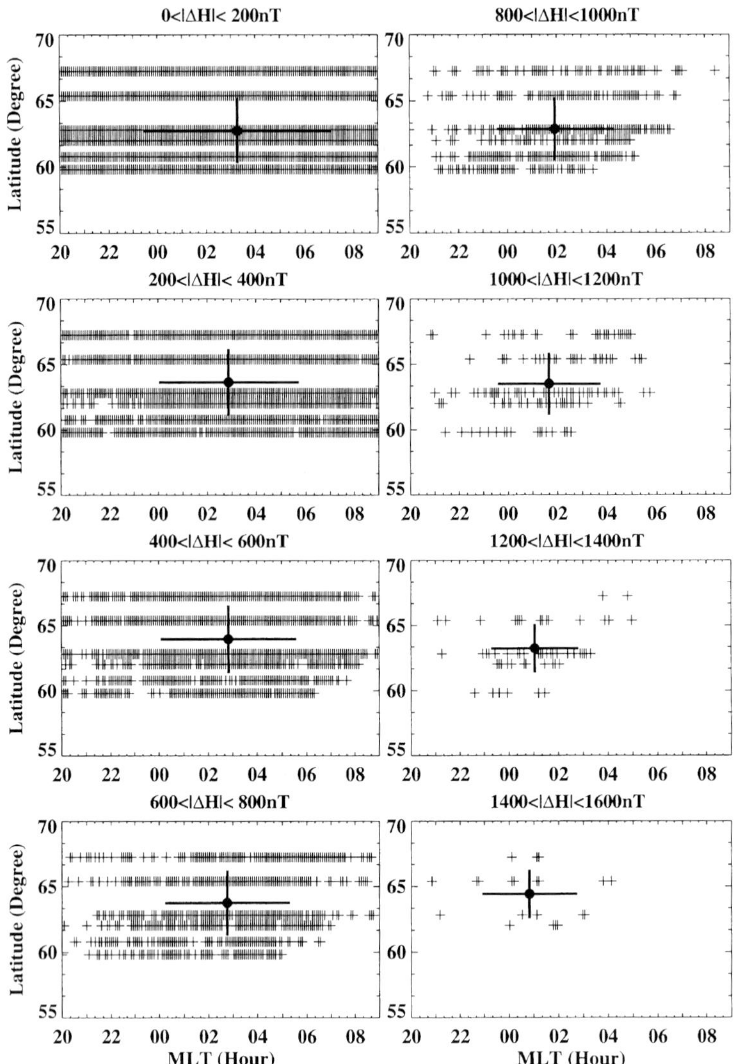

Figure 2. Same as Figure 1 but for Alaska chain of magnetometers.

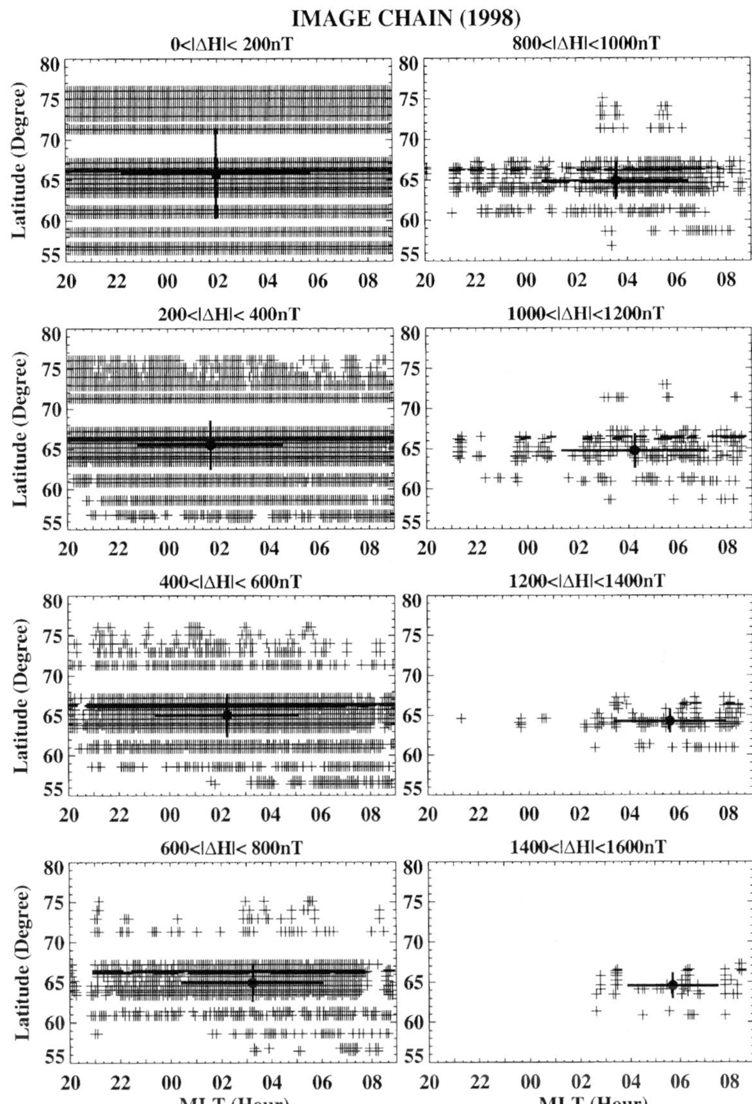

Figure 3. Same as Figure 1 but for IMAGE chain of magnetometers.

152

the extremely disturbed period, e.g., $1400 < |\Delta H| < 1600$ nT, the center is located at $64.55° \pm 1.59°$ and 5.64 ± 1.84 MLT. This latitude of the center is comparable to that of CANOPUS chain for the same activity level. Since IMAGE chain covers all the way down to $56.85°$ in terms of latitude, it is ideal to check whether the westward electrojet flows along such a lower latitudinal region during extremely disturbed periods, the region that neither Alaska nor CANOPUS chain covers. It is noted, however, that the center does not seem to shift steadily below the latitudinal range of either Alaska or CANOPUS chain but rather to approach $63\text{-}64° \pm \approx 2.0°$, regardless of magnetic activity. On the other hand the center shifts toward late local time sector, about three or four hours, compared to either the Alaska or CANOPUS chains. In any case it should be pointed out that the peak occurrence rates of the H maximum consistently occur in the post-midnight region in all three magnetometer chains, indicating that the peak is associated with the convection driven DP-2 current system rather with the DP-1 substorm electrojet.

3. The Relatonship Between the Westward Electrojet and Auroral Image

During intense geomagnetic storms aurora was occasionally reported even over Texas. It is an interesting question to be asked whether the westward electrojet belt also shifts to such low latitude together with the bright auroral image. Plate 1 shows two Polar VIS images taken at 0714 and 0723 UT during the maximum phase of a magnetic storm occurred on October 22, 1999 with the Dst index recording -237 nT at 0700 UT. Overlapped with the images are seven geomagnetic stations: St. Johns ($57.6°$), Ottawa ($58.5°$), Pinawa ($60.7°$), Meanook ($62.6°$), New Port ($55.0°$), Victoria ($54.1°$), and Sitka ($59.8°$). A map is also included to show the locations of those ground magnetic stations, see Figure 4. A color code for the luminosity scale of aurora is adopted. One can note that there are approximately three classes of color in the images; purple, deep blue and green or brighter, representing the low, intermediate and bright regions of luminosity. Overlapped are the latitude circles of $80°$, $70°$, $60°$ and $50°$ and the dawn-duck and noon-midnight meridians.

An extraordinary bright region was located slightly poleward of the southern boundary of the auroral oval in the midnight sector stretching east-west direction at 0714 UT. The oval becomes narrower towards both afternoon and post-midnight sectors. It is interesting to note that magnetic stations, New Port and Victoria, which were embedded in or close to the brightest auroral region recorded horizontal magnetic disturbances of -201 and -62 nT, respectively, less intense compared to those recorded at the other five stations located in the poleward side. Higher disturbances were recorded rather away from the bright region. The highest ΔH disturbance, -872 nT, was record at Ottawa. Pinawa and Meanook also recorded relatively high ΔH disturbance, -660.0 and -662nT, respectively. It is a strong indication that the intensity of the auroral electrojet is not necessarily proportional to the auroral luminosity.

We examined ΔH and ΔZ of the seven magnetic stations during the first half-day of October 22, 1999. The Z-component can be used to determine the relative location of a magnetic station with respect to the center of the auroral electrojet, [Rostoker and Hughes, 1979]. Only two out of the seven stations, Pinawa and Meanook, recorded positive Z-component changes during the period covered by the two images, indicating that the center of the westward electrojet was located equatorward of the two stations. In other words, it was located poleward of the rest of the stations during the entire period. Thus the center of the electrojet seemed to be located somewhere between Meanook and New Port in the pre-midnight sector. At about 0723 UT the overall brightness of the entire auroral image was reduced further while the level of overall magnetic disturbance persisted at a similar level of the previous epoch. A close examination of the auroral image reveals that there seems to be a boundary between two regions in terms of brightness of aurora,   dividing a deep blue colored zone occupying the equatorward side from a

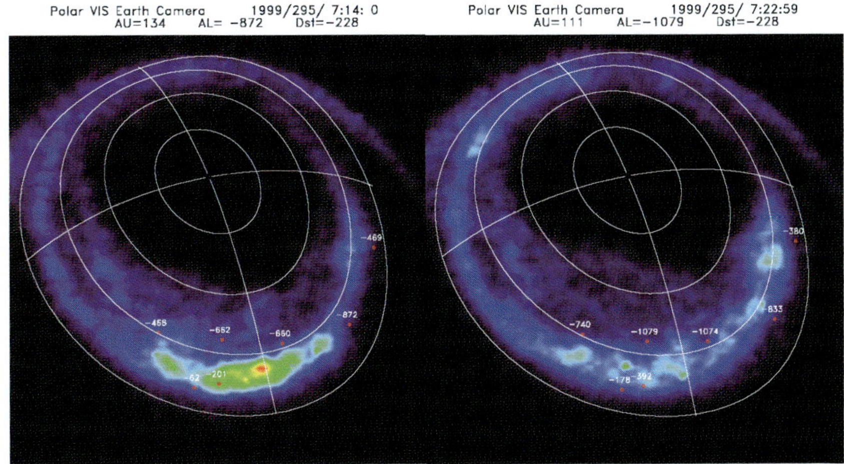

Plate 1. Polar VIS auroral images taken at 0714 and 0723 UT on October 22, 1999. Superimposed are ΔH variations of the seven magnetic stations referred in Figure 4. Also shown are the latitude circles of 80°, 70°, 60° and 50° along with the dawn-dusk and noon-midnight meridians. The AU, AL and Dst indices during each epoch are also indicated

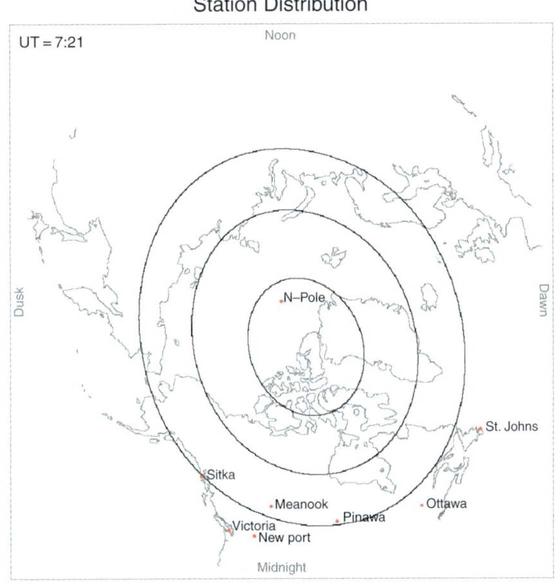

Figure 4. A map showing the location of the seven magnetic stations, St. Johns, Ottawa, Pinawa, Meanook, New Port, Victoria and Sitka.

154

purple colored zone in the poleward side. It is likely that the demarcation in the auroral brightness delineates the center of the westward auroral electrojet. Note that three stations, Pinawa, Meanook and Sitka, located along the demarcation line recorded relatively high magnetic disturbances.

High ionospheric conductivity would be expected over a bright auroral region and it is also the region of a weak electric field [Kamide et al., 1986]. Therefore, an intense auroral electrojet would flow in the region where both the ionospheric conductivity and the electric field are considerably high. The center of the auroral electrojet seems to correspond to the boundary that separates the brighter auroral region of the equatorward portion from the wide but less bright area of the poleward portion. Note that although they are under the brigher part of the auroral image, the magnetic disturbances recorded at New Port and Victoria during the epoch (0723 UT) were lower than those of Pinawa (-1074 nT), Meanook (-1079nT) and Sitka (-740 nT), all located in the less bright region. It is particularly noteworthy that Pinawa recorded a significantly high ΔH and also a significantly positive ΔZ, indicating that the center of the westward electrojet is at least a few degrees equatorward of it.

Although it is a subauroral station and under rather less bright auroral region, Ottawa recorded relatively high magnetic disturbance. At the same time, it recorded a considerably large negative Z-component, indicating that the center of the westward electrojet flows far poleward of the station. It can be interpreted that the electric field plays an important role in intensifying the auroral electrojet in the region. St. Jones, which was far east of Ottawa, showed the same tendency. On the other hand, Sitka, under less bright region (purple), recorded a significantly high disturbance, indicating that electric field also plays an important role. Enhancements in the electric field and ionospheric conductivity seem to be complementary of each other, implying that if the electric field is strong in a region, the ionospheric conductivity would be rather weak and vice versa. However, the electric field appears to play a leading role in intensifying the westward electrojet, while the ionospheric conductivity enhancement is also essential but seems to play a secondary role. We have seen that large magnetic disturbances are not associated with the brightest aurora region. It is an indirect indication that the electric field plays a major role in intensifying the auroral electrojet. Thus an enhancement in the AL index is, in general, more closely associated with electric field intensification rather than impulsive ionospheric conductivity enhancement associated with auroral brightening.

During the two epochs shown in Plate 1, Pinawa and Meanook recorded positive Z-component while the other five stations recorded negative Z-component, suggesting that the center of the auroral electrojet seems to be located equatorward of the two stations, Meanook and Pinawa, and poleward of the rest of the five stations. Considering the magnetic latitude of the seven stations, we can infer that the latitude of the westward electrojet center is at around 60° in the night hemisphere during these epochs. It would be approximately the same lowest possible latitude of the westward electrojet center, noted from the three magnetic meridian chain data shown in Figures 1, 2 and 3.

4. Summary

Based on an extensive magnetic disturbance database obtained from CANOPUS, Alaska and IMAGE chains of magnetometers we examined equatorward shift of the auroral electrojet during very disturbed periods. As expected, the auroral electrojet tends to shift equatorward with magnetic activity. But the shift does not seem to exceed a certain latitudinal range even during severely disturbed periods. The lowest possible latitude of the center of the westward electrojet seems to be at around 60° in corrected geomagnetic latitude. Such a trend is also apparent from an event study of a major magnetic storm during which a bright aurora was observed as low as the northern part of continental USA. Interestingly the intense part of the westward electrojet during the storm period does not seem to be located near the equatorward boundary of the auroral oval, where the aurora is brightest, but along the demarcation line, which divides bright and less bright regions. In other words, an intense auroral electrojet is not necessarily collocated with a bright aurora region. Admitting that the auroral electrojet is a combined result of the enhancement of the ionospheric conductivity and the electric field, it is reasonable to say that a bright aurora alone cannot activate an intense electrojet. Actually the intense westward electrojet during the storm period seemed to flow along the demarcation line between the bright auroral region and a less bright one or along the poleward boundary of the bright auroral image. It is the region where both the conductivity and electric field are expected to be considerably high. Since an intense auroral electrojet is collocated with less luminous auroral region, the electric field seems to play a more important role in intensifying the auroral electrojet than the ionospheric conductivity does.

Employing a realistic conductivity distribution inferred from bremsstrahlung X ray image data, Ahn et al. [1989] examined the spatial relationship between the conductivity enhancement and auroral electrojet. They noted that the poleward half of the westward electrojet in the post-midnight sector is dominated by the electric field, while the equatorward half is largely governed by the ionospheric conductivity. The demarcation line seems to delineate the boundary between the bright and less bright aurora regions. They further confirmed that the center of the intense westward electrojet is approximately collocated along the demarcation line between the electric field- and conductivity-dominant regions. It was also noted that the enhanced conductivity region even at the nightside auroral latitude does not necessarily accompany an enhanced ionospheric current. Such a tendency is clearly reflected in the ionospheric conductivity model by Ahn et al. [1998]. According to them the ionospheric conductivity is significantly higher in the equatorward half of the auroral electrojet than the poleward half for a given level of magnetic activity.

While the bright aurora can shift toward lower latitudes during severe magnetic storm, the center of the westward electrojet does not seem to shift accordingly but approaches to a limiting latitude, about 60°. By examining magnetic records from College (65.1°) and Sitka (59.8°) during the International Geophysical Year (one of the most disturbed periods ever recorded during the last century), Weimer et al. [1990] found that perturbations in the Z-component from Sitka almost always shows negative values, indicating that the westward electrojet was always located poleward of Sitka. Therefore the current AE network, which covers as low as about 62°, does not have any serious problems in monitoring the auroral electrojet, particularly the westward electrojet even during intense geomagnetic storm as long as the lowest AE station is located in the dark sector. Weimer et al. [1990] also confirmed that the electrojet appears to remain poleward of 60° during substorms. Ahn et al. [2000] noted that the AL index tends to be underestimated during severely disturbed periods. It is particularly the case during early UT hours when the westward electrojet flows equatorward of Narssarssuag (68.9°) and Leirvogur (66.8°) in the midnight-early morning sector.

Acknowledgments

This work was supported by grant (R05-2003-000-10059-0) from the Basic Research Program of the Korea Science Engineering Foundation. G. X. Chen is partly supported by National Natural Science Foundation of China (Grant No. 40236058). The Alaska meridian chain data were provided by the Geophysical Institute of University of Alaska. The CANOPUS instrument array constructed, maintained and operated by the Canadian Space Agency, provided

Agency, provided the data used in this study. The IMAGE magnetometer data are collected as a Finnish-German-Norwegian-Polish-Russian-Swedish project. Magnetograms of St. Johns, Ottawa, New Port, Meanook and Victoria were provided by INTERMAGNET.

References

Ahn, B.-H., Y. Kamide, and S.-I. Akasofu, Global ionospheric current distribution during substorms, *J. Geophys. Res., 89*, 1613, 1984.

Ahn, B.-H., Y. Kamide, and S.-I. Akasofu, Electrical changes of the polar ionosphere during magnetospheric substorms, *J. Geophys. Res., 91*, 5737, 1986.

Ahn, B.-H., H. W. Kroehl, Y. Kamide, and D. J. Gorney, Estimation of ionospheric electrodynamic parameters using ionospheric conductance deduced from Bremsstrahlung X ray image data, *J. Geophys. Res., 94*, 2565, 1989.

Ahn, B.-H., A. D. Richmond, Y. Kamide, H. W. Kroehl, B. A. Emery, O. de la Beaujardiere, and S.-I. Akasofu, An ionospheric conductance model based on ground magnetic disturbance data, *J. Geophys. Res., 103*, 14769, 1998.

Ahn, B.-H., H. W. Kroehl, Y. Kamide, and E. A. Kihn, Universal time variations of the auroral electrojet indices, *J. Geophys. Res., 105*, 267, 2000.

Akasofu, S.-I., Relationship between the AE index and Dst indices during geomagnetic storms, *J. Geophys. Res., 86*, 4820, 1981.

Feldstein, Y. I., A. Grafe, L. I. Gromova, and V. A. Popov, Auroral electrojets during geomagnetic storms, *J. Geophys. Res., 102*, 14223, 1997.

Kamide, Y., J. D. Craven, L. A. Frank, B.-H. Ahn, and S.-I. Akasofu, Modeling substorm current systems using the conductivity distribution inferred from DE auroral images, J. Geophys. Res., 91, 11235, 1986.

Rostoker, G., and T. J. Hughes, A comprehensive model current system for high latitude magnetic activity, *Geophys. J.R. Astron. Soc., 58*, 571, 1979.

Sun, W., Y. Kamide, J. R. Kan, and S.-I. Akasofu, Inversion of the auroral electrojets from magnetometer chain data based on the flexible tolerance method, *J. Geomag. Geoelec., 45*, 1151, 1993.

Weimer, D. R., L. A. Reinleitner, J. R. Kan, L. Zhu, and S.-I. Akasofu, Saturation of the Auroral Electrojet Current and the Polar Cap Potential, *J. Geophys. Res., 95*, 18981, 1990

Multiscale Coupling of Sun-Earth Processes
A.T.Y. Lui, Y. Kamide and G. Consolini, editors.

GLOBAL DAYSIDE IONOSPHERIC RESPONSE TO INTERPLANETARY ELECTRIC FIELDS: PLASMA UPLIFT AND INCREASES IN TOTAL ELECTRON CONTENT

A. J. Mannucci, B. T. Tsurutani, B. A. Iijima
Jet Propulsion Laboratory, California Institute of Technology,
4800 Oak Grove Drive, Pasadena CA 91109

W.D. Gonzalez, F. L. Guarnieri
Instituto Nacional de Pesquisas Espaciais, São José dos Campos, SP, Brazil

K. Yumoto
Space Environment Research Center, Kyushu University, Japan

Abstract

The interplanetary shock/electric field event of 5-6 November 2001 is analyzed using ACE interplanetary data. The consequential ionospheric effects are studied using GPS receiver data from the CHAMP satellite and data from ~100 ground-based GPS receivers. The dawn-to-dusk interplanetary electric field was initially ~ 33 mV/m just after a forward shock (IMF B_Z= -48 nT), reaching a peak value of ~54 mV/m one hour and 40 minutes later (B_Z= -78 nT). The electric field was ~45 mV/m (B_Z=-65 nT) two hours after the shock and a magnetic storm intensity of D_{ST} = -275 nT was reached. The dayside satellite GPS receiver data plus ground-based GPS data indicate that the entire equatorial and midlatitude (up to ± 50° MLAT) dayside ionosphere was uplifted, significantly increasing the column-integrated electron content at altitudes above the CHAMP orbital altitude of 430 km. Based on the total electron content (TEC), this uplift appeared to peak ~2 1/2 hours after the shock passage. The effect of the uplift leading to TEC increase lasted for 4 to 5 hours. We suggest that the interplanetary electric field "promptly penetrated" to the ionosphere, and the dayside plasma was convected (by $\vec{E} \times \vec{B}$ drift) to higher altitudes. Upward plasma transport led to a ~ 55-60% increase in equatorial ionospheric TEC measured from altitudes of ~430 km (at 1930 LT). This inferred vertical transport plus photoionization of atmospheric neutrals at lower altitudes caused a 21% TEC increase in equatorial ionospheric TEC at ~14 LT (from ground based measurements). Data from an equatorial magnetometer confirms the electric field behavior inferred from the TEC changes on the dayside. Seven to nine hours after the beginning of the electric field event, the dayside ionospheric TEC above ~430 km decreased to values approximately 45% lower than quiet day values. The ground-based equatorial ionospheric TEC decrease was ~16%. This decrease occurred both at midlatitudes and at the equator. We presume that thermospheric winds and neutral composition changes produced by the storm-time frictional heating and disturbance dynamo electric fields are responsible for these decreases.

1. Introduction

It is well-known that the interconnection between southward interplanetary magnetic fields (IMFs) and the Earth's magnetic fields leads to strong dawn-to-dusk electric fields and an overall increase in magnetospheric convection. This convection, in turn, causes intense ring current buildup and magnetic storms [*Gonzalez and Tsurutani*, 1987; *Gonzalez et al.*, 1994; *Kozyra et al.*, 1997; *Kamide et al.*, 1998]. These same dawn-to-dusk electric fields can have dramatic effects on the Earth's ionosphere. If the electric fields can penetrate to the ionosphere before shielding builds up [*Tanaka and Hirao*, 1973], they can modify equatorial ionospheric electrodynamics. Dawn-to-dusk directed electric fields will be eastward in the daytime and westward at night. Thus, such electric fields will cause upward **ExB** dirft in the equatorial ionosphere during the day and downward drift at night.

Magnetospheric convection electric fields have another effect. The plasmasheet is convected earthward forming a ring current, and a magnetic storm main phase commences, as indicated by the *Dst* index [*Tsurutani and Gonzalez*, 1988]. The resultant large increase in energy and momentum deposition in the high latitude ionosphere and thermosphere causes a chain of events. Large-scale high-latitude convection of ionospheric plasmas by magnetospheric electric fields leads to neutral winds with velocities in excess of 1 km/s, through the mechanism of ion-neutral collisions. Frictional dissipation within the high latitude thermosphere drives large scale wind surges. These winds alter the midlatitude ionosphere by moving plasma along field lines to regions of altered composition [*Proelss*, 1997; *Fuller-Rowell et al.*, 1997; *Buonsanto*, 1999]. The flow from high to low latitudes, through continuity, causes upwelling of the neutral gas and an increase of nitrogen molecules relative to oxygen atoms. This in turn leads to increased ion recombination rates and ionospheric plasma density reductions.

In this paper we study an interplanetary event with an unusually strong southward (negative B_z in GSM coordinates) interplanetary magnetic field and measure the ionospheric consequences throughout the magnetic storm. At equatorial and mid-latitude regions, significant changes occur to the ionospheric TEC (measured by dual-frequency GPS receivers). Little is presently known about the global dayside features of such modifications during intense magnetic storms. It appears that drastic changes in ionospheric vertical TEC can be produced by intense disturbance electric fields originating from the magnetosphere-ionosphere interaction. The disturbance electric fields at low latitudes have been identified principally in two forms, as: (a) prompt penetration zonal electric field often observed in the equatorial latitudes [*Sastri*, 1988; *Fejer and Scherliess*, 1995; *Abdu et al.*, 1995; *Sobral et al,.* 1997; *Sobral et al.*, 2001; *Sastri et al.*, 2002; *Kelley et al.*, 2003] and/or (b) delayed electric fields produced by the disturbance dynamo driven by frictional heating due to storm energy input at high latitudes [*Blanc and Richmond*, 1980; *Richmond and Lu*, 2000; *Scherliess and Fejer*, 1997]. Zonal electric fields from these two causes, depending on their polarity and duration, could cause large uplifts or downdrafts of the ionospheric plasma leading to large-scale local time dependent enhancements or decreases of the vertical TEC. The effects at low latitudes are especially important because the equatorial plasma fountain is highly responsive to such disturbance electric fields and the latter can strongly affect space-based communications and GPS application systems. We note that downdrafts of the ionosphere can lead to net TEC reductions and uplifts can lead to net TEC enhancements, by a simple mechanism discussed later.

In this paper we will address the interesting problem of the global scale ionospheric response to an intense solar wind dawn-to-dusk electric field. The resulting equatorial disturbance electric field variations appear to control a major part of the observed global TEC features. Effects in both the storm main phase and recovery phase will be discussed. We will use low-Earth orbiter and ground-based GPS data primarily. A companion paper [*Tsurutani et al.*, 2004b, in press] includes Brazilian digital ionosonde data and magnetometer data from the Pacific sector in the analyses. The interplanetary event will be analyzed using ACE solar wind data.

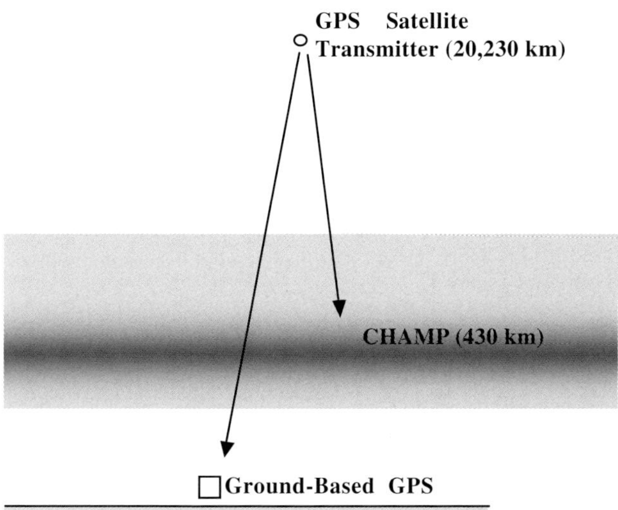

Figure 1: A schematic of GPS satellite and low altitude satellite and ground-based GPS receivers. The system is used to measure multipoint ionospheric path-integral total electron content. The shaded area represents the ionosphere.

2. Data and Analysis Method

We used data from the ACE satellite upstream of the Earth at a GSM position of (1.4×10^6 km, 1.2×10^5 km, -2.0×10^5 km) to measure the solar wind properties causing the geomagnetic disturbance. In the GSM coordinate system, $\hat{x}$ is in the direction from Earth to the Sun, $\hat{y} = \dfrac{\hat{x} \times \hat{M}}{\left| \hat{x} \times \hat{M} \right|}$, where $\hat{M}$ is the north magnetic pole, and $\hat{z}$ completes the right hand system. The plasma detector and the magnetometer are described in *McComas et al.* [1998] and *Smith et al.* [1998], respectively. Due to the contamination of energetic particles, the ACE SWEPAM plasma data are provided with reduced time resolution [*R. Skoug*, personal communication, 2003].

Our measurements of the column density of electrons, referred to as total electron content (TEC) were obtained from the Global Positioning System (GPS) satellite signals as received by both ground-based receivers (~100 distributed around the globe) and satellites in low-Earth orbit. The locations of both the satellites and receivers are known to a few meters, and since straight line propogation is essentially correct for the signals, the electron content is measured along a geographically well-located line-of-sight between GPS satellite and receiver.

A schematic of the TEC measurement system is shown in Figure 1. There were 28 GPS satellites located in circular Earth orbit at an altitude of 20,200 km. For simplicity, only one GPS satellite is illustrated in the Figure. Each receiver simultaneously tracks multiple GPS satellite signals. The relative delay between the two GPS transmission frequencies is directly related to the

the total number of electrons along the line of sight [*Mannucci et al.*, 1999; *Mannucci et al.*, 1998].

The satellite orbital parameters are given in Table 1. CHAMP was in a polar orbit (87° inclination) at an altitude of ~430 km with an orbital period of 91 minutes. Data for elevation angles within 50° of the vertical were used exclusively in this study. Otherwise no other data deletions have been performed. To normalize the measurements obtained at multiple elevation angles, the slant TEC data have been used to estimate the vertical TEC directly above the LEO, assuming a simple geometrical factor accounts for the difference between slant and vertical TEC. In this case, the vertical distribution of density that we assume is a spherical shell ionosphere of uniform (horizontally stratified) density 700 km thick above the CHAMP altitude. For elevation angles greater than 50° as used in this study, the error from this simplifying assumption is not a significant factor for our analysis (see Figure 4).

Table 1: CHAMP Orbital Parameters

Altitude	Inclination	Orbital Period	Local Time Equatorial Crossings (Nov. 6th, 2001)
430 km	87°	91 min	7AM - 7PM

The ground based GPS data set used is composed of ~100 stations from the International GPS Service data centers [*Moore*, 2001]. The obliquity function used to estimate the vertical TEC is computed modeling the ionosphere as a spherical slab of uniform electron density between 450 and 650 km altitude. The latitude and longitude at which the ground-to-satellite line-of-sight intersects the ionosphere is computed using a spherical shell at 450 km altitude. Details concerning the removal of instrumental offsets for both ground based receivers and satellite receivers is beyond the scope of this paper. If the reader is interested in further information on the topic of extracting TEC measurements from ground-based GPS receivers, see *Mannucci et al.* [1998].

3. Results

Measured parameters pertaining to the interplanetary event of 5-6 November 2001 are shown in Figure 2. The top six panels are interplanetary parameters taken by the ACE spacecraft located at 1.4×10^6 km upstream of the Earth. From top to bottom, the parameters are: the solar wind speed, the proton temperature, the solar wind ram pressure, the proton density, the magnetic field magnitude and the magnetic field B_z component in GSM coordinates. Using the measured solar wind speed of ~700 km/s, a convection delay time of ~34 min from ACE to the magnetosphere was estimated. The solar wind data has therefore been shifted by this time in the Figure to match the ground based AE and Dst index data (the bottom two panels).

Nov. 5 - 6, 2001 Interplanetary Event

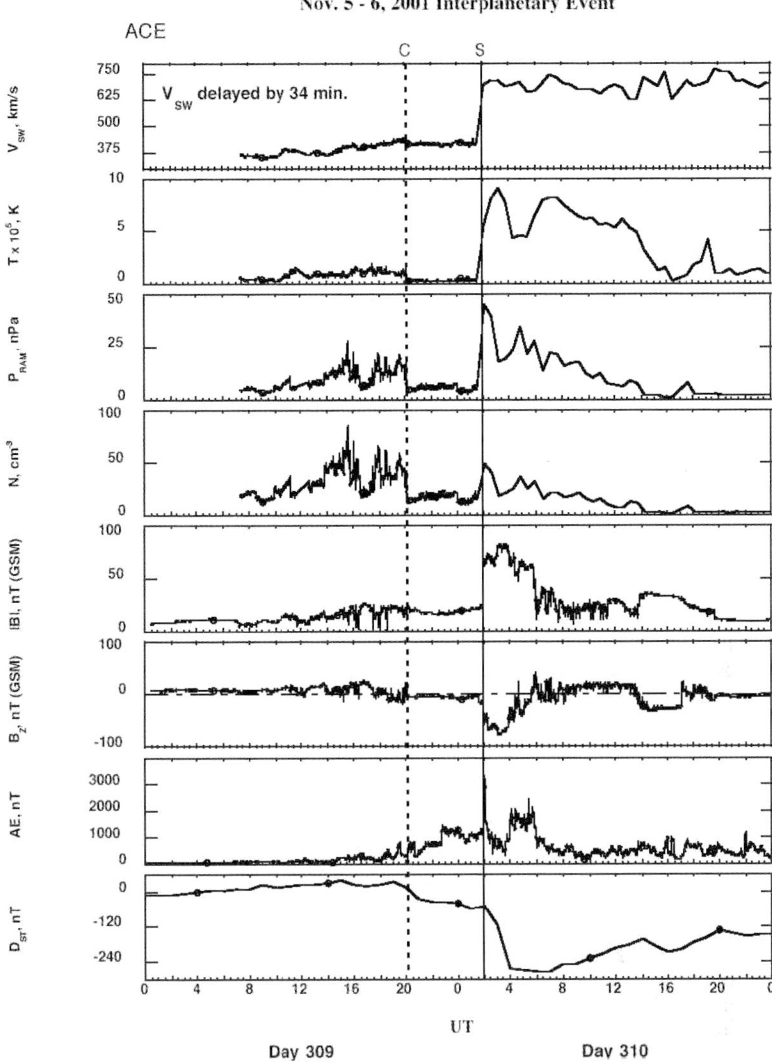

Figure 2: The interplanetary event of 5-6 November 2001. The interplanetary data is taken by the ACE spacecraft. The magnetic field is plotted in GSM coordinates. The time delay of the solar wind and magnetic field convection from ACE to the magnetopause is ~34 min. Thus the interplanetary shock should impinge upon the magnetosphere at ~0154 UT. A strong dawn-to-dusk electric field is imposed on the magnetosphere at this time. An AE ~ 3000 nT substorm onset and a Dst = -275 nT magnetic storm onset also start at the time of the shock arrival.

The dashed vertical line labeled "C" indicates the start of a magnetic cloud [*Klein and Burlaga*, 1982]. The speed is ~420 km/s and is thus a "slow" cloud (see *Tsurutani et al.* [2004a] for discussion of slow cloud properties). It is identified by quiet magnetic fields with the general absence of large amplitude Alfvén waves [*Tsurutani et al.*, 1988] and very low proton temperatures, ~2.5x10^4 K. The plasma density is ~18 cm^{-3} and |B| ~18 nT.

The solid vertical line labeled "S" is a fast forward shock. The shock occurred at ~0120 UT at ACE. It is identified by an abrupt solar wind speed increase from ~420 km/s to ~700 km/s, a proton temperature increase to ~8 x 10^5 K, a density increase to ~48 cm^{-3}, and a magnetic field increase to ~70 nT. The magnetic field magnitude reached a maximum value of 80 nT ~1 hr and 40 min after the shock passage.

The shock interaction with the upstream slow magnetic cloud has a profound effect on the resultant geomagnetic activity at Earth. The cloud has a steady B_Z = -7 nT (southward) field. The interplanetary shock compresses this preexisting interplanetary negative B_Z [*Tsurutani et al.*, 1988] to B_Z = -48 nT. At the peak field strength, ~1 hr 40 minutes after the shock, the B_Z component reached –78 nT. Two hours after the shock, B_Z was ~ -65 nT. B_Z remained at negative values until ~3+ hrs after the shock. This intensely negative IMF B_Z feature of several-hour duration is the cause of the main phase of the magnetic storm with $D_{ST} \cong$ -275 nT. Such intense IMF B_Z events lasting of order hours are always present during major magnetic storms [*Gonzalez and Tsurutani*, 1987; *Gonzalez et al.*, 1994].

The sudden IMF negative B_Z increase at the shock caused a sharp decrease in the Dst value. Dst reached a minimum value of -275 nT. For the remainder of the paper, we will discuss the sudden intense interplanetary electric field at (and after) the shock and its effect on the Earth's ionosphere. This electric field onset occurs at ~0120 UT at ACE and assuming a time shift of 34 minutes, it should have been imposed on the magnetosphere at ~0154 UT. It should also be remembered that there is a small but important "precursor" interplanetary electric field associated with the negative B_Z (due to the slow magnetic cloud ahead of the shock) and a concomitant moderate storm, occurring prior to the shock electric field event.

The large ~3000 nT AE increase just after shock passage is presumably a substorm expansion phase (this is beyond the present paper and is not addressed in detail). Zhou and Tsurutani (2001) and Tsurutani and Zhou (2003) have statistically studied interplanetary shocks and auroral activity using POLAR UV imaging data. The authors conclude that shocks may indeed trigger substorms under the proper interplanetary conditions. Some suggested mechanisms are shock compressional effects or sudden onsets of dawn-to-dusk cross-tail electric fields. We refer the interested reader to those papers and schematics.

Figure 3 shows a multi-day time series of the "verticalized" TEC above the polar orbiting CHAMP satellite for two days, 4 November 2001 (top panel) and 6 November 2001 (bottom panel). Since CHAMP has a nearly identical ground track every 2 days, we use the 4 November TEC data as a quiet "baseline" to identify the ionospheric features in response to the November 6 interplanetary shock. CHAMP was at a nearly constant altitude of ~430 km and was at a nearly constant longitude at midlatitudes and at the equator during these two days. CHAMP crossed the equator at ~7 am LT and at dusk, ~7 pm LT. The ~7 pm LT track is the one that is of most interest for this paper. The vertical line labeled by "S" in the bottom panel (~0154 UT) indicates the time of shock arrival at the Earth's magnetosphere.

There are two important features to note in the Figure. First, a significant TEC increase is observed in the ~7 pm near-equatorial crossing within about one hour of the shock reaching the magnetosphere. At ~0259 UT, the TEC is ~125 TECU compared with ~80 TECU over approximately the same landmass 2 days earlier (a TEC unit [TECU] is 10^{16} electrons/m^2). On the next 7 pm pass, the ionospheric electron enhancement above CHAMP increases to ~160 TECU at ~0425 UT, whereas it was ~100 TECU two days earlier. Thus the TEC above 430 km at ~7 pm

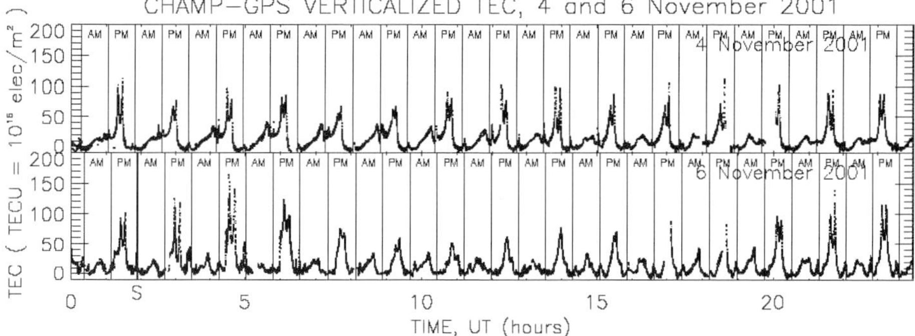

Figure 3: The interplanetary event of 5-6 November 2001. The interplanetary data is taken by the ACE spacecraft. The magnetic field is plotted in GSM coordinates. The time delay of the solar wind and magnetic field convection from ACE to the magnetopause is ~34 min. Thus the interplanetary shock should impinge upon the magnetosphere at ~0154 UT. A strong dawn-to-dusk electric field is imposed on the magnetosphere at this time. An AE ~ 3000 nT substorm onset and a Dst = -275 nT magnetic storm onset also start at the time of the shock arrival.

increased by 55-60% as soon as ~2 hrs after the shock. A second feature in the Figure is a significant decrease in the TEC above CHAMP well after the southward IMF B_z event ended at ~0400 UT. The minimum TEC value at the equatorial anomaly is 55 TECU at ~1050 UT compared with ~95 TECU on the quiet baseline day. The decrease is ~45%. Another noticeable feature several hours after shock passage is the suppression of the dual peak TEC structure that is observed at this local time throughout the quiet day.

Variability in total electron content is also observed on the quiet day (November 4), although the storm day has more variability and larger TEC values. The largest TEC value recorded by CHAMP on November 4 is 110 TECU, whereas the TEC reaches 125 and 163 TECU in the two passes following the shock. The largest TEC increase also occurs on November 6, approximately 38 TECU over 1.5 hours comparing the two post-shock CHAMP passes. On November 4, the largest TEC increase pass-to-pass is 18 TECU.

Figure 4 shows an expanded view of the CHAMP verticalized TEC data during the 7 pm passes plotted against magnetic latitude (it should be noted that for such a plot, the time tick marks are different from panel to panel). The different plots in each panel correspond to successive ascending GPS satellite tracks. The top panel (starting at 1856 UT on November 5) shows the two equatorial anomaly peaks located at ± 15°-20° MLAT, with a local minimum (~40 TECU) as the satellite traverses the magnetic equator at about 1916 UT. After the shock, the TEC in the anomaly peaks increases (see discussion for Figure 3) and the equatorial TEC minimum decreases. These structural changes are consistent with an enhanced equatorial fountain effect induced by disturbance eastward electric fields originating in the magnetosphere that penetrate to equatorial latitudes. At ~0300 UT, the equatorial TEC was 25 TECU compared with ~40 TECU at ~1916 UT. At ~0434 UT, the equatorial TEC increases to 50 TECU and then to 75 TECU by ~0608 UT, with the crests retreating back towards the equator. With the anomaly peak intensities decreasing, the dual peak structure begins to disappear, which is consistent with a westward electric field, possibly from a disturbance dynamo of thermospheric origin [Fuller-Rowell et al., 1997]. By ~0742 UT, there is just a single broad peak from -20° to +20° MLAT.

164

An important aspect of the ionospheric response to the shock is exceptionally large fractional changes in TEC at midlatitudes, for example at ~0425 UT and ~0559 UT. The TEC values are ~60-70 TECU at both of these times in southern magnetic latitudes of ~ -40 degrees, in comparison with values of between 20-25 TECU at the same universal times on November 4 (not shown here). Thus the shock results in midlatitude TEC values that are 250-350% greater than quiet time values.

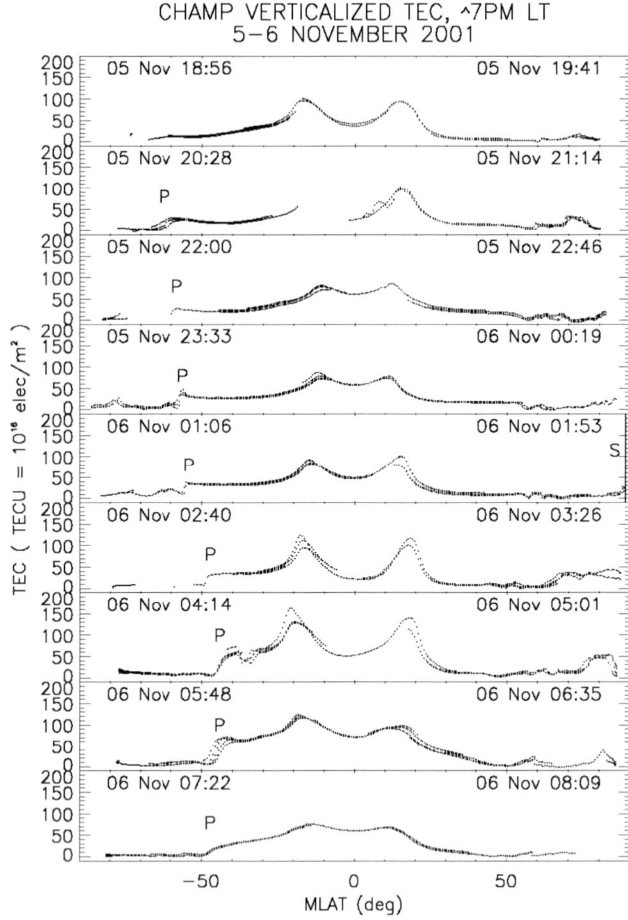

Figure 4: The CHAMP verticalized TEC data for 6 November 2001 and for a few passes on 5 November 2001 (the latter for comparative purposes). The equatorial anomaly maxima at ±15°-20° increased by 55% to 60% within 3 hours after the shock passage at ~0154 UT 6 November. A substantial (~50%) decrease to values below quiet time levels occurs ~9 hours after the shock passage. The TEC difference between the traces in each plot is due to errors in the scaling of slant TEC to vertical TEC above the satellite. Because the traces in each panel follow each other relatively closely, we believe the scaling error is not a significant factor in the analysis presented here.

The CHAMP data capture a sharp TEC decrease with higher latitudes, or a TEC "shoulder" feature. We note that the behavior of this shoulder suggests the TEC observations mirror the signature of plasmapause motion during a geomagnetic storm. At ~2030 UT we label the shoulder feature with a "P". This occurs at a MLAT of ~ -60°. This feature then moves slowly to lower latitudes with increasing time. At ~0115 UT, it has moved equatorward to ~ -54° MLAT (this initial slow inward motion may be associated with the "precursor" negative B_Z prior to shock arrival). After the shock onset at 0154 UT, the shoulder moves equatorward quite rapidly, reaching ~ -43° MLAT by ~0425 UT. If associated with the plasmapause, this motion of the feature towards lower latitudes is due to plasmasphere erosion as the region of enhanced magnetospheric convection expands towards the Earth and maps to ionospheric midlatitudes. Consistent with this interpretation, the shoulder eventually moves poleward again, and is once again at ~ -47° at ~ 0732 UT. It is also noted that the plasma content equatorward of the "shoulder" increases as the equatorward motion takes place. This will be discussed later.

A synoptic view of the dayside TEC response is available from ground-based TEC measurements. Figure 5 shows the ground-based data with satellite track TEC data superposed. The data is plotted as a function of local time and magnetic latitude. Figure 5a shows the 4 November 2001 "baseline" data from 0409 to 0456 UT. Each colored point in the figure represents an estimate of vertical TEC from a link between a GPS ground receiver and a GPS satellite. About ten data points are recorded along each satellite-receiver link during the 45 minute period of the map. In addition, each receiver tracks numerous satellites, typically a number between four and ten.

Figure 5b shows the TEC distribution after the shock event on 6 November 2001, from 0414 to 0500 UT (some 2 to 3 hours after the shock). The satellite data are also shown. The satellite data correspond to verticalized TEC above ~430 km at ~7 pm (CHAMP). All satellite and ground GPS data are plotted to the same TEC scale. The color scale is such that the anomaly trough is not clearly visible.

Figure 5a shows a typical pattern of global TEC which is maximum on the dayside due to solar UV irradiation, and centered at approximately 2 pm local time. The area of the largest TEC intensities, the red area, extends from – 30° MLAT to +20° MLAT and is slightly offset in latitude from the equator (~ -10° MLAT). Such a latitudinal offset would be expected for northern wintertime (November) due to the influence of the prevailing circulation from south-to-north.

Figure 5b shows the post-shock TEC distribution is markedly different to that of Figure 5a. The region of enhanced TEC (> 100 TECU) over the Earth is much larger. The most obvious feature is the greater latitudinal extent, ranging from +50° MLAT to almost -50° MLAT. The westward and eastward extent of the red areas is approximately the same: ~9 am to ~9 pm LT.

From a detailed comparison of Figures 5a and 5b at 14 LT, where the maximum TEC occurs at the equatorial region, it is noted that on 6 November the ground-based TEC has increased from ~145 TECU (4 November) to ~170-180 TECU. Thus, there is an absolute increase in TEC at this local time by ~21%. Figures 5a and 5b also contain the CHAMP upward viewing TEC data (shown earlier in Figures 3 and 4). The TEC values above CHAMP are the same as those above the ground within an uncertainty of 10%. Thus, at ~7 pm, nearly all of the ionospheric plasma is above ~430 km altitude.

4. Interpretation

In this section, we will attempt to interpret the data and draw conclusions on the major points.

1. The interplanetary electric field was initially ~33 mV/m immediately after the shock and then

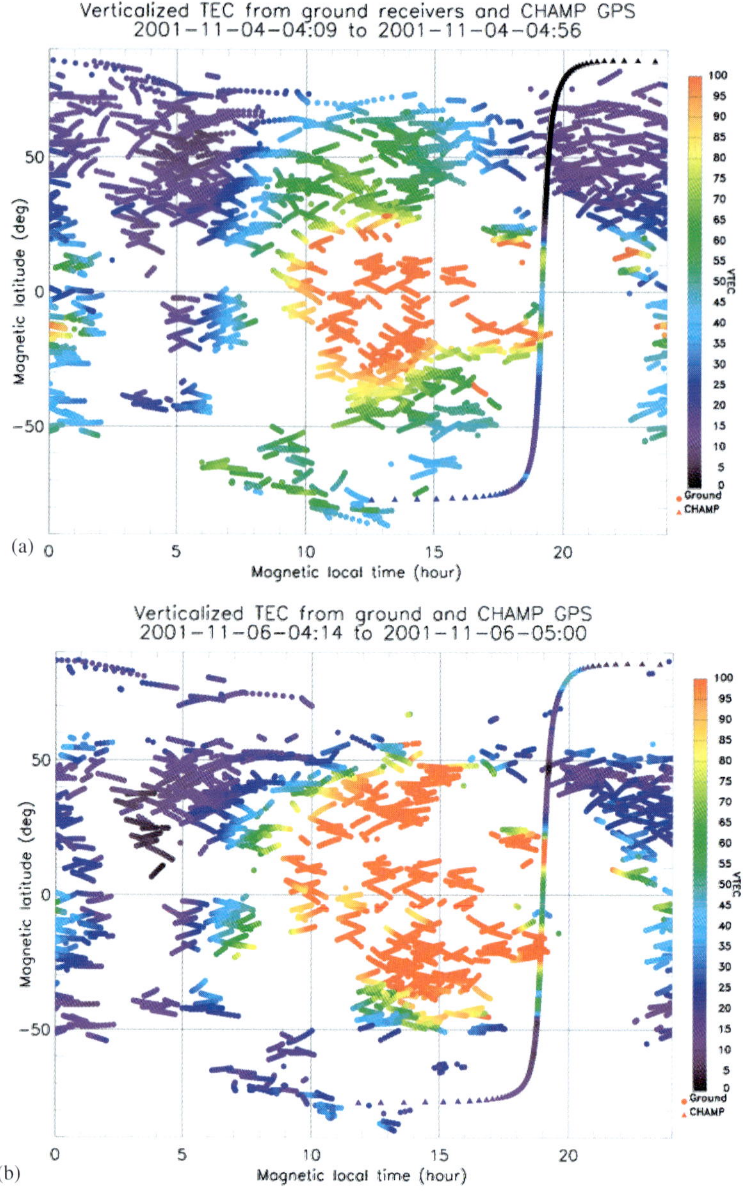

Figure 5: The ~100-station ground-based TEC data for 4 November from 0409 to 0456 UT (background) in panel (a) and for 6 November from 0414 to 0500 UT (post shock event) in panel (b). Various satellite TEC data are also shown (to the same intensity scale). The dayside post-shock region of enhanced TEC is much broader in latitudinal extent, ranging from +50 ° MLAT to almost -50 ° MLAT.

increased to a value of ~54 mV/m one hour and 40 minutes later. The southward IMF event lasted over 3 hours. The dawn-to-dusk directed reconnection electric field convected the plasmasheet into the magnetosphere, forming the storm-time ring current resulting in a large magnetic storm main phase (Dst = -275 nT). The prompt penetration of the interplanetary electric field to the dayside ionosphere had a major impact that is the major focus of this paper. On the dayside, the dawn-to-dusk electric field will be eastward in direction, and the $\vec{E} \times \vec{B}$ plasma drift associated with this eastward electric field will lift the ionospheric plasma to higher altitudes where recombination rates are low. Solar UV radiation will form new electron-ion pairs at lower altitudes,

2. Large TEC increases in the dusk sector were measured by CHAMP. Thus it appears that the "daytime" TEC enhancement extends at least to the 7 pm LT sector. A dusk TEC increase results from longitudinal extension to the dusk sector of the large dayside disturbance eastward electric field. An eastward disturbance electric field at 1900 LT is consistent with the results of *Nopper and Corovillano* [1978], which contains a calculation of equatorial electric field direction assuming an average ionospheric conductance pattern. The maximum electric field intensity, as inferred from the increased depth of the EIA crest intensity (in Figure 4) observed by CHAMP, occurred at ~03 UT. Independent magnetic field data obtained from the low latitude magnetometer at Yap (discussed below) suggests that the increased crest depth occurred nearly simultaneously with the peak eastward disturbance electric field inferred from magnetometer data near noon local time [*Tsurutani et al.*, 2004b].

3. On the basis of the shock-associated event sequences of Figure 2, the evolution of the dayside TEC can be explained as follows: after the shock passage and the imposition of the interplanetary electric field (with the associated AE activity enhancement of ~3000 nT), large eastward electric fields were present over the dayside. The eastward electric field was responsible for the uplift of the dayside (and evening sector) ionosphere that contributed, due to the reduced loss rate of the elevated ionosphere and the presence of the ongoing ion production by solar radiation, to the large scale buildup of the TEC over the dayside hemisphere. A decrease in the eastward electric field, or even its subsequent possible reversal to westward, will not produce a prompt (immediate) decrease of the TEC due to the very slow F layer plasma decay rate by chemical recombination and the continuing ionization production by the solar radiation (even in the 7 pm LT sector where ionization by direct solar radiation cannot operate, the TEC decay time should be of the order of hours at F-region altitudes). By ~06 UT the AE activity shows recovery (Figure 2) which may have produced equatorward penetrating electric fields of westward polarity on the day sector, because of overshielding, or a westward disturbance dynamo electric field in that sector due to thermospheric winds. The resulting "downdraft" of plasma results in TEC decrease over the equatorial region seen in the CHAMP data several hours after the shock. Persistence of the disturbance dynamo westward electric field is a possible driving force for the reduced/inhibited EIA that ensued in the following hours.

4. Another new feature is that the dayside ionospheric TEC increases extended to ±50° MLAT. The increases at these latitudes exceed ~80% compared to quiet time values. One obvious possibility is that the prompt electric field for these intense interplanetary events not only affects the equatorial ionosphere but midlatitude regions as well. This "superfountain effect" will transport electrons from lower to higher magnetic latitudes.

The "shoulders" or plasma boundaries at -54° MLAT (and more equatorward) are a dayside feature that has never been noted before. The equatorward motion of this boundary during the main phase of the magnetic storm is analogous to the inward motion of plasmaspause. The retreat of this boundary towards the pole is analogous to the expansion of the plasmasphere in the storm recovery phase. Similar results were observed recently by the Radio Plasma Imager (RPI) on board the IMAGE satellite [*Reinisch et al.*, 2004]. It will be interesting to make a detailed comparison of this ionospheric data to IMAGE results for the same event to determine if plasmaspheric erosion (and the plasmapause) can be identified by ionospheric TEC data.

It is noted that the TEC measured above CHAMP altitudes is increasing during the same period when the "shoulder" feature (or plasmapause) moves equatorwards. Plasmaspheric erosion associated with enhanced magnetospheric convection is probably occurring at the same time that the superfountain effect is causing transport of equatorial plasma to higher magnetic latitudes. These two physically distinct processes are possibly contributing to the sharpness of the shoulder feature. It should be noted that possible indications of a superfountain effect similar to that reported here may have been observed during the great storm of March 1989. A TEC depletion at the equatorial anomaly crest location [*Batista et al.*, 1991] was accompanied by concurrent large enhancements at higher magnetic latitudes [*Abdu*, 1997].

5. The dusk ionospheric TEC decreases (~45%) well after the storm main phase are observed by CHAMP at mid-latitudes. We note that the ground based peak values for the daytime equatorial region at ~14 LT decreased by ~16% during this period (see Tsurutani et al., [2004b]). This decrease is almost as impressive as the increase. The coexistence of both positive and negative storm phases in TEC have been previously noted by *Lu et al.* [2001]. The most likely explanation for the depletions in TEC at midlatitudes are neutral compositional changes [*Proelss*, 1997; *Fuller-Rowell et al.*, 1997; *Buonsanto*, 1999]. At lower latitudes, the long lived depletions can be driven by the westward disturbance dynamo electric fields as noted previously [*Blanc and Richmond*, 1980; *Scherliess and Fejer*, 1997; *Fuller-Rowell et al.*, 2002].

Figure 6 shows the equatorial electrojet (EEJ) current intensity over the Pacific equatorial station Yap (9.3° N, 138.5° E, dip angle: -0.6°), which was obtained by subtracting the diurnal range of the horizontal (northward) component of magnetic field intensity (_H) over a non-EEJ station Guam (13.58° N, 144.87° E, dip angle: 9°) from that of Yap. During the interplanetary shock event and consequential storm, Yap was in the midday sector. The large increase of the EEJ intensity over the Pacific sector (bottom panel) is evidence for the very large disturbance penetration electric field of eastward polarity produced on the dayside by the shock event. This eastward polarity of the penetration electric field on the dayside is in general agreement with theoretical expectations and model results (Senior and Blanc, 1984; Spiro et al., 1988; Fejer et al., 1990).

5. Conclusions

The dawn-dusk interplanetary electric field has been invoked to cause a myriad of effects: 1) convection of the plasmasheet Earthward to create the ring current and the storm main phase, 2) uplifting of the dayside equatorial and midlatitude ionosphere and suppression of the nightside equatorial ionosphere, and 3) convection of the auroral zone ionospheric plasmas such that ion-neutral collisions lead to thermospheric winds that propagate to low latitudes.

The intense penetration electric fields strongly affect the dayside midlatitude ionospheric plasma densities. The abnormally high TEC values caused by the expansion of the EIA encounter TEC decreases associated with plasmaspheric erosion caused by the magnetospheric convection electric fields, leading to a sharp "shoulder" feature we believe is indicative of the storm-time plasmapause. Such a signature may be also present (but less noticeable) during smaller magnetic storm events, and thus GPS data might provide another tool to identify plasmaspheric erosion events and a measure of magnetospheric convection electric fields. At this time we don't know how this GPS signature is related to Subauroral Polarization Streams (SAPS: Foster and Vo, 2002; Foster et al., 2002]. Also we remain with the question: why does the plasmapause signature occur in only one hemisphere?

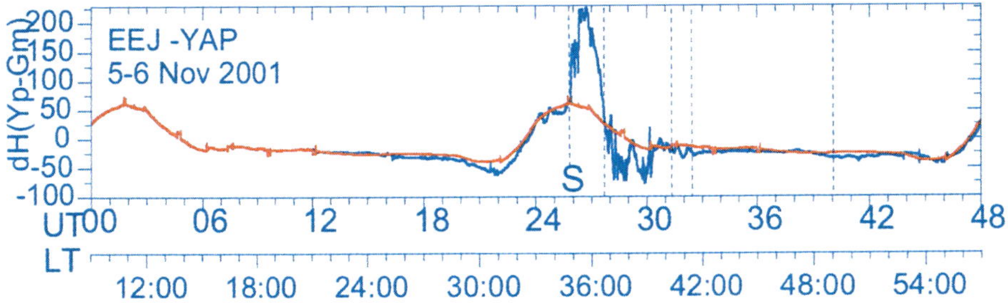

Figure 6: The Equatorial Electrojet –EEJ current intensity over the Yap magnetometer station in the midday sector is shown. The reference day curve for Yap (in red) is taken from 5 Nov. The starting time in the figure is 00 UT 5 Nov. The LT at Yap (UT plus 9 hrs) is shown at the bottom. Vertical line 1 indicates the onset of the shock. The interval between lines 1 and 2 is the interval of strong prompt penetration eastward electric field over Yap in the midday sector. The time interval 2-3 indicates an interval of westward electric field over Yap. The electric field polarity is uncertain during the interval 3-4 due to the sunrise effect. Yap does not show any effect due to the absence of the EEJ at night (lines 4-5).

The results reported here raise the issue of shielding effectiveness at low latitides during these extreme events [*Fejer*, 2002]. For many storms, the direct penetration of magnstospheric electric fields to the low latitude ionosphere is found to be suppressed within about one hour [*Scherliess and Fejer*, 1997], and this mechanism has not been emphasized as a major factor in global plasma redistribution during geomagnetic storms. Our GPS data suggest that the prompt penetration fields are a major driver of ionospheric response lasting for several hours. One possible explanation for underestimating the impact of shielding is the dissimilar nature of the ionosphere during day from night. The higher ionospheric conductivity during the day may lead to considerably longer shielding time scales in this region [*Southwood*, 1977; *Southwood and Wolf*, 1978]. Past studies may have emphasized nighttime response of the ionosphere.

We find that the interplanetary electric field has an enormous effect on the dayside ionosphere. We are only beginning to understand the full ramifications.

Acknowledgements

Portions of the research for this paper were performed at the Jet Propulsion Laboratory, California Institute of Technology under contract with NASA. FLG would like to thank CAPES/MEC (Brazil) for the fellowship allowing an extended stay at JPL. We wish to thank R. S. Skoug for her kind help in providing the ACE SWEPAM (plasma) data. The authors are grateful for excellent referee comments that led to significant manuscript improvements.

References

Abdu, M. A., I. S. Batista, G. O. Walker, J. H. A. Sobral, N. B. Trivedi, and E. R. de Paula, Equatorial ionospheric electric field during magnetospheric disturbances: local time/longitude dependences from recent EITS campaigns, *J. Atmos. Terr. Physics, 57,* 1065, 1995.

Abdu, M. A., Major phenomena of the equatorial ionosphere-thermosphere system under disturbed conditions, *J. Atmosph. Solar Terr. Physics, 59, (13),* 1505, 1997.

Batista, I.S.; de Paula, E.R.; Abdu, M.A.; Trivedi, N.B.; "Ionospheric Effects of the 13 March 1989 Magnetic Storm at Low Latitudes", *J. Geophys. Res., 96,* 13943, 1991.

Blanc, M., and A. D. Richmond, The ionospheric disturbance dynamo, *J. Geophys. Res., 85*, 1669, 1980.

Buonsanto, M. J., Ionospheric storms – A review, *Space Science Reviews, 88*, 563, 1999.

Fejer, B. G., and L. Scherliess, Time dependent response of equatorial ionospheric electric fields to magnetospheric disturbances, *Geophys. Res. Letts., 22*, 851, 1995.

Fejer, B. G., Low latitude storm time ionospheric electrodynamics, Journal of Atmospheric and Solar-Terrestrial Physics, 64, 1401-1408, 2002.

Foster, J. C., P. J. Erickson, A. J. Coster, J. Goldstein and F. J. Rich, Ionospheric signatures of plasmaspheric tails, *Geophys. Res. Lett., 29(13)*, 10.1029/2002GL015067, 2002.

Foster, J. C. and H. B. Vo, Average characteristics and activity dependence of the subauroral polarization stream, *J. Geophys. Res., 107 (A12)*, 1475, doi:10.1029/2002JA009409, 2002.

Fuller-Rowell, T.M., M.V. Codrescu, R G. Roble, and A.D. Richmond, How does the thermosphere and ionosphere react to a geomagnetic storm?, in *Magnetic Storm*, edited by B.T. Tsurutani, W.D. Gonzalez, Y. Kamide, J.K. Arballo, American Geophysical Union, Washington D.C., *98*, 203, 1997.

Fuller-Rowell, T. M., G. H. Millward, A. D. Richmond and M. V. Codrescu, Storm-time changes in the upper atmosphere at low latitudes, *J. Atmos. And Solar Terr. Phys., 64*, 1383, 2002.

Gonzalez, W. D. and B. T. Tsurutani, Criteria of interplanetary parameters causing intense magnetic storms (Dst < -100nT). *Planetary Space Science, 35(9)*: 1101, 1987.

Gonzalez, W. D.; Joselyn, J. A.; Kamide, Y.; Kroehl, H. W.; Rostoker, G.; Tsurutani, B. T.; Vasyliunas, V. M. What is a geomagnetic storm? *J. Geophys. Res., 99(A4)*: 5771, 1994.

Kelley, M. C., J. J. Makela, J. L. Chau, and M. J. Nicolls, Penetration of the solar wind electric field into the magnetosphere/ionosphere system, *Geophys. Res. Letts., 30*, 1158, 2003.

Klein, L.W., and L.F. Burlaga, Inter-planetary magnetic clouds at 1-AU, *Journal of Geophysical Research-Space Physics, 87* (NA2), 613, 1982.

Lu, G., A. D. Richmond, R. G. Roble and B. A. Emery, Coexistence of ionospheric positive and negative storm phases under northern winter conditions: A case study, *J. Geophys. Res., 106*, 24493, 2001.

Mannucci, A.J., B.D. Wilson, D.N. Yuan, C. M. Ho, U.J. Lindqwister, T. F., Runge. A global mapping technique for GPS-derived ionospheric total electron content measurements. *Radio Science 33* (3), 565, 1998.

Mannucci, A.J., Iijima, B. A., Lindqwister, U. J., Pi, X., Sparks, L., Wilson, B.D., GPS and Ionosphere, published in *URSI Reviews of Radio Science, 1996-1999*, Oxford University Press, August 1999.

Moore A.W., A review of currently available IGS network summaries, Phys. And Chem. of the Earth – Part A, **26** (6-8), 591-594, 2001.

Nopper, R.W., and R.L. Carovillano, Polar-Equatorial Coupling During Magnetically Active Periods, *Geophysical Research Letters*, 5 (8), 699-702, 1978.

Proelss, G. W., Magnetic storm associated perturbations of the upper atmosphere, in *Magnetic Storms*, edited by B.T. Tsurutani, W.D. Gonzalez, Y. Kamide and J. K. Arballo, Amer. Geophys. Un. Press, Wash. D.C., *98*, 227, 1997.

Reinisch B. W., Huang X., Song P., Green J. L., Fung S. F., Vasyliunas V. M., Gallagher D. L., and B. R. Sandel, Plasmaspheric mass loss and refilling as a result of a magnetic storm, *J. Geophys. Res., 109*, Art. No. A01202 JAN 6, 2004.

Richmond, A. D., and G. Lu, Upper-atmospheric effects of magnetic storms: a brief tutorial, *J. Atmos. Solar-Terrest. Phys., 62*, 1115, 2000.

Sastri, J. H., Equatorial electric fields of the disturbance dynamo origin, *Annales Geophysicae 6*, 635, 1988.

Sastri, J. H., K. Niranjan, and K.S.V Subbarao, Response of the equatorial ionosphere in the Indian (midnight) sector to the severe magnetic storm of July 15, 2000, *Geophys. Res. Lett., 29*, No. 13, 10.1029/2002GL015133, 2002.

Scherliess, L., and B. G. Fejer, Storm time dependence of equatorial disturbance dynamo zonal electric field, *J. Geophys. Res., 102*, 24037, 1997.

Smith, C. W., M. H. Acuna, L. F. Burlaga, J. L'Heureux, N. F. Ness and J. Scheifele, The ACE Magnetic Field Experiment, *Space Science Reviews, 86*, 613, 1998.

Sobral, J. H. A., M. A. Abdu, W. D. Gonzalez, I. Batista, A. L. Clua de Gonzalez, Low-latitude ionospheric response during intense magnetic storms at solar maximum, *J. Geophys. Res. 102*, 14305, 1997.

Sobral, J. H. A., M. A. Abdu, W. D. Gonzalez, C. S. Yamashita,A. L. Clua de Gonzalez, , I. Batista and C. J. Zamlutti, Responses of the low latitude ionosphere to very intense geomagnetic storms, *J. Atmos. Solar Terr. Phys., 63*, 965, 2001.

Southwood, D. J., Role of hot plasma in magnetospheric convection, *J. Geophys. Res., 82*, 5512, 1977.

Southwood, D. J. and R. A. Wolf, Assessment of roles of precipitation in magnetospheric convection, *J. Geophys. Res., 83*, 5227, 1978.

Tanaka, T. and K. Hisao, Effects of an electric field on the dynamical behavior of the ionospheres and its application to the storm time disturbances of the F-layer, *J. Atmos. Terr. Phys., 35*, 1443, 1973.

Tsurutani, B.T., W.D. Gonzalez, F. Tang, S.I. Akasofu, and E.J. Smith, Origin of Interplanetary Southward Magnetic-Fields Responsible for Major Magnetic Storms near Solar Maximum (1978-1979), *Journal of Geophysical Research-Space Physics, 93* (A8), 8519, 1988.

Tsurutani, B. T., and X. Y. Zhou, Interplanetary shock triggering of substorms: WIND and POLAR, *Adv. Space Res., 31*, 1063, 2003.

Tsurutani, B. T., W. D. Gonzalez, X. –Y. Zhou, R. P. Lepping, and V. Bothmer, Properties of slow magnetic clouds, *J. Atmosph. Solar Terr. Physics, 66*, 147, 2004a.

Tsurutani, B. T., A. J. Mannucci, B. A. Iijima, M. A., Abdu, J. H. A. Sobral, W. D. Gonzalez, F. L. Guranieri, T. Tsuda, A. Saito, K. Yumoto, B. G. Fejer, T. Fuller-Rowell, J. U. O. Kozyra, J. C. Foster, A. J. Coster, V. M. Vasyliunas, Global Dayside Ionospheric Uplift and Enhancement Associated With Interplanetary Electric Fields, *Journal of Geophysical Research-Space Physics*, in press, 2004b.

Zhou, X.Y., and B.T. Tsurutani, Interplanetary shock triggering of nightside geomagnetic activity: Substorms, pseudobreakups, and quiescent events, *Journal of Geophysical Research-Space Physics, 106* (A9), 18957, 2001.

CHAPTER THREE

MAGNETOSPHERIC SUBSTORMS

176

poised to collapse. It was then suggested that the "loaded" magnetosphere is required to progress to the original ground-state configuration irrespective of the detailed initiation mechanism, which is localized by its nature.

Another area of controversy regarding important aspects in modeling phenomenological substorm sequences relates to the question of how the intensity of substorms is determined for a given level of loaded energy. Observationally, the substorm intensity ranges from some 100 nT to 5000 nT in terms of the magnetic effects of the auroral electrojets, which is a commonly used parameter for monitoring the development and decay of substorms. *Kamide and Akasofu* [1974] showed that the substorm intensity has a wide dynamic range from less than 10^5 A to more than 10^6 A in the total current of the auroral electrojets. If there were a critical energy level in the magnetosphere for substorm generation and substorms were produced by a process that removes the excess energy from that level toward the energy level of the ground state, then we would always have the same substorm intensity.

It seems likely that we have tacitly assumed that the substorm intensity is controlled by the intensity of southward interplanetary magnetic field (IMF) impinging effectively upon the magnetopause, because it determines the efficiency of merging between the IMF and earth's magnetic field, through which process solar wind energy enters the magnetosphere. Figure 1, from *Caan et al.* [1975], is an example of such expectation, trying to show the close relationship between the magnetic flux of the IMF preceding a substorm and the intensity of substorms, which is expressed, by the amplitude of the westward electrojet in the midnight sector. From this level of correlation (the authors did not show the corresponding correlation coefficient), it is difficult to access physical implications between the parameters. If one believes in the existence of a good relationship between the two, it would be stated that the two quantities are highly correlated, but if one does not, their correlation can be seen to be rather low in the scattered points.

By reviewing statistical characteristics of substorms that may limit the choices of theories of

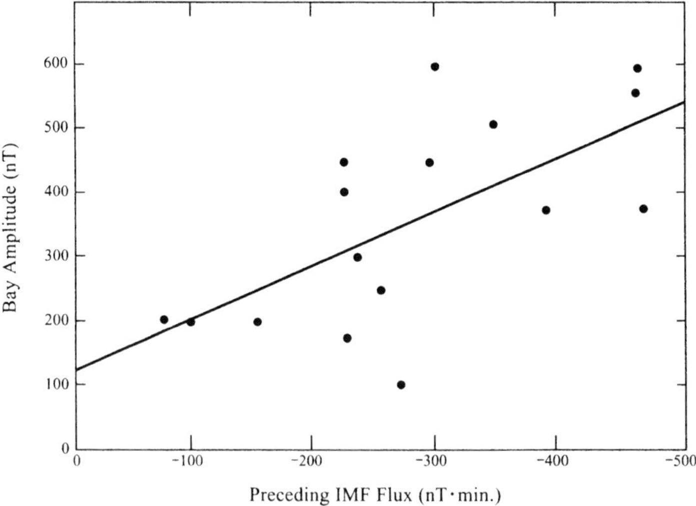

Figure 1. Plot of the flux of the interplanetary magnetic field (IMF) between the southward turning of the IMF and the time of the associated substorm expansion onset versus the intensity of H component magnetic perturbations (designated as Bay Amplitude). A least squares fit is shown by the solid line. After *Caan et al.* [1977].

substorm generation mechanisms, this paper discusses (1) the best available parameters that quantify the intensity of substorms, (2) their mutual relationships, and (3) the possibility of observing a variety of substorm intensities for a given input energy from the solar wind.

2. The Intensity of Substorms: Its Latitude Dependence

Before addressing what determines the intensity of substorms, it is quite natural to ask ourselves whether or not we understand what the intensity is, and how we are able to measure the intensity. As listed by *McPherron and Baker* [1993] and *Lui* [1993], some of the possible parameters are: the standard auroral electrojet indices, the total current of the auroral electrojets, the latitude of the initial auroral breakup, mid-latitude magnetic perturbations, the size of the auroral oval, the brightness of auroras, and the degree of changes in the magnetotail configuration. Note that all these parameters are not mutually exclusive but in fact are closely related each other in a statistical sense. Their correlations are not necessarily 100%, however, implying perhaps that each of the parameters represents only one specific facet of the whole substorm process. We do not necessarily limit ourselves to one particular parameter but in the present paper, we refer mostly to the auroral electrojets to discuss the magnitude of substorms.

2.1   Auroral and Polar Magnetic Substorms

The aurora has continuously been the subject of curiosity for space physicists. It has been well documented that on a global scale, the aurora behaves in rather a systematic manner with characteristic features that recur under similar conditions in terms of the degree to which the solar wind interacts with the magnetosphere [*Akasofu*, 1964]. This recurring process, or cycle, was coined as an auroral substorm. The auroral substorm involves a brightening of auroral arcs in the midnight sector with their subsequent intensifications and poleward motion. The brightening of auroras initiated in the midnight sector proceeds westward as the westward traveling surge, which is one of the dominant signatures of substorms [*Rostoker et al.*, 1980]. All these auroral phenomena represent non-linear energy coupling processes in the magnetosphere-ionosphere system.

During an auroral substorm, the earth's magnetic field undergoes considerable perturbations as well, particularly at high latitudes. This is commonly referred to as a polar magnetic substorm, where the magnetic perturbations are produced by the auroral electrojets that are activated by an enhancement in the ionospheric conductivities caused by auroral precipitation. Because of the one-to-one correspondence between an auroral substorm and a polar magnetic substorm and because of the availability of extensive data from a number of magnetic observatories on the earth's surface, studies of magnetospheric substorms have relied heavily on ground-based magnetometer observations.

2.2   Dynamic Range of Substorm Intensity

Substorms can occur anywhere in the dark sector [*Clauer and McPherron*, 1974], and expansion onsets of substorms take place anywhere between 55° and 67° in geomagnetic latitude [*Kamide and Winningham*, 1977]. Figures 2a and b demonstrate clearly this nature from two independent statistical studies on the basis of two completely different data. In Figure 2a, the magnitude of substorms is defined by the intensity of mid-latitude positive bays observed typically in the midnight sector, which is one of the signatures to identify substorms. From an early statistical study using satellite-viewed imagery data, *Lui et al.* [1975] defined the most probable location of the auroral oval, which is indicated in Figure 2b. It is noticed that the probability of seeing the substorm-associated auroral ovals is shifted equatorward by, at least, several degrees, compared to the quiet-time ovals. *Craven and Frank* [1991] contended also that an initial

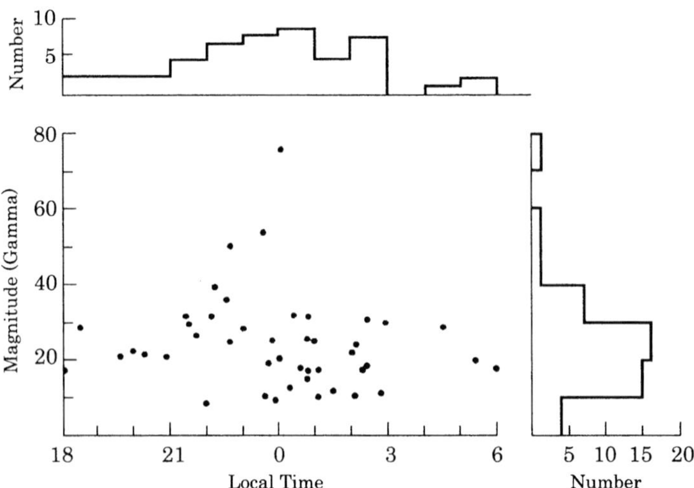

Figure 2a.   Distribution of substorms in local time and magnitude.   The top histogram shows the local time distribution of the central meridian of all substorms without regard to magnitude, while the right histogram shows the distribution in magnitude without regard to local time.   After *Clauer and McPherron* [1974].

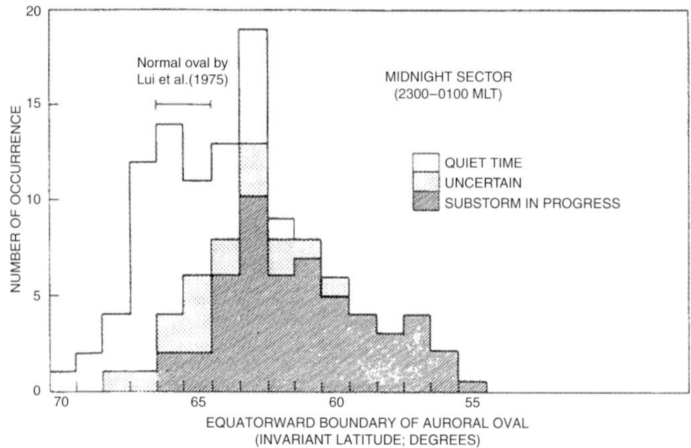

Figure 2b.   Histogram showing the latitudinal distribution of the auroral oval in the midnight sector (measured by the equatorward boundary of diffuse electron precipitation).   Symbols for different categories of substorm activity are defined in the inset.   The latitudinal area of the "normal" auroral oval defined by *Lui et al.* [1975] is also indicated.   After *Kamide and Winningham* [1977].

brightening of auroras leading to the substorm expansion phase can occur anywhere between 2000 and 0200 MLT, although the most prominent local time is 2200-2300 MLT: see Figure 3.   All this supports the idea that no particular choices exist with regard to the location where the magnetosphere-ionosphere system suddenly becomes unstable.

While substorms can take place anywhere at auroral latitudes, the location of the substorm expansion onset seems to be related to the intensity of substorms and also to the polarity of the IMF.   That is, substorms occur at lower latitudes when the southward IMF is more intense. Figure 4 shows the latitude of the electrojet center during the maximum phase of substorms as a function of the total current intensity.   It is evident that the region of the auroral electrojet shifts systematically equatorward as the total current intensity increases.   An increase in substorm intensity from less than 10^5 A to more than 10^6 A in terms of the total current intensity corresponds to an equatorward shift, on average, from 70° to 67°.   Note that in individual substorms, latitudinal motions of the electrojet are more dynamic.

It is important to note that there is a difference of more than 100 in the amount of dissipated energy, which is proportional to the square of the currents, between weak and intense substorms. This indicates that the parameters that control the total energy of substorms, such as the magnetic energy $B^2/8\pi$ at the location of the substorm origin in the magnetotail, must have a similar dynamic range.

The magnitude of substorms in terms of the total current of the auroral electrojet is correlated with the IMF as well.   *Kamide and Akasofu* [1974] showed statistically that the total electrojet, and thus the energy of substorms dissipated in the polar ionosphere, is more intense when the southward component of the IMF is stronger.   It was also revealed that the magnetosphere reaches the "ground" state after a prolonged period of northward IMF, and under such conditions, the auroral oval contracts to its minimum size [*Akasofu*, 1975].   The midnight portion of the ground-state auroral oval contracts to invariant latitudes of 72° - 73°, implying that no substorm expansions take place beyond this latitude.

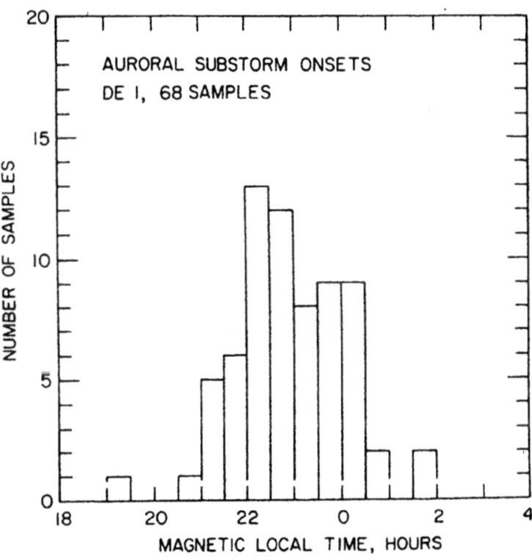

Figure 3.   Histogram showing the number of auroral breakups in half-hour increment of magnetic local time, identified with the Dynamics Explorer-1 spacecraft.   After *Craven and Frank* [1991].

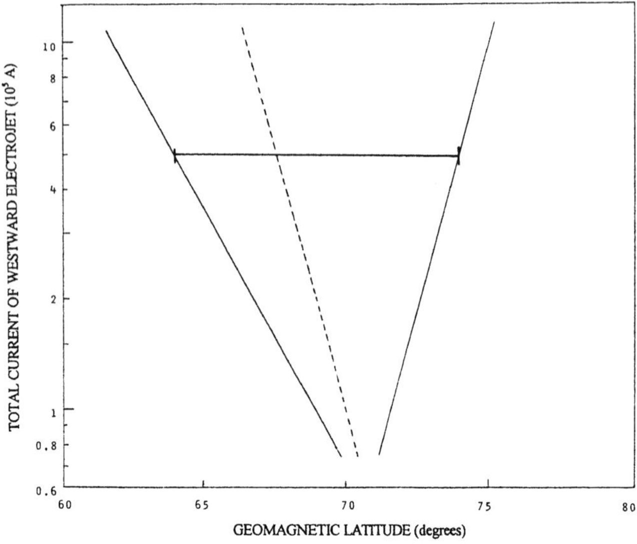

Figure 4. Location in geomagnetic latitude and total current intensity of the westward auroral electrojet at the maximum epoch of substorms. The half-width as well as the center of the substorm westward electrojet are shown. Modified after *Kamide and Akasofu* [1974].

2.3 Probability of Substorm Occurrence

Does substorm occurrence probability depend on the IMF, or, equivalently, on the size of the polar cap? If the probability has a strong dependence on the IMF direction, it can be stated that the occurrence frequency of substorms has something to do with the amount of energy stored in the magnetotail, since there is a definite relationship between the north-south component of IMF and the magnetic energy stored in the magnetotail [e.g., *Fairfield and Lepping*, 1981]. If, on the contrary, the IMF does not influence in any obvious way the triggering of the breakup of auroral forms and the associated electrojet intensifications, substorm expansions can be regarded as resulting from a random process. That is, one could assume that substorms occur randomly at any time regardless of the amount of energy that the magnetosphere gains from the solar wind. This dependence, or no dependence, has in fact been difficult to examine critically, simply because of the lack of continuous observations for the entire polar region.

The amount of energy released in the high-latitude ionosphere during a single substorm is very small, when compared to free energy in the magnetosphere, which is powered by the solar wind-magnetosphere dynamo. It is therefore possible to continue to pursue the possibility that an impulsive substorm-expansion process is not led by a specific IMF signature, but is a rather random internal process occurring in the magnetosphere [*Akasofu et al.*, 1973]. However, studies by, for example, *Kamide et al.* [1977], showed that substorms do not occur randomly but that the magnetosphere is an efficient system in both extracting solar wind energy into the magnetosphere and consuming that energy in substorm processes on a time scale as short as a few hours. In other words, input from the solar wind and output for substorms are both controlled by the solar wind.

Figure 5 shows the probability of seeing substorms and that of seeing quiet times as a function

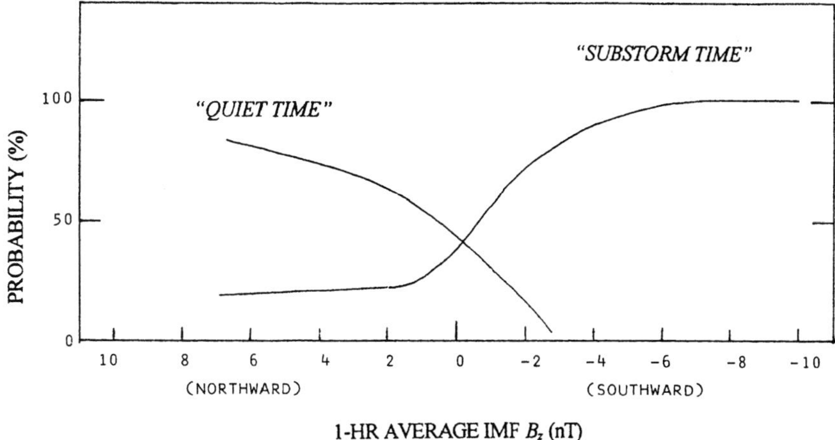

1-HR AVERAGE IMF B_z (nT)

Figure 5. Probability of observing substorm times and quiet times as a function of the north-south component of the IMF. For the definition of substorm probability, see *Kamide et al.* [1977].

of IMF B_Z values. It is clear that the substorm probability increases as B_Z decreases; if the magnitude of the southward B_Z is greater than 5 nT, the substorm probability reaches essentially 100%. Note that here we use the average value for one hour preceding substorm expansions. Figure 5 clearly demonstrates that substorms do not occur with an equal probability for different values of the IMF B_Z component. Since the magnetospheric substorm is a process by which the magnetosphere sporadically removes the excess energy in the magnetotail, it is reasonable to assume that the substorm occurrence probability is closely related to the amount of energy stored in the magnetotail.

These statistical results restrict our options in seeking initiation mechanisms for magnetospheric substorms. Both the location of main substorm actions and the probability of substorm occurrence as well as the strength of substorms are all strongly controlled by the IMF. This accords with the idea that the polar cap, whose size is a measure of the amount of energy stored in the magnetotail, expands and shrinks efficiently according to the B_Z component of the IMF.

3. How Does the Intensity Substorms Relate to Other Parameters?

3.1   How an Auroral Breakup Begins?

There are many varieties of substorms in terms of their strength and the location at which they are triggered. There are also so-called isolated substorms and continuous substorm activations. There are periodic substorms as well [*Borovsky et al.*, 1993]. Regardless of such complications, it is important to emphasize that common threads exist with regard to the phenomenology of most, if not all, substorms. One of these common features is the equatorward shift of the polar cap before substorm expansion takes place. This enlargement of the polar cap has been observed most extensively in terms of equatorward motions of auroral forms prior to a breakup of an auroral arc [e.g., *Snyder and Akasofu*, 1972].

During periods of expansion of the polar cap, there seem to be no obvious and consistent signatures in auroras that generate an auroral breakup. It seems unlikely that a specific signal leads to an expansion onset of an auroral substorm. Which particular brightening actually causes the effective expansion into the entire sky cannot be ascertained since the process appears to be utterly random. Some of the brightening spots are just fluctuations whereas some others may be pseudobreakups. Auroral brightening occurs at least several times at various places in the sky before the "grand opening" begins. Occasionally, no breakup results even though the polar cap has expanded significantly and several increases in the luminosity have occurred. Once the first "sign" of an explosive auroral display begins, however, a chain of brightening actions propagates to all directions in the sky. This may be very similar to a train of explosive powder. This observation is probably consistent with what *Baker et al.* [1999] describes as "a local trigger with global consequences" and what *Chapman et al.*'s [1998] avalanche model dictates: see also *Chang* [1992], *Sharma* [1995], *Klimas et al.* [1996], *Consolini and Chang* [2001], *Chapman and Watkins* [2001], and *Chang et al.*, [2003]. *Lui et al.* [2000] have used a large number of satellite images of auroras to determine statistics of size as a measure of the energy output of the magnetosphere, finding that the dynamic magnetosphere resembles a simple avalanche, or sandpile, system.

In spite of various initial signatures of the breakup process, the strength and the location in which that initial brightening begins must follow the empirical relationships shown in Figure 4. It is also interesting to speculate that the ionosphere-magnetosphere system may cause a substorm expansion onset as long as it is not at the "ground" state, and as long as the polar cap is expanding equatorward, subject, of course, to the probability shown in Figure 5. In other words, the system is always unstable. All that is necessary are minor triggering actions and the conditions that nurture further development of the explosion process. Here we define the ground state of the magnetosphere-ionosphere system as the state where the auroral oval reaches its minimum size and the substorm occurrence probability is literally zero [*Akasofu and Kamide*, 1976]. It is not the state where the solar wind disappears. It is not the state of the IMF being intensely northward in which the effects of magnetic reconnection are expected to be minimal: see *Tsurutani and Gonzalez* [1995].

There is no doubt that the energy released during substorms comes from magnetic energy stored in the magnetotail; therefore, substorm energy must be stored in the tail prior to expansion onset. By "prior" it is meant not only a typical growth phase period of approximately 30 min to 1 hour but also a wider time range, say, from 10 min. to more than a few hours, even during previous substorms in the case of continuous substorm activity.

There is another important issue to consider in seeking the mechanism of how the intensity of substorms is determined. That is, even for the same amount of stored energy, there are large varieties of substorms in terms of the released energy at the polar ionosphere [*Akasofu and Kamide*, 1976]. Figure 6 shows a scattered plot of the total electrojet current as a function of the latitude, where each point represents one substorm: note that the vertical line is in a logarithmic scale. It is clear that the scattered points are well bounded by two curves. The two curves are obtained from a thought experiment by assuming that 1%, or 20%, of free energy available for substorms is consumed as the electrojet energy. This wide dynamic range in the electrojet intensity for a given "stored" energy can be attributed to the following causes: Even once substorm expansion has begun, the process may not continue until all of the available free energy is depleted. A choking mechanism acts to suppress further development of the substorm expansion process. A pseudobreakup may be a type of such imperfect burning of energy. In such a case the energy released during the substorm does not reach a maximum intensity commensurate with the size of the "expanded" polar cap. The poleward motion of the broken-up aurora is then suppressed.

Summarizing the complications discussed above, auroral evolution prior to a breakup is shown schematically in Figure 7. Within several minutes prior to the breakup, auroral brightening

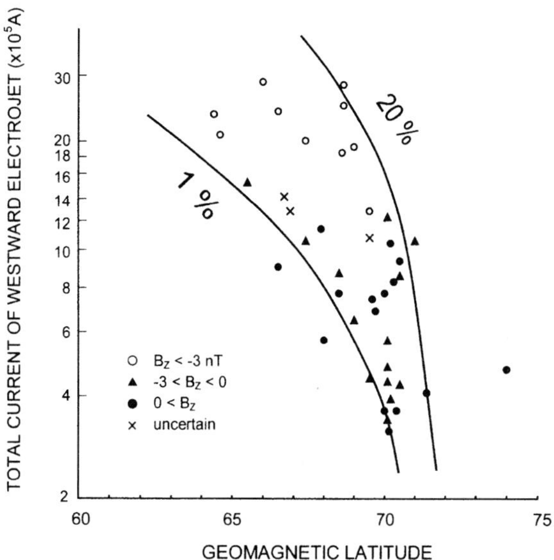

Figure 6. Relationship between the total current of the westward auroral electrojet and geomagnetic latitude of the electrojet center for various values of the B_z component of the IMF, as indicated by different symbols. After *Akasofu and Kamide* [1976].

(marked by x) occurs at least several times at various places in the sky before the "grand opening" (marked by X) begins. As well documented, the auroral arcs may, at times, dim before breakup: see *Pellinen et al.* [1982], *Pellinen and Heikkila* [1984], *Safargalleev et al.* [1997], and *Kauristie et al.* [1997].

Occasionally, no breakup occurs even though the polar cap has expanded significantly and several on-and-offs in the luminosity have occurred. Once the first "sign" of an explosive auroral display, marked by X, begins, however, a chain of brightening signatures propagates to all directions in the whole sky. It is practically impossible which one of "x"s in the unstable region becomes X.

It is important to realize that an auroral breakup begins within a preexisting auroral arc, which has marched from the northern horizon over several tens of minutes [*Snyder and Akasofu*, 1972; *Samson*, 1994]. The breakup does not begin suddenly from "nothing" in the dark sky [*Lui and Murphree*, 1998). This sequence of auroras that is repeatable for every isolated substorm must properly be taken into account when modeling substorm initiation. This feature appears to be missing in the current models of near-earth reconnection in the magnetotail.

This line of thoughts about the unstable magnetosphere can perhaps be best described according to an analogy of a leaky, fragile water bucket shown in Figure 8. The total amount of water in the bucket, i.e., the free energy stored in the magnetotail, is controlled in general by variable solar wind-magnetosphere interactions. It is assumed that the tank is so fragile that water leaks can occur anywhere and at any time unless the tank is completely dry, and that the occurring probability of leak becomes higher with a higher water level. If a small hole at A near the bottom of the bucket happens to develop explosively, the volume of water above the A level will be dumped. If, on the other hand, a hole at B near the water surface grows rapidly, the amount of tipped water will not be enormous. This analogy demonstrates that although the

184

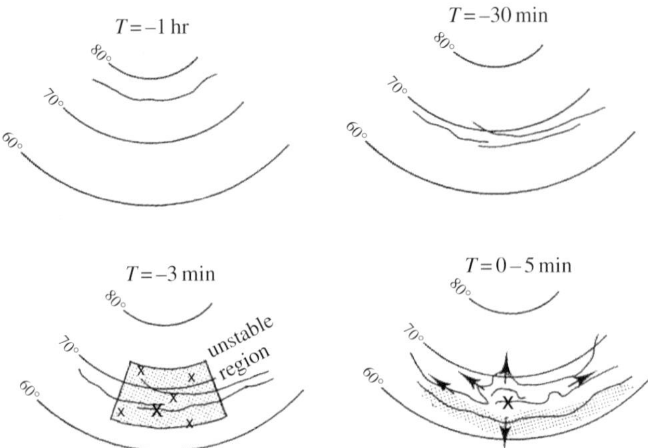

Figure 7. Schematic illustration of auroral evolution prior to an auroral breakup. Within several minutes preceding the auroral breakup, several bright spots (shown by 'x's) appear at various places, but it is not predictable which spot will lead to the real breakup: X marks the location where a breakup has in fact taken place. After *Kamide* [2001].

precise location of the hole can never be predicted with accuracy, a clear relationship between the trigger point and the amount of expelled water can be determined. Note that the height of the bucket corresponds to geomagnetic latitude at the polar ionosphere: the substorm triggered at A (at a lower latitude) is more intense than that at B (at a higher latitude), as shown in Figure 8.

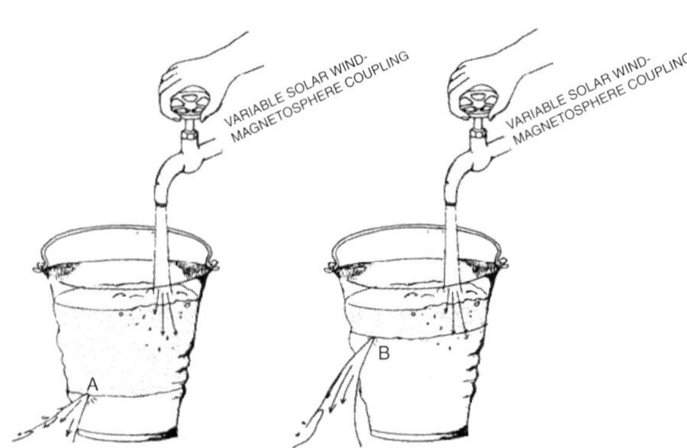

Figure 8. A water tank model showing how the unstable magnetosphere works for substorm initiation. A scenario is shown for two possible cases, A and B, for a given state of solar wind-magnetosphere coupling.

3.2   Magnetotail Variations Associated With Substorm Expansions

Among a number of working models proposed to account for triggering substorm expansions, the near-earth neutral line is thought to be one of the prime mechanisms.   Where is then the ionospheric projection of the near-earth neutral line?   According to *Nagai et al.* [1998a, b], the near-earth reconnection from which strong plasma flows most likely forms at $X = -25\ R_E$ ($\pm 5\ R_E$). On the basis of a number of substorm case studies, *Baumjohann et al.* [1999] conducted a superposed epoch analysis, and concluded that the near-earth neutral line is located between –21 and –26 R_E.   However, it is difficult to accept that, regardless of the high variability in the substorm intensity, a near-earth neutral line is formed almost always near –25 R_E and, within a few minutes, a current wedge associated with dipolarization is generated well inside –10 R_E: as shown in Figure 7, at the latitudes auroral arcs ready for a breakup have long existed.

These studies also appear to ignore the varying intensities of substorms: according to Figure 4, stronger substorms originate at radial distances closer to the earth, and vice versa, as far as we refer to the standard magnetospheric models which are commonly being used.   Reconnection modeling must account consistently for the latitude of auroral breakups, the location of reconnection, dipolarization, formation of the substorm wedge current, and braking of high-speed flows in the magnetosphere, all of which are supposed to occur within a few minutes at such a great distance away, if all the reported observations are correct [*Shiokawa et al.*, 1997; *Birn et al.*, 1999; *Sergeev et al.*, 2000].   No studies have ever shown a systematic relationship between the locations of near-earth neutral lines and the intensity of the corresponding substorms.

It was therefore hypothesized that even though the initial signatures of the breakup process are not identified, the strength of substorms and the location where the initial brightening begins are required to follow the statistical relationship shown in Figure 4.   Most recently, *Miyashita et al.* [2004] have approached this question by reexamining substorm data from the GEOTAIL spacecraft and by dividing nearly 400 substorm cases into two classes: intense and weak substorms, on the basis of the corresponding mid-latitude magnetic perturbations which were utilized as the intensity of substorms.   For the two classes, Figure 9 shows changes in the B_Z component just a few minutes after expansion onset of substorms as a function of the distance from the earth.   The "demarcation" point between the positive and negative perturbations, which can be taken as the location of the X neutral line, is located at –17 R_E and –23 R_E, for intense and weak substorms, respectively.   Having examined not only the magnetic changes but also other available

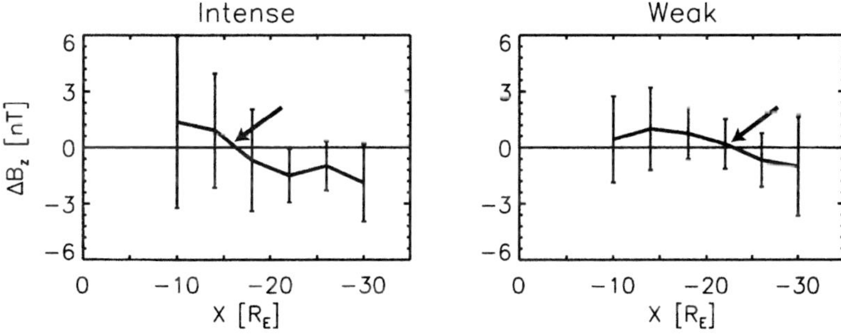

Figure 9.   Average distribution for the distance from the earth of the north-south component of the magnetic field in the magnetotail a few minutes after the expansion onset of substorms, for intense (left) and weak (right) substorms.   The location of the X neutral point is shown by an arrow for each category.   See *Miyashita et al.* [2004] for definition of the intensity of substorms.

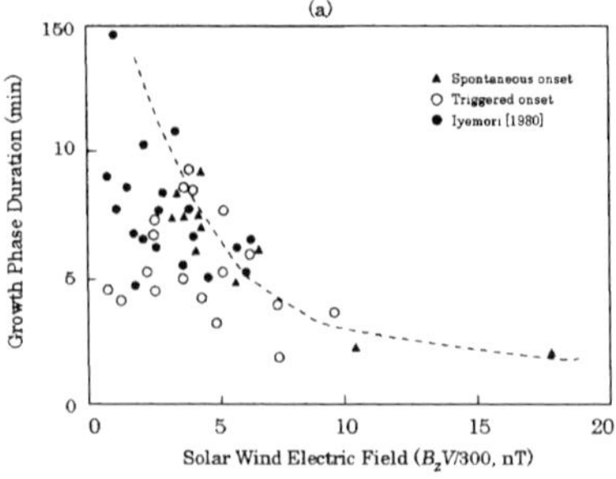
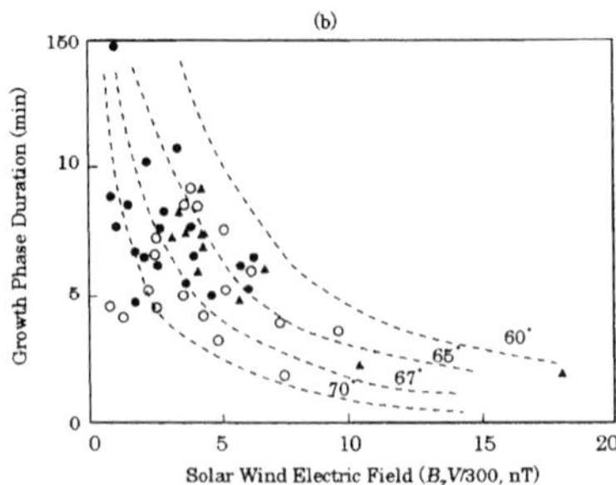

Figure 10. (a) Relationship between the duration of the substorm growth phase (*T*) and the strength of the interplanetary dawn-dusk electric field during the growth phase (*E*).   The dashed line represents the $T \cdot E$ = const. relation for spontaneous onset substorms (triangles).   After *Dmitrieva and Serggev* [1983].   (b) For the same data set, four curves of the $T \cdot E$ = const. Relation are drawn for various latitudes of the substorm expansion onset.

parameters, such as electric field and pressure and energy fluxes, it is suggested that the magnetic reconnection site approaches earthward for intense substorms.

3.3   Spontaneous Substorms and Triggered Substorms

Dmitrieva and Sergeev [1983] put forwarded an idea of separating substorms into two categories according to how their expansion onsets initiate: spontaneous onset and induced onset. The release of the accumulated energy, i.e., auroral breakup, begins either (1) spontaneously after reaching an energy threshold or (2) suddenly under the influence of a change in external conditions, such as a northward turning of IMF or a sudden change in dynamic pressure of the solar wind: e.g., *Rostoker* [1983], *Lyons* [1995], and *Zhou and Tsurutani* [2001].

Thirty isolated substorms were identified by *Dmitrieva and Sergeev*.   It was found that for cases where no noticeable variations of any kind in the solar wind existed within 10 min. of the expansion onset, the growth phase duration was inversely proportional to the average dawn-dusk electric field in the solar wind.   This category forms the outer envelope in Figure 10a.

One can argue, however, that the conclusion of *Dmitrieva and Sergeev* [1983] ignores the statistical result for the location of the substorm expansion and the intensity of substorms.   That is, they disregard the well-established results shown in Figure 4 of this paper.   Is there really a need of separating substorms into the spontaneous substorms and the externally triggered substorms particularly if the magnetosphere is always unstable to substorm expansions?   At each energy level or substorm intensity, or at each latitude (of breakup) range, an inverse relationship similar to the *Dmitrieva and Sergeev* curve for spontaneous substorms can be drawn: see Figure 10b.   In this way, there is no need of separating all the substorms into the two categories.

4. Discussion and Concluding Remarks

The present paper has intended to consider how the intensity of magnetospheric substorms is determined.   Undoubtedly this question is very important for understanding quantitatively the energy budget in the solar wind-magnetosphere system, but, for some unknown reasons, it is a missing element in current substorm theories.   Researchers in the field may simply assume that the substorm intensity is determined by input from the solar wind.   Is this really the case?   Why do a variety of substorms in size occur even for a given solar wind and IMF condition?

We all are aware that an auroral breakup is the first indicator of substorm expansion onset. It is thus intuitive to use auroral observations prior to breakups to time various substorm phenomena in the magnetosphere. It is also important to realize that an auroral breakup emanates from a preexisting auroral arc, which has shifted equatorward slowly during the growth phase of the substorm.

As this paper has contended, the probability of substorm occurrence does increase with an increase in available energy. That is, the probability tends to increase with an increase in the southward component of the IMF and, thus, the corresponding increase in the size of the polar cap. In other words, the expanding polar cap, or more practically, the shift of the poleward-most auroral forms can be regarded as a necessary condition for the occurrence of an expansion onset. However, this is not a sufficient condition to guarantee that substorm expansions will indeed occur.

The triggering mechanism has something to do with an increase in magnetic energy stored in the magnetotail. This increase can be accomplished by raising either the energy density or the volume of the magnetotail, or both, but the actual triggering action presumably follows the growth of local plasma instabilities. If there were a certain critical energy level and substorms were produced by a process that removes all excess energy toward the energy level of the ground state, then substorm intensity would always be the same for a given latitude. In fact, differences of more than 100 in the amount of substorm energy can occur at any given latitude: see Figure 6.

4.1 What is the Intensity of Substorms?

Although it appears that the a whole variety of substorm intensities can occur, previous studies indicated that, at least, for isolated substorms, there is a statistical relationship between the substorm intensity and the size of the auroral oval along which substorms take place: the oval size is determined by the IMF polarity. Figure 4 clearly shows such a relationship, although it is from ground magnetic observations. Figure 11 demonstrates that this is really the case for auroras [*Zverev and Starkov*, 1979]. The lower line indicates the latitude of auroral breakups and the upper line indicates the latitude, which is reached by discrete auroral forms at the maximum of substorm expansions, both against the largest southward component that IMF assumed before the breakup. This diagram clearly shows that the IMF not only leads to a shift of auroras toward lower latitudes before the expansion phase of substorms but also influences the maximum latitude, which is attained by active auroras at the peak of substorms.

An extensive study is in progress concerning the intensity of substorms on the basis of the ground magnetic perturbations for a large number of substorm events. Its motivation is to attempt to answer the question of whether the substorm intensity, i.e., how deep the horizontal component will decrease at high latitudes, is deterministic at the beginning of each substorm, or it is simply proportional to the duration of the expansion phase, or has no physical link with any measurable parameters because of random processes.

From a most recent study of a somewhat similar nature, Figure 12 shows the relationship between the intensity and the rise time (the duration of substorm expansions in our case) for storm sudden commencements (SSCs) [*Araki et al.*, 2004]. The upper envelope of the points delineates a clear inverse relationship between the two quantities, but the intensity does not correlate well with the rise time when the rise time is short. In the case of SSCs, their intensity (the amplitude in Figure 12) is determined by the strength of the interplanetary shocks or discontinuities whereas the rise time reflects how fast the shocks sweeps the magnetosphere. Thus, the upper envelope in Figure 12 only indicates that weak shocks tend to occur under the slow solar wind, taking long rise times, and that strong shocks are accompanied by fast speeds. In substorms, the situation is more complicated in the sense that substorms are not a result of something clearly identifiable sweeping the magnetosphere.

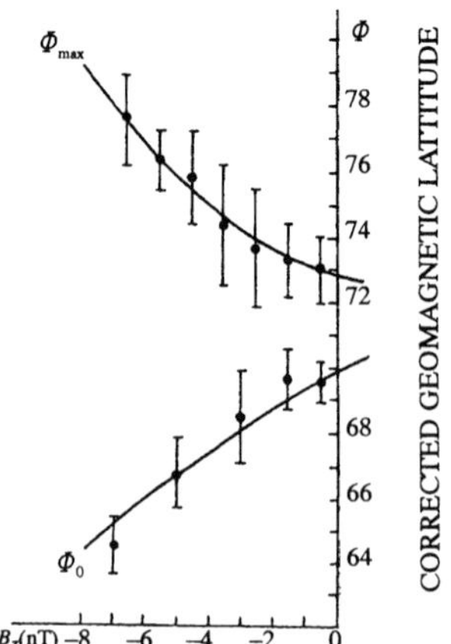

Figure 11. Latitudinal dependence of auroral breakup locations and the location of auroras at the maximum epoch of substorms, as functions of the largest southward component of the IMF preceding the expansion onset. After *Zverev and Starkov* [1979].

The study underway takes a very simple technique to address this basic question of what determines the substorm intensity. After computer-selecting a number of substorms from the *AL* index in 1991 – 1995 according to the algorithm shown in Figure 13, we examine the intensity of substorms (ΔH) as functions of the duration of the expansion phase (ΔT), the duration of the growth phase and the latitude of the auroral belt when a breakup occurs. It is quite possible that not all *AL* enhancements identified in this way are substorms [e.g., Tsurutani et al., 2004], but it is

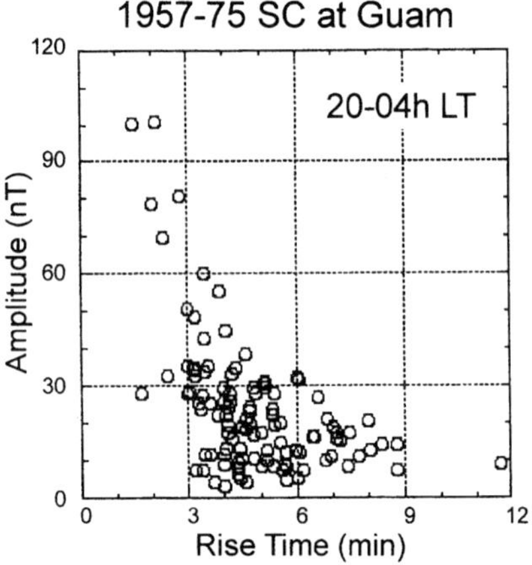

Figure 12. Scattered plot for the amplitude vs. the rise time of geomagnetic sudden commencements (SSCs) observed in the nighttime, covering a total of 17 years. After *Araki et al.* [2004].

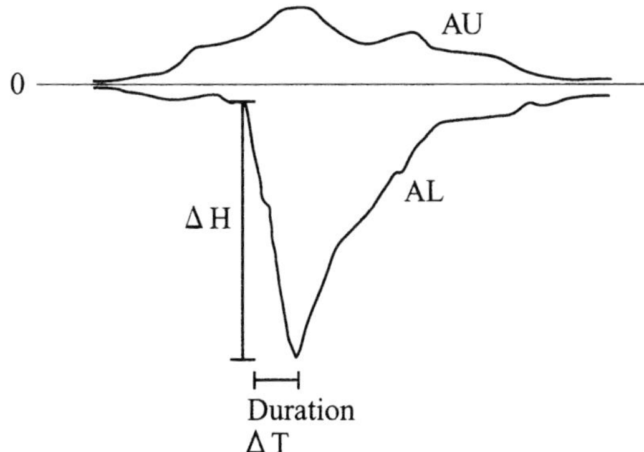

Figure 13. Schematics showing our algorithm selecting substorm events from *AL* index variations. The intensity of substorms, Δ*H*, and the duration of the expansion phase, Δ*T*, are defined.

assumed here that the number of non-substorms is relatively small, thus not influencing too much to our statistics.

Figure 14 shows the log-log distribution of all the 10,850 substorms identified for this study, representing seemingly a typical self-organized criticality phenomenon. This distribution of events has been sorted according to the north-south component of IMF. The result is shown in Figure 15 for all the substorm events, and for intense substorms only, against the B_z value averaged for 30 minutes preceding the expansion onset of substorms. We have found that changing the averaging intervals does not change the overall characters of Figure 15.

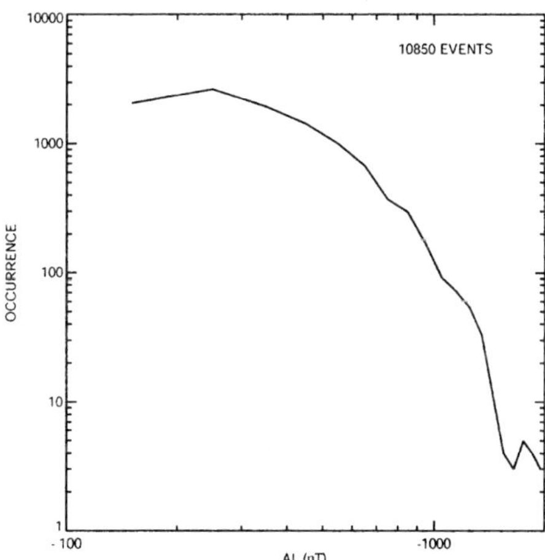

Figure 14. A log-log plot of the occurrence frequency of substorms which have the corresponding *AL* values. This covers 10,850 computer-selected events in 1991 - 1995.

190

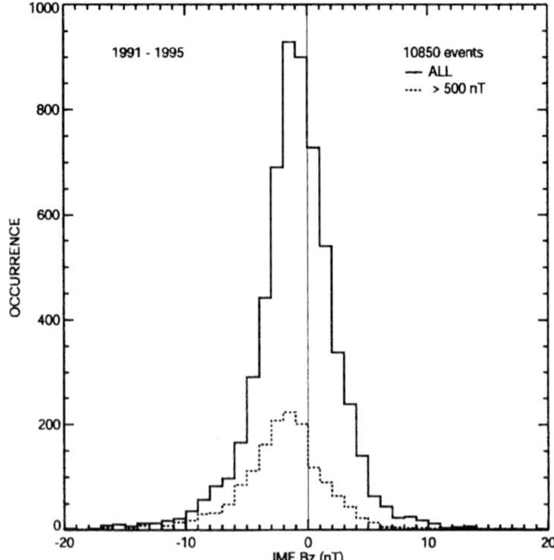

Figure 15. Occurrence frequency of substorm events sorted for the north-south component of the IMF. Two distributions are shown for all substorm events and only for intense substorms, whose peak values in absolute *AL* values exceed 500 nT.

As expected, the peak B_z values are shifted to southward IMF by 1 - 2 nT regardless of the intensity of substorms. Figure 16 shows the duration-intensity distribution, grouped into two categories according to the IMF B_z values. Although the result is still preliminary, it seems likely

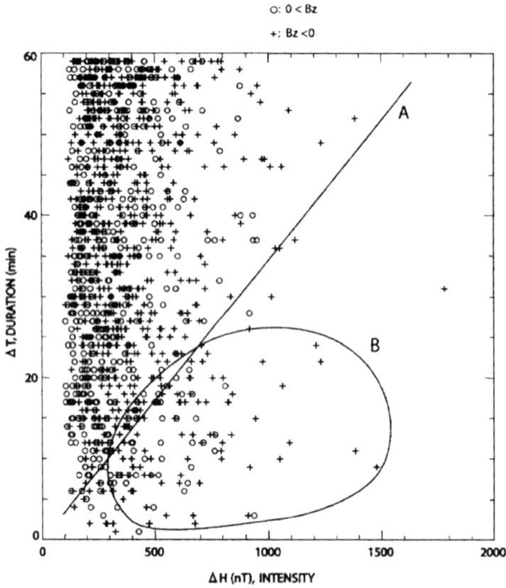

Figure 16. Scattered plots of all substorm events in the duration of the expansion phase and the intensity of substorms. The plots can be seen in terms of two groups, A and B. For the implications of A and B, see the text.

that for weak substorms, there are no particular choices for the duration of the expansive phase. That is, to reach the peak of the electrojet, the duration ranges from 5 min to 60 min. Intense substorms associated with large southward IMF, however, seem to have two subsets: A and B. One subset of intense substorms, A, results simply from the long duration, located at the upper-right corner of the scattered points, while the other subset, B, is in accordance with the duration of the expansion phase shorter than 20 minutes. There can be seen no clear-cut boundary between the two subsets, as is usually the case in this type of statistical treatment of scattered points. It seems likely, however, that in this preliminary analysis for subset A, the intensity of intense substorms greater than 500 nT tends to increase with an increase in the duration, as indicated by the straight line in Figure 16.

4.2 A Variety of Substorms

There is no doubt that substorm energy must be stored in the tail prior to expansion. The amount of energy released in the high-latitude ionosphere during a single substorm is relatively small when compared to the free energy in the magnetosphere, which is powered by the solar wind-magnetosphere dynamo. This leaves open the possibility that individual substorm processes are not led by the southward IMF, but the impulsive expansion process is instead a random internal process. As the present paper demonstrates, however, the magnetosphere, is indeed an "efficient" system, in that the energy budget is ordered by the solar wind. The "expanding" polar cap is one of the necessary conditions for the occurrence of an expansion onset. A reasonable compromise is that the solar wind provides the magnetosphere with first-order, global conditions, including the probability of substorm occurrence and the rough magnitude of substorms (if they occur), but where and how exactly a substorm initiates depends on the local conditions of the magnetosphere-ionosphere system. It would be impossible on the basis of the presently available observational techniques, or even with a perfect network of observations, to predict exactly where a chain of substorm expansion actions will begin.

As far as auroral features are concerned, it seems unlikely that a specific signal leads to an expansion onset of an auroral substorm. Several brightening spots are normally observed during the polar cap expansion, as seen in Figure 7. Which particular brightening actually causes the effective expansion into the entire sky cannot be ascertained since the process appears to be utterly random. Some of the brightening spots are simply fluctuations, which are not related to a breakup.

Another important issue to consider in seeking the generation mechanism of substorms pertains to the great variety of substorms in terms of the released energy at the polar ionosphere. In this paper we have used the auroral electrojet to measure the substorm intensity or energy. It is not too difficult to argue, however, that it represents only a small portion of the total energy of magnetospheric substorms and naturally we need to substantiate what we have discussed in this paper on the basis of other parameters. Most of the substorm energy may in fact be released in the form of the ring current in the inner magnetosphere and plasmoids that leave the magnetosphere. It is quite possible that the partition of energy dissipated into the polar ionosphere and that deposited in the magnetosphere are not always in a constant proportion, so that it may be a mistake to evaluate the size of magnetospheric substorms relying on the quantity that is measured at polar ionosphere.

4.3 Pseudobreakups or Simply Noise?

A number of studies [e.g., *Koskinen et al.*, 1993; *Ohtani et al.*, 1993] have attempted to understand the nature of pseudobreakups. According to these studies, a pseudobreakup commonly exhibits the local signatures of an auroral expansion onset and produces a weak electrojet system in the midnight sector, but they do not develop to a full-scale substorm expansion.

192

However, all these suppressed signatures are practically the same as those of weak substorms. It is argued that we should differentiate pseudobreakups and truly weak substorms in terms of their generation mechanisms.

Generically, weak substorms are weak simply because the energy delivered from the solar wind to the magnetosphere is minuscule or minor. Even weak substorms have all the observable signatures that intense substorms have. It is thus a matter of quantity (or size), not of quality. As evident from Figure 4, weaker substorms tend to occur at higher latitudes, along the contracted auroral oval. These weak substorms cannot be easily identified as pseudobreakups. On the other hand, pseudobreakups appear to be weak because a choking mechanism works, or perhaps because the magnetosphere-ionosphere system has not yet been ready to accept a full substorm, even though the energy extracted from solar wind-magnetosphere coupling is great.

It is also quite possible that scientists have confused pseudobreakups with fluctuations (or noise) during auroral activity. Auroras, as natural phenomena, are almost always active and fluctuate except during extremely quiet periods. This implies that auroral brightening or fluctuations almost always occur, not only before substorm expansion, but even after expansion onset as well. It is inappropriate to refer to minor auroral fluctuations or noise that occur during the growth phase and throughout the duration of the event as "pseudobreakups."

Acknowledgements

I would like to thank my substorm colleagues for their critical discussions during the preparation of this manuscript. I note, however, that I was not able to subscribe consistently to all of their comments because of mutual "conflict" in their views. I particularly thank G. Consolini and B. T. Tsurutani for the helpful discussions in finalizing the manuscript. I also would like to acknowledge the able assistance of Y. Kadowaki in the statistical analysis. The work presented in this paper was supported in part by the Ministry of Education, Culture, Sports, Science and Technology under a Grant-in-Aid for Scientific Research.

References

Akasofu, S.-I., The development of the auroral substorm, *Planet. Space Sci., 12*, 273, 1964.

Akasofu, S.-I., The roles of the north-south component of the interplanetary magnetic field on large-scale auroral dynamics observed by the DMSP satellites, *Planet. Space Sci., 23*, 1349, 1975.

Akasofu, S.-I., and Y. Kamide, Substorm energy, *Planet. Space Sci., 24*, 223, 1976.

Akasofu, S.-I., P. D. Perreault, F. Yasuhara, and C.-I. Meng, Auroral substorms and the interplanetary magnetic field, *J. Geophys. Res., 78*, 7490, 1973.

Araki, T., T. Takeuchi, and Y. Araki, Rise time of geomagnetic sudden commencements – Statistical analysis of ground magnetic data, *Earth Planet Space, 54*, in press, 2004.

Baker, D. N., T. I. Pulkkinen, J. Büchner, and A. J. Klimas, Substorms: A global instability of the magnetosphere-ionosphere system, *J. Geophys. Res., 104*, 14601, 1999.

Baumjohann, W., M. Hesse, S. Kokubun, T. Mukai, T. Nagai, and A. A. Petrukovich, Substorm depolarization and recovery, *J. Geophys. Res., 104*, 24995, 1999.

Birn, J., M. Hesse, G. Haerendel, W. Baumjohann, and K. Shiokawa, Flow braking and the substorm current wedge, *J. Geophys. Res., 104*, 19895, 1999.

Borovsky, J. E., R. J. Nemzek, and R. D. Belian, The occurrence rate of magnetospheric-substorm onsets: Random and periodic substorms, *J. Geophys. Res., 98*, 3807, 1993.

Caan, M. N., R. L. McPherron, and C. T. Russell, Substorm and interplanetary magnetic field effects on the geomagetotail lobes, *J. Geophys. Res., 80*, 191, 1975.

Chang, T., Low-dimensional behavior and symmetry breaking of stochastic systems near criticality - can these effects be observed in space and in the laboratory?, *IEEE Trans. of*

Plasma Sci., 20, 691, 1992.

Chang, T., S. W. Y. Tam, C. C. Wu, and G. Consolini, Complexity, forced and/or self-organized criticality, and topological phase transitions in space plasma, *Space Sci. Rev., 107*, 425, 2003.

Chapman, S. C., and N. W. Watkins, Avalanching and self-organized criticality, a paradigm for geomagnetic activity?, *Space Sci. Rev., 95*, 293, 2001.

Chapman, S. C., N. W. Watkins, R. O. Dendy, P. Helander, and G. Rowlands, A simple avalanche model as an analogue for magnetospheric activity, *Geophys. Res. Lett., 25*, 2398, 1998.

Clauer, C. R., and R. L. McPherron, Variability of mid-latitude magnetic parameters used to characterize magnetospheric substorms, *J. Geophys. Res., 79*, 2898, 1974.

Consolini, G., and T. Chang, Magnetic field topology and criticality in geotail dynamics: relevance to substorm phenomena, *Space Sci. Rev., 95*, 309, 2001.

Craven, J. D., and L. A. Frank, Diagnosis of auroral dynamics using global auroral imaging with emphasis on large-scale evolutions, in *Auroral Physics*, edited by C.-I. Meng, M. J. Rycroft, and L. A. Frank, p. 275, Cambridge Univ. Press, 1991.

Dmitrieva, N. P., and V. A. Sergeev, The spontaneous and induced onset of the expansive phase of a magnetospheric substorm and the duration of its preliminary phase, *Geomag. Aeronomy* (Eng. Trans.), *23*, 380, 1983.

Fairfield, D. H., and R. P. Lepping, Simultaneous measurements of magnetotail dynamics by IMP spacecraft, *J. Geophys. Res., 86*, 1396, 1981.

Kamide, Y., and S.-I. Akasofu, Latitudinal cross section of the auroral electrojet and its relation to the interplanetary magnetic field polarity, *J. Geophys. Res., 79*, 3755, 1974.

Kamide, Y., and J. D. Winningham, A statistical study of the 'instantaneous' nightside auroral oval: The equatorward boundary of electron precipitation as observed by the Isis 1 and 2 satellites, *J. Geophys. Res., 82*, 5589, 1977.

Kamide, Y., P. D. Perreault, S.-I. Akasofu, and J. D. Winningham, dependence of substorm occurrence probability on the interplanetary magnetic field and on the size of the auroral oval, *J. Geophys. Res., 82*, 5521, 1977.

Kauristie, K., T. I. Pulkkinen, A. Huuskonen, R. J. Pellinen, H. J. Opgenoorth, D. N. Baker, A. Korth, and M. Syrjäsuo, Auroral precipitation fading before and at substorm onset: ionospheric and geostationary signatures, *Ann. Geophys., 15*, 967, 1997.

Klimas, A. J., D. Vassiliadis, D. N. Baker, and D. A. Roberts, The organized nonlinear dynamics of the magnetosphere, *J. Geophys. Res., 101*, 13089, 1996.

Koskinen, H. E., R. E. Lopez, R. J. Pellinen, T. I. Pulkkinen, D. N. Baker, and T. Bösinger, Pseudobreakup and substorm growth phase in the ionosphere and magnetosphere, *J. Geophys. Res., 98*, 5801, 1993.

Lui, A. T. Y., A synthesis of magnetospheric substorm models, *J. Geophys. Res., 96*, 1993, 1991.

Lui, A. T. Y., What determines the intensity of magnetospheric substorms?, *J. Atmos. Terr. Phys., 55*, 1123, 1993.

Lui, A. T. Y., and J. S. Murphree, A substorm model with onset location tied to an auroral arc, *Geophys. Res. Lett., 25*, 1269, 1998.

Lui, A. T. Y., C. D. Anger, and S.-I. Akasofu, The equatorward boundary of the diffuse aurora and auroral substorms as seen by the Isis 2 auroral scanning photometer, *J. Geophys. Res., 80*, 3603, 1975.

Lui, A. T. Y., S. C. Chapman, K. Liou, P. T. Newell, C.-I. Meng, M. Brittnacher, and G. K. Parks, Is the dynamic magnetosphere an avalanching system?, *Geophys. Res. Lett., 27*, 911, 2000.

Lyons, L. R., A new theory for magnetospheric substorms, *J. Geophys. Res., 100*, 19069, 1995.

McPherron, R. L., and D. N. Baker, Factors influencing the intensity of magnetospheric substorms, *J. Atmos. Terr. Phys., 55*, 1091, 1993.

Miyashita, T., Y. Kamide, S. Machida, K. Liou, T. Mukai, Y. Saito, C.-I. Meng, and G. K. parks, The difference between intense and weak substorms in magnetotail variations, *J. Geophys. Res.*, submitted, 2004.

first-principle models has many limitations. They often require too much computer resources and yet remain imprecise, for instance, in determining the timing of the substorm onset.

Figure 1. Representation of the magnetosphere in a data-derived approach. Input parameters are represented here by the solar wind speed v, southward component of the interplanetary magnetic field B_s, and dynamical pressure P_{dyn}, while the output is represented by the auroral indices AL and AE reflecting the substorm activity, and the disturbance storm-time index D_{st}.

One can propose another approach to complex system modeling. It is a more pragmatic approach without recourse to the tools of classical physics. For instance, although the dynamics of a human body can in principle be described in terms of quantum mechanics and electrodynamics, physicians and sociologists quite seldom use that type of description in their practice. Let us consider instead our system, the magnetosphere, as a black box with some input and output (Figure 1) and let us try to create a model of system's dynamics directly from data, using the techniques of signal processing, nonlinear dynamics and statistics. This is an empirical or *data-derived approach*. It often provides the optimum level of resolution and strongly complements the first-principle models being more robust and efficient in many applications.

2. Linear Filters

The very first look at the input and output time series of substorm data, an example of which is shown in Figure 2, reveals a clear correlation between the solar wind input (inductive electric field parameter vB_s) and the magnetospheric output (AL index). This clear correlation gave rise to the first family of data-derived models of the solar wind-magnetosphere coupling, namely the linear filters [*Burton et al.* 1975, *Iyemori et al.*, 1979; *Clauer et al.*, 1983; *Bargatze et al.*, 1985].

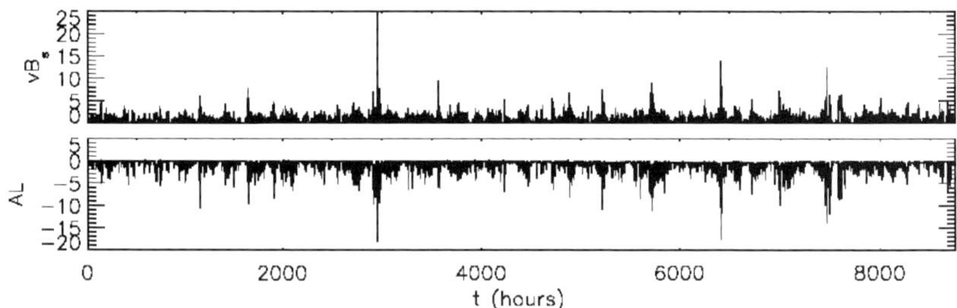

Figure 2. vB_s - AL time series with 1 minute resolution compiled by *Weigel et al.* [2001] for the whole year 1998.

In linear filters the output is supposed to be a linear combination of the input with different time delays

$$O(t) = \int_0^\infty d\tau f(\tau) I(t-\tau). \tag{2.1}$$

If the response is indeed linear, the filter function f will be independent of the activity level. However, already the early studies [*Bargatze et al.*, 1985] revealed the significant dependence of linear filters on the activity level and thus suggested the nonlinearity of the magnetospheric response to the solar wind loading.

3. Dynamical Chaos Hypothesis

Nonlinearity often results in phase divergence, as it takes place, for example, in a nonlinear pendulum. Such a divergence in bounded systems may be a mechanism of dynamical chaos [e.g., *Tabor*, 1989]. The hypothesis that the magnetosphere may be in a regime of dynamical chaos appeared to explain the seeming randomness of the auroral index time series. Note here that most dynamical chaos models deal with autonomous systems, which is not the case for magnetospheric activity represented by the auroral indices, as is clear from Figure 2. However, the closer examination [e.g., *Kamide and Baumjohann*, 1993] reveals that the substorm current system at the ionospheric level, which contributes to the auroral index dynamics, consists of two subsystems, DP1 and DP2. While the DP2 system is actually directly driven by the solar wind input, the DP1 system shows rather autonomous "unloading" behavior.

It was suggested [*Vassiliadis et al.*, 1990; *Roberts*, 1991; *Shan et al.*, 1991a, 1991b; *Pavlos et al.*, 1992, 1994; *Sharma et al.*, 1993] that the complex dynamics of the magnetosphere is *self-organized* and as a result it is controlled by a few effective degrees of freedom, while its complexity arises from the dynamical chaos effects. If that is true than the form of the attractor of the magnetospheric dynamics may be reconstructed using the time delay technique. It can be elucidated by the simple chaotic map model, given by the recurrent relation

$$x_{n+1} = Kx_n \mod 1, \tag{3.1}$$

where $mod1$ denotes taking the fractional part of the number Kx_n, and K is a constant number. If $K<1$, the relation (3.1) describes the sequence, which converges to zero. In contrast, if $K>>1$, then for most initial values x_0, the sequence (3.1) looks quite random as shown in Figure 3a. This type of behavior is also called deterministic chaos because of the deterministic rule (3.1). The rule can be inferred from the series x_n if one plots x_n versus x_{n-1} as shown in Figure 3b

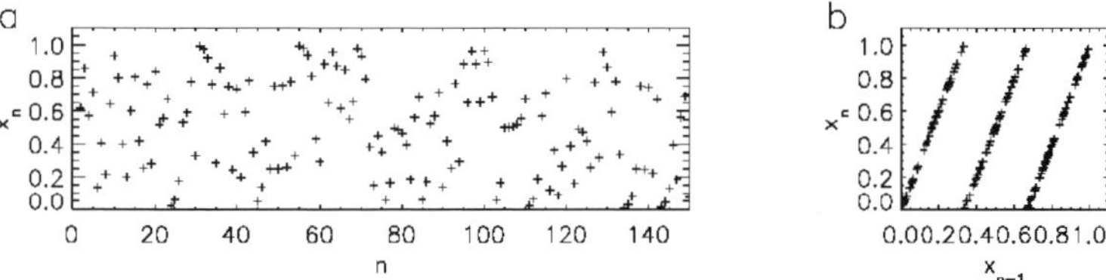

Figure 3. Pseudo-random series x_n generated by the map (3.1) with $x_0 = \sqrt{15}$ and $K=3$ (a), and the reconstruction of the deterministic rule (3.1) using time delays (b).

This simple example has several important implications. First, it shows that the seemingly random dynamics of a nonlinear low-dimensional system can be reconstructed using the *time-delay embedding technique*, that is by plotting the original series in an artificial phase space formed by the original variable and others constructed from the same variable using various time delays. Such an approach was proposed by *Packard et al.* [1980] and *Takens* [1981]. Second, the regular structure of the reconstructed attractor in Figure 3b suggests that the time series x_n as a function of the discrete "time" n can be predicted using local approximation of the attractor. The

198

corresponding technique of *locally linear filters* was elaborated [e.g., *Casdagli*, 1992] and then applied to forecasting of storms and substorms [*Vassiliadis et al.*, 1995; *Valdivia et al.*, 1996] along with other classes of low-dimensional deterministic models [*Baker et al.*, 1990; *Klimas et al.*, 1992; *Hernandes et al.*, 1993; *Horton and Doxas*, 1998; *Weigel et al.*, 1999]. First estimates of the effective dimension D of the magnetosphere, the so-called correlation dimension [*Grassberger and Procaccia*, 1983], gave very optimistic results with D ranging between 2 and 4 [*Vassiliadis et al.*, 1990; *Roberts*, 1991; *Shan et al.*, 1991]. Moreover, the fact that D was a noninteger number suggested the "strange" character of the attractor [*Lorenz*, 1963], which might explain, in turn, the multiscale features of the substorm structure and dynamics.

4. Self-organized Criticality Models

The dynamical chaos hypothesis was based on one important assumption of low effective dimension of the magnetosphere. However, soon after the first estimates of the correlation dimension of the magnetosphere, *Prichard and Price* [1992] showed that those earlier estimates were caused by the long autocorrelation times of the system rather than by the low-dimensional dynamics. It was shown in particular that the modified correlation integral [*Theiler*, 1986] with excluded pairs of points, which differ in time less than by the autocorrelation time, does not converge for AE index time series. Two possible explanations of that negative result were proposed. There might be strong influence of the solar wind input (most of the dynamical chaos concepts were formulated for autonomous systems) or truly multiscale behavior of the magnetosphere with many excited degrees of freedom. The latter point of view in the form of another paradigm of the so-called *self-organized criticality* (SOC) was strongly motivated by the power-law spectra of the magnetospheric activity. They were first inferred from *AE* index data by *Tsurutani et al.* [1990]. Then *Ohtani et al.* [1995] revealed similar properties of magnetic field fluctuations in the geomagnetotail, consistent with intermittent energy transport phenomena studied by *Angelopoulos et al.* [1994]. The examples of power-law spectra in the form of the burst life time probability distribution [*Freeman et al.*, 2000] and the singular spectrum of solar wind and magnetospheric data [*Sitnov et al.*, 2000] are shown in Figure 4.

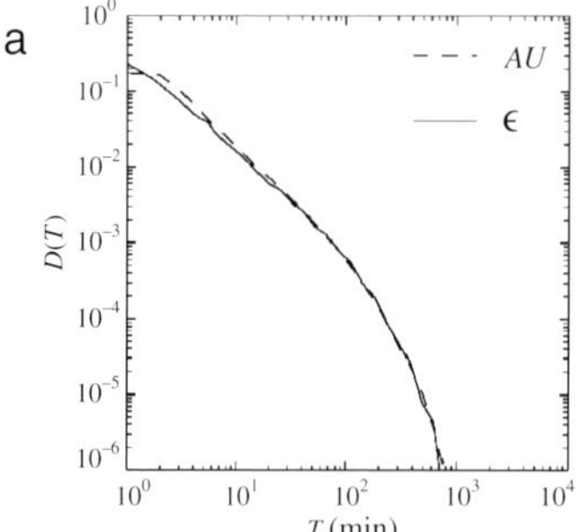
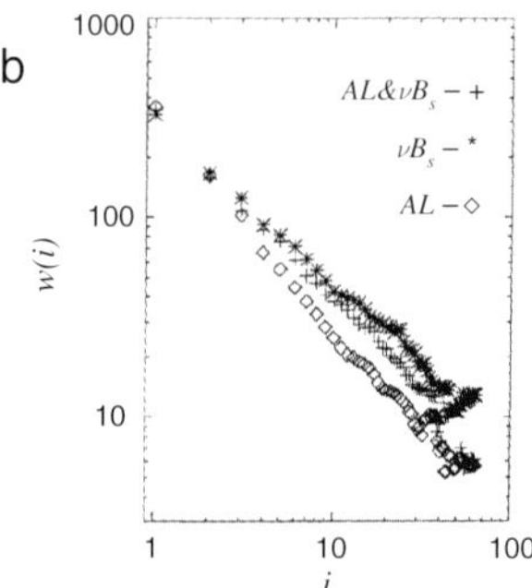

Figure 4. Power-law spectra of solar wind and magnetospheric activity in the forms of (a) life time T probability distribution $D(T)$ of the auroral index AU and Akasofu parameter ε [*Freeman et al.*, 2000] (a burst lifetime, T, is the duration for which the measurement, $AU(t)$ exceeds a given threshold value) and (b) the singular spectrum of AL and vB_s time series as well as their combined set [*Sitnov et al.*, 2000] inferred from first 15 intervals of *Bargatze et al.* [1985] database (for details see equation (5.2) and related discussions in section 5).

Power-law spectra usually indicate that fluctuations of all scales give comparable contributions to dynamics and the smaller scales cannot be simply averaged out. It is interesting to note here that stochastic time series with $1/f^\alpha$ power spectra, known as the colored noise, have a finite correlation dimension $D = 2/(\alpha - 1)$ [*Osborne and Provenzale*, 1989] and thus may mimic the dynamical chaos effects in the magnetosphere [*Prichard and Price*, 1992]. To resolve this uncertainty *Theiler* [1991] proposed the method of surrogate data, in which the correlation dimension of the original data is compared with the stochastic time series with the same power spectrum and autocorrelation. The surrogate time series can be formed, for instance, by randomizing the phases of the Fourier transform of the original time series. Using this technique the conclusion on the absence of a converged modified correlation dimension of the magnetosphere inferred from the AE time series has been confirmed [*Price and Prichard*, 1992; *Prichard and Price*, 1992]. Note however that *Roberts* [1991] and *Pavlos* [1992] showed that the data trajectories of AE time series in embedding space may still be drastically different from those of the color noise (for details and further discussions see [*Pavlos et al.*, 1999] and refs. therein, [*Prichard*, 1995] and reviews [*Sharma*, 1995] and [*Klimas et al.*, 1996]).

In 1987 Bak, Tang and Wiesenfeld proposed an elegant model (hereafter BTW), which appeared to explain the ubiquitous nature of power-law fluctuations in different systems. That model as well as its modifications have been later applied by *Consolini* [1997], *Chapman et al.* [1998], and *Uritsky and Pudovkin* [1998] to explain the power-law spectra in the magnetosphere. The original BTW model [*Bak et al.*, 1987] represents a mathematical sandpile, that is a grid of cells, each of which may accumulate some amount of grains Z due to the external loading. When the parameter Z exceeds some threshold value Z_c, which is determined by either the critical slope or energy accumulated in the cell, 4 sand units are transferred to 4 closest cells (the number of cells, 4, is for the specific 2D BTW model):

$$z_{t+1}(x,y) = z_t(x,y); \qquad z \le z_c, \qquad (4.1)$$

$$z_{t+1}(x \pm 1, y \pm 1) = z_t(x \pm 1, y \pm 1) + 1; \quad z_{t+1}(x,y) = z_t(x,y) - 4, \qquad z > z_c, \qquad (4.2)$$

Bak et al. [1987] argued that this model reveals the power spectra distributions of avalanches in their size and energy independent of the details of the driving process. To highlight the latter property and distinguish the process from conventional critical phenomena [e.g., *Stanley*, 1971] also known as the second order phase transitions, they called it the self-organized criticality. SOC has been documented in various systems, including rice piles, type II superconductors, droplet formation and a multitude of computer models [e.g., *Jensen*, 1998 and refs. therein].

SOC as a new paradigm for geomagnetic activity has been further tested by *Freeman et al.* [2000], *Consolini and Chang* [2001], *Chapman and Watkins* [2001], and *Watkins et al.* [2001]. In particular, *Consolini and Chang* [2001] discussed a modification of the SOC concept, taking into account the variable solar wind input (the so-called forced self-organized criticality). *Freeman et al.* [2000] and *Chapman and Watkins* [2001] discussed the possibility to separate the autonomous SOC behavior reflected in power-law spectra of auroral indices from its part, possibly related to the DP2 system, which is directly driven by the solar wind. *Watkins et al.* [2001] elucidated that SOC is not an aspect of the global magnetosphere but relevant more locally to the magnetotail. *Takalo et al.* [1999a, 1999b], *Klimas et al.* [2000] and *Uritsky et al.* [2002] studied a class of continuum running avalanche models of the tail current sheet based on a resistive MHD model.

Due largely to the amazing simplicity and seeming universality of the original BTW model, the SOC concept has become very popular in complexity studies. However, it was soon realized that the BTW model is not as ubiquitous as expected and it does not describe in particular the sand dunes themselves. It has been shown [*Nagel*, 1992] that sand behaves in a manner more reminiscent of a first order phase transition than of a second one. Similarly, in the magnetopshere, in spite of the scale-invariant behavior of some parameters, there are clear signatures of characteristic time scales associated with storms (~ days) and substorms (~ hours). More rigorous results were obtained by *Borovsky et al.* [1993], *Prichard et al.* [1996], and *Smith et al.* [1996], who showed, in particular, that the intensity and inter-substorm interval for one-half of substorms have the probability distribution with a well-defined mean. This apparent contradiction between scale invariance in index spectra on the one hand and the appearance of time and magnitude scales in substorm data on the other hand stimulated *Chapman et al.* [1998] to propose for explanation of the magnetospheric activity an advanced avalanche model elaborated by *Dendy and Helander* [1998]. In contrast to the classical BTW cellular automation, the new model was continuous in height and slope with the threshold condition governed by a probabilistic rule. It demonstrated two types of avalanches, internal that did not reach the edges of the system, and systemwide ones. Internal avalanches reproduced the classical SOC picture with the absence of an intrinsic scale. In contrast, the systemwide discharges had the probability distribution of the time intervals between each discharge and the next with a well-defined mean. Those systemwide avalanches were interpreted by *Chapman et al.* [1998] as substorms.

It is worth noting here that the fundamental physical processes in magnetospheric plasmas underlying their complex dynamical behavior may not allow scale-invariance in the form of avalanches. For instance, in contrast to current disruption phenomena [e.g., *Lui*, 2002 and refs. therein], which indeed resemble avalanches in a ricepile, the processes of magnetic reconnection [e.g., *Birn et al.*, 2001] reveal scale-invariance only near the singular X-line, whereas the minimum size L of a plasmoid formed by reconnection is too large $L \geq \sim 4R_E$ [*Ieda et al.*, 1998] to be utilized in any granular multiscale model of the magnetospheric activity.

Another important feature of magnetospheric dynamics, which distinguishes it from SOC, is its strong dependence on the solar wind input. This dependence is evident from Figure 1 and is confirmed by the successful use of linear and nonlinear filters in forecasting the magnetospheric activity as discussed in more detail in the sections 2 and 3. Moreover, even the multiscale features of this activity are strongly affected by the solar wind input. In particular, such parameters as the fractal dimension [*Uritsky and Pudovkin*, 1998] or κ-exponent in the sign-singular analysis [*Consolini and Lui*, 1999] show clear dependence on the substorm phase, which is controlled in turn by the solar wind loading. Thus, the original 1987 SOC paradigm does not seem to be quite useful for modeling and particularly forecasting of the magnetospheric activity. It introduces however a very promising *probabilistic description* of this activity and highlights its scale-invariant features, which were explicitly demonstrated by *Lui et al.* [2000] and *Uritsky et al.* [2002] on the basis of Polar data. The most important lesson of SOC, according to *Jensen* [1998], is that, in a great variety of systems, it is misleading to neglect the fluctuations.

5. Data-derived Picture of Solar Wind-Magnetosphere Coupling: Analogy with Phase Transitions

Another class of magnetospheric activity models, which reconciles the hypothesis of dynamical chaos and the seemingly alternative SOC interpretation, resulted from the attempts to improve the low-dimensional description of the solar wind-magnetosphere interaction and to elucidate in particular the role of the solar wind driving. In contrast to the previous works based on the assumption that the dynamics of the magnetosphere is autonomous [*Vassiliadis et al.*, 1990; *Roberts*, 1991; *Shan et al.*, 1991b; *Sharma et al.*, 1993], an attempt was made [*Sitnov et al.*, 2000] to construct a low-dimensional input-output relationship, or improve it as compared to its simplest form of linear filters. For that purpose the data were studied in the time-delay space involving both the output of the system (the auroral index AL) and its solar-wind input (the parameter vB_s)

$$\mathbf{X} = \left(\mathbf{O}_t^T, \mathbf{I}_t^T\right) = \left(O_t,...,O_{t-(M-1)}, I_t,...,I_{t-(M-1)}\right), \quad I_t = vB_s(t)/\sigma_{vB_s}, \quad O_t = AL(t)/\sigma_{AL}, \quad (5.1)$$

with both input and output parameter being normalized by their standard deviations σ. The resulting expanded time series were additionally sorted using the singular spectrum analysis (SSA) [*Broomhead and King*, 1986] to reveal their linear combinations, which are essential to reproduce the dynamics of the system adequately. In the SSA the matrix X is expanded in a series of projections P_i through the singular value decomposition [e.g., *Press et al.*, 1992]

$$\mathbf{X} = \mathbf{UWV}^T, \quad P_i = \left(\mathbf{XV}\right)_i, \quad (5.2)$$

where $\mathbf{W}$ is the diagonal matrix, whose values w_i reflect the significance of projections P_i so that the spectrum w_i resembles the corresponding Fourier and wavelet spectra with the data-derived basis functions [e.g., *Preisendorfer*, 1988].

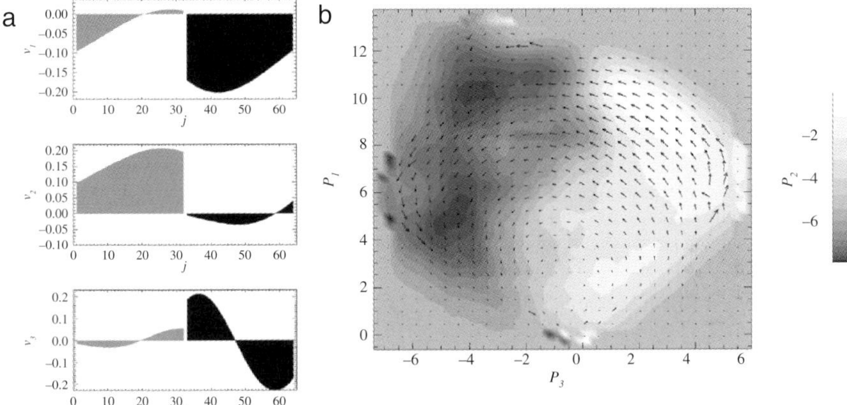

Figure 5. Eigenvectors corresponding to three largest SSA eigenvalues w_i, i=1,2,3 for embedding (5.1) with M=32 inferred from first 20 intervals of *Bargatze et al.* [1985] database (a) and the 2D manifold of the magnetospheric dynamics in the space of the corresponding projections P_i. The arrows show the data-derived circulation flows of the system ($\dot{P}_1, \dot{P}_3$).

Sitnov et al. [2000] found that, in contrast to the correlation dimension, and in spite of the fact that the SSA spectrum is a power-law, the more general dimension estimate, the so-called coast-line dimension D_f [e.g., *Abarbanel et al.*, 1993] saturates as a function of the embedded dimension $2M$ in the region $D_f \in (2,3)$, and this saturation persists with an increase in the number of the data points for the largest scales. This implies that the magnetosphere trajectory in the embedded input-output phase space lies close to some two-dimensional surface. Subsequent attempts to plot that surface as well as the evolution of the magnetosphere on it gave surprisingly interesting results shown in Figure 5.

The substorm dynamics of the magnetosphere appears in this data-derived picture as a counter-clockwise motion of a point on a two-level surface in the 3D space formed by the papameters P_1, P_2, amd P_3. The corresponding projectors V_1, V_2, and V_3 of the matrix V are shown in Figure 5a. Together with the general structure (5.1) of the matrix $\mathbf{X}$ they demonstrate that the subspace (P_1, P_2, P_3) is formed by two functions of the input P_1 and P_3 (integrated and differential parameter vB_s, top and bottom panels in Figure 5a, respectively) and a function of the output (integrated AL, middle panel). Figure 5b reveals two interesting features of substorm dynamics. First, the substorm onset (the transition from the upper level to the lower one) is associated with the reduction of the solar wind-magnetosphere coupling (saturation of the input P_1 and transition from positive to negative values of the parameter P_3) as suggested and interpreted in detail by *Lyons et al.* [2003]. Second, the whole surface in Figure 5b resembles the temperature-pressure-density diagram of conventional (non-self-organized) water-steam phase transitions [*Stanley*, 1971] shown in Figure 6a. This diagram has a special *critical point* where density jumps disappear, although jumps or singularities are observed for the second derivatives of the thermodynamic potentials such as the heat capacity (Figure 6b). It is also known [e.g., *Stanley*, 1999; *Kadanoff*, 1999] that near the critical point the fluctuations of the system become multiscale with power-law spectra.

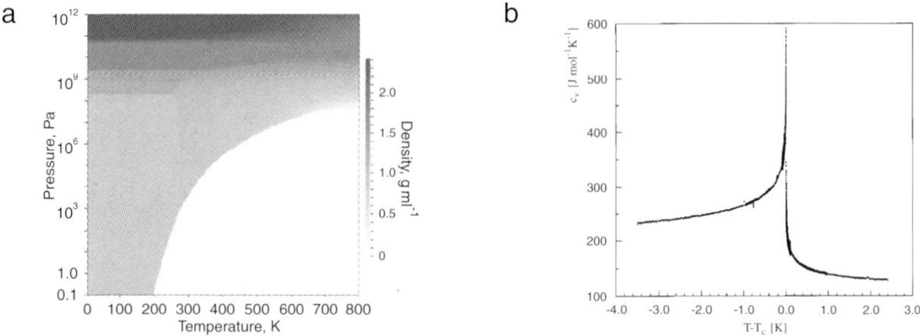

Figure 6. (a) Temperature-pressure-density diagram for water-steam transitions with the critical point (T, P, n)=(647K, 22Mpa, 322 kg/m$_3$) (http://www.lsbu.ac.uk/water/images/den.gif). (b) Specific heat C_V of SF_6 near the critical point [*Haupt and Straub*, 1999].

An important feature of the second order phase transitions, which distinguishes them from other scale-invariant processes and in particular those in the BTW model of SOC, is a series of scale-invariant relationships different from the scale distributions such as AU and AL spectra shown in Figure 4. They relate different parameters of the system such as the temperature fluctuations near the critical point and the corresponding density fluctuations

$$n - n_c \propto \left| T / T_c - 1 \right|^{\beta}, \quad T / T_c - 1 \to 0 -, \tag{5.3}$$

Here the parameter β is a critical exponent. *Sitnov et al.* [2001] computed an analog of the critical exponent near the effective critical point $\left(P_1, P_2, P_3 \right) = \left(0, 0, 0 \right)$ in the form of the lower envelope of the derivative dP_2 / dt as

$$\min \left(dP_2 / dt \right) \propto P_1^{\beta_*}, \tag{5.4}$$

This envelope is shown in Figure 7. It was shown also that within the framework of the simplest mean-field model of phase transition dynamics, the so-called dynamical Ising model [*Zheng and Zhang*, 1998], the exponent β_* is connected to the classical exponent β by the relation $\beta_* = 3\beta$. Note here that the existence of the multiscale input –output relationship of the type of (5.4) is consistent with the results of *Freeman et al.* [2000] and *Sitnov et al.* [2000] shown

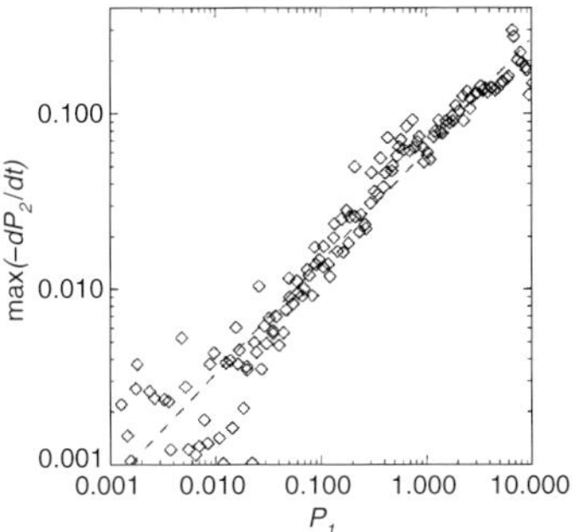

Figure 7. Scale invariant input-output relation of the type of (5.3) with $\beta_* \approx 0.64$ inferred from the SSA analysis of substorm dynamics [*Sitnov et al.*, 2001].

in Figure 4, which reveal close resemblance between solar wind and magnetospheric fluctuation spectra. Although the results of *Freeman et al.* [2000] and similar results by *Hnat et al.* [2002] reveal nearly same spectra of solar wind and auroral indices suggesting that the magnetosphere works as an exhaustive pipe for solar wind and the input-output exponent β should be simple rational number, the results of *Sitnov et al.* [2000] (Figure 4b) as well as the extended analysis of *Hnat et al.* [2002] performed in [*Hnat et al.*, 2003] reveal quite considerable difference between the input and output spectra.

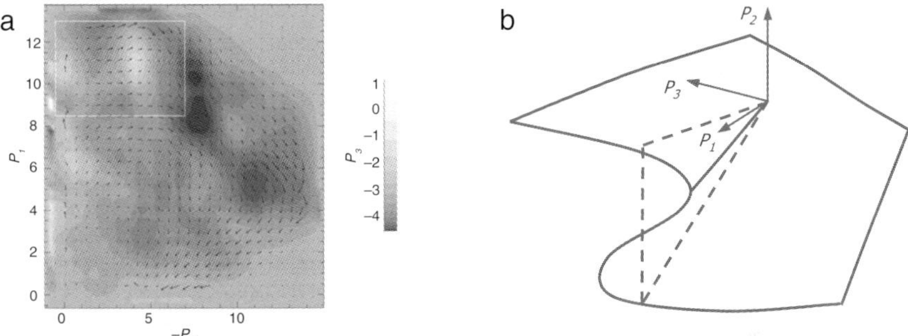

Figure 8. (a) Hysteresis phenomenon (white rectangle) inferred from the SSA analysis of substorm data [*Sitnov et al.*, 2001]. (b) Interpretation of Figure 7a as a phase separation surface of nonequilibrium phase transitions; dashed lines show its analog in the equilibrium case similar to Figure 5a.

The dynamical and non-equilibrium nature of magnetospheric phase transitions during substorms is reflected by the *hysteresis* phenomena, when the system may be in two or more different states under the same quasi-static conditions such as, for instance, its temperature and

204

pressure. Then the analog of the temperature-pressure-density diagram differs from the equilibrium one shown in Figure 6a and resembles more the one shown in Figure 8b. The original projection of the substorm data shown in Figure 5b cannot reveal hysteresis as the corresponding intervals in data would be folded. In fact, those intervals were removed from the main data set to reveal the two-level surface; however, we show below that the analog of Figure 5b can be obtained without that rather artificial procedure. Another projection shown in Figure 8a reveals the hysteresis phenomena in substorm dynamics.

Hysteresis is closely related to the *metastability* property of the system, because one of many states available at given parameters of the system is often metastable and becomes unstable in presence of fluctuations. The fact that the magnetosphere during substorms behaves as a metastable system is reflected in many models of the substorm onset in the form of the corresponding plasma instabilities and catastrophes [*Goertz and Smith*, 1989; *Sitnov et al.*, 1997; *Hurricane et al.*, 1998; *Birn et al.*, 2004].

The apparent controversy between the data-derived picture of the magnetospheric dynamics described above and the classical SOC model [*Bak et al.*, 1987] has been resolved recently within the framework of the SOC theory itself. *Vespignani and Zapperi* [1998] noted that the BTW model contains a condition, which is quite unusual for real open systems. Specifically, each new grain of sand is added to the pile after the completion of all avalanches induced by previous grains. This makes the loading rate the smallest parameter in the model. According to *Vespignani and Zapperi* [1998], elimination of the aforementioned condition almost restores the conventional picture of phase transitions for both classical SOC models, sandpile and forest fire model. There exists, in particular, an analog of the temperature-pressure-density diagram similar to Figure 5b (Fig.1 in [*Vespignani and Zapperi*, 1998]) as well as the input-output multiscale relationship similar to (5.3) and (5.4) and Figure 7 (Figs.2 and 3 in [*Vespignani and Zapperi*, 1998]). According to *Vespignani and Zapperi* [1998], the genuine SOC regime arises only in the limit when both driving and dissipation rates tend to zero. On the other hand, the behavior of cellular automata beyond the SOC regime still has an important feature, which distinguishes them from conventional water-steam transitions. Having driving and dissipation rates as the control parameters instead of the state parameters such as the pressure or temperature, these transitions turn out to be inherently non-equilibrium.

These results have been further substantiated by *Consolini and De Michelis* [2001], who studied a revised forest-fire model driven by a 1D coupled map, which shows on-off intermittency with quiet and turbulent periods and thus is able to mimic the important features of the solar wind driving of the Earth's magnetosphere. They report in particular the criticality, which is "forced" rather than "self-organized" and is very similar to ordinary critical phenomena with first and second order phase transitions. A transition from the avalanche regime characteristic of SOC to another, "continuous flow" regime was also reported by *Corral and Paczuski* [1999] for the so-called "Oslo" model, which is widely used in SOC studies. In contrast to the avalanche regime, where the width of the active zone diverges with system size, in the continuous flow regime the active zone width is independent of the system size. The transition to the continuous flow regime occurs when the time between grain additions becomes comparable with the mean avalanche time. A seemingly different result on the robustness of the SOC behavior under strong driving conditions has been reported by *Watkins et al.* [1999]. However, *Watkins et al.* [1999] used in their analysis a modified sandpile model [*Dendy and Helander*, 1998; *Chapman et al.*, 1998], in which the probability distribution already contains both the SOC-like power-law constituent and a non-SOC part with a characteristic scale.

6. Phase Transition Analogy Applications: New Generation of Forecasting Tools

The phase transition analogy revealed in the analysis of the correlated solar wind-magnetosphere data made several important contributions to our understanding of modeling solar wind-magnetosphere interaction. First, it showed that global coherent and multiscale phenomena may co-exist like first- and second order transitions do in the water-steam system. Second, it revealed the importance of averaging in the reconstructed phase space necessary to reveal the "mean-field" dynamical features (figures 5b and 8a). Third, it confirmed that the mutiscale features of the magnetospheric dynamics may depend on the solar wind conditions. Therefore, the probabilistic description of those multiscale features must have the form of *conditional* probabilities. The utilization of these new concepts have resulted in creating a new generation of tools to model and forecast the magnetospheric activity.

6.1 Mean-field dimension

In the mean-field approach the dynamics of the system is approximated by the motion of the center of mass of its *NN* nearest neighbors in the embedded and SSA-ordered phase space

$$\mathbf{x}_t^{cm} = \sum_{i=1}^{NN} \mathbf{x}_i, \quad \mathbf{x}_t = \left(I_t, ..., I_{t-(M-1)}, O_t, ..., O_{t-(M-1)} \right)\left(\mathbf{v}_1, ... \mathbf{v}_D \right), \quad \left\| \mathbf{x}_i - \mathbf{x}_t \right\| < \varepsilon(NN), \tag{6.1}$$

where $\mathbf{v}_i$, $i=1,...D$, is the ith column of the orthogonal projection matrix V in the singular value decomposition (5.2), and $\varepsilon(NN)$ is the minimum radius of an n-dimensional sphere containing the nearest neighbors. In particular, the next step in time is given by the formula

$$O_{t+1}^{mf} = F\left(\mathbf{x}_t^{cm} \right) = \left\langle O_{k+1} \right\rangle, \quad \mathbf{x}_k \in NN, \tag{6.2}$$

The minimum embedding dimension D in these equations is determined by the condition of a regular behavior of the function F. The point is that the averaging procedure in multiscale systems often becomes incorrect as it takes place, for instance, in case of the phase transitions near the critical point [*Kadanoff*, 1999]. On the other hand, the theory of critical phenomena [e. g., *Kadanoff*, 1999 and refs. therein] as well as the dynamical system theory [e.g., *Casdagli*, 1992] suggest that the mean-field description based on the averaging procedure may be correct if the dimension is high enough. This is the dimension we search for. In particular, as suggested by *Ukhorskiy et al.* [2003], given the specific number of *NN* and accuracy ε_F we calculate the minimum dimension $D=D_t$, for which the following condition holds

$$\left| O_{t+1}^{mf} - \widetilde{O}_{t+1}^{mf} \right| < \varepsilon_F, \tag{6.3}$$

where $\widetilde{O}_{t+1}^{mf}$ differs from O_{t+1}^{mf} given by (6.2) in that the set of the nearest neighbors of $\mathbf{x}_t$ includes now the point $\mathbf{x}_t$ itself. Then the optimum *mean-field dimension* D_{mf} of the system for a given pair *(NN, ε_F)* is found as a cut-off of the probability distribution function $P(D_t)$. This definition of dimension ensures the regular character of the forecasting process (6.2). Its dependence on the averaging level *NN* is also quite natural as in practice the dimension does depend on the region of scales considered (one can propose as an example a seashell, which is 1D at scales more them 10cm, 2D at 1 cm scale and 3D at smaller scales). The effective dimension $D_{mf}=3$, consistent with the SSA images in Figures 5b and 8a, has been found for *NN=300* [*Ukhorskiy et al.*, 2003]. Moreover, the new probabilistic approach allows us to provide the analog of Figure 5b without removing hysteresis intervals from the original data set. The hysteresis phenomena responsible for the folding of the surface shown in Figure 8a make the distribution of trajectory points $P(P_2)$ double peaked. However, one of two peaks of that distribution is usually larger than the other, and the former peak determines the value of the parameters P_2 for the corresponding equilibrium system (for details see, for instance [*Gilmore*, 1993]). The maximum probability plots $\widetilde{P}_2(P_1,P_3)$, $P(\widetilde{P}_2)=\max P(P_2)_{|P_1,P_2}$ for two different data sets [*Bargatze et al.*, 1985] and [*Weigel et al.*, 2001] are shown in Figure 9. Note that the former data set includes only non-storm substorms of different averaged levels of activity, while the latter one represents just one year record of AL and vB_s parameters including both storm and non-storm substorms. The amazing similarity of plots in Figure 9 with conventional temperature-pressure-density diagram of water-steam transitions (Figure 6) strongly supports the hypothesis of the phase transition-like behavior of the magnetospheric index AL during substorms.

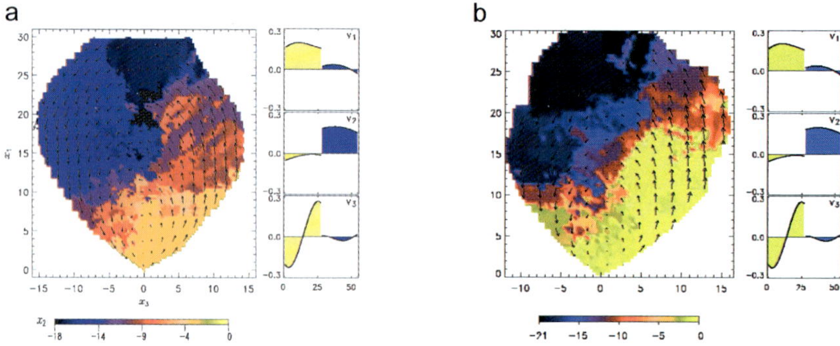

Figure 9. Substrom dynamics of the magnetosphere inferred from the singular spectrum analysis of AL-vB_s data for *Bargatze et al.* [1985] (a) and *Weigel et al.* [2001] (b) data sets. Different colors code maxima of the conditional probability distribution $P(x_2|x_1,x_3)$. Structure of the SSA eigenvectors v_n, which determine the relation of the variables x_n, to the original time series AL and vB_s through (6.1), are shown in yellow-blue panels.

6.2 Conditional probability approach to forecasting of substorm activity

Owing to equation (6.2) the mean-field description provides the forecasting algorithm similar to that of nonlinear filters, which is superior to the linear filter approach. Moreover, in contrast to nonlinear filters and similar to linear ones, the optimization of this nonlinear algorithm involves only one global parameter D_{mf} and therefore significantly improves deterministic predictions. This type of prediction becomes possible in spite of the power-law spectra of the output fluctuations as the system is strongly driven by the solar wind input. In particular, *Ukhorskiy et al.* [2004a] have shown that the *conditional probability* distributions $P(O|I)$, which constitute the almost power-law marginal distribution $P(O)$ in case of the solar wind-magnetosphere system described by AL and vB_s are much more localized and non-power-law. This fact not only justifies the mean-field deterministic forecasting algorithm (6.2) but allows to further improve predictions by supplying each specific value of the deterministic forecast O_{t+1} at given time $t+1$ with the conditional probability bar $P(O_{t+1}|I)$

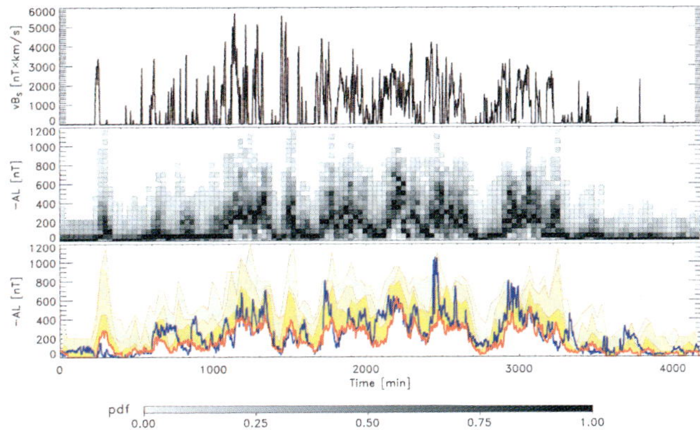

Figure 10. Mean-field and conditional probability (gray and color shading) predictions of AL index (blue) using the solar wind data input (top panel) for the 27th interval of *Bargatze et al.* [1985] database. Middle panel shows the probability $P(O_{t+1}|x_t)$ of AL with the optimum choice $(D_{mf}, NN)=(5, 50)$. Bottom panel shows the mean-field forecast O_{t+1}^{mf} (red) and the conditional probability $P\left(O_{t+1} > O_{t+1}^{mf} \mid \mathbf{x}_t\right)$ of deviations from O_{t+1}^{mf} with $P \leq 0.7$ (dark yellow), 0.9 (yellow) and 1 (light yellow).

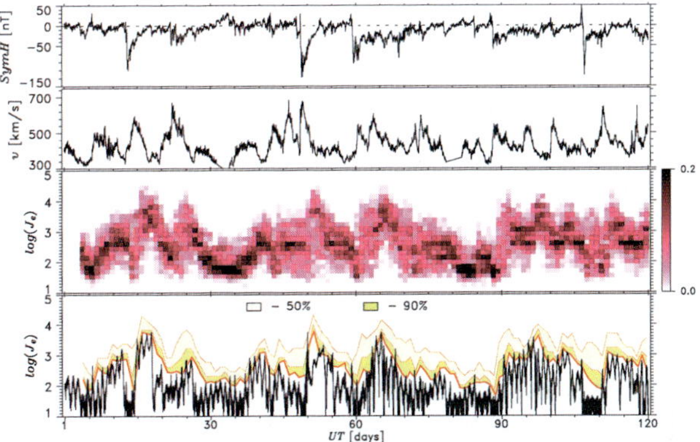

Figure 11. One-day predictions of relativistic electron intensity at geosynchronous orbit $(M=3, NN=300)$ for a part of of 1999. Inputs: the *SymH* index (top panel) and the solar wind velocity (upper middle panel). Outputs: the conditional probabilities $P(O_{t+1}|x_t)$ (lower middle panel), predicted $\log\left(J_e^{max}\right)$ (red, bottom panel), contours of probability $P(\delta_{t+1}|x_t)$ (yellow shading, bottom panel) for the actual $\log(J_e)$ (black, bottom panel) to exceed the predicted $\log\left(J_e^{max}\right)$ under the given solar wind and magnetospheric conditions.

208

$$P\left(O_{t+1}\mid\mathbf{x}_t\right)\simeq P\left(O_{k+1}\right),\quad \mathbf{x}_k\in NN,\tag{6.4}$$

showing the predicted distribution of deviations from that deterministic value O_{t+1} due to the multiscale properties of the system.

An example of such predictions for *Bargatze et al.* [1985] data set is shown in Figure 10. It shows in particular that the mean-field model (6.2) forecast (bottom panel, red curve) is quite efficient in reproducing the average dynamics of *AL*. In contrast to earlier locally linear filter techniques [*Vassiliadis et al.*, 1995; *Valdivia et al.*, 1996], the new method does not require any tuning of the filter parameters, as demonstrated by *Ukhorskiy et al.* [2002], as long as the mean-field dimension is D_{mf} determined using the procedure described in the previous subsection. The complementary conditional probability forecasts are given in Figure 10 in two forms. First, the middle panel shows the probability $P(O_{t+1}|x_t)$ of *AL* with the optimum choice $(D_{mf}, NN)=(5, 50)$. The bottom panel gives the conditional probability $P\left(O_{t+1}>O_{t+1}^{mf}\mid\mathbf{x}_t\right)$ of deviations from the mean-field forecast O_{t+1}^{mf}. The probability distributions are computed at each time step using only those events that correspond to AL values greater than the mean-field model output. In particular, the 100% contour includes all deviations from the meanfield prediction, including the sharpest peaks.

6.3 Conditional probability forecasting of radiation belt electron fluxes

The unified data-derived description based on the combination of mean-field dynamical model and conditional probability approach provides an efficient technique for modeling and forecasting various magnetospheric time series. *Ukhorskiy et al.* [2004b] used this technique for one-day predictions of the MeV electron fluxes at geosynchronous orbit. The parameters of the model were derived from the correlated database of solar wind parameters, geomagnetic indices, and daily maxima of relativistic (> 2 MeV) electron flux observed by GOES 7 and 8 satellites during years 1995 through 2000. The logarithm of the flux daily maxima was taken as the output of the model, i.e $O_t=\log\left(J_e^{\max}\right)$. The model was driven with daily maxima of the input parameters

$$\left(I_t^{(1)},...I_t^{(5)}\right)=\max\left(v,-SymH,vB_s,P_{dyn},AsyH\right),\tag{6.5}$$

where *SymH* and *AsyH* are longitudinally symmetric and asymmetric mid-latitude geomagnetic indices, respectively, with the former one being often considered as a high-resolution analog of the *Dst* index.

The small-scale dynamics of electrons in the outer radiation belt have a multiscale nature due to stochastic interactions with various wave fields. In a first principle approach averaging over these small scales yields a set of diffusion equations that govern the large-scale dynamics of the electron fluxes. To derive the global dynamical constituent *Ukhorskiy et al.* [2004b] used the mean-field dynamical approach based on the ensemble averaging of similar states in the reconstructed phase space of mean-field dimension. In particular, it was shown that the mean-field model yields the average prediction efficiency of 0.77, which exceeds the predictability of diffusion-based models [*Li et al.*, 2001] by about 0.2.

To describe the constituent of the MeV electron flux time series not captured by the deterministic model the conditional probability approach was used. The conditional probabilities were calculated using (6.4). Since the mean-field model often underestimates the fluxes,

$P(O_{t+1} \mid \mathbf{x}_t)$ was used to evaluate the probability $P(\delta_{t+1} \mid \mathbf{x}_t)$ of the observed value of the flux O_{t+1} to exceed its predicted value $\hat{O}_{t+1}$, where $\delta_t = O_t - \hat{O}_t$, for $O_t > \hat{O}_t$, and 0 for $O_t \leq \hat{O}_t$.

At each time step the output of the deterministic model was supplemented with the corresponding probability contours δ_t^C

$$\int_0^{\delta_t^C} P(\delta \mid \mathbf{x}_{t-1})\,d\delta = C, \tag{6.6}$$

which were used in risk analysis forecasting. For example, $C=1$ corresponds the highest possible value of the flux, while $C=0.7$ corresponds to the flux level which is not exceeded in 70% of the cases.

An example of one-day prediction of the combined model given by the mean-field and conditional probability descriptions together is shown in Figure 11. Taking into account the relatively limited amount of data, the calculation of the mean-field dimension was replaced in this case by the straightforward optimization of the mean-field evolution equation (6.2) over the parameters M and NN. In particular, the optimal set ($M=3$, $NN=300$) used to Figure 11 gives the highest Prediction Efficiency $PE=1-MSE/VAR$, where MSE is the mean squared error and VAR is the variance of the observed time series.

The most relevant input parameters v and $SymH$ are shown in top panels of the figure. The time series of $\log(J_e)$ are shown in black in the bottom panel and the one-day predictions of $\log(J_e^{max})$ are shown in red. The large daily variations of $\log(J_e^{max})$ are caused by the satellite rotation around the Earth. The time series of predicted $\log(J_e^{max})$ has one-day time scale and provides an upper envelope for the observed variations of the flux. The conditional probabilities of $\log(J_e^{max})$ calculated at each time step are shown in the third panel and these are used to calculate the probability contours shown in different shades of yellow above the predicted value of $\log(J_e^{max})$ in the bottom panel. This way the model yields risks of deviations from deterministic predictions as a function of the average solar wind and magnetosphere conditions.

To validate the probabilistic approach one must answer the following question. Being calculated using in-sample statistics, how does $P(\delta_t \mid x_{t-1})$ perform out-of-sample? In other words, by the construction δ_t^C probability contours embrace $C \cdot 100\%$ of in-sample events, will it still hold for out-of-sample predictions? To answer the first question we calculate the following average of (6.6) over all $\mathbf{x}_{t-1}$

$$C = \sum_t P(\mathbf{x}_{t-1}) \int_0^{\delta_t^C} P(\delta \mid \mathbf{x}_{t-1})\,d\delta, \tag{6.7}$$

for out-of sample predictions. Note that for in sample predictions (6.7) becomes an identity by virtue of the definition (6.6). On the other hand, using the definition of the conditional probability, the right hand side of (6.7) can be re-written in the form, which does not contain the unknown function $P(\mathbf{x}_{t-1})$

$$\sum_t P(\mathbf{x}_{t-1}) \int_0^{\delta_t^C} P(\delta \mid \mathbf{x}_{t-1})\,d\delta = \int_0^d P(\delta')\,d\delta', \tag{6.8}$$

where $\delta' = \delta / \delta_t^C$.

The integrals in the right hand side of (6.8) calculated for out-of-sample predictions of log fluxes for years 1995-2000, are listed in Table 1. The average error is only *0.03*, which means that the in-sample training set provides the sufficient coverage of the dynamic range of the flux and therefore can be used for out-of-sample probabilistic predictions. There are two major sources of the observed uncertainty. First, the events lying outside of the dynamic range of the training set (i.e. most extreme events like the Bastille day 2000 storm), cannot be predicted based on this training set. This explains the fact that $C=1$ contour does not embrace 100% of events. Second, due to the coarse partitioning of the flux range used in the calculations the contours corresponding to different values of C can coincide at some places, which explains why a given probability contour generally embraces a slightly higher number of events.

Table 1. Integrals (6.8) calculated for out-of-sample predictions of log fluxes for years 1995-2000

C	0.50	0.60	0.70	0.80	0.90	1.0
$\int_0^1 P(\delta')d\delta'$	0.55	0.64	0.73	0.83	0.91	0.97

A more conventional way of estimating the probability contours is based on the marginal probability density function $P(\delta_t)$ calculated with all events in the training set. The resultant contours δ_t^C do not depend on any control parameters and therefore are fixed in time $\delta_t^C \to \delta^C$. The discrepancy between the two approaches can be quantified in terms of the standard deviation $\sigma_C^2 = (1/N)\sum_{t=1}^{N}(\delta_t^C / \delta^C - 1)^2$, which also shows the variability of the conditional probability functions. The average σ_C calculated for 50-100% contours for years 1995-2000 is as high as 0.5. This means that the shape of conditional probability function strongly depends on the solar wind and magnetospheric conditions and that the probability contours derived from the marginal probability density function in most cases strongly underestimate or overestimate the risks in the system.

7. Conclusion

We have reviewed three types of models of geomagnetic activity, based on the concepts of dynamical chaos, self-organized criticality and nonequilibrium phase transition physics. These concepts are shown to provide important new techniques improving the simplest approach to modeling solar wind-magnetosphere interactions based on linear filters, including (I) time delay embedding, (II) averaging over the nearest neighbors and (III) conditional probability approach. The resulting model of the magnetosphere as a dynamical system reveals interesting new features of substorm dynamics such as an analog of the phase transition diagram, hysteresis, and input-output multiscale relationships. Importantly, these features are consistent with the generalized SOC theory, which treats the BTW regime as a limiting case of the specific class of nonequilibrium phase transitions. This also results in a new generation of forecasting tools with deterministic predictions of the global component of magnetospheric dynamics and probabilistic predictions of its multi-scale features.

Acknowledgements

The research was supported by NSF grants ATM-0119196 and 0318629.

References

Abarbanel, H. D., R. Brown, J. J. Sidorovich, and T. S. Tsimring, The analysis of observed chaotic data in physical systems, *Rev. Mod. Phys., 65,* 1331, 1993.

Angelopoulos, V., C. F. Kennel, F. V. Coroniti, R. Pellat, M. G. Kivelson, R. J. Walker, C. T. Russel, W. Baumjohann, W. C. Feldman, and J. T. Gosling, Statistical characteristics of bursty bulk flow events, *J. Geophys. Res., 99,* 21257, 1994.

Bak, P., C. Tang, and K. Wiesenfeld, Self-organized criticality: An explanation of 1/f noise, *Phys. Rev. Lett., 59,* 381, 1987.

Baker, D. N., A. J. Klimas, R. L. McPherron, and J. Buchner, The evolution from weak to strong geomagnetic activity: An interpretation in terms of deterministic chaos, *Geophys. Res. Lett., 17,* 41, 1990.

Bargatze, L. F., D. N. Baker, R. L. McPherron, and E. W. Hones Jr., Magnetospheric impulse response for many levels of geomagnetic activity, *J. Geophys. Res., 90*(A7), 6387-6394, 1985.

Birn, J., J. F. Drake, M. A. Shay, B. N. Rogers, R. E. Denton, M. Hesse, M. Kuznetsova, Z. W. Ma, A. Bhattacharjee, A. Otto, and P. L. Pritchett, Geospace Environmental Modeling (GEM) Magnetic Reconnection Challenge, *J. Geophys. Res., 106,* 3715, 2001.

Birn, J., J. C. Dorelli, M. Hesse, K., Schindler, Thin current sheets and loss of equilibrium: Three-dimensional theory and simulations, *J. Geophys. Res., 109,* A02215, doi:10.1029/2003JA010275, 2004.

Borovsky, J. E., R. J. Nemzek, and R. D. Balian, The occurrence rate of magnetospheric-substorm onsets: Random and periodic substorms, *J. Geophys. Res., 98,* 3807 1993.

Broomhead, D. S., and G. P. King, Extracting qualitative dynamics from experimental data, *Physica D, 20,* 217, 1986.

Burton, R. K., R. L. McPherron, and C. T. Russell, An emphirical relationship between interplanetary conditions and Dst, *J. Geophys. Res., 80,* 4204, 1975.

Casdagli, M., A dynamical approach to modeling input-output systems, in *Nonlinear Modeling and Forecasting,* Casdagli, M., and S. Eubank, editors, p.265, Addison-Wesley, 1992.

Chapman, S., and N. Watkins, Avalanching and self-organized criticality, a paradigm for geomagnetic activity? *Space Sci. Rev., 95,* 293, 2001.

Chapman, S. C., N. W. Watkins, R. O. Dendy, P. Helander, and G. Rowlands, A simple avalanche model as an analogue for magnetospheric activity, *Geophys. Res. Lett., 25,* 2397, 1998.

Clauer, C. R., R. L. McPherron, and C. Searls, Solar wind control of the low-latitude asymmetric magnetic disturbance field, *J. Geophys. Res., 88,* 2123, 1983.

Consolini, G., Sandpile cellular automata and magnetospheric dynamics, in *Proc. Vol. 58, "Cosmic Physics in the Year 2000",* edited by S. Aiello, N. Iucci, G. Sironi, A. Traves, and U. Villante, SIF Bologna, Italy, 1997.

Consolini, G., and T. S. Chang, Magnetic field topology and criticality in geotail dynamics: Relevance to substorm phenomena, *Space Sci. Rev., 95,* 309, 2001.

Consolini, G., and P. De Michelis, A revised forest-fire cellular automation for the nonlinear dynamics of the Earth's magnetotail, *J. Atmos. Solar Terr. Phys., 63,* 1371, 2001

Consolini, G., and A. T. Y. Lui, Sign-singular analysis of current disruption, *Geophys. Res. Lett., 26,* 1673, 1999.

Corral, A., and M. Paczuski, Avalanche merging and continuous flow in a sandpile model, *Phys. Rev. Lett., 83,* 572, 1999.

Dendy, R. O. and P. Helander, On the appearance and non-appearance of self-organised criticality in sandpiles, *Phys. Rev. E, 57,* 3641, 1998.

Freeman, M. P., N. W. Watkins, and D. J. Riley, Evidence for a solar wind origin of the power law burst lifetime distribution of the AE index, *Geophys. Res. Lett., 27,* 1087, 2000.

Gilmore, R., *Catastrophe Theory for Scientists and Engineers*, Dover, Mineola, New York, 1993.

Goertz, C. K., and R. A. Smith, The thermal catastrophe model of substorms, *J. Geophys. Res.*, *94*, 6581, 1989.

Grassberger, P. and I. Procaccia, Measuring the strangeness of strange attractors, *Physica D, 9*, 189, 1983.

Haupt, A., and J. Straub, Evaluation of the isochoric heat capacity measurements at the critical isochore of SF_6 performed during the German Spacelab Mission D-2, *Phys. Rev. E, ,* 1795, 1999.

Hernandes, J. V., T. Tajima, and W. Horton, Neutral net forecasting for geomagnetic activity, *Geophys. Res. Lett., 20,* 2707, 1993.

Hnat, B., S. C. Chapman, G. Rowlands, N. W. Watkins, M. P. Freeman, Scaling in solar wind ε and the AE, AL and AU indices as seen by WIND, *Geophys. Res. Lett., 29(10),* 1446, doi:10.1029/2001GL014597, 2002.

Hnat, B., S. C. Chapman, G. Rowlands, N. W. Watkins, M. P. Freeman, Scaling in long term data sets of geomagnetic indices and solar wind ε as seen by WIND spacecraft, *Geophys. Res. Lett., 30(22),* 2174, doi:10.1029/2003GL018209, 2003.

Horton, W., and I. Doxas, A low-dimensional dynamical model for the solar wind driven geotail-ionosphere system, *J. Geophys. Res., 103,* 4561, 1998.

Hurricane, O. A., B. H. Fong, S. C. Cowley, F. V. Coroniti, C. F. Kennel, and R. Pellat, Substorm detonation – the unification of substorm trigger mechanisms, in *Substorms-4: International Conference on Substorms-4, Lake Hamana, Japan, March 9-13, 1998*, edited by S. Kokubun and Y. Kamide, p.373,, Terra Sci., Tokyo, 1998.

Ieda, A., S. Machida, T. Mukai, Y. Saito, T. Yamamoto, A. Nishida, T. Terasawa, S. Kokubun, Statistical analysis of the plasmoid evolution with Geotail observations, *J. Geophys. Res., 103,* 4453, 1998.

Iyemori, T., H. Maeda, and T. Kamei, Impulse response of geomagnetic indices to interplanetary magnetic field, *J. Geomagn. Geoelectr, 6,* 577, 1979.

Jensen, H. J., *Self-Organized Criticality: Emergent Complex Behavior in Physical and Biological Systems*, Cambridge Univ. Press, Cambridge, 1998.

Kadanoff, L. P., *Statistical physics: statics, dynamics and renormalization*, World Scientific Publishing Co. Pte. Ltd, Singapore, 1999.

Kamide, Y., and W. Baumjohann, *Magnetosphere-Ionosphere Coupling*, Springer-Verlag, Berlin Heidelberg, 1993.

Klimas, A. J., D. N. Baker, D. A. Roberts, D. H. Fairfield, and J. Buchner, A nonlinear dynamical analogue model of geomagnetic activity, *J. Geophys. Res., 97,* 12,253, 1992.

Klimas, A. J., D. Vassiliadis, B. N. Baker, and D. A. Roberts, The organized nonlinear dynamics of the magnetosphere, *J. Geophys. Res., 101,* 13,089, 1996.

Klimas, A. J., J. A. Valdivia, D. Vassiliadis, D. N. Baker, M. Hesse, and J. Takalo, The role of self-organized criticality in the substorm phenomenon and its relation to localized reconnection in the magnetospheric plasma sheet, *J. Geophys. Res., 105,* 18,765, 2000.

Li, X., M. Temerin, D. N. Baker, G. D. Reeves, and D. Larson, Quantitative prediction of radiation belt electrons at geostationary orbit based on solar wind measurements, *Geophys. Res. Lett., 28,* 1887, 2001.

Lorenz, E. N., Determining nonperiodic flow, *J. Atmos. Sci., 20,* 130, 1963.

Lui, A. T. Y., Multiscale phenomena in the near-Earth magnetosphere, *J. Atmos. Solar Terr. Phys., 64,* 125, 2002.

Lui, A. T. Y., S. C. Chapman, K. Liou, P. T. Newell, C. I. Meng, M. Brittnacher, G. K. Parks, Is the dynamic magnetosphere an avalanching system?, *Geophys. Res. Lett., 27,* 911, 2000.

Lyons, L. R., S. Liu, J. M. Ruohoniemi, S. I. Solovyev, J. C. Samson, Observations of dayside convection reduction leading to substorm onset, *J. Geophys. Res., 108(A3),* 1119, doi:10.1029/2002JA009670, 2003.

Nagel, S. R., Instabilities in a sandpile, *Rev. Mod. Phys., 64*, 321, 1992.

Ohtani, S., T. Higuchi, A. T. Y. Lui, and K. Takahashi. Magnetic fluctuations associated with tail current disruption: Fractal analysis, *J. Geophys. Res., 100*, 19,135, 1995.

Osborne, A. R. and A. Provenzale, Finite correlation dimensions for stochastic systems with power-law spectra, *Physica D, 35*, 357, 1989.

Packard, N., J. Crutchfield, D. Farmer, and R. Shaw, Geometry from a time series, *Phys. Rev. Lett., 45*, 712, 1980.

Pavlos, G. P., G. A. Kyriakov, A. G. Rigas, P. I. Liatsis, P. C. Trochoutsos, and A. A. Tsonis, Evidence for strange attractor structures in space plasmas, *Ann. Geophys., 10*, 309, 1992.

Pavlos, G. P., D. Diamandidis, A. Adamopoulos, A. G. Rigas, I. A. Daglis, and E. T. Sarris, Chaos and magnetospheric dynamics, *Nonlin. Proc. Geophys., 1*, 124, 1994.

Pavlos, G. P., M. A. Athanasiu, D. Diamandidis, A. G. Rigas, and E. T. Sarris, Comments and new results about the magnetospheric chaos hypothesis, *Nonlin. Proc. Geophys., 6*, 99, 1999.

Preisendorfer, R. W., *Principal Component Analysis in Meteorology and Oceanography*, Elsevier, New York, 1988.

Press, W. H., B. P. Flannery, S. A. Teukolsky, and W. V. Vetterling, *Numerical Recipes: The Arth of Scientific Computing*, 2nd ed., Cambridge Univ. Press, New York, 1992.

Price, C. P., and D. Prichard, The chaotic or non-chaotic behabior of global geomagnetic processes, in: *Physics of Space Plasmas*, SPI Conf. Proc. And Reprint Ser., No.12, Scientific Publ., Inc., Cambridge, MA, pp.265-277, 1992.

Prichard, D., Comment on "Chaos and magnetospheric dynamics" by G. P. Pavlos, D. Diamandidis, A. Adamopoulos, A. G. Rigas, I. A. Daglis, and E. T. Sarris, *Nonlin. Proc Geophys., 2*, 58, 1995.

Prichard, D., and C. P. Price, Spurious dimension estimates from time series of geomagnetic indices, *Geophys. Res. Lett., 19*, 1623, 1992.

Prichard, D., J. E. Borovsky , P. M. Lemons, C. P. Price, Time dependence of substorm recurrence: An information-theoretic analysis, *J. Geophys. Res., 101*, 15,359, 1996.

Roberts, D. A., Is there a strange attractor in the magnetosphere?, *J. Geophys. Res., 96*, 16,031, 1991.

Shan, L. H., P. Hansen, C. K. Goertz, and R. A. Smith, Chaotic appearance of the ae index, *Geophys. Res. Lett., 18*, 147, 1991a.

Shan, L. H., C. K. Goertz, and R. A. Smith, On the embedding-dimension analysis of ae and al time series, *Geophys. Res. Lett., 18*, 1647, 1991b.

Sharma, A. S., assessing the magnetosphere's nonlinear behavior: Its dimension is low, its predictability, high, *Rev. Geophys. Suppl.*, pp.645-650, 1995.

Sharma, A. S., D. V. Vassiliadis, and K. Papadopoulos, Reconstruction of low-dimensional magnetospheric dynamics by singular spectrum analysis, *Geophys. Res. Lett., 20*, 335, 1993.

Sitnov, M. I., H. V. Malova, and A. T. Y. Lui, Quasi-neutral sheet tearing instability induced by electron preferential acceleration from stochasticity, *J. Geophys. Res., 102*, 163, 1997.

Sitnov, M. I., A. S. Sharma, K. Papadopoulos, D. Vassiliadis, J. A. Valdivia, A. J. Klimas and D. N. Baker, Phase transition-like behavior of the magnetosphere during substorms, *J. Geophys. Res., 105*, 12,955, 2000.

Sitnov, M. I., A. S. Sharma, K. Papadopoulos, and D. Vassiliadis, Modeling substorm dynamics of the magnetosphere: From self-organization and self-organized criticality to nonequilibrium phase transitions, *Phys. Rev. E, 65*, 016116, 2001.

Smith, A. J., M. P. Freeman, and G. D. Reeves, Postmidnight VLF chorus events, a substorm signature observed at the ground near $L = 4$, *J. Geophys. Res., 101*, 24,641, 1996

Stanley, H. E., *Introduction to Phase Transition and Critical Phenomena*, Oxford Univ. Press, New York, 1971.

H.E. Stanley, Scaling, universality, renormalization: three pillars of modern critical phenomena, *Rev. Mod. Phys. 71*, S358, 1999.

Tabor, M., *Chaos and Integrability in Nonlinear Dynamics*, John Wiley & Sons, 1989.

Takalo, J., J. Timmonen, A. Klimas, J. Valdivia, and D. Vassiliadis, Nonlinear energy dissipation in a cellular automation magnetic field model, *Geophys. Res. Lett., 26*, 1813, 1999.

Takalo, J., J. Timmonen, A. Klimas, J. Valdivia, and D. Vassiliadis, A coupled-map model for the magnetotail current sheet, *Geophys. Res. Lett., 26*, 2913, 1999.

Takens, Detecting strange attractors in turbulence, in: *Lecture Notes in Mathematics 898*, D. A. Rand and L. S. Young, eds., Springer, Berlin, pp. 366-381, 1981.

Theiler, J., Spurious dimension from correlation algorithms applied to limited time series data, *Phys. Rev. A, 34*, 2427, 1986.

Theiler, J., Some comments on the correlation dimension of $1/f^\alpha$ noise, *Phys. Lett. A, 155*, 480, 1991.

Tsurutani, B. T., M. Sugiura, T. Iyemori, B. E. Goldstein, W. D. Gonzalez, S. I. Akasofu, and E. J. Smith, The nonlinear response of AE to the IMF Bs driver: A spectral break at 5 hours, *Geophys. Res. Lett., 17*, 279, 1990.

Ukhorskiy, A. Y., M. I. Sitnov, A. S. Sharma, and K. Papadopoulos, Global and multi-scale aspects of magnetospheric dynamics in local-linear filters, *J. Geophys. Res., 107*(A11), 1369, doi: 10.1029/2001JA009160, 2002.

Ukhorskiy, A. Y., M. I. Sitnov, A. S. Sharma, and K. Papadopoulos, Global and multi-scale features in a description of the solar wind-magnetosphere coupling, *Ann. Geophys., 21*, 1913, 2003.

Ukhorskiy, A. Y., M. I. Sitnov, A. S. Sharma, and K. Papadopoulos, Combining global and multi-scale features of the solar wind-magnetosphere coupling: From modeling to forecasting, *Geophys. Res. Lett., 31*, L08802, doi:10.1029/2004GL018932, 2004a.

Ukhorskiy, A. Y., M. I. Sitnov, A. S. Sharma, B. J. Anderson, S. Ohtani, and A. T. Y. Lui, Data-derived forecasting model for relativistic electron intensity at geosynchronous orbit, *Geophys. Res. Lett., 31*, L09806, doi:10.1029/2004GL019616, 2004b.

Uritsky, V. M., and M. I. Pudovkin, Low frequency 1/f-like fluctuations of the AE index as a possible manifestation of self-organized criticality in the magnetosphere, *Ann. Geophys., 16(12)*, 1580, 1998.

Uritsky, V. M., A. J. Klimas, and D. Vassiliadis, Multiscale dynamics and robust critical scaling in a continuum current sheet model, *Phys. Rev. E, 65*, 046113, 2002.

Uritsky, V. M., A. J. Klimas, D. Vassiliadis, D. Chua, G. Parks, Scale-free statistics of spatiotemporal auroral emissions as depicted by POLAR UVI images: Dynamic magnetosphere is an avalanching system, *J. Geophys. Res., 107(A12)*, 1426, doi:10.1029/1001JA000281, 2002.

Valdivia, J. A., A. S. Sharma, and K. Papadopoulos, Prediction of magnetic storms by nonlinear models, *Geophys. Res. Lett., 23*, 2899, 1996.

Vassiliadis, D., A. S. Sharma, T. E. Eastman, and K. Papadopoulos, Low-dimensional chaos in magnetospheric activity from AE time series, *Geophys. Res. Lett., 17*, 1841, 1990.

Vassiliadis, D., A. J. Klimas, D. Baker, and D. A. Roberts, A description of the solar wind-magnetosphere coupling based on nonlinear filters, *J. Geophys. Res., 100*, 3495, 1995.

Vespignani, A., and S. Zapperi, How self-organized criticality works: A unified mean-field picture, *Phys. Rev. E, 57*, 6345, 1998.

Watkins, N. W., S. C. Chapman, R. O. Dendy, and G. Rowlands, Robustness of collective behaviour in strongly driven avalanche models: Magnetospheric implications, *Geophys. Res. Lett., 26*, 2617, 1999.

Watkins, N. W., M. P. Freeman, S. C. Chapman, and R. O. Dendy, Testing the SOC hypothesis for the magnetosphere, *J. Atm. Sol. Terr. Phys., 63*, 1435, 2001.

Weigel, R.S., W. Horton, T. Tajima, and T. Detman, Forecasting auroral electrojet activity from solar wind input with neural networks, *Geophys. Res. Lett., 26*, 1353, 1999.

Weigel, R.S., A.J. Klimas, D. Vassiliadis, V. Uritsky, and W. Horton, Features of auroral zone magnetometer time series during undriven relaxation, Spring AGU meeting, Boston, MA, May 29-June 2, Abstract SM61B-06, 2001.
Zheng, G. P., and J. X. Zhang, Determination of dynamical critical exponents from hysteresis scaling, *Phys. Rev. E, 58*, 1187, 1998.

Multiscale Coupling of Sun-Earth Processes
A.T.Y. Lui, Y. Kamide and G. Consolini, editors.

SUBSTORM ONSET CAUSED BY THE TWO STATE TRANSITION OF THE MAGNETOSPHERE

T. Tanaka

Department of Earth and Planetary Sciences, Graduate School of Sciences, Kyushu University, Fukuoka 812-8581, Japan

Abstract. The substorm process is simulated by resistive magneto-hydrodynamic (MHD) equations. Starting from a stationary solution under a northward interplanetary magnetic field (IMF) condition with non-zero IMF B_y, changes in magnetospheric convection following the southward turning of the IMF are investigated. The simulation results first show the progress of plasma sheet thinning in the tail. This thinning is caused by the drain of closed flux from the near-earth plasma sheet coupled with the enhanced ionospheric convection. In this stage, the flux transported from the near-earth tail to the dayside cannot be replaced by the flux from the midtail, because the reclosure process of open field lines in the midtail plasma sheet is still slow due to the control by the remnant of northward IMF. The substorm onset occurs as an abrupt shift of the pressure peak in the plasma sheet from -12 Re to -8 Re and an intrusion of convection flow into the inner magnetosphere. After the onset, the simulation results reproduce both the dipolarization in the near-earth tail and the formation of near-earth neutral line (NENL) at the midtail, together with plasma injection into the inner magnetosphere and an enhancement of the nightside field-aligned current (FAC). During the dipolarization process, the static magnetospheric force balance changes from the z-direction-dominated state to the x-direction-dominated state. Thus, the dipolarization is not a mere pile up of the flux ejected from the NENL. The pressure distribution in the near-earth plasma sheet changes so as to self-adjust the restored magnetic tension associated with the establishment of force balance in the x direction. It is concluded that the direct cause of the onset is the state (phase-space) transition of the convection system from a thinned state to a dipolarized state.

Keywords. substorm, state transition.

1. Introduction

In the earliest magnetospheric model founded by *Chapman and Ferraro* [1931], the Earth's magnetic field is confined to a comet-shaped configuration by the dynamic pressure of the solar wind. This model results in an absolutely quiet magnetosphere with no plasma motion (= no electric field), because the solar wind plasma and magnetospheric plasma (if present) would not mix in ideal magnetohydrodynamic (MHD) processes under the frozen-in principle. In the real magnetosphere, however, a large scale plasma motion is induced through non-ideal MHD processes such as magnetopause instability or reconnection together with the penetration of solar wind plasma into the magnetosphere. This convection process proceeds involving the ionospheric plasma as well and forms the most fundamental mechanism in transferring the free energy of disturbances in the magnetosphere-ionosphere (M-I) system. One of the most important problems in such dynamic magnetosphere concerns the cause-effect relationship which controls the substorm process. The concept of substorm was first introduced by *Akasofu* [1964]. It is a consequence of northward-to-southward turning of the IMF that enhances dayside reconnection and subsequent convection due to its antiparallel-merging characteristic.

218

When the interplanetary magnetic field (IMF) is southward for about 1 hour, three phases of the substorm can be identified: the growth phase, the expansion phase, and the recovery phase. Observations have revealed that the progress of magnetospheric convection accompanies a thinning of the plasma sheet and enhanced two-cell convection in the ionosphere with a gradual increase of intensity in tail current, field-aligned currents (FACs), and ionospheric currents. This enhanced convection in the M-I system represents well-known features of the growth phase [*Baker et al.*, 1996]. The expansion phase starts with a sudden brightening of the equatorward-most pre-onset arc, accompanied by an enhancement of electrojet activities. After the initial brightening, the region of intensified aurora moves poleward and westward [*Akasofu*, 1964; *Elphinstone et al.*, 1996]. In the near-earth plasma sheet where the ground onset position is mapped back, the expansion onset is signified by the dipolarization of the tail field [*Lopez and Lui*, 1990; *Lui*, 1996]. The dipolarization accompanies the plasma injection into the inner magnetosphere, the current disruption (CD), and the formation of the current wedge. When a spacecraft is located in the near-earth tail during the dipolarization, a high level of magnetic-field fluctuation is observed that is associated with the CD events [*Takahashi et al.*, 1987; *Ohtani et al.*, 1998]. In the middle tail, fast tailward plasma flows threaded by the southward magnetic field are often observed beyond 20-25 Re [*Nagai et al.*, 1994]. These observations suggest the tail reconnection located at 20-25 Re. Flows in the near-earth region between 9 and 19 Re are observed as a form of bursty bulk flow (BBF) indicating that the tail reconnection is patchy-bursty [*Baumjohann et al.*, 1990; *Angelopoulos, et al.*, 1992].

This paper shows that the substorm process is a natural consequence of the development in magnetospheric convection that is inevitable during the condition of southward IMF. As shown in *Tanaka* [2003b], the magnetospheric convection is the basic process for the generation of disturbance phenomena in the earth's magnetosphere. It represents a process in which plasma and magnetic field in the magnetosphere and the ionosphere realize a quasi-circular motion at all altitudes, with momentum transfer between the altitudes connected by the FAC. More basically, it is the fundamental process to self-consistently organize almost all physical structures in the M-I system including magnetospheric currents, ionospheric currents, FACs, dynamo and plasma population regimes [*Tanaka*, 2003b]. Analyzing the development of convection, this paper gives new interpretations for two key features of the substorm, namely the plasma sheet thinning during the growth phase and the dipolarization associated with the expansion phase. The most important question addressed in this paper is "What is responsible for the discontinuous behavior of the M-I system that characterizes the substorm onset", since the magnetic disturbance is not called as substorm if it starts gradually [*Yahnin et al.*, 1994; *Sergeev et al.*, 1996].

2. Numerical Method

Almost all analytic models cannot be completed to treat the convection in the M-I system, because its structure depends on the topologies specific to the M-I system [*Tanaka*, 2003b]. A crucial method that can overcome this difficulty is the recent solar wind-magnetosphere-ionosphere (S-M-I) interaction model by the three-dimensional (3-D) MHD simulations coupled with a model ionosphere [*Ogino et al.*, 1986; *Tanaka*, 1994, 1995, 1999, 2000a, 2000b, 2001, 2003a, 2003b; *Crooker et al.*, 1998; *Fedder et al.*, 1998; *Siscoe et al.*, 2000]. Recent MHD simulations [*Tanaka*, 1995, 2003b] have revealed the relationship between the convection and FAC to have advanced our understanding for the mutual relations between plasma structure, current system, and convection. It is actually proved from MHD models that the magnetospheric and ionospheric convections construct a coupling system regulated by the exchange of FAC. For the generation of FAC in a realistic magnetosphere and the projection of convection onto the ionosphere, MHD simulations provide a powerful tool. However, application of MHD simulation had not been so straightforward because the magnetosphere and the ionosphere exhibit large differences in their characteristics. Three major difficulties arise in the numerical calculation of MHD when adapted to the S-M-I coupling simulation.

The first difficulty comes from the situation that the respective sizes of the magnetosphere and the ionosphere are extremely different. To overcome this difficulty and to numerically project the convection onto the ionosphere, MHD simulations using unstructured grids are being studied [*Tanaka*, 1995; *Fedder et al.*, 1998; *Siscoe et al.*, 2000; *Gombosi et al.*, 2000]. The second difficulty concerns non-ideal MHD effects and numerical resistivity imbedded in the scheme [*Gombosi et al.*, 2000]. In the MHD regime, non-ideal MHD effects are considered through the transport coefficients. In almost all MHD models, however, the distribution of transport coefficients is due to the numerical dissipation. In this paper, high resolution schemes such as the total-variation diminishing (TVD) scheme are used to reduce background numerical resistivity. Then, actual resistivity is controlled by a given model [*Tanaka*, 2000b] in which resistivity increases in the downtail direction. It was shown by *Tanaka* [2000b] that this kind of resistivity model induces a reconnection that generates a realistic substorm solution. A wide range in the magnitude of magnetic field causes the third difficulty. The magnitude of the dipole magnetic field is about 30,000nT in the ionosphere near the earth, while it diminishes rapidly in the magnetosphere to about 10 nT. Therefore, the ratio of variable to intrinsic components of the magnetic field becomes extremely small in the ionosphere. In order to avoid this difficulty, the MHD equations are reconstructed by dividing $\mathbf{B}$ as $\mathbf{B}=\mathbf{B}_0+\mathbf{B}_1$, where $\mathbf{B}_0$ and $\mathbf{B}_1$ are known intrinsic and unknown variable components, respectively [*Tanaka*, 1994]. In the S-M-I interaction problem, $\mathbf{B}_0$ adopts a dipole field. It was shown by *Tanaka* [1994] that the TVD scheme can be organized even for the reconstructed equations treating $\mathbf{B}_1$. By adopting this method, *Tanaka* [1995] first reproduced the FAC numerically and reconstructed the 3-D structure of the FAC system. This success was the first step to discuss the self-consistent configuration of the M-I convection system.

The following part of this paper will mainly focus on the self-consistent picture of convection system obtained from above MHD simulations employing unstructured grids, TVD scheme and reconstructed equations. In the presentation of simulation results in this paper, the x axis is pointing toward the sun, the y axis is pointing toward the opposite direction of the earth's orbital motion, and the z axis is pointing toward the north. In the MHD simulation, the Ohm's law in the ionosphere acts as an inner boundary condition. Ionospheric conductance is determined as a sum of sunlight-induced part that is calculated as a function of solar zenith angle and auroral-particle-induced part that is given as a function of pressure, temperature and FAC projected from the inner boundary along the magnetic field lines [*Tanaka*, 2000b]. In the inner boundary, the number of fixed variables must coincide with the number of characteristic lines that are directed toward the calculation domain. The Ohm's law fixes two velocity components through the determination of potential. Up to three more variables are fixed in the inner boundary; radial component of the magnetic field, density, and pressure. However, this number depends on the flow direction in the inner boundary (depends on the polar wind case or precipitation case). If the flow at inner boundary is inward (precipitation case), density and pressure are determined by the projection of variables from the upper region. These processes must be constructed along the principle of characteristic line for hyperbolic equation.

3. Numerical Substorm

The most general configuration that satisfies the quiet time geospace condition may be the magnetospheric configuration shown by *Tanaka* [1999] (under northward IMF condition with nonzero IMF B_y). Starting from this solution and changing the IMF input from northward to southward, observed sequence of substorm were numerically reproduced as shown in Figure 1. This figure shows time sequences of simulated pressure (normalized by the solar wind pressure P_{sw}) and $-V_x$ ($-x$ velocity) distributions (normalized by the solar wind sound velocity C_s) in the meridional plane of the magnetosphere. The time is from top to bottom.

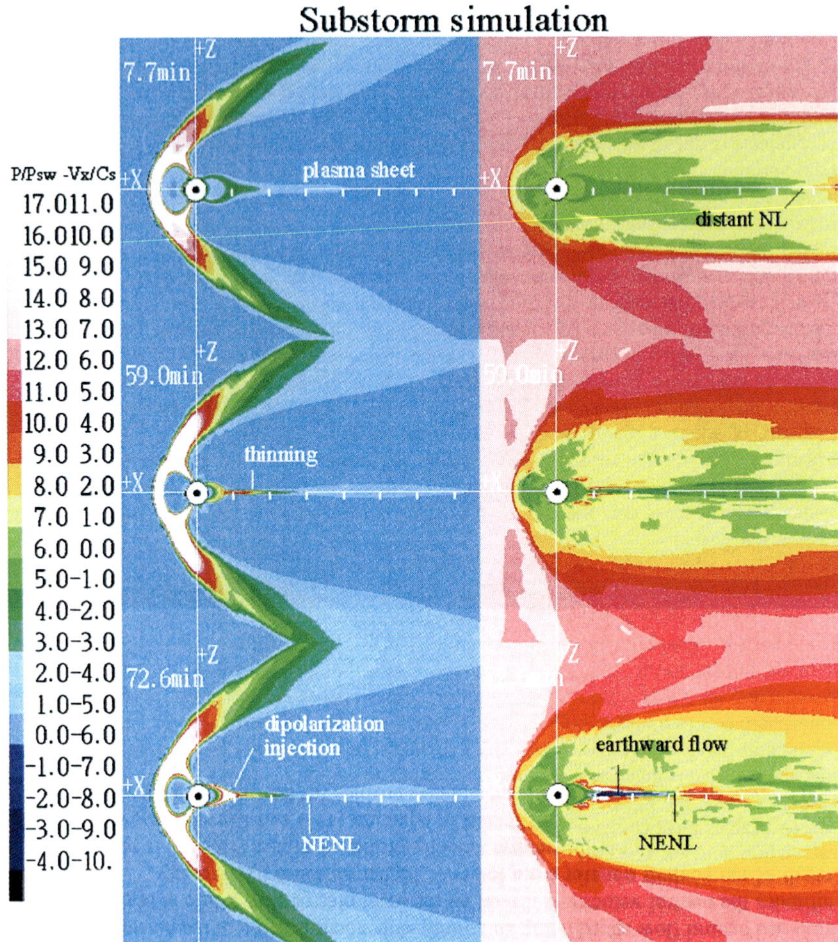

Figure 1. Time sequences of simulated pressure (P) and velocity ($-V_x$) distributions in the meridian plane of the magnetosphere. The pressure values are normalized by the solar wind pressure (P_{sw}=47pPa) and the velocity values are normalized by the solar wind sound velocity (C_s=53km/sec).The size of the earth is shown by black spheres. Marks on the $-x$ axis are 10 Re apart. Time is measured after the southward turning of the IMF. Thinning and dipolarization sequence is observable in this figure.

The top left panel of Figure 1 at t=7.7 minute illustrates the magnetospheric configuration soon after an IMF southward turning. At this time, a thick and low-pressure plasma sheet is still observed without a noticeable effect of the southward IMF. The flow structure in the top right panel of Figure 1 indicates that the x line is situated beyond x=-60 Re shortly after the southward turning of the IMF. This structure, which is generally called the distant neutral line, is a continuation of the tail structure under the northward IMF condition shown in *Tanaka* [1999].

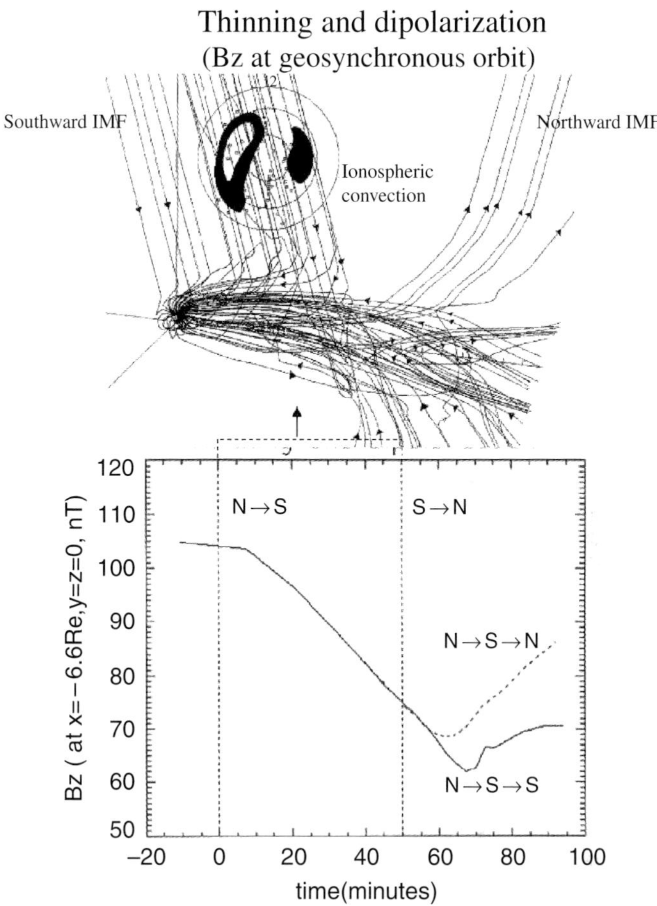

Figure 2. Magnetic configuration during the growth phase at t=30 minute (upper) and development of B_z at the midnight geosynchronous orbit showing the thinning and dipolarization process (lower). The upper inset in the upper panel shows the ionospheric convection. It is seen in this inset that the two-cell convection is developed to a certain extent.

222

At this time, sunward flow in the plasma sheet is still fairly slow (less than 50 km/sec). In Figure 1, the thinning of the near-earth plasma sheet (growth phase) develops continuously until t=59.0 minute shown in the middle panels. As thinning proceeds, earthward flow increases its speed in the near-earth tail. During this interval, the blue area in the dayside magnetosphere seen in the pressure distribution tends to shrink slightly (12.5 %) and the flaring angle of the tail lobe tends to increase considerably, due to an erosion effect. In the growth phase (middle panels of Figure 1), the pressure maximum in the plasma sheet exists around x=-12 Re. The x dependence of the pressure profile in the plasma sheet inside x=-20 Re does not change severely during the growth phase, even though the absolute value increases according to the development of thinning.

After the middle panels of Figure 1, some qualitative changes occur in the tail configuration. At first, a remarkable tailward flow appears in the midtail at t=61.8 minute. While the flow inside x=-60 Re is earthward throughout the growth phase, flow around x=-39 Re changes tailward at this time (t=61.8 minute). Associated with this tailward flow, negative B_z appears around x=-34 Re showing the reconnection in the plasma sheet. The next change is a start of the dipolarization. To show the thinning and dipolarization sequence, the solid curve in the lower panel of Figure 2 illustrates the development of the B_z component at x=-6.6 Re in the midnight equatorial plane (at the midnight geosynchronous orbit). A noticeable thinning starts after about 10 minutes from a southward turning of the IMF. As shown in Figure 2, the B_z component at the midnight geosynchronous orbit decreases continuously during the growth phase. This growth-phase signature continues for about 50 minutes until a dipolarization which starts at t=67.2 minute. After t=67.2 minute, decreasing tendency of B_z at the midnight geosynchronous orbit changes to increasing. We tentatively define the onset by the start time of the B_z increase at the midnight geosynchronous orbit at t=67.2 minute. Under a continuously southward IMF, dipolarization tends to saturate after 10 minutes. In the lower panel of Figure 2, dashed curve shows the case of northward re-turning of IMF at t=50.0 minutes. In this case, dipolarization continues without saturation and, in addition, the onset time becomes earlier indicating the substorm triggering by a northward turning of the IMF. The dipolarization is hastened 5 minutes compared with the steady southward IMF case. A northward re-turning of the IMF results in the deceleration of the ionospheric convection. The result in Figure 2 suggests that the deceleration of convection tends to promote the substorm onset occurring in the M-I convection system [*Tanaka*, 2000b].

The bottom panels of Figure 1 at t=72.6 minute show the pressure and velocity distributions after the onset. At this time, the pressure peek in the plasma shifts to x=-8 Re and remains there afterward. In this way, the position of the peak plasma pressure shifts from x=-12 to x=-8 Re in a short time. These variations show the signature of the dipolarization and injection occurring in the inner magnetosphere. From the development of the plasma sheet configuration in Figure 1, a plasmoid formation is observed after the onset, with the near-earth neutral line (NENL) formed around x=-30 Re. A closeup of this NENL structure can be seen in *Tanaka* [2000c]. During the expansion phase, this NENL persists in the midtail, generating fast earthward flows inside (earthward of) x=-32 Re and tailward flows outside.

4. Growth Phase Mechanism

The upper panel of Figure 2 at t=30 minute shows a snapshot of the magnetospheric configuration during the growth-phase interval together with the ionospheric convection. While open field lines generated from the southward IMF through the dayside reconnection wrap the outer part of the lobe, the core part still consists of open field lines generated from the northward IMF. In the thinning stage during the growth phase, therefore, the reclosure process in the plasma sheet (at x=-60 Re in Figure 1) is still slow, since it is controlled by the remnants of the northward IMF. From numerical results shown by *Tanaka* [1999], it is well understood that the magnetic

panel). In this model, the onset proceeds from the inner magnetosphere to the tail (steps 1, 2, and 3 in Figure 6, top). The CD is a kinetic process and must also be an instability. Supporters of this model associate the severe magnetic oscillations accompanying dipolarization with non-MHD processes [*Takahashi et al.*, 1987; *Ohtani et al.*, 1998], and believe that there may even be a global-scale slip between the magnetic field and plasma (non-ideal MHD process). Therefore, this model is constructed under the EJ paradigm in which **E** is regarded as the cause of **V** and **J** is regarded as the cause of **B**. The CD model is consistent with the brightening glow order of the quiet arcs. Both of the NENL and CD models attribute the substorm onset to local plasma processes. In these models, global magnetospheric structure is controlled by local processes. On the contrary, the two-state transition model considers the change of global force balance associated with the convection motion as the main controlling factor for magnetospheric development. In this model, onset is generated retaining $\mathbf{J} \times \mathbf{B} = -\nabla P$ in the near-earth-tail.

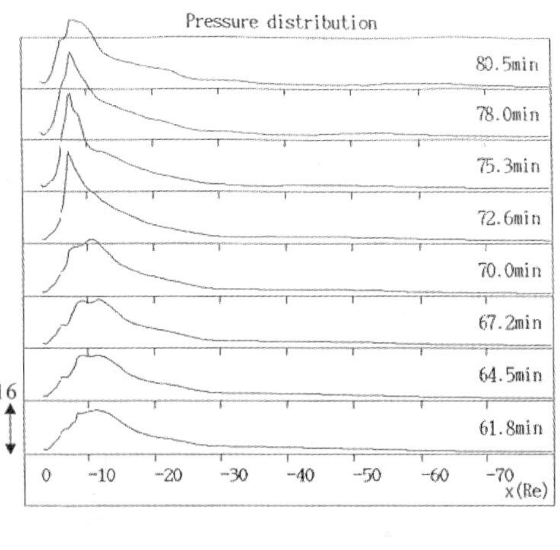

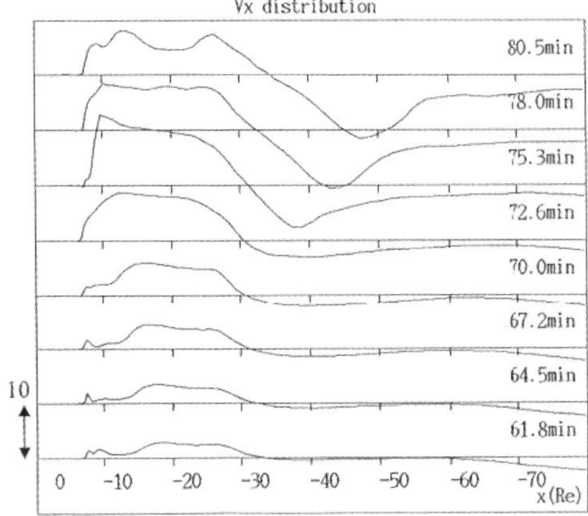

Figure 5. Time sequence of P and V_x distributions along the $-x$ axis around the onset. Pressure and velocity is normalized by solar-wind pressure and solar-wind sound velocity. Sudden transition is observable between t=70.0 minute and 72.6 minute.

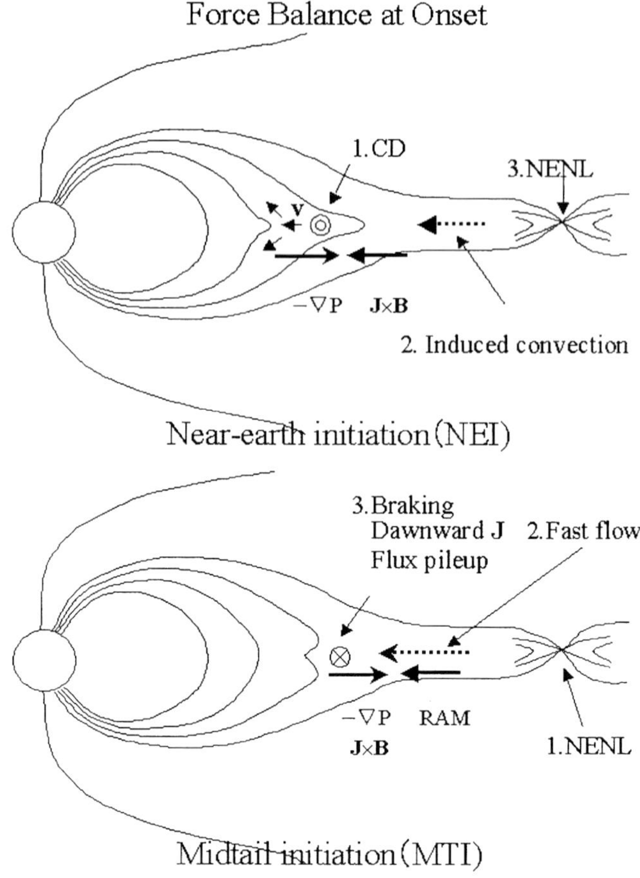

Figure 6. Schematic diagrams showing two different concepts for substorm initiation. Top and bottom panels represent near-earth-initiation and mid-tail-initiation models, respectively. If a primary cause for the onset is assumed, the resulting effects occur in sequence as show by numbers. Solid arrows show the force balance. In the lower panel (mid-tail initiation), dominance of kinetic energy is unavoidable in the near-earth region.

6. Mechanical Balance

In the near-earth region, plasma sheet is filled with plasma even during the growth phase. This plasma supports the **JxB** force in the x direction and constructs the convection system including the plasma population regimes [*Tanaka*, 2003b]. In the region between −12 and −30 Re, the distribution of plasma pressure changes little even after the onset, whereas a drastic change of plasma pressure occurs in the near-earth region inside −12 Re (Figure 5). These results strictly coincide with observations given by *Kistler et al.* [1992]. In the simulation result shown in Figure

1, therefore, the NENL is never a disconnected system from the earth but an attached system with the supporting plasma under the force balance. In this situation under the large $-\nabla P$ force, large magnetic tension and small acceleration in the x direction, the convection in the plasma sheet is in the subsonic regime and nearly incompressible. Actually, flow surrounding the slant slow shock is almost incompressible [*Lee*, 1995]. In the incompressible convection system, the flow configuration must change as a whole. Only a portion of the convection cannot become fast, since in such a case flow convergence is required at the front of the fast flow and flow divergence is required at the rear. Briefly speaking, a fluid element in the incompressible flow cannot move until a fluid element in the front position moves aside. The concept of state transition in the substorm onset has been suggested by *Atkinson* [1991], followed by *Sitonov et al.* [2000], *Shao et al.* [2003] and *Tanaka* [2000b]. The state transition model by *Tanaka* [2000b] regards the onset as a natural consequence of development process in convection.

During the growth phase, the dominant element in the force balance is that between plasma pressure and magnetic pressure in the z direction, under a small contribution of pressure variation in the x direction. The dipolarization is considered to be a change of this force balance. The contraction of tail-like field lines to a dipolar form compresses the plasma as stress balance in the x direction becomes important. After the dipolarization process, the increased tension balances with newly established pressure which is, in turn, a result of the work done by the plasma sheet convection that directs from low-pressure region to high-pressure region against $-\nabla P$ force [*Tanaka*, 2000b, 2003b]. Thus, the energy conversion between the magnetic and internal energies acts as a kind of self-adjustment system. In other words, restored magnetic tension confines the plasma to enable a convergence of convection. At the onset, the magnetosphere and ionosphere are partially out of synchronization since a compressional motion cannot be mapped down onto the ionosphere. The change in force balance from the z direction to x direction enables the motion even under this condition (non-shear motion). It is natural to observe earthward flow before the onset (Figure 5) because under the frozen-in condition the magnetic configuration cannot change without preceding flow which carries the magnetic field. The kinetic aspect of reconnection may be conspicuous in the pathway between the states. However, the discontinuous behavior at the onset is primarily attributed to a realization of initial and destination states in the MHD regime. The sudden dipolarization is interpreted as a self-organization phenomenon in a nonlinear system [*Tanaka*, 2000b]. In this view, the onset is a bifurcation process from a tail-like state to a dipolar state through the change of force balance. At the onset, magnetically driven motion due to the restored magnetic tension along the dipolarizing field lines accumulates plasma by transporting it along the magnetic field lines without a compression of magnetic field. This kind of motion corresponds to the slow mode variation. Recently such motion is confirmed from the satellite observations [*Nakamizo and Iijima*, 2003]. The development of pressure distribution within the plasma sheet shown in Figure 5 is similar to those observations by *Kistler et al.* [1992], and generates a maximum within 10 Re after onset.

These variations are rapid, and within 1 minute, a transition of pressure distribution occurs from the growth phase to the expansion phase. In this process, the dipolarization is a restoration of tension and never a relaxation to the potential field. In the NENL model of the substorm, on the contrary, the magnetic energy is converted to the kinetic energy through the NENL formation. The ground onset is attributed to the braking of this kinetic energy in the inner magnetosphere [*Baker et al.*, 1996]. On the other hand, the state transition model can generate the substorm onset without the dominance of kinetic energy everywhere in the tail. It generates the onset along a natural extension of the incompressible convection in the M-I coupling system. The NENL model and the CD model both assume that there must be a central player initiating the onset. In contrast, there is no central player in the state transition model. The state transition model resembles the economic model of major depressions, the Ising model for magnetization, or the avalanche model for substorms [*Chapman et al.*, 1998]. These models are all based on cooperative phenomena, and no central player exists. Instead, in these models, many similar elements coexist and interact with one another. The state transition of the interacting systems corresponds to the conditions of a major depression and of magnetization. The onset of substorm is the occurrence of state transition in the interacting system, as shown in Figure 5. The substorm does not involve as many elements as the economic model or the model for magnetic bodies. However, all elements in substorm have its characteristic topologies. Since the conditions differ significantly from those of typical complex systems found in economic models and models for magnetic bodies, we refer to the M-I system as a compound system. The state transition in the near-earth region explains quite well why the onset starts from the equatorwardmost preonset arc. In general, the state transition requires the existence of multiple solutions for one boundary condition. As a consequence, it requires a nonlinearity of the system. On the other hand, the relation between the NENL solution and inflow and outflow boundary conditions is in one-to-one correspondence which generates no state transition [*Lee*, 1995]. The state transition model can also explain the triggering effect caused by the northward turning of IMF. The deceleration of the ionospheric convection reduces the flux exiting the plasma sheet to the dayside, creating a region of flow convergence. The triggering effect by northward turning of IMF is also explained by *Lyons* [1995]. However, Lyons' theory attempts to explain the triggering effect as originating from the deceleration of convection caused by the penetration of the weakened solar wind electric field into the magnetosphere; this leads to the mistake of EJ paradigm pointed out by *Parker* [2000].

The pseudo breakup and steady magnetospheric convention (SMC, or convection bay) are phenomena resembling a substorm but which are grouped in a different category. The pseudo breakup displays a common morphology with substorms until the onset. However, it does not p produce poleward expansion or a westward traveling surge (WTS), and terminates before changes are propagated to the whole magnetosphere [*Pulkkinen et al.*, 1998]. The magnitude of disturbance in an SMC is comparable to that of the substorm, but an SMC progresses to the expansion phase without displaying a clear discontinuity, or onset [*Yahnin et al.*, 1994; *Sergeev et al.*, 1996]. In an SMC, a strong stationary two-cell ionospheric convection is realized. The pseudo breakup can be explained in terms of the NENL model as a mid-course termination of an initiated reconnection. In contrast, in the state transition model, it can be interpreted as follows. Normally, state transition is initiated in the inner magnetosphere with the extinction of the core part (distant-tail neutral line) shown in the upper panel of Figure 2, and the substorm proceeds to the expansion phase. However, if state transition occurs before the extinction of the core part (distant-tail neutral line) shown in Figure 2, the discontinuity at onset is realized, but the NENL cannot be formed at this stage due to the presence of the core structure. This results in a failure to proceed to the expansion phase, and instead, the event ends as a pseudo breakup. If convection is promoted without state transition or after the state transition has been initiated, and the state of no distant-tail neutral line is maintained, then the event is an SMC. In this way, the state transition model provides simple explanations for the pseudo breakups and SMC.

7. Conclusion

Starting from a stationary solution under a northward IMF condition with non-zero IMF B_y, changes in magnetospheric convection following the southward turning of the IMF are investigated by the resistive MHD. The simulation results first show the progress of plasma sheet thinning in the tail. Then, the substorm onset occurs as the depolarization accompanying an abrupt shift of the pressure peak in the plasma sheet from -12 Re to -8 Re and an intrusion of convection flow into the inner magnetosphere. These improved solutions are obtained by the resistivity that is given so as to increase as it goes further downtail. During the dipolarization process, the static magnetospheric force balance changes from the z-direction-dominated state to the x-direction-dominated state. Unlike the MHD substorm model based on the NENL concept, in which the onset occurs through the reconnection with a threshold value, the onset in this model is caused through the nonlinearity of MHD system itself. Thus the traditional concept that attributes the onset to local instabilities is not the conclusion of this paper. Any instability must occur to realize a new state that is described by the MHD. Consequently, simultaneous realization of initial and final states as the solution of MHD is essential, although the kinetic aspect may be important in the pathway between the states. It is concluded that the direct cause of the onset is the state (phase-space) transition of the convection system from a thinned state to a dipolarized state.

References

Akasofu, S. -I., The development of the auroral substorm, *Planet. Space Sci.*, *12*, 273, 1964.

Angelopoulos, V., W. Baumjohann, C. F. Kenel, F. V. Coroniti, M. G. Kivelson, R. Pellat, R. J. Walker, H. Luhr, and G. Paschmann, Bursty bulk flows in the inner central plasma sheet, *J. Geophys. Res.*, *97*, 4027, 1992.

Atkinson, G., A magnetosphere WAGS the tail model of substorms, in *Magnetospheric substorms, Geophys. Monogr. Ser.*, vol. 64, edited by J. R. Kan, T. A. Potemra, S. Kokubun, and T. Iijima, p. 191, AGU, Washington, D.C., 1991.

Baker, D. N., T. I. Pulkkinen, V. Angelopoulos, W. Baumjohann, and R. L. McPherron, Neutral line model of substorms: Past results and present view, *J. Geophys. Res.*, *101*, 12,975, 1996.

Baumjohann, W., G. Paschmann, and H. Luhr, Characteristics of high-speed ion flows in the plasma sheet, *J. Geophys. Res.*, *95*, 3801, 1990.

Chapman, S., and V. C. A. Ferraro, A new theory of magnetospheric storms, Part 1, The initial phase, *J. Geophys. Res.*, *36*, 77, 1931.

Chapman, S. C., N. W. Watkins, R. O. Dendy, P. Helander, and G. Rowlands, A simple avalanshe model as an analogue for magnetospheric activity, *Geophys. Res. Lett.*, *25*, 2397, 1998.

Crooker, N. U., J. G. Lyon, and J. A. Fedder, MHD model merging with IMF By: Lobe cells, sunward polar cap convection, and overdraped lobes, *J. Geophys. Res.*, *103*, 9143, 1998.

Elphinstone, R. D., J. S. Murphree, and L. L. Cogger, What is a global auroral substorm, *Rev. Geophys.*, *34*, 169, 1996.

Fedder, J. A., S. P. Slinker, and J. G. Lyon, A comparison of global numerical simulation results to data for the January 27-28, geospace environment modeling challenge event, *J. Geophys. Res.*, *103*, 14,799, 1998.

Gombosi, T. I., K. G. Powell, and B. van Leer, Comment on "Modeling the magnetosphere for northward interplanetary magnetic field: Effects of electrical resistivity" by Joachim Raeder, *J. Geophys. Res.*, *105*, 13,141, 2000.

Hashimoto, K., T. Kikuchi, and Y. Ebihara, Response of the magnetospheric convection to sudden IMF change as deduced from the evolution of partial ring currents, *J. Geophys. Res.*, *107*(A11), 1337, doi:10.1029/2001JA009228, 2002.

Kistler, L. M., E. Mobius, W. Baumjohann, and G. Paschmann, Pressure changes in the plasma sheet during substorm injection, *J. Geophys. Res.*, *97*, 2973, 1992.

Kivelson, M. G., and D. J. Southwood, Ionospheric traveling vortex generated by solar wind buffering of the magnetosphere, J. Geophys. Res., 96, 1661, 1991. Lee, L. C., A review of magnetic reconnection: MHD models, in *Physics of magnetopause, Geophys. Monogr. Ser.*, vol. 90, edited by P. Song et al., p. 139, AGU, Washington, D. C., 1995.

Lee, L. C., A review of magnetic reconnection: MHD models, in *Physics of magnetopause, Geophys. Monogr. Ser.*, vol. 90, edited by P. Song et al., p. 139, AGU, Washington D. C., 1995.

Lopez, R. E., and A. T. Y. Lui, A multi-satellite case study of the expansion phase of a substorm current wedge in the near-Earth magnetotail, *J. Geophys. Res.*, 95, 8009, 1990.

Lui, A. T. Y., Current disruption in the earth's magnetosphere: Observations and models, *J. Geophys. Res.*, *101*, 13,067, 1996.

Lyons, L. R., A new theory for magnetospheric substorms, *J. Geophys. Res.*, 100, 19,069, 1995.

Lyons, L. R., G. T. Blanchard, J. C. Samson, R. P. Lepping, T. Yamamoto, and T. Moretto, Coordinated observation demonstrating external substorm triggering, *J. Geophys. Res.*, *102*, 27,039, 1997.

Nakamizo, A., and T. Iijima, A new perspective on magnetotail disturbances in terms of inherent diamagnetic processes, *J. Geophys. Res. 108*(A7), 1286, doi:10.1029/2002JA009400, 2003.

Nagai, T., K. Takahashi, H. Kawano, T. Yamamoto, S. Kokubun, and A. Nishida, Initial geotail survey of magnetic substorm signatures in the magnetotail, *Geophys. Res. Lett.*, *25*, 2991, 1994.

Ogino, T., R. J. Walker, M. Ashour-Abdalla, and J. M. Dawson, An MHD simulation of the effects of interplanetary magnetic field B_y component on the interaction of the solar wind with the earth's magnetosphere during southward interplanetary magnetic field, *J. Geophys. Res.*, *91*, 10,029, 1986.

Ohtani, S., K. Takahashi, T. Higuchi, A. T. Y. Lui, H. E. Spence, and J. F. Fennell, AMPTE/CCE-SCATHA simultaneous observations of substorm-associated magnetic fluctuations, *J. Geophys. Res.*, *103*, 4671, 1998.

Parker, E. N., The alternative paradigm for magnetospheric physics, J. Geophys. Res., 101, 10,587, 1996.Parker, E. N., Newton, Maxwell, and Magnetospheric Physics, in *Magnetospheric current systems, Geophys. Monogr. Ser.*, vol. 118, edited by S. Ohtani et al., p.1, AGU, Washington, D. C., 2000.

Pulkkinen, T. I., D. N. Baker, M. Wiltberger, C. Goodrich, R. E. Lopez, and J. G. Lyon, Pseudobreakup and substorm onset: Observations and MHD simulations compared, *J. Geophys. Res.*, *103*, 14,847, 1998.

Sergeev, V. A., R. J. Pellinen, and T. I. Pulkkinen, Steady magnetospheric convection: A review of recent results, *Space Sci. Rev.*, *75*, 551, 1996.

Shao, X, M. I. Sitnov, S. A. Sharma, K. Papadopoulos, C. C. Goodrich, P. N. Guzdar, G. M. Milikh, M. J. Wiltberger, J. G. Lyon, Phase transition-like behavior of magnetospheric substorms: Global MHD simulation results, *J. Geophys. Res.*, *108*(A1), doi:10.1029/2001JA009237, 2003

Siscoe, G. L., N. U. Crooker, G. M. Erickson, B. U. O. Sonnerup, K. D. Siebert, D. R. Weimer, W. W. White, and N. C. Maynard, Global geometry of magnetospheric currents inferred from MHD simulations, in *Magnetospheric current systems, Geophys. Monogr. Ser.*, vol. 118, edited by S. Ohtani et al., p.41, AGU, Washington, D. C., 2000.

Sitnov, M. I., A. S. Sharma, K. Papadopoulos, D. Vassiliadis, J. A. Valdivia, A. J. Klimas, and D. N. Baker, Phase transition-like behavior of the magnetosphere during substorms, *J. Geophys. Res.*, *105*, 12,955, 2000.

Takahashi, K., L. J. Zanetti, R. E. Lopez, R. W. McEntire, T. A. Potemra, and K. Yumoto, Disruption of the magnetotail current sheet observed by AMPTE/CCE, *Geophys. Res. Lett.*, *14*, 1019, 1987.

Tanaka, T., Finite volume TVD scheme on an unstructured grid system for three-dimensional MHD simulation of inhomogeneous systems including strong background potential fields, *J. Comput. Phys.*, *111*, 381, 1994.

Tanaka, T., Generation mechanisms for magnetosphere-ionosphere current systems deduced from a three-dimensional MHD simulation of the solar wind-magnetosphere-ionosphere coupling processes, J. Geophys. Res., *100*, 12,057, 1995.

Tanaka, T., Configuration of the magnetosphere-ionosphere convection system under northward IMF condition with non-zero IMF By, *J. Geophys. Res.*, *104*, 14,683, 1999.

Tanaka, T., Field-aligned current systems in the numerically simulated magnetosphere, in *Magnetospheric current systems, Geophys. Monogr. Ser.*, vol. 118, edited by S. Ohtani et al., p.53, AGU, Washington, D. C., 2000a.

Tanaka, T., The state transition model of the substorm onset, *J. Geophys. Res.*, *105*, 21,081, 2000b.

Tanaka, T., Development of a global MHD simulation model for the S-M-I coupling process in the substorm prediction program at CRL, *Adv. Space Res.*, *26*(1), 151, 2000c.

Tanaka, T., IMF B_y and auroral conductance effects on high-latitude ionospheric convection, *J. Geophys. Res.*, *106*, 24,505, 2001.

Tanaka, T., Finite volume TVD scheme for magnetohydrodynamics on unstructured grids, in *Space plasma simulation, Lecture notes in physics*, edited by J. Buchner et al., p. 275, Springer, 2003a.

Tanaka, T., Formation of magnetospheric plasma population regimes coupled with the dynamo process in the convection system, *J. Geophys. Res.*, *108*(A8), 1315, doi:10.1029/2002JA009668, 2003b.

Watanabe, M., G. J. Sofko, D. A. Andre, T. Tanaka, and M. R. Hairson, Polar cap bifurcation during steady-state northward interplanetary magnetic field with $|B_y| \sim B_z$, *J. Geophys. Res.*, *109*(A1), A01215, doi:10.1029/2003JA009944, 2004.

Yahnin, A., M. V. Malkov, V. A. Sergeev, R. J. Pellinen, O. Aulamo, S. Vennerstrom, E. Friis-Christensen, K. Lassen, C. Danielsen, J. D. Craven, C. Deehr, and L. A. Frank, Features of steady magnetospheric convection, *J. Geophys. Res.*, *99*, 4039, 1994

Multiscale Coupling of Sun-Earth Processes
A.T.Y. Lui, Y. Kamide and G. Consolini, editors.

EXPLOSIVE INSTABILITIES AND SUBSTORM INTENSIFICATIONS IN THE EARTH'S MAGNETOTAIL

J.C. Samson and P. Dobias

Department of Physics, University of Alberta, Edmonton, Alberta, T6G-2J1, Canada.

Abstract. We show that during the substorm growth phase, the near Earth magnetotail, between about 8 and 12 R_E , evolves toward a configuration that is potentially explosively unstable to nonlinear magnetohydrodynamic instabilities. Our model requires the existence of an auroral arc on field lines mapping to the Earthward edge of the plasma sheet and we base our model of this arc on a shear Alfvén, field line resonance model. In evaluating the nonlinear stability or instability of the region of the near Earth magnetotail we use an analysis based on a Lagrangian-Hamiltonian approach for ideal magnetohydrodynamics and a magnetotail equilibrium model based on the Grad-Shafranov equations.

Keywords. Substorms, nonlinear magnetohydrodynamics, explosive instabilities

1. Introduction

Following a growth phase with typical durations between 30 minutes to over an hour, the initiation of the substorm intensification begins in an arc embedded within a region of energetic proton precipitation at the equatorward border of the auroral zone [Samson et al., 1992a]. At least some of these arcs are produced by ULF (1-4 mHz) field line resonances (FLRs) [Samson et al., 2003]. The initiation of a global substorm expansive phase is generally preceded by a number of pseudobreakups, with small scale (100s of km) auroral vortex structuring [Voronkov et al., 2000]. We shall call these pseudobreakups "precursors". Eventually one of the small scale vortex events initiates a true substorm intensification and expansive phase with a global reconfiguration of the magnetotail, including dipolarization. The evolution of the expansive phase is very rapid with time scales of 10s of seconds [Friedrich et al., 2001]. These times scales are typical Alfvénic time scales when mapped from the ionosphere to the equatorial magnetosphere, that is time scales of the order L/V_A where L is the characteristic length and V_A is the Alfvén speed. Consequently these observations suggest that the substorm expansive phase begins as an explosive magnetohydrodynamic (MHD) instability. These explosive instabilities are nonlinear and evolve in time as $(t-t_o)^{-\alpha}$ (α is positive) which is much faster than the exponential growth of linear instabilities [Cowley and Artun, 1997]. Cowley and Artun [1997] give a clear explanation of why explosive instabilities must be nonlinear, particularly in systems with slow adiabatic storage of energy, as is found in the substorm growth phase, thus drawing into doubt any models of the substorm intensification based on linear instabilities, including linear ballooning modes. The finite time singularities in these explosive modes indicate that eventually new physics must come into play. One possible candidate for this new physics is the cross field current instability advocated by Sitnov and Lui [1999].

Even though we shall not address the later evolution of the substorm intensification and expansive phase, a few comments are in order. The observations of Friedrich et al. [2001] show that the signature of lobe flux reconnection in the ionosphere follows rapidly (within minutes) after the initiation of the substorm intensification near the equatorward edge of the auroral region. These conclusions are based on observations of 630.0 nm emissions to mark the border of open

236

field lines. Consequently, the ionospheric data suggest that the near Earth plasma sheet might play the important role in the initiation of substorm intensifications, with the process evolving tailward. The synthesis of Lui [1991] gives a reasonable scenario for this. Conversely, a number of models suggest that the near Earth neutral line (NENL) at 20-30 R_E starts the process, with destabilization or instabilities in the near Earth plasma sheet following, possibly as a result of enhanced Earthward flows due to the NENL [Pu et al., 1999]. In our minds, a tailward progression of the events, as discussed in Lui [1991], gives the simplest and most plausible model. Most models, whether NENL or near Earth plasma sheet initiation, consider some type of ballooning mode to be the likely process for the near Earth instabilities [Ohtani and Tamao, 1993; Liu, 1997]. Unfortunately, most ballooning models are based on linear instabilities. A major weakness of any scenario advocating a NENL initiation of the substorm intensification is the lack of adequate explanations of why the substorm intensification in the ionosphere must start within a preexisting auroral arc, and the lack of any mechanisms to produce this arc, particularly the organized topology of the arc. Note that this arc is very thin (typically 10 km in latitude in the ionosphere), is relatively stable in time (over many minutes to tens of minutes), and extends over many time zones. Furthermore this arc exists long before (10s of minutes) the substorm intensification is seen in the ionosphere. In our analysis in this paper we will show that the auroral arc and the associated equatorial fields are essential to the initiation of the substorm intensification. Furthermore, we will give a mechanism to produce these arcs through the FLR process.

Cherry [1959, 1968] has shown that the presence of resonances in a Hamiltonian system can have destabilizing effects for the system as a whole. In line with Cherry's results, we believe that the low frequency, long wavelength, more global modes (FLRs in the range 1-4 mHz) are a generic feature required for the initiation of these types of nonlinear instabilities in plasma systems with a slow adiabatic storage of energy. In essence, these global modes break the symmetry in the system, allowing azimuthal structuring of the topology and pressure gradients of the system (e.g. the magnetosphere). A similar process occurs in high-β disruptions in Tokamaks [Park et al., 1995; Fredrickson et al, 1996]. In the Tokamak case, toroidally localized, high-n (toroidal) ballooning modes are driven unstable in limited regions by toroidally localized changes in the pressure gradient due to low frequency, low-n modes. These ballooning modes evolve nonlinearly, producing a strong localized pressure bulge, destroying flux surfaces, followed by a thermal quench and the formation of magnetic islands. A comparison of the Tokamak instability with the substorm instability suggests similarities in the mechanism of the two processes, noting however, that in the case of high-β disruptions the instability is driven by local modifications in the pressure gradients, whereas in the substorm the instability is possibly driven by modification of field line curvature by the FLR, as we shall discuss later.

We suggest that the sequence of events for the substorm is:

(i)   The slow adiabatic storage of energy: the growth phase (10s of minutes),
(ii) Azimuthal symmetry breaking: arc formation and FLRs (minutes),
(iii) The precursor phase: ballooning and pseudobreakups (minutes),
(iv) Initiation of the explosive MHD instability: the substorm intensification and explosive
       ballooning modes (10s of seconds),
(v)   Destruction of flux surfaces: possible tailward propagation of a rarefaction wave and
       initiation of tearing modes or cross field current instabilities leading quickly to the beginning
       of dipolarization (10s of seconds),
(vi) magnetic "island" formation: lobe flux reconnection with the NENL now at 20-30 R_E.
       (minutes).

The sequence of events we outlined above is essentially the same as that outlined by Lui [1991] but we have used a different terminology to be more compatible with our discussion of the generic features of explosive instabilities and plasma relaxation. The study outlined in our paper brings us as far as (iv) without a full description of the evolution of the ballooning modes.

In the following section we outline the use of the Lagrangian of ideal MHD to determine nonlinear stability or stability and the derivation of this Lagrangian to fourth order (section 2). In section (3) we review the equations and fields associated with a FLR. Section (4) outlines our equilibrium model for the near Earth magnetotail. Section (5) shows results from one substorm event including equilibrium models and an evaluation of nonlinear instability in the presence of FLRs. Discussion and conclusions follow.

2. The Lagrangian to Fourth Order

Our derivation of the Lagrangian follows the derivation in Dobias et al. [2004]. We combine the variational (energy based) approach for the Lagrangian formulation of MHD, which has a long history in stability analysis problems [Bernstein et al., 1958; Low, 1958; Arnold, 1969; Pfirsch and Sudan, 1993], with the component formalism of differential geometry [Misner et al., 2002; Shultz, 1980]. The methods of differential geometry in MHD [Zeitlin and Kambe, 1993] greatly simplify the derivation of the fourth order Lagrangian and give results that are much easier to understand and to use compared to those found using standard vector tensor formalisms. A comparison of the third order term that we derive below with that of Pfirsch and Sudan [1993], who use standard vector tensor methods, serves to illustrate our point.

We consider displacements from equilibrium in the form of a coordinate change

$$\hat{\mathbf{r}} = \mathbf{r} + \xi(\mathbf{r},t), \tag{2.1}$$

where $\mathbf{r}$ is the equilibrium coordinate, $\hat{\mathbf{r}}$ is the displaced coordinate, and ξ is the perturbation displacement. The Lagrangian of ideal MHD is [Pfirsch and Sudan, 1993]

$$L = \int dV \left(\frac{1}{2}\rho v^2 - \frac{p}{\gamma-1} - \frac{1}{2}\mathbf{B}^2 \right) \tag{2.2}$$

In (2.2), V is volume, $\mathbf{v}$ is velocity, $\mathbf{B}$ is the magnetic field, p is the plasma pressure, ρ is the plasma density, and γ is the polytropic index of adiabatic compression. Lagrangian (2.2) can be expressed in terms of the plasma displacement by using suitable constraints including:
Conservation of mass

$$\int \rho dV = const \tag{2.3}$$

giving

$$\hat{\rho}(\xi) = J^{-1}(\xi)\rho, \tag{2.4}$$

conservation of magnetic flux

$$\int \mathbf{B} \cdot d\mathbf{S} = const \tag{2.5}$$

giving

$$\mathbf{B}(\xi) = \mathbf{A}^{-1}(\xi) \cdot \mathbf{B}, \tag{2.6}$$

and the adiabatic constraint

the metric tensor q_{ij}=diag(1,1,1), and vectors and covectors are equivalent. Assuming a harmonic time dependence of the form $e^{-i\omega t}$, the wave equation becomes

$$\rho\omega^2\xi^i = -\partial^i\left(\left(\gamma p + B^2\right)\partial_l\xi^l - B_j B^p \partial_p\xi^j\right)$$
$$- B^l \partial_l \partial_p\left(\xi^i B^p - \xi^p B^i\right)$$
$$- \left(\partial_j B^i\right)\partial_p\left(\xi^j B^p - \xi^p B^j\right). \tag{3.2}$$

We choose a coordinate system (x,y,z) where x corresponds to the radial direction in the equatorial plane, y is the azimuthal direction, and z is in the direction of the magnetic field. We also assume

$$\xi(\mathbf{r}) = \xi(x)\exp\left(i\left(k_y y + k_z z\right)\right) \tag{3.3}$$

and that the gradients in plasma parameters are only in the x-direction. Solving for the radial displacement vector, ξ_x, gives the equation

$$\rho\left(\omega^2 - k_z^2 V_A^2\right)\xi_x = -\partial_x\left(F(x)\partial_x\xi_x\right) \tag{3.4}$$

where

$$F = \frac{\rho_0\left(\omega^2 - k_z^2 V_A^2\right)}{G}, \tag{3.5}$$

$$G(x) = \frac{\omega^2\left[\omega^2 - \left(k_y^2 + k_z^2\right)\left(V_A^2 + V_S^2\right)\right] + k_z^2\left(k_y^2 + k_z^2\right)V_S^2 V_A^2}{\omega^2\left(V_A^2 + V_S^2\right) - k_z^2 V_S^2 V_A^2}, \tag{3.6}$$

$V_A^2(x) = B^2/\rho(x)$, $V_S^2(x) = \gamma p(x)/\rho(x)$, and V_A and V_S are respectively the Alfvén and sound speed. This equation has two turning points x_T, $(G(x_T) = 0)$, and two resonances at positions, x_R, $(F(x_R) = 0)$, corresponding to the shear Alfvén resonance ($\omega^2 - k_z^2 V_A^2(x) = 0$) and the cusp resonance ($\omega^2\left(V_A^2(x) + V_S^2(x)\right) - k_z^2 V_S^2(x)V_A^2(x) = 0$) where compressional energy of the fast mode couples to the slow mode. We will consider only the shear Alfvén resonance.

Near the shear Alfvén resonance, the equation has the approximate form

$$\frac{d^2\xi_x}{dx^2} + \frac{1}{x - x_R + i\eta}\frac{d\xi_x}{dx} - k_y^2\xi_x = 0 \tag{3.7}$$

where $\eta = \operatorname{Im}\varepsilon(x_R)/(d\operatorname{Re}\varepsilon/dx)_{x=x_R}$, $\operatorname{Re}\varepsilon = \omega^2/V_A^2(x) - k_z^2$ and the imaginary component of ε takes into account losses. The solution is

$$\xi_x = \xi_0 \ln\left(k_y\left(x - x_R + i\eta\right)\right). \tag{3.8}$$

The solutions for ξ_y and E_x, for example

$$E_x(x) = \frac{ik_y E_0}{\left(\text{Re}\,\varepsilon - k_y^2\right)\left(x - x_R + i\eta\right)} \tag{3.9}$$

have a singularity at the resonance if $\eta = 0$.

If ionospheric conductivities are high, the azimuthal velocity and displacement of a fundamental mode FLR have an approximate node at the ionosphere and an antinode or maximum at the magnetic equator. Profiles of fields associated with FLRs along dipole magnetic field lines can be found in Samson et al. [1996]. Profiles for a stretched magnetotail topology can be found in Samson et al. [2003]. Based on the results of Samson et al. [1996] and observations of azimuthal velocities in FLRs by Mitchell et al. [1990], in the equatorial plane the maximum of ζ_y in the FLR can be greater than 1 R_E. Though ζ_y dominates in the FLR at the resonance, a finite ζ_x is required for the possibility of nonlinear instabilities (see (2.11)-(2.14)). The ratio $|\zeta_y|/|\zeta_x|$ depends on the azimuthal wavenumber k_y. For typical azimuthal wavenumbers (see for example [Samson et al., 2003]) $|\zeta_y|/|\zeta_x| \approx 10$. Figure 1 shows a profile of $\zeta_\varphi(\zeta_y)$ and $\zeta_r(\zeta_x)$ in the equatorial

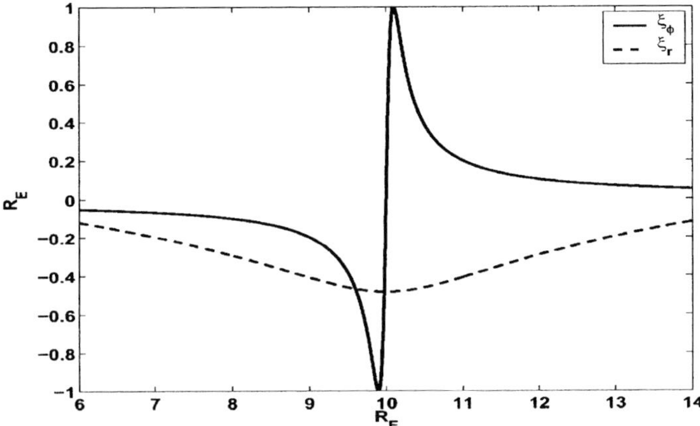

Figure 1. Radial (x-component) profiles of the displacements $\zeta_\varphi(\zeta_y)$ and $\zeta_r(\zeta_x)$ in a FLR. Bottom axis: radial distance. Left axis: displacement.

plane for the FLR model we will use in this study. This corresponds to the time of maximum displacement. Note the abrupt 180 degree phase shift at the resonance position. Earthward of 10 R_E the azimuthal displacement is +1 R_E and anti-Earthward the azimuthal displacement is -1 R_E. This leads to strong radial shears in the magnetic field directions, and substantial changes in the distortion of the curvature of the magnetic fields. These features are a contributing factor to the possibility of nonlinear instabilities, and, in fact, our calculations show that the curvature term in the third order Lagrangian, Ω_{BC} grows most rapidly as the amplitude of the displacement in the FLR increases and probably plays the major role in the evolution toward nonlinear instability. Though there is some modification of the radial pressure profile due to ζ_x, these changes seem to play only a minor role.

We note at this point that the shear Alfvén FLR model we use above has no compressional component, but these FLRs are driven by monochromatic compressional modes, possibly compressional eigenmodes in the magnetotail [Liu et al., 1995] These compressional modes couple to shear Alfvén modes in regions with large Alfvén velocity gradients perpendicular the ambient magnetic fields. Preferred regions are the plasma sheet boundary layer, and the inner edge of the plasma sheet [Liu et. al., 1995], within the region of energetic proton precipitation (the field lines that map to the H_β, region in the ionosphere). In our study, the perturbation displacement in the FLR only serves to show that these large scale modes can lead to conditions that allow possible nonlinear instabilities. Others (for example, Crabtree et al. [2003]) consider compressional modes as possible instabilities leading to substorm processes near the Earth.

4. The Equilibrium Model of the Magnetotail

In our analysis of possible nonlinear instabilities during and after the substorm intensification, we will use an equilibrium model based on the Grad-Shafronov (GSh) equation. We develop a model with two free parameters and use CANOPUS H_β data [Samson et al., 1992a; Wanliss et al., 2000] to estimate these parameters and to construct equilibrium models of the near Earth magnetotail during the substorm growth phase, and just after the substorm intensification and expansive phase. We then use these models and our fourth order Lagrangian to determine whether explosive, nonlinear instabilities are possible in the near Earth magnetotail (8-12 R_E).

The GSh equation describes equilibrium configurations in a magnetized plasma. For a system with azimuthal symmetry the GSh equation can be written in terms of the magnetic flux function ψ as

$$\partial_{rr}\psi + \frac{\sin\theta}{r^2}\partial_\theta \frac{1}{\sin\theta}\partial_\theta \psi = 4\pi^2 r^2 \sin^2\theta \partial_\psi p \tag{4.1}$$

where r and θ are the radius and polar angle in spherical coordinates, and $p=p(\psi)$. The pressure, p, remains constant along a magnetic field line and consequently p is a function of the magnetic flux, which is also constant along a field line. The magnetic field is related to ψ by

$$\mathbf{B} = -\frac{1}{2\pi}\nabla\psi \times \nabla\phi \tag{4.2}$$

where ϕ is the azimuthal angle and $\nabla\phi = \hat{\phi}/r$ where $\hat{\phi}$ is a unit vector in the azimuthal direction. In the case where pressure is constant, the GSh equation has a dipolar solution

$$\psi_0(r,\theta) = 2\pi \frac{M}{r}\sin^2\theta \tag{4.3}$$

where M is the magnetic moment. Another solution is

$$\psi_1 = -\pi B_0 r^2 \sin^2\theta . \tag{4.4}$$

If the plasma pressure is a linear function of ψ, then one further solution is

$$\psi_2 = 0.4\pi^2 \partial_\psi p r^4 \sin^2\theta . \tag{4.5}$$

Although ψ_1 and ψ_2 diverge at infinity, they are extremely useful in modeling the nightside magnetosphere, at least as far as the position of the neutral line or x-line.

A combination of these three solutions to the GSh equation has a dipole structure near the Earth and an x-line or neutral line in the equatorial plane at a distance from the Earth. The solution has only two free parameters, which we shall estimate from CANOPUS H_β data. Now the equation for ψ is

$$\psi(r,\theta) = 2\pi \frac{M}{r}\left[1 + \frac{1-\alpha}{2}\left(\frac{r}{R_x}\right)^3 + \frac{\alpha}{4}\left(\frac{r}{R_x}\right)^5\right]\sin^2\theta \qquad (4.6)$$

where α accounts for the pressure gradient and R_x is the position of the neutral line. This solution is not valid over all space, but gives very reasonable models out to about 20 R_E.

The plasma pressure consists of a constant part, p_0, and an inhomogeneous part that is proportional to the coefficient α, with

$$p(r,\theta) = p_0 + \frac{5}{4\pi}\frac{\alpha M}{R_x^5}\psi(r,\theta). \qquad (4.7)$$

Since R_x is the radial position of the minimum of ψ, α must be positive to ensure an Earthward directed pressure gradient in the region $r < R_x$. The equation for p also allows us to determine the azimuthal, diamagnetic current.

The position of the x-line is defined as a line in the equatorial plane where the B-field perpendicular to the equatorial plane, $\theta = \pi/2$, is zero. In the equatorial plane,

$$B_\theta = -\frac{1}{2\pi r}\partial_r\psi = \frac{M}{r^3}\left[1 - (1-\alpha)\left(\frac{r}{R_x}\right)^3 - \alpha\left(\frac{r}{R_x}\right)^5\right] \qquad (4.8)$$

clearly illustrating that $r = R_x$ is the position of the neutral line.

5. A Case Study of a Substorm

To estimate the two parameters, α and R_x, during the evolution of a substorm we use the CANOPUS H_β data. The band of H_β emissions just equatorward of the diffuse 630.0 nm emissions is due to the precipitation of relatively energetic precipitating protons, with energies greater than about 5 keV at the poleward edge of the band and increasing Earthward due to adiabatic heating. The equatorward edge of the band is determined by the trapping boundary Samson et al. [1992a] show some comparisons of DMSP proton data with CANOPUS H_β emissions. The precipitating protons are due to nonadiabatic scattering of the protons in regions of strong magnetic field curvature in the equatorial plane of the magnetotail. Buchner and Zelenyi [1987] showed that when $1 < K < 3$, where $K^2 = R/r_g$, R is the radius of curvature of the field lines, and r_g is the proton gyroradius, nonadiabatic scattering will occur. In the near Earth plasma sheet, K increases Earthward. To determine our parameters for the equilibrium models we assume that the equatorward border of the H_β emissions is on field lines mapping to an equatorial plasma sheet where $K=3$. We place the poleward border of the emissions on field lines

244

where the precipitating protons have an energy of 5 keV. This energy was determined empirically by comparing CANOPUS H_β data with DMSP particle data. Note that there is proton precipitation poleward of the poleward boundary of this H_β emission band. The energies (typically 1 keV or so) are not large enough to produce observable emissions. We determine the latitudinal boundaries of the H_β emissions by fitting a Gaussian to the emission data (see Wanliss et al. [2000]) and choosing the boundaries to be at 0.3 of the latitudinal maximum of the emission intensity. Parameters α and R_X are determined by computing the equilibrium model in which field lines at the latitude of the equatorward border map to a $K=3$ region in the equatorial plane, and field lines at the poleward border map to the 5 keV position (computed from the equilibrium pressure profile and assuming a proton number density of 1/cc in the equatorial plane).

The CANOPUS meridian scan, H_β data for the substorm we shall study is shown in Figure 2. This event has also been studied by Lui et al. [1998]. Further details on this event, including an example of the fit of a Gaussian profile to the equatorward H_β band, can be found in Friedrich et al. [2001]. The two times we consider are 04:35 UT, just minutes before the substorm intensification (occurring at 04:37 UT), and 04:40 UT, just minutes after the intensification when the near Earth magnetotail has reached a lower energy equilibrium configuration. The equilibrium, magnetic field models, radial magnetic field profiles in the equatorial plane, and pressure profiles in the equatorial plane are shown in Figures 3, 4, and 5 respectively.

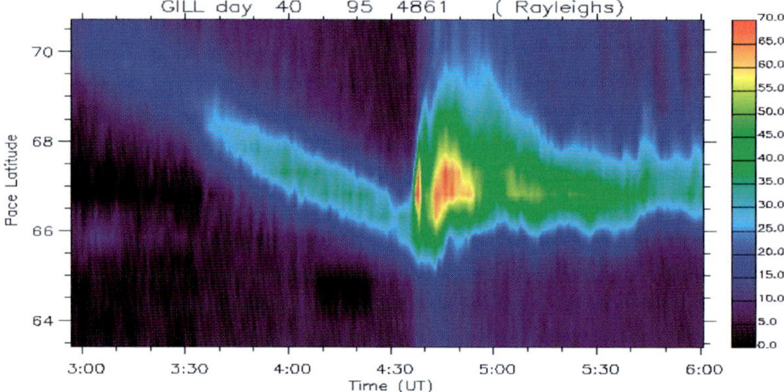

Figure 2. H_β data from the CANOPUS meridian scanning photometer at Gillam for the substorm event on 9 February, 1995 (see also Friedrich et al. [2001]). Eccentric dipole latitude (EDL) is Pace Latitude – 3 degrees.

The equilibrium models show much enhanced stretching of the magnetotail just prior to the substorm intensification (at 04:37 UT), with a marked reduction in stretching after the intensification. This reflects the explosive return to a lower energy state, but, of course, the magnetosphere does not return to the lowest energy state, a dipole. The models show a pronounced dipolarization of field lines inside about 10 R_E, following the substorm intensification. The profiles in Figure 4 show that the fields at geosynchronous orbit relax to an almost dipolar configuration after the intensification. We will show below that at 04:40 UT, the

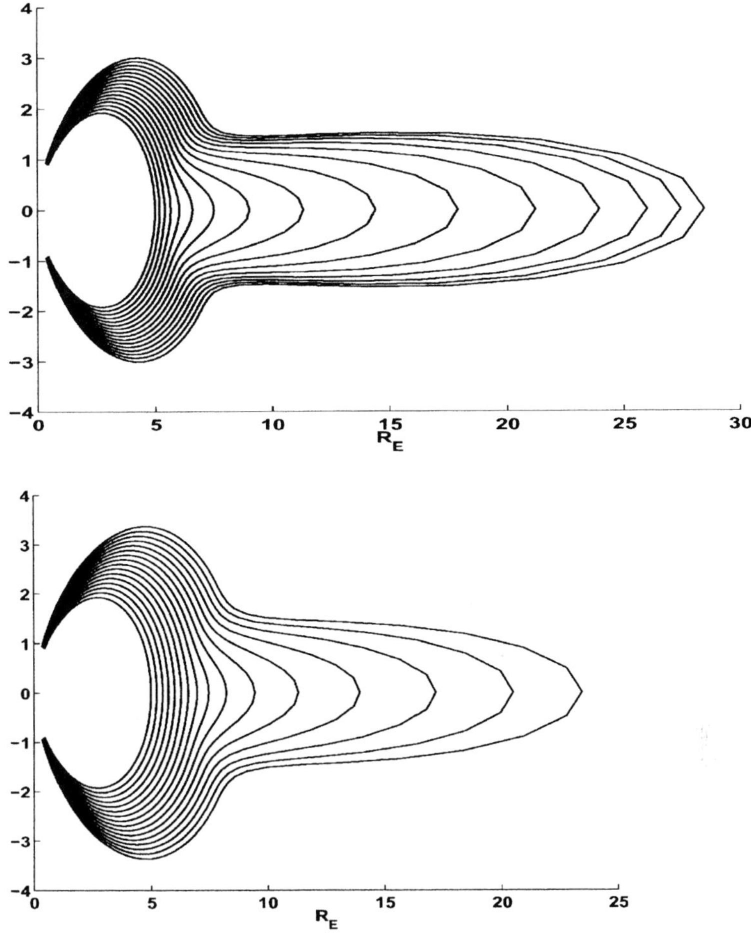

Figure 3. Top: Equilibrium magnetic field model for 04:35 UT, 9 February, 1995. Field lines range in latitude from 63.4 to 70.2. Bottom: Equilibrium magnetic field model for 04:40 UT, 9 February, 1995. Field lines range in latitude from 63.4 to 70.2 degrees.

inner plasma sheet has returned to a configuration that is not susceptible to nonlinear instabilities. At 04:35 UT, prior to the intensification, our analysis shows that the inner plasma sheet, near 10 R_E has become nonlinearly unstable in the presence of an FLR. The plasma pressure profiles show a maximum of 7 nP at less than 6 R_E at 04:35, and a maximum β of 50 at 12 R_E. At 04:40 UT, the maximum pressure has dropped markedly, to approximately 2.5 nP at 7.5 R_E.

While many more tests must be done to determine the accuracy and robustness of our equilibrium models, some possibly surprising results are suggested here. Perhaps the most remarkable observation is the short time required for magnetic field reconfiguration and for very

246

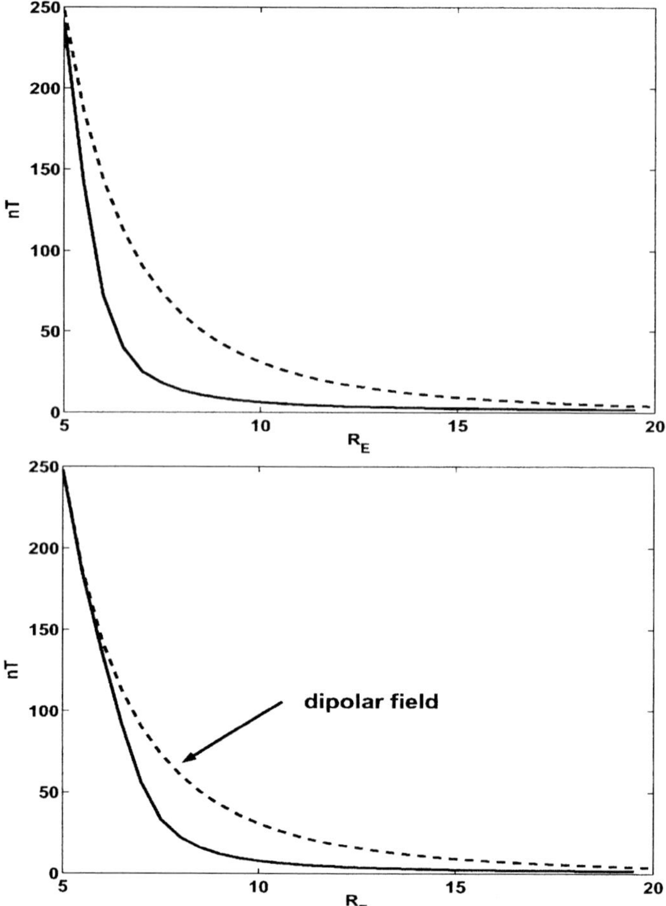

Figure 4. Radial magnetic field profiles for the equatorial plane. Top: 04:35 UT. Bottom: 04:40 UT.

substantial reductions in plasma pressure. Noting that the intensification was at 04:37 UT and that the H_β band had made its maximum northward excursion within 1 or 2 minutes after the intensification [Friedrich et al., 2001] indicates that the reconfiguration occurs on time scales of 10s of seconds, with the whole process lasting on the order of two minutes. This requires Alfvén speeds for the disturbance to propagate over the substantial distances involved (10s of R_E). Furthermore, whatever this explosive process is, it must substantially cool or remove hot plasma over a large region, starting at about 6 R_E and extending beyond 15 R_E down the magnetotail within this time frame. The scenario we outlined in section 1 might be able to do this, but, of course, we have to fill in a lot more physics. Just how decelerated Earthward flows from a NENL at 20-30 R_E might do this is not clear to us (see Pu et al., [1999] and references therein).

The radial profiles for the potential energies for the Lagrangian (2.11) using the FLR in Figure 1 are shown in Figure 6. In determining possible nonlinear instabilities we used the following criteria). If the displacement in the FLR brings a

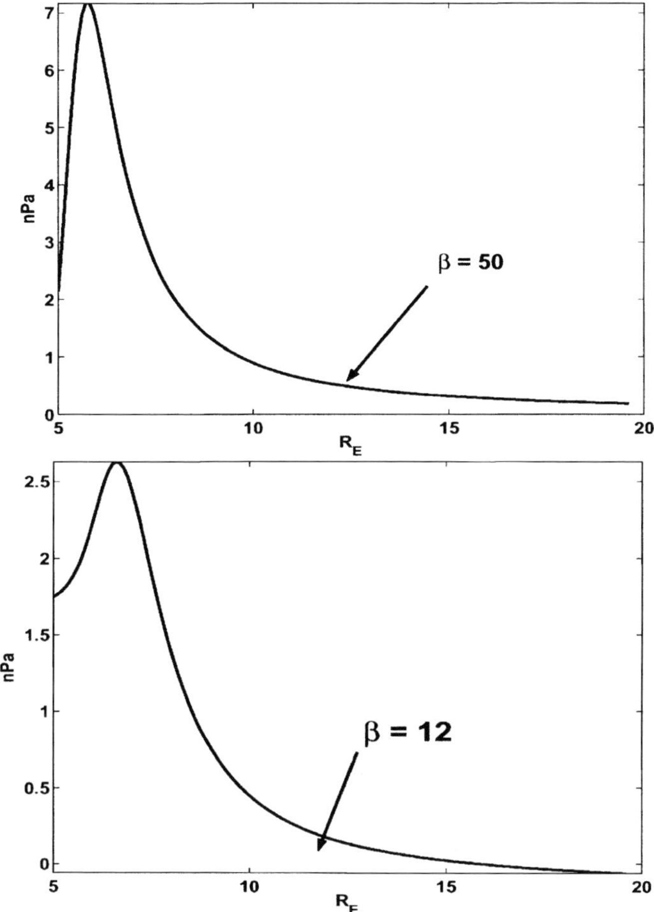

Figure 5. Radial pressure profiles in the equilibrium model. Top: 04:35 UT. Bottom: 04:40 UT.

violation of the ordering $|\Omega^{(2)}| \ll |\Omega^{(3)}| \ll |\Omega^{(4)}|$, then a nonlinear instability is possible if $|\Omega^{(3)}|$ is the dominant term. If $|\Omega^{(4)}|$ is the dominant term, then the nonlinear instability, if it occurs, will rapidly saturate with no appreciable growth in amplitude. If $|\Omega^{(2)}|$ is the dominant term, then the system behaves as a linear system.

Inspection of Figure 6 shows that the region near $10\ R_E$ has become susceptible to a nonlinear instability by the later stages of the growth phase (04:35 UT). This nonlinear destabilization threshold is passed when the amplitude of the azimuthal displacement of the FLR has reached 0.5 R_E. Below that amplitude, the second order term dominates. Without a FLR with displacement amplitudes well beyond 0.5 R_E at 04:35 UT, there would likely be no nonlinear instabilities and the growth phase would continue with further tail-field stretching. Presumably the required amplitude for a FLR to cause nonlinear instabilities would continue to decrease as the growth phase continued. Figure 6 also shows that the region near 10 RE should not allow further evolution of a nonlinear instability after the intensification, as now $|\Omega^{(4)}|$ is the dominant term.

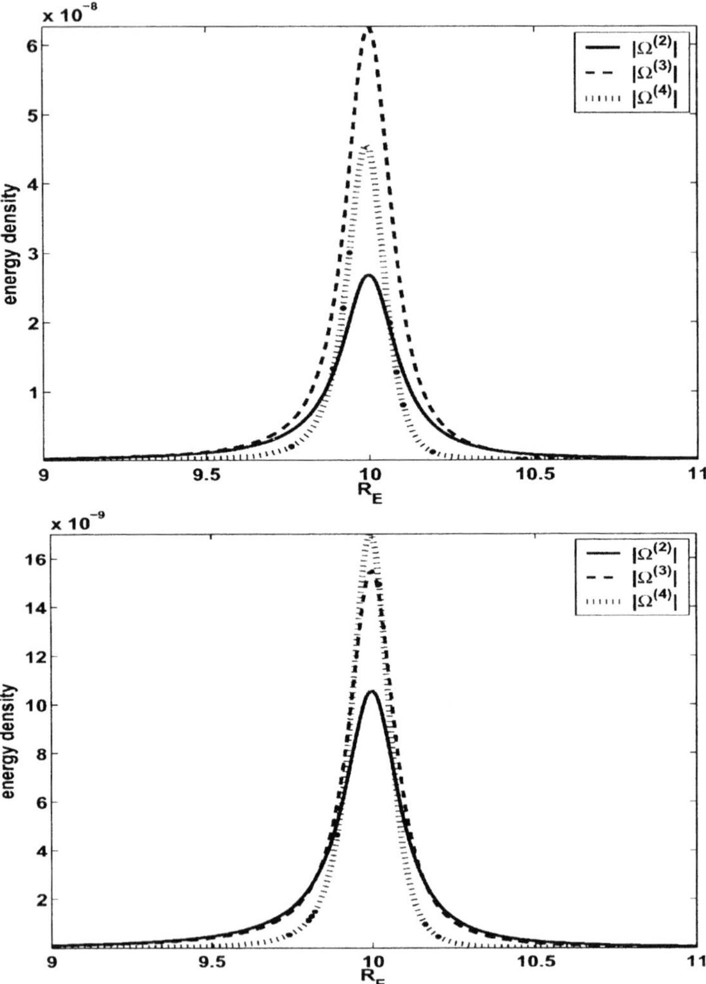

Figure 6. Equatorial, radial profiles of the magnitudes of the second, third and fourth order terms for the potential energy in the Lagrangian (2.11). Top: 04:35 UT. Bottom: 04:40 UT. The FLR fields are given in Figure 1. The FLR, azimuthal displacement is 1 R_E.

The third order curvature term in Ω_{BC} grows the fastest and plays the major role in the eventual evolution toward possible nonlinear instability. This is expected, as the pressure distribution in our model is axially symmetric and the azimuthal component of the displacement in the FLR is dominant (see Figure 1). This axisymmetry is a reasonable approximation for the dusk to local midnight sector of the magnetosphere, but may not be appropriate near local midnight where there could be substantial azimuthal gradients in the plasma pressure due to the gradient and curvature drift of the adiabatically heated ions in the near Earth region (less than 10 R_E). These azimuthal gradients would enhance the possibility of a nonlinear instability, as the azimuthal displacements in the FLRs would also lead to modifications in the pressure gradients. Consequently we might expect that the region just before local midnight would be most susceptible to substorm intensifications.

6. Discussion and Conclusions

The material presented here, and in Dobias et al. [2004], offer the first proof that the during the substorm growth phase the near Earth magnetotail might reach a configuration that is nonlinearly and explosively unstable. In fact, we believe that our conclusions are relatively conservative on this issue. FLRs allow other nonlinear effects, such as the formation of Kelvin-Helmholtz (KH) instabilities in the equatorial plane [Samson et al., 1992b; Rankin et al., 1993; Samson et al., 1996]. The evolution of these KH instabilities can substantially modify pressure profiles in the pre-midnight magnetosphere leading to a further enhancement of the possibility of nonlinear instabilities, particularly through the shear flow - ballooning mode [Voronkov et al., 1997, 2000]. This is another area that deserves further study, using variational methods.

An important element of our model is the need for an auroral arc associated with magnetospheric fields that drive the nonlinear instability. In our case we have chosen a FLR model, and it seems to work, but other avenues need to be explored. These FLRs seem to be a common feature in the magnetosphere. Still, much remains to explain their origin. We understand the physics of FLRs but the origin of the monochromatic compressional modes that drive them is not yet resolved. The possibility that they are driven by global MHD modes in the magnetosphere is compelling, but there are problems. The FLRs often occur at discrete and predictable frequencies, and these frequencies can be stable over long time intervals [Fenrich et al. 1995]. The magnetosphere is a very dynamic system, particularly during substorm times, and consequently it is difficult to understand how stable frequencies might exist, but they do.

The model we have presented here offers a possible explanation for pseudobreakups or precursors, which we suspect are localized ballooning modes, as well as the initiation of the substorm intensification. In essence, the pseudobreakups and the beginning of the initiation of the substorm intensification both evolve as a localized ballooning mode at about 8-12 R_E. We expect this to occur in the midnight or premidnight sector of the magnetosphere where radial pressure gradients are likely to be larger due to ion drift in the electric field of the magnetotail and gradient-curvature drift of the adiabatically heated ions moving Earthward in the magnetotail near the inner edge of the plasmasheet [Lyons and Samson, 1992].

When the substorm intensification occurs, these ballooning modes rapidly evolve as an explosive, nonlinear instability, leading eventually (within 10s of seconds) to more global reconfigurations of parts of the nightside magnetosphere. As we discussed in Section 1, following the initial ballooning mode, we need to look at other processes that allow diffusion of magnetic field lines and the eventual formation of a NENL at about 20-30 R_E.

In addition to suggesting some possibilities for the physical processes leading to the substorm intensification, we believe our study can be useful in a number of other areas, some of them quite practical. A better understanding of the substorm process will clearly require computer models, particularly in light of the possibly complex topology of the process after the intensification and the nonlinear evolution of the instability. The method we have outlined allows the construction of a equilibrium magnetotail configuration that can be shown to be nonlinearly unstable near 10 R_E. These equilibrium models can be used to initiate computational models with ideal and Hall MHD and more. The possibility of using our method to predict substorm intensifications by monitoring the auroral ionosphere is very appealing, and we are looking at this problem. In addition to using 486.1 nm emissions, as we have done here, we hope to use 630.0 nm emissions to identify the boundary of open field lines (Friedrich et al., [2001] and references therein) allowing one further constraint for an equilibrium model that includes the magnetotail x-line. We are also exploring the use of ground based magnetometer data to monitor the ionospheric electrojets an the nightside. Initial studies indicate that it might be possible to delineate the approximate H_β boundaries from these data. This would allow a backup to the optical data that would not be available during cloudy conditions.

Acknowledgements

This research was funded by the Natural Sciences and Engineering Research Council (NSERC) of Canada and the Canadian Space Agency (CSA). We would like to thank V.T. Tikhonchuk for his help in developing the Grad-Shafranov model and I. Voronkov for many helpful suggestions on the nature of the ballooning modes associated with substorm intensifications.

References

Arnold, V.I., On apriori estimates in the theory of hydrodynamical stability, *Amer. Math Soc. Transl., 79*, 267, 1969.

Bernstein, I.B., E.I. Frieman, M.D. Kruskal, and R.M. Kulsrud, An energy principle for hydromagnetic stability problems, *Proc. R. Soc. London, A17*, 244, 1958.

Buchner, J., and L.M. Zelenyi, Regular and chaotic charged particle motion in magnetotaillike field reversals 1. Basic theory of trapped motion, *J. Geophys. Res., 94*, 11821, 1989.

Cherry, T.M., The patology of differential equations, *J. Austral. Math. Soc. 1*,1,1959.

Cherry, T.M., Asymptotic solutions of analytic Hamiltonian systems, *J. Differential Equations, 4*, 142, 1968.

Cowley, S.C., and M. Artun, Explosive instabilities and detonation in magnetohydrodynamics, *Phys. Rep., 283*, 185, 1997.

Crabtree, C., W. Horton, H.V. Wong, and J.W. Van Dam, Bounce-averaged stability of compressional modes in geotail flux tubes, *J. Geophys. Res., 108*(A2), doi:10.1029/2002JA009555, 2003.

Dobias, P., I. Voronkov, and J.C. Samson, On nonlinear plasma instabilities during the substorm expansive phase onset, *Phys. Plasmas, 11*, 2048, 2004.

Fenrich, F.R., J.C. Samson, G. Sofko, and R.A. Greenwald, ULF high and low-*m* field line resonances observed with SuperDARN, *J. Geophys. Res., 100*, 21535, 1995.

Fredrickson, E.D., K.M. McGuire, Z.Y. Chang, A. Janos, J. Manickam, G. Taylor, S. Mirnov, I. Semenov, D. Kislov, and D. Martynov, Ballooning instability precursors to high β disruptions on the Tokamak fusion test reactor, *Phys. Plasmas, 3* (7), 2620, 1996.

Friedrich, E., J.C. Samson, I. Voronkov, and G. Rostoker, Dynamics of the substorm expansive phase, *J. Geophys. Res., 106*, 13145, 2001.

Fong, B.H., S.C. Cowley, and O.A. Hurricane, Metastability of magnetically confined plasmas, *Phys. Rev. Lett., 82*, 4651, 1999.

Liu, Physics of the explosive growth phase: Ballooning instability revisited, *J. Geophys. Res., 102*, 4927, 1997.

Liu, W.W., B.-L. Xu, J.C. Samson, and G. Rostoker, Theory and observation of auroral substorms: A magnetohydrodynamic approach, *J. Geophys. Res.*, *100*, 79, 1995.

Low, F.E., A Lagrangian formalism of the Boltzman-Vlasov equations for plasmas, *Proc. Roy. Soc. London, A248*, 282, 1958.

Lui, A.T.Y., A synthesis of magnetospheric substorm models, *J. Geophys. Res.*, *96*, 1849, 1991.

Lui, A.T.Y, et al., Multipoint study of a substorm on February 9, 1995, *J. Geophys. Res.*, *103*, 17333, 1998.

Lyons, L.R., and J.C. Samson, Formation of the stable auroral arc that intensifies at substorm onset, *Geophys. Res. Lett.*, *19*, 2171, 1992.

Mitchell, D.G., M.F. Engebretson, D.J. Williams, C.A. Cattell, and R. Lundin, Pc5 pulsations in the outer dawn magnetosphere, *J. Geophys. Res.*, *95*, 967, 1990.

Misner, C.W., Thorne, K. S., and J.A. Wheeler, *Gravitation*, W.H. Freeman and Company, New York, 2002.

Ohtani, S.I., and T. Tamao, Does the ballooning instability trigger substorms in the near-Earth magnetotail?, *J. Geophys. Res.*, *98*, 19369, 1993.

Park, W., E.D. Fredrickson, A. Janos, J. Manickam, and W. M. Tang, High-β disruptions in Tokamaks, *Phys. Rev. Lett.*, *75* (9), 1763, 1995.

Pfirsch, D., and R.N. Sudan, Nonlinear ideal magnetohydrodynamic instabilities, *Phys. Fluids, 7*, 2052-2061, 1993.

Pu, Z.Y., K.B. Kang, A. Korth, S.Y. Fu, Q.G. Zong, Z.X. Chen, M.H. Hong, Z.X. Liu, G.G. Mouikis, R.W.H. Fridel, and T. Pulkkinen, Ballooning instability in the presence of a plasma flow: A synthesis of tail reconnection and current disruption models for the initiation of substorms, *J. Geophys. Res.*, *104*, 10235, 1999.

Rankin, R., B.G. Harrold, J.C. Samson, and P. Frycz, The nonlinear evolution of field line resonances in the Earth's magnetosphere, *J. Geophys. Res.*, *98*, 5839, 1993.

Samson, J.C., L.L. Cogger, and Q.Pao, Observations of field line resonances, auroral arcs, and auroral vortex structures, *J. Geophys. Res.*, *101*), 17373, 1996a.

Samson, J.C., L.R. Lyons, P.T. Newell, F. Creutzberg, and B. Xu, Proton aurora and substorm intensifications, *Geophys. Res. Lett.*, *19*, 2167, 1992a.

Samson, J.C., R. Rankin, and V.T. Tikhonchuk, Optical signatures of auroral arcs produced by field line resonances: comparison with satellite observations and modeling, *Annales Geophys.*, *21*, 933, 2003.

Samson, J.C., D.D. Wallis, T.J. Hughes, F. Creutzberg, J.M. Ruohoniemi, and R.A. Greenwald, Substorm intensifications and field line resonances in the nightside magnetosphere, *J. Geophys. Res.*, *97*, 8495, 1992b.

Schultz, B.F., *Geometrical methods of mathematical physics*, Cambridge University Press,1980.

Sitnov, M.I., and A.T.Y. Lui, Cross-field current instability as a catalyst of the explosive reconnection in the magnetotail, *J. Geophys. Res.*, *104*, 6941, 1999.

Voronkov, I., E.F. Donovan, B.J. Jackel, and J.C. Samson, Large-scale vortex dynamics in the evening and midnight auroral zone: Observations and simulations, *J. Geophys. Res.*, *105*, 18505, 2000.

Voronkov, I., R. Rankin, P. Frycz, V.T. Tikhonchuk, and J.C. Samson, Coupling of shear flow and pressure gradient instabilities, *J. Geophys. Res.*, *102*, 9639, 1997.

Wanliss, J.A., J.C. Samson, and E. Friedrich, On the use of photometer data to map dynamics of the magnetotail current sheet during substorm growth phase, *J. Geophys. Res.*, *105*, 27673-27684, 2000.

Zeitlin , V., and T. Kambe, Two-dimensional ideal magnetohydrodynamics and differential geometry, *J. Phys. A: Math Gen.*, *26*, 5025, 1993.

Multiscale Coupling of Sun-Earth Processes
A.T.Y. Lui, Y. Kamide and G. Consolini, editors.

253

SUBSTORM AURORAL MORPHOLOGY DURING GEOMAGNETIC STORMS ON 19-20 APRIL 2002

X.-Y. Zhou

Jet Propulsion Laboratory, California Institute of Technology, 4800 Oak Grove Drive, Pasadena, CA 91109-8099, USA

S. B. Mende

University of California, Space Sciences Laboratory, 7 Gauss Way, Berkeley, CA 94720-7450, USA

Abstract. In this paper, we report on a study of the auroral activity during three substorm expansion phases that occurred prior to, during and after the magnetic storm main phase on April 20 during which the SYM-H index decreased from -75 to -185 nT. The first substorm actually occurred in the recovery phase of a previous storm on April 19. The IMAGE WIC (the Wideband Imaging Camera) data showed that auroral morphologies of the three events were almost identical. In the first $10 - 20$ min after substorm expansion phase onsets, there were significant auroral brightenings and abrupt expansions. In each case, double auroral oval occurred during the later substorm expansion phases and early substorm recovery phases. South-north orientated auroral patches and torches filled in the space between the two ovals. It was found that even with very similar auroral patterns, the three substorm expansion phases may have been very different in terms of auroral electrojet intensity and energy transport from the solar wind into the magnetosphere/ ionosphere system. The first substorm event may have been generated by the release of energy stored in the magnetotail, although the trigger was an external IMF northward turning. The second substorm was more likely driven by the intense solar wind dawn-dusk electric field. The third substorm might have been internally triggered by a steady-loading process. Our argument is that the similar auroral morphologies may have been manifestations of different substorm processes, i.e., loading-unloading and/or being directly driven.

Keywords. Auroral morphology, substorm expansion phase, geomagnetic storm

1. Introduction

A Chapman conference was held in India in March 2001 to discuss the most recent results on the storm-substorm relationship. To show the importance of substorm, Akasofu (pioneer of substorm physics) emphasized again that no storms occur without substorms (Akasofu et al., 2003). On the other hand, some studies argued that substorm injections have nothing to do with ring current intensification (Grande et al., 2003, Clauer et al., 2003). In addition, Lui et al. (2001), Tsurutani et al. (2003), and Zhou et al. (2003a) have shown that there could be a lack of substorm expansion phases during magnetic storm main phases that were caused by a smooth southward interplanetary magnetic field (IMF) and/or by magnetic clouds. Although it was still unclear after the meeting how storms and substorms influence each other, it was generally accepted that substorms, at the very least, bring ions out from the ionosphere during ring current intensification. We have known that intense auroral activity always accompanies substorm expansions. The purpose of this paper is to review the results of our study of the storm-time auroral patterns that may have something to do with how substorms influence the ring current buildup, since auroral patterns are related to the ion outflow intensity and distribution. While it is unknown how auroral forms and

254

brightening are related to the ion outflow intensity, it has been found that that the brighter the aurora the more intense the ion outflow (Wilson et al., 2001; Mende et al., 2003).

There was an interval of intense solar activity during 10 – 22 April 2002, which caused a series of heliosphere/magnetosphere/ionosphere disturbances. The geomagnetic Dst index reached -125 nT (or lower) three times on 18, 19, and 20 April. The space storms in this interval have been extensively studied as a series of Sun-Earth Connection (SEC) events. For example, two SEC workshops in Augusts 2002 and 2003 have been focused on these case studies. A special issue on JGR that will contribute to this SEC event investigation will be published next year. This paper will focus on the auroral morphology in the latter magnetic storm, during which there was continued IMAGE Wideband Imaging Camera (WIC) observation from 0100 to 0800 UT. The imaging data perfectly showed the auroral evolution of three substorm expansion phases.

It has been found that substorm expansion phases can be triggered externally by sudden changes in the solar wind (Burch, 1972; Kokubun et al., 1977; McPherron et al., 1986; Lyons, 1995; Zhou and Tsurutani, 2001), and internally by strong-steady-loading processes, similar to the catastrophic collapse of a sand dune (Baker et al., 1990; Baker et al., 1999). The three substorms on 20 April 2002 may have been caused by those external and internal processes. It is interesting to know whether the auroral pattern and distribution were the same when the substorm triggers and processes were different. We will also discuss whether substorm expansion phases within a magnetic storm main phase are different from those outside the main phase in terms of the auroral electrojet and the energy transportation.

2. Observations

2.1 Solar Wind And Geomagnetic Parameters

The solar wind and geomagnetic activity discussed in this paper are part of a series of Sun-Earth Connection events. An intense interplanetary fast forward shock was detected by the ACE spacecraft at ~1030 UT on 17 April, followed by a very disturbed sheath region from 1030 to 1830 UT. In this interval, the Dst index decreased from ~0 to -100 nT. This interplanetary shock was driven by a magnetic cloud that occurred from 2130 UT on 17 April to 0800 UT on 19 April (Zhang et al., 2004). During the cloud, the most intense solar wind dawn-dusk electric field was ~7 mV/m, from 0100 to 0900 UT on 18 April, which caused a further ring current intensification as indicated by a decrease in Dst from ~ -80 to -130 nT. Another fast forward shock occurred at the end of the cloud and led to another disturbed sheath region, from ~0800 UT on 19 April to ~0430 UT on 20 April, which caused a double magnetic storm with Dst decreases from ~ -40 to -125 nT during 1000 – 1900 UT on 19 April, and from -87 to -148 nT during 0300 - 0700 UT on 20 April. Another magnetic cloud occurred from 0430 UT on 20 April to 1730 UT on 21 April (Zhang et al., 2004), which mainly lengthened the storm recovery phase. The second part of the double storm will be studied in this paper because there was very good coverage in the IMAGE WIC data for this interval, so that the auroral morphology could be examined.

Figure 1 shows the ACE solar wind observation and the geomagnetic SYM-H index on 20 April 2002. Panels, from top to bottom, are the solar wind ram pressure (ρV^2), the total IMF value and the IMF B_z component (in gray), the solar wind speed, the proton density, the plasma β, the solar wind dawn-dusk field E_{VB}, and the ground SYM-H index, which is a high time resolution (1 min) Dst index. There was no AE (AL, AU) data downloading service at the time when this paper was drafted, so we could not scale it to this figure, but it is shown in Figure 3. The solar wind data have been shifted 40 min, which is the estimated solar wind propagation time form ACE to the nose of the magnetopause (assumed at 10 R_E). The shifted time that is shown at the bottom of the figure will be used in the rest of the paper.

The ACE spacecraft was in the upstream solar wind at (224, 32, -10 R_E) in GSM coordinates during the storm main phase (shaded in gray) from ~0300 to 0600 UT. The solar wind ram pressure was intense and highly variable from ~0130 to 0730 UT. The highest ram pressure reached ~15 nPa at 0450 UT. During the storm main phase, the IMF value was high at ~20 nT at the beginning, then decreased to ~15 nT in the middle stage of the main phase when the IMF turned almost completely southward. The IMF decreased to 7 nT and maintained this level the rest of the day. The solar wind was at high speed, 650 to 750 km/s, during the main phase. The speedy dense plasma caused a high and varying solar wind ram pressure and a high plasma β region. The dawn–dusk electric field E_{VB} was ~8 mV/m from 0300 to 0500 UT, except for a narrow low E_{VB} interval (~30 min) in the middle when the IMF B_z turned to less southward. This intense solar wind E_{VB} field caused the SYM–H index to decrease from ~ -80 to -180 nT during 0300 – 0600 UT. Three substorm expansion onsets were observed by the IMAGE WIC instrument and are marked in this figure by three arrows in the third panel. The first two substorm expansion phases might have been triggered by the IMF B_z northward turnings at ~0145 and 0340 UT, respectively. It seems that there was no clear trigger in the IMF for the third substorm expansion phase, but there was a solar wind ram pressure fluctuation from 11 to 5, then back to 8 nPa. It should be noted that the magnetic cloud that started from ~0520 UT did not lead to a further intensification of the ring current. Instead, the storm recovery phase started at ~0600 UT, soon after the cloud embraced the magnetosphere. However, with E_{VB} slowly varying between 0 – 4 mV/m, this cloud did maintain the SYM-H index at the level of -100 nT for about 7 hr from ~1200 to 1900 UT.

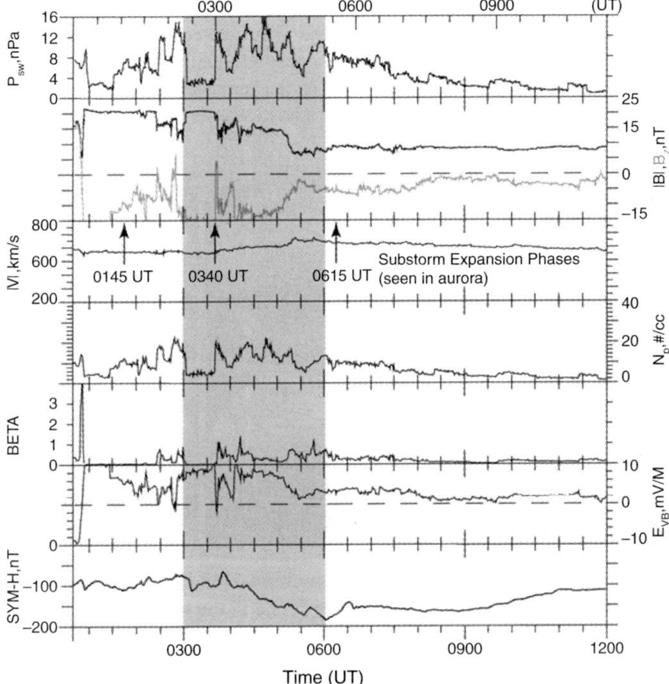

Figure 1. The ACE solar wind observation and the geomagnetic SYM-H index on 20 April 2002. The shaded area shows the storm main phase. The solar wind data have been shifted 40 min. The time on the top of the figure is the universal time at ACE before the shift.

258

Around 0606 UT, the auroral oval recovered to the quiet level similar to that preceding the second substorm expansion phase. Meanwhile, the storm recovery phase started. The aurora broke up again between 0614 and 0616 UT at 01 MLT. Figure 2c shows the substorm expansion phase after the third auroral breakup. The aurora had a significant westward expansion, as shown in the image at 0618:20 UT. As this westward expansion continued, for next ~10 min, the poleward and eastward expansions occurred. The aurora was very bright during the expansion. The dawnside auroral oval from 05 – 07 MLT was not affected before 0700 UT. After this time, there was some cutoff of the dawnside image. The lowest AL index was ~ -600 nT before 0700 UT. The AL index reached to -1000 nT at ~0715 UT, but this decrease was not related to the substorm expansion phase. The aurora at 0715 UT was very similar to that in the image at 0713:35 UT. This is another example that a low AL (high AE) (at 0715 UT) index does not necessarily indicate a substorm expansion phase. There was very bright aurora along the duskside auroral poleward boundary after 0649:02 UT. Then a double auroral oval occurred. The two ovals were connected by the south-north orientated auroral patches and/or torches.

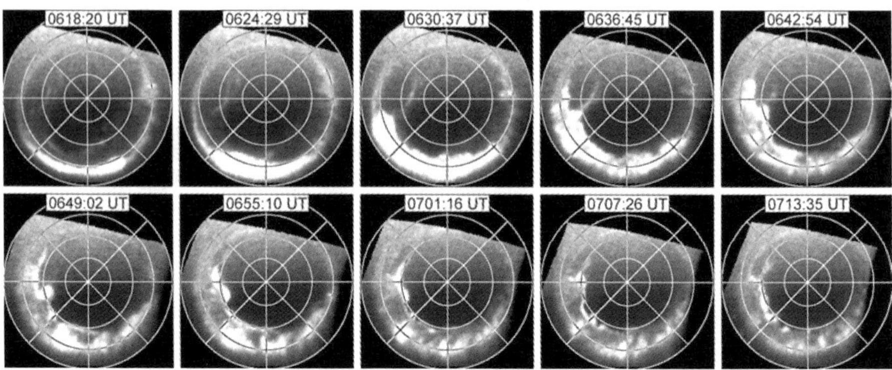

Figure 2c. The substorm expansion phase with onset at ~0615 UT on 20 April 2002. The images have the same format as Figure 2a.

3. Discussion and Summary

We have shown auroral images for three substorm expansion phases that occurred, under different solar wind conditions, within and outside of a magnetic storm main phase. The three substorm expansion phases occurred during very disturbed solar wind and geomagnetism. The IMF was mainly southward. The solar wind speed was higher than 650 km/s. The proton density was high and perturbed. Therefore, the solar wind ram pressure was very intense and highly variable between 5 to 15 nPa. As a result, the SYM-H index was lower than -100 nT for most of the day.

Corresponding to the three substorm expansion phases, three dispersionless energetic particle injections were observed by the LANL 1990-095 satellite (not shown in this paper) when it was in the nightside plasma sheet. Accordingly, significant magnetic field dipolarizations were observed by GOES satellites. The quasi-periodic substorm signature that was observed at the geosynchronous orbit has been discussed extensively in the recent literature (e.g., Reeves et al., 2003, Lui et al., 2004).

The common characteristics of the aurora during the three substorm expansion phases include expansion onsets near local midnight, very bright auroral illuminations during abrupt eastward, westward and poleward auroral expansions, and occurrences of a double auroral oval in late substorm expansion phases and early recovery phases. South-north orientated auroral patches and auroral torches filled in between the double oval in each case. The auroral morphology of the three substorm expansion phases is very similar except the poleward auroral boundary of the first event reached to very high latitude (at ~82°) as a result of the substorm expansion phase. This significant polar cap contraction was due to the decrease of the dayside magnetic reconnection when the IMF turned to less southward, meanwhile the tail open flux was abruptly reduced during the substorm expansion phase. In general, substorm expansion phases can cause a decrease of the lobe magnetic flux by ~20 to 30% (Sauvaud 1992). However for this event, the reduction of the lobe flux was very high at ~60%, according to the contraction of the polar cap. As a result, this northward turning reduced the energy input into the magnetosphere and the ionosphere (Meng and Makita 1986; Kamide et al., 1999). Therefore, this substorm expansion phase did not lead to a further ring current intensification.

Although the auroral morphology of the three events was almost identical, no conclusion can be drawn whether the substorms were also identical in terms of the intensity of the eastward and westward auroral electrojets. The auroral electrojets are represented by the AL and AU indices. We show the AL, AU, and AE indices on 20 April 2002 in the top two panels of Figure 3. The CANOPUS local electrojet indices CU and CL are shown in the bottom panel. During the three substorms, the CANOPUS magnetometer chain was in the 18 – 00 LT sector, which provided a favorable opportunity for the electrojet observation. For the first substorm expansion event, the magnitude of both global and local westward electrojet indices (AL and CL) increased significantly after the substorm expansion phase onset when the aurora abruptly expanded poleward, westward and eastward. But this energy dissipation only lasted ~30 min. CU increased from ~300 to 1000 nT, which was not seen in AU. The abrupt change in CU and CL was very well correlated with the violent nightside auroral activity.

In the second event before 0400 UT, both AL and CL showed no response to the substorm expansion onset. This was because the auroral oval was at low magnetic latitudes, due to a large polar cap. The first row of Figure 2b shows that in the first 20 min of the substorm expansion phase, the auroral oval from 18 to 00 LT was mainly lower than 65° MLAT. However, most CANOPUS stations are higher than 65° MLAT. The geomagnetic X-component decreased ~500 nT at the three low MLAT CANOPUS stations, PINA, MCMU, and ISLL, when the aurora expanded poleward. It might be due to the same reason the CL magnitude was lower than the AL magnitude during the rest of the storm main phase. The interval for AL < -1000 nT was ~1 hr, from ~0520 to 0620 UT. The auroral activity lasted as long as the storm main phase, from ~0330 to 0600 UT. These observations showed that the second substorm had an extended expansion phase that was 3 times longer than the maximum expansion phase duration defined by Akasofu (1964).

Although the third substorm event was not noticeable in AL or AE, the event was clearly shown by CL. This difference may be simply due to the improper coverage of the AE stations during that interval. Similar to the first substorm event, the third one had energetic particle injection and tail magnetic field depolarization. However, the third substorm did not reduce the tail open magnetic flux at an efficiency that was higher than the dayside magnetic reconnection. Therefore, the polar cap was about the same size after 0630 UT. Similar to the second substorm event, the third substorm expansion broke up at low MLAT. However, the substorm did not lead to a further ring current intensification because the solar dawn-dusk electric field was relatively weak. Therefore, the magnetospheric convection driven by this electric field would have been reduced, which would lead to the storm recovery phase.

The observations have shown that although the auroral intensity was almost the same for the three events, the electrojet intensity was different. One of the reasons, as discussed above, was because the magnetometers were not under the auroral electrojets when the oval was at lower magnetic latitudes. Another reason could be that auroral brightening may not relate to the auroral electrojet. The electrojet intensity is controlled not only by ionosphere conductance, but also by the electric field that could be affected by both solar wind dawn-dusk electric field and ram pressure (Hairston et al., 1999). Therefore, with similar intensity in the solar wind electric field, the higher the ram pressure, the larger the AL value, which partially explains the intense and weak auroral electrojet in the second and third substorm events.

Figure 3 shows that during the first substorm expansion phase there was only a westward electrojet spike at 0200 UT. Later in that hour, the eastward electrojet increased as shown by several enhancements in the AU index, which were consistent with the variations in the solar wind density and therefore the ram pressure. From 0230 to 0300 UT, the AE variation was mainly contributed by AU. This observation shows that during northward IMF or less southward IMF, the AE index is actually controlled by the solar wind density, which is consistent with what has been found by Kamide et al. (2003). It has been found that high and sudden increases in solar wind ram pressure cause dayside auroras (Zhou and Tsurutani, 1999; Frey et al., 2002; Hubert et al., 2003; Zhang et al., 2003; Zhou et al., 2003b), therefore increases the AU index. This phenomenon has been named a "dayside substorm" by Kamide et al. (2003).

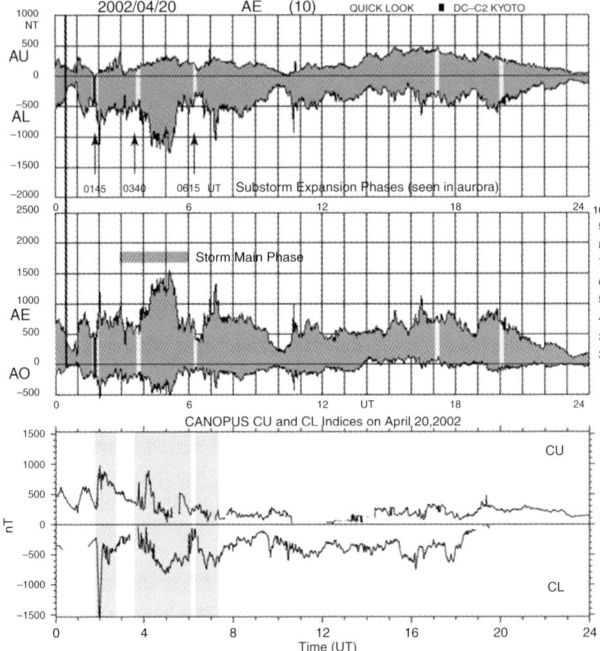

Figure 3. Global and CANOPUS local electrojet indices on 20 April 2002. The duration of the magnetic storm main phase is shown by a horizontal grey bar in the middle panel. Shadings in the bottom panel are the intervals for which the auroral imaging data are shown in Figure 2. The second shading interval is actually consistent with the storm main phase.

The intense solar wind dawn-dusk electric field during 0300 – 0500 directly drove the second substorm event, although it was actually triggered by either the sudden IMF northward turning (Lyons, 1995), or by the abrupt enhancement of the solar wind ram pressure (Zhou and Tsurutani, 2001). The westward electrojet was very intense and lasted more than an hour from ~0410 – 0510 UT as shown by the intense AL, at lower than -1100 nT. While the solar wind density was very high during the first event, it did not cause high eastward electrojets, due to the southward IMF. Instead, the AU index was lower than that during 0200 – 0300 UT. Comparing the first and second events, it seems that the first auroral expansion was due to the injection of an excess amount of energy that has been stored in the magnetotail before the onset of the substorm expansion phase, but the second substorm expansion was caused by the energy dumping into the open magnetic field from the solar wind. The third substorm expansion phase was almost unnoticeable in the AL (AE) index. The value of AL and AU was very low during 0600 – 0700 UT compared to what one would expect based on the corresponding WIC images. The reason of the inconsistency between the space-based auroral observations and the ground-based geomagnetic indices is unknown. One possible explanation could be that the AE station coverage was not very good during this interval.

The discrepancy between the auroral morphology and the electrojet variation shown by the comparison between the imaging data (Figure 2) and the AL index (Figure 3) has an interesting implication, i.e., the most important factor affecting substorm and storm intensity was how long the substorm expansion phase endures, but not how bright and violent the aurora becomes, or how intense the electrojet becomes in the first 10 to 20 minutes of expansion phases. When there was an intense and long lasting dawn-dusk electric field, substorm expansion phases were maintained for a long time, and only during that time the ring current intensified.

In summary, although the three substorm expansion phases occurred within and outside of the storm main phase on 20 April 2002, their auroral morphology was identical. However, this does not necessarily mean that the three substorm events were the same in terms of the energy transport from the solar wind into the magnetosphere/ionosphere system. The similar auroral morphologies could have been manifestations of the different substorm processes of loading-unloading and/or being directly driven (Akasofu, 1979; Rostoker et al., 1987). Further study, such as the calculation of the loading and driven components (Sun et al., 1998), is needed to understand the energy transportation mechanism of the three substorm expansion phases.

Acknowledgements

Portions of this paper represent work done at the Jet Propulsion Laboratory, California Institute of Technology, Pasadena, under contract with the National Aeronautics and Space Administration. We would like to acknowledge the Coordinated Data Analysis Web (CDAWeb) for providing access to the ACE solar wind data, and WDC-C2, at Kyoto, Japan for providing the ground-based AL, AU, and AE indices. The CANOPUS instrument array constructed, maintained and operated by the Canadian Space Agency, provided the CU and CL indices used in this study.

References

Akasofu, S.-I., The development of the auroral substorm, *Planet. Space Sci., 12,* 273, 1964.
Akasofu, S.-I., What is a magnetospheric substorm?, in *Dynamics of the Magnetosphere*, p.447, D. Reidel., Norwell, Mass., 1979.
Akasofu, S.-I., W. Sun, and B.-H. Ahn, Comments on some long-standing problems in storm/substorm studies, in *Disturbances in Geospace: The Storm-Substorm Relationship*, Geophysical Monograph Series, Vol. 142, edited by A.S. Sharma et al., AGU Washington DC, pp.243, 2003.
Baker, D.N., A.J. Klimas, R.L. McPherron, J. Buechner, The evolution from weak to strong geomagnetic activity: an interpretation in terms of deterministic chaos, *Geophys. Res. Lett., 17,* 41, 1990.
Baker, D.N., T.I. Pulkkinen, J. Buchner, and A.J. Klimas, Substorms: A global instability of the magnetosphere-ionosphere system, *J. Geophys. Res., 104,* 14601, 1999.

Burch, J.L., Preconditions for the triggering of polar magnetic substorms by storm sudden commencements, *J. Geophys. Res., 77,* 5629, 1972.

Clauer, C.R., M.W. Liemohn, J.U. Kozyra, and M.L. Reno, The relationship of storms and substorms determined from mid-latitude ground-based magnetic maps, in *Disturbances in Geospace: The Storm-Substorm Relationship*, Geophysical Monograph Series, Vol. 142, edited by A.S. Sharma et al., AGU Washington DC, pp.143, 2003.

Frey, H.U., S. B. Mende, T. J. Immel, S.A. Fuselier, E.S. Claflin, J.-C. Gérard, B. Hubert, Proton aurora in the cusp, *J. Geophys. Res., 107(A7),* 10.1029/2001JA900161, 2002.

Grande, M., C.H. Peery, A. Hall, J. Fennell, R. Nakamura, and Y. Kamide, What is the effect of substorms on the ring current ion population during a geomagnetic storms, in *Disturbances in Geospace: The Storm-Substorm Relationship*, Geophysical Monograph Series, Vol. 142, edited by A. S. Sharma et al., AGU Washington DC, pp.75, 2003.

Hairston, M. R., D. R. Weimer, R. A. Heelis, and F. Rich, Analysis of the ionospheric cross polar cap pothetial drop and electrostatic potential distribution patterns during the January 1997 CME event using DMSP data, *J. Atmos. Solar Terr. Phys., 61,* 195, 1999.

Hubert, B., J. C. Gérard, S. A. Fuselier, and S.B. Mende, Observation of dayside subauroral proton flashes with the IMAGE-FUV imagers, *Geophys. Res. Lett., 30,* 10.1029/2002GL016464, 2003.

Kamide, Y., S. Kokubun, L. F. Bargatze, and L. A. Frank, The size of the polar cap as an indicator of substorm energy, *Phys. Chem. Earth., 24,* 119, 1999.

Kamide, Y., J.-H. Shue, and M. Brittnacher, Effect of solar wind density on the auroral electrojets and global auroras during geomagnetic storms, in *Disturbances in Geospace: The Storm-Substorm Relationship*, Geophysical Monograph Series, Vol. 142, edited by A.S. Sharma et al., AGU Washington DC, pp.15, 2003.

Kokubun, S., R.L. McPherron, and C.T. Russell, Triggering of substorms by solar wind discontinuities, *J. Geophys. Res., 82,* 74, 1977.

Lui, A.T.Y., R.W. McEntire, K.B. Baker, A new insight on the cause of magnetic storms, *Geophys. Res. Lett., 28,* 3413, 2001.

Lui, A.T.Y., T. Hori, et al., Magnetotail behavior associated with "sawtooth injections" during magnetic storms in April 2002, appear to *J. Geophys. Res.,* 2004.

Lyons, L.R., A new theory for magnetospheric substorms, *J. Geophys. Res., 100,* 19069, 1995.

McPherron, R.L., T. Terasawa, and A. Nishida, Solar wind triggering of substorm onset, *J. Geomag. Geoelectr., 38,* 1089, 1986.

Mende, S.B., *et al.*, IMAGE FUV and in situ FAST particle observation of substorm aurora, *J. Geophys. Res., 108,* 1145, 10.1029/2002JA009413, 2003.

Meng, C.-I, and K. Makita, Dynamic variation of the polar cap, in *Solar Wind-Magnetospheric Coupling*, edited by Y. Kamide and J.A. Slavin, pp. 605-631, Terra Sci., Tokyo, 1986.

Reeves, G.D. M.G. Henderson, et al., IMAGE, POLAR, and geosynchronous observations of substorm and ring current ion injection, in *Disturbances in Geospace: The Storm-Substorm Relationship*, Geophysical Monograph Series, Vol. 142, edited by A.S. Sharma et al., AGU Washington DC, pp.91, 2003.

Rostoker, G., S.-I. Akasofu, W. Baumjohann, Y. Kamide, and R.L. McPherron, The roles of direct input of energy from the solar wind and unloading of stored magnetotail energy in driving magnetospheric substorms, *Space Sci. Rev., 46,* 93, 1987.

Sun, W., W.-Y. Xu, and S.-I. Akasofu, Mathematic separation of directly driven and unloading components in the ionospheric equivalent currents during substorms, *J. Geophys. Res., 103,* 11,695, 1998.

Tsurutani, B.T., X.-Y. Zhou, and W.D. Gonzalez, A lack of substorm expansion phases during magnetic storms induced by magnetic cloud, in *Disturbances in Geospace: The Storm-Substorm Relationship*, Geophysical Monograph Series, Vol. 142, edited by A.S. Sharma et al., AGU Washington DC, pp.23, 2003.

Wilson, G.R., D.M. Ober, G.A. Germany, and E.J. Lund, The relationship between suprathermal heavy ion outflow and auroral electron energy deposition: Polar/Ultraviolet Imager and Fast auroral Snapshot/Time-of-Flight Energy Angle Mass Spectrometer observations, *J. Geophys. Res., 106*, 18981, 2001.

Zhang, Y., L.J. Paxton, T. J. Immel, H.U. Frey, and S.B. Mende, Sudden solar wind dynamic pressure enhancements and dayside detached auroras: IMAGE and DMSP observations, *J. Geophys. Res., 108*, 10.1029/2002JA009355, 2003.

Zhang, J.-C., M.W. Liemohn, J.U. Kozyra, B.J. Lynch, and T.H. Zurbuchen, A statistical study of the geoeffectiveness of near-Earth magnetic clouds during high solar activity years, submitted to *J. Geophys. Res.*, 2004.

Zhou, X.-Y., and B.T. Tsurutani, Rapid intensification and propagation of the dayside aurora: Large scale interplanetary pressure pulses (fast shocks), *Geophys. Res. Lett., 26*, 1097, 1999.

Zhou, X.-Y. and B.T. Tsurutani, Interplanetary shock triggering of nightside geomagnetic activity: Substorms, pseudobreakups, and quiescent events, *J. Geophys. Res., 106*, 18957, 2001.

Zhou, X.-Y., B. T. Tsurutani, G. Reeves, G. Rostoker, W. Sun, J.M. Ruohoniemi, Y. Kamide, A.T.Y. Lui, G.K. Parks, W.D. Gonzalez, and J.K. Arballo, Ring current intensification and convection-driven negative bays: Multisatellite studies, *J. Geophys. Res., 108*, doi:10.1029/2003JA09881, SMP 13-1, 2003a.

Zhou, X.-Y., R.J. Strangeway, P.C. Anderson, D.G. Sibeck, B.T. Tsurutani, G. Haerendel, H.U. Frey, and J.K. Arballo, Shock-aurora: FAST and DMSP observations, *J. Geophys. Res., 108*, doi:10.1029/2002JA009701, 2003b.

Multiscale Coupling of Sun-Earth Processes
A.T.Y. Lui, Y. Kamide and G. Consolini, editors.

PERIODIC SUBSTORMS: A NEW PERIODICITY OF 2-3 HOURS IN THE MAGNETOSPHERE

Chao-Song Huang

Haystack Observatory, Massachusetts Institute of Technology, Westford, MA 01886, USA

Geoff D. Reeves

Los Alamos National Laboratory, Los Alamos, NM 87545, USA

Abstract. An outstanding problem in magnetospheric physics is whether substorms have some specific periodicity. We will present observations of substorms during magnetic storms. In two storm cases, the Geotail satellite was located in the near tail between X_{GSM} = -20 and -30 R_E and detected periodic southward turnings of the magnetospheric magnetic field; these southward turnings represent periodic magnetic reconnection in the tail. Geosynchronous satellites measured sawtooth-like injections of energetic charged particle fluxes. The periodic reconnection in the near tail is well correlated with the sawtooth injections at geosynchronous orbit. These measurements indicate the occurrence of periodic substorms. A prominent feature of these storm-time substorms is the well-defined periodicity of 2-3 hours. The periodic substorms can continue for 6-10 cycles over an interval of up to 24 hours under stable and southward interplanetary magnetic field (IMF) conditions. Each substorm onset does not have to be triggered by an IMF northward turning or by a solar wind pressure impulse. The periodic substorms can also occur when the IMF B_z component is small and fluctuating between southward and northward, and the period is nearly the same as that during strongly and continuously southward IMF. The observations show that the period of substorms is controlled neither directly by variations in the solar wind nor indirectly by the energy transferred to the magnetosphere from the solar wind. We suggest that substorms have an intrinsic period of 2-3 hours. This period is an eigenmode of the magnetosphere during magnetic storms; each cycle of the magnetospheric oscillations may result in the generation of a substorm onset.

Keywords. Substorms, periodic substorms, magnetic storms.

1. Introduction

Substorms are the fundamental process of magnetospheric dynamics [*Akasofu*, 1968; *McPherron*, 1995]. Although significant progress has been made in understanding the physics of substorms, there is serious controversy on a number of important issues. One of the issues is what process is responsible for the onset of substorms. Magnetic reconnection at a near-Earth neutral line (NENL) is a widely accepted mechanism [*Hones*, 1984; *Baker et al.*, 1996, 1999]. In the NENL model, magnetic reconnection onset occurs in the near tail on the closed field lines of the plasma sheet between -20 and -30 R_E. The reconnection causes the formation of plasmoids and the release of the energy stored in the tail into the ionosphere. A statistical study of Geotail measurements provides evidence of the reconnection location and support for the NENL model [*Nagai et al.*, 1998]. The current disruption, which occurs in the near-Earth magnetosphere, is another candidate mechanism [*Lui*, 1991, 1996]. In the current disruption model, the substorm onset location is near the boundary between the tail-like and the dipolar-like field regions, and

1984, 1993]. The magnetic field B_z component became negative in the first seven cycles and showed a decrease after 1800 UT. Figure 2b shows the ion density, ion energy, ion pressure, and ion velocity X (earthward) component. The ion velocity was generally negative (tailward), consistent with the location of Geotail in the lobe. Since Geotail was in the lobe on the open field lines, the periodic variations in the plasma particle parameters were not so clear as in the magnetic field. Figure 2c shows the energetic electron fluxes measured by the geosynchronous satellites LANL 1990-095 (MLT = UT - 2.5 hours) and 1991-080 (MLT = UT - 8 hours). The vertical dotted lines in Figure 2c indicate the times of the southward turnings of the magnetic field B_z identified from Figure 2a.

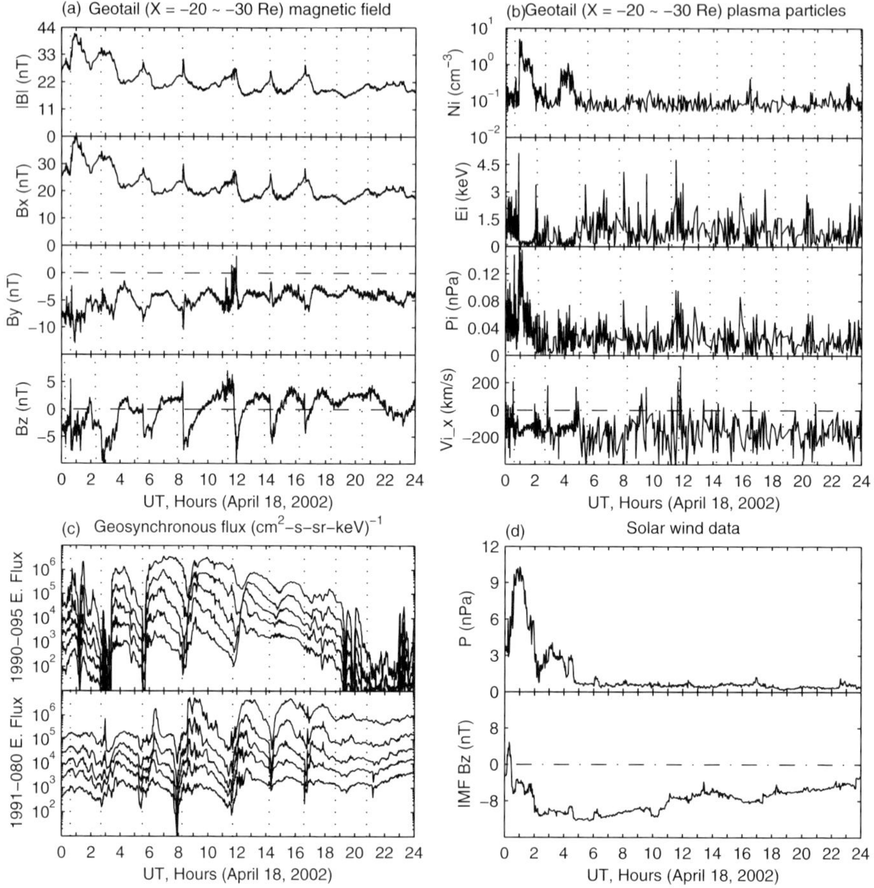

Figure 2. (a) The magnitude and components of the magnetospheric lobe magnetic field and (b) ion density, ion energy, ion pressure, and ion velocity measured by the Geotail satellite, (c) the electron fluxes measured by the LANL 1990-095 and 1991-080 geosynchronous satellites, and (d) the solar wind pressure and IMF B_z component measured by the Wind satellite on 18 April 2002. The electron energy channels are 50-75, 75-105, 105-150, 150-225, and 225-315 keV. The vertical dotted lines indicate substorm onsets.

A profound characteristic of the flux variations is the excellent correlation between the reconnection onsets in the near tail and the energetic flux increases at geosynchronous orbit, and a one-to-one correspondence exists clearly. The flux increases represent particle injections. When energetic plasma particles are injected to the inner magnetosphere, the charged particles collide with neutral atoms and become neutrally-charged and cause enhanced emissions from the newly produced energetic neutral atoms. The Imager for Magnetopause-to-Aurora Global Exploration (IMAGE) satellite indeed measured enhancements of the neutral emissions and auroral brightenings after the sudden flux increases [*Huang et al.*, 2003a], providing unambiguous evidence that the flux increases detected by geosynchronous satellites were true energetic plasma particle injections from the tail into the inner magnetosphere at substorm onsets.

In order to see whether the period of the substorms was determined by the solar wind, we plot in Figure 2d the solar wind pressure and IMF B_z component measured by the Wind satellite which was located at $X_{GSM} = 9 \sim 12\ R_E$, $Y_{GSM} = 168 \sim 200\ R_E$, and $Z_{GSM} = 40 \sim 120\ R_E$ on 18 April 2002. Since Wind was far from the Sun-Earth line, we checked the solar wind data of the ACE satellite and found that the measurements of ACE and Wind were in good agreement. A solar wind pressure impulse was detected by Wind at ~0010 UT. The solar wind pressure impulse might have triggered the first cycle of the periodic substorms. It is important to note that the solar wind pressure was rather stable after the impulse and that the IMF B_z was continuously negative and very stable on this day. However, the substorms continued to occur for many cycles. The results suggest that the periodic occurrence of the substorms was not controlled by the solar wind.

We now provide an interpretation of the observations. We plot in Figure 3 an illustration of the plasmoid formation during substorms. Figures 3a-c represent the magnetospheric magnetic topology in the growth phase of a substorm, at the onset, and in the expansion phase, respectively.

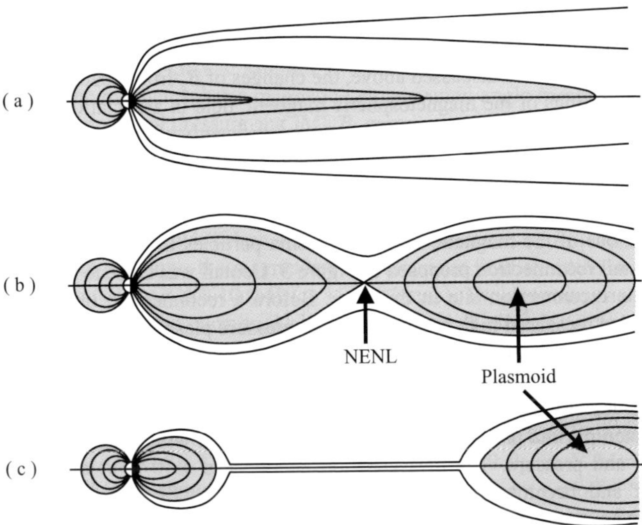

Figure 3. Illustration of the magnetospheric magnetic topology (a) during the growth phase of a substorm, (b) at the substorm onset, and (c) during the substorm expansion phase. NENL represents the near-Earth neutral line. The region with closed field lines earthward of the NENL becomes dipolar during the expansion phase.

-30 R_E for 24 hours in 6 cases (including the 2 cases shown in Figures 2 and 4) and detected the signatures of periodic formation of the NENL and plasmoids. Periodic magnetic reconnection in the near tail is always correlated with sawtooth-like injections at geosynchronous orbit.

4. Periodic Substorms During Stable or Fluctuating IMF

We present more examples of periodic substorms that occurred under different IMF conditions. Figure 5a shows the energetic electron fluxes measured by the geosynchronous satellites LANL 1989-046 and 1991-080 on October 4, 2000. There is no measurement in the tail in this case. Because we have demonstrated in the previous cases that the sudden flux increases correspond to the near-tail reconnection and represent particle injections at substorm onsets, the flux variations in Figure 5a are interpreted as the signature of periodic substorms. The vertical dotted lines indicate the flux increases (injections) at substorm onsets. The sudden decreases in the AL index of this case shown in the upper-right panel of Figure 1 correspond well to the flux injections at geosynchronous orbit.

We want to find whether the periodic substorms were related to solar wind variations. Figure 5b shows the solar wind pressure and IMF B_z measured by the Wind and Geotail satellites. On this day, Wind was located at $X_{GSM} = 32\ R_E$, Y_{GSM} varied between -245 and -211 R_E, and Z_{GSM} varied between 50 and 135 R_E. Geotail was very close to the magnetosphere and moved from (26, 2, 3) R_E GSM at 0000 UT to (9, 12, 0) R_E GSM at 2400 UT. The IMF B_z data of Wind and Geotail are in good agreement. The IMF B_z changed to negative around 0400 UT and remained continuously southward and very stable for the rest of the time of this day. The vertical dotted line at 0610 UT indicates the time at which the first flux injection was detected by the 1989-046 satellite.

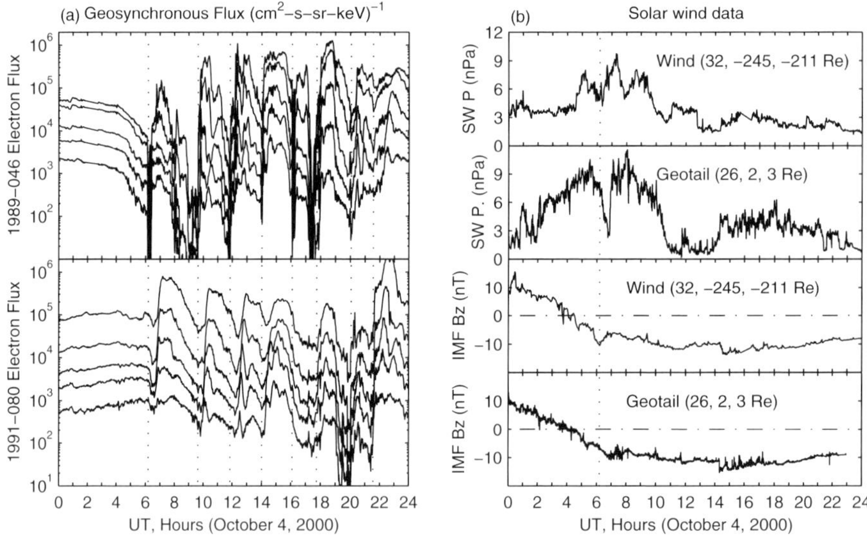

Figure 5. (a) The electron fluxes measured by the LANL 1989-046 and 1991-080 satellites and (b) the solar wind pressure and IMF B_z component measured by the Wind and Geotail satellites on October 4, 2000. The vertical dotted lines indicate substorm onsets.

In Figure 5b, there is some difference in the solar wind pressure measurements of the two satellites. Large perturbations in the solar wind pressure were detected by both satellites around 0600 UT, and these perturbations might be related to the first cycle of the periodic substorms. Then some irregular variations in the solar wind pressure occurred, but these irregular variations were not correlated with the periodic occurrence of the substorms.

An additional case of periodic substorms during stable IMF is shown in Figure 6. The energetic fluxes measured at geosynchronous orbit showed periodic variations, with gradual decreases and rapid increases. The characteristics of the flux variations are similar to those in the previous cases and therefore interpreted as the flux injections at substorm onsets. The solar wind data measured by the Wind and Geotail satellites were in good agreement. The IMF B_z was strongly and continuously southward. The solar wind pressure decreased rapidly before 2100 UT on 10 August and became small since then. There were a solar wind pressure impulse measured by Wind at ~2030 UT and a spike in the IMF B_z measured by Geotail at ~2100 UT. It is not certain whether the solar wind pressure impulse and/or the IMF B_z spike triggered the first cycle of the periodic substorms. However, the solar wind pressure and the IMF B_z did not show any periodic variations that could be related to the periodic substorms.

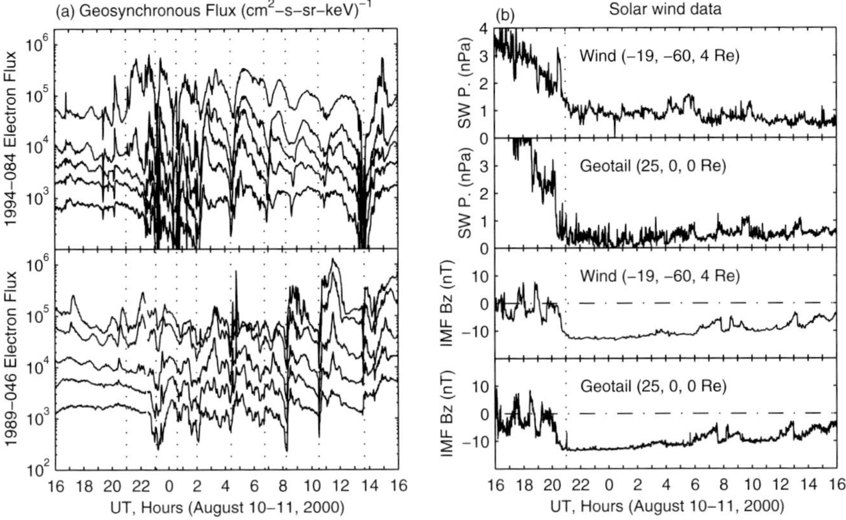

Figure 6. (a) The electron fluxes measured by the LANL 1994-084 and 1989-046 satellites and (b) the solar wind pressure and IMF B_z component measured by the Wind and Geotail satellites on August 10-11, 2000. The vertical dotted lines indicate substorm onsets.

We finally examine the periodic substorms when the IMF B_z was small and fluctuating between southward and northward. Figures 7a and 7b show the electron fluxes measured by the geosynchronous LANL 1994-084 and 1990-095 satellites and the solar wind pressure and IMF B_z measured by the Wind satellite on 4 November 2002. Figures 7c and 7d show the same data but on 25 October 2002. The electron fluxes showed quasi-periodic variations, with a mean period of ~2.8 hours. In these two cases, Geotail was upstream of the magnetosphere at $X_{\mathrm{GSM}} = $ ~20 R_E but not in the magnetotail, and there was no measurement of near-tail reconnection and plasmoids.

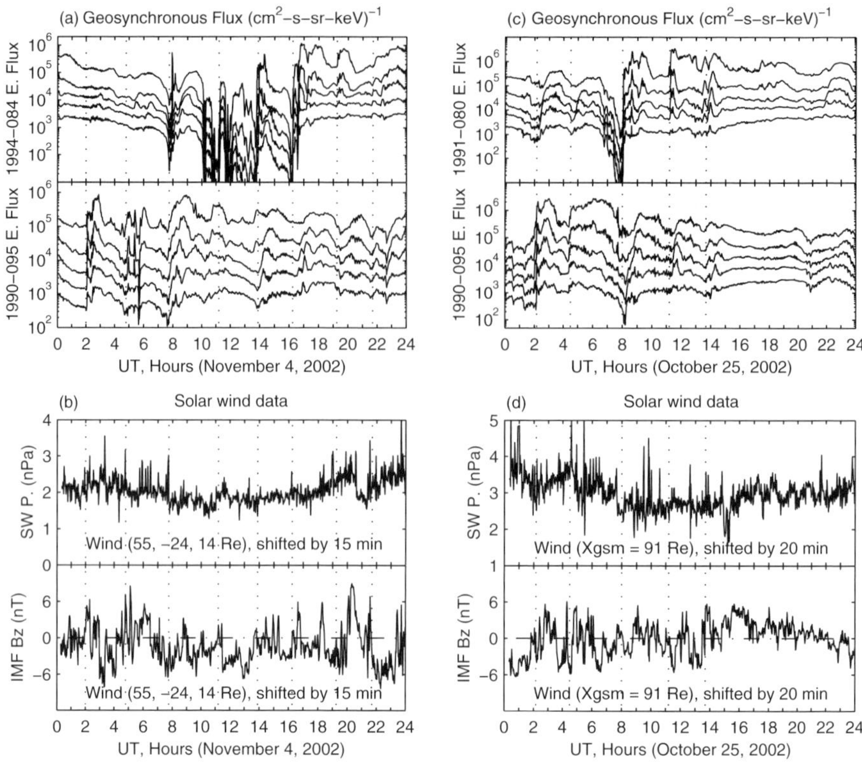

Figure 7. (a) The electron fluxes measured by the LANL 1994-084 and 1990-095 satellites and (b) the solar wind pressure and IMF B_z component measured by the Wind satellite on November 4, 2002. The vertical dotted lines indicate substorm onsets. Figures 7c and 7d show the same data but on October 25, 2002.

We checked the measurements of the CANOPUS meridian scanning photometers. Auroral intensifications occurred around 0800 and 1100 UT on 4 November 2002 and around 0200 and 0400 UT on 25 October 2002 when the photometers were located near midnight and when the data were available (The photometer plots are not shown here). These auroral intensifications are typical signature of substorm onsets and were well correlated with the energetic flux injections at geosynchronous orbit in these cases. Because all flux increases have the same characteristics, the flux variations in Figure 7 are interpreted as the signature of periodic substorms.

The solar wind parameters measured by the Wind satellite are plotted in Figures 7b and 7d. A very important characteristic of the IMF is that the IMF B_z was small and fluctuating between southward and northward. Although the IMF B_z fluctuated very rapidly, the substorm onsets still occurred every 2-3 hours and did not follow the rapid IMF northward turnings. The significance of the observations in Figure 7 is that substorm onsets occurred every ~3 hours, no matter whether the IMF was stable or fluctuating. A northward turning of the IMF after an interval of southward IMF does not necessarily trigger a substorm onset. For example, a substorm onset occurred at 0745 UT on 4 November 2002, and an IMF northward turning at 0845 UT did not cause an onset. Similarly, an onset occurred at 1615 UT, and no onset occurred at 1810 UT at which there was a large northward spike in the IMF B_z.

5. Discussion

The most remarkable characteristic of the substorms presented in this paper is the periodic occurrence. In all the cases, the substorms showed periods of 2-3 hours. Now the question is whether the periodic substorms represent an intrinsic eigenmode of the magnetosphere or a forced oscillation driven by an external source.

If the periodic occurrence of the substorms is driven by an external oscillation source, the most likely source is the solar wind. Previous studies found that isolated substorms could occur coincidently with a sudden change in the solar wind [*Cann et al.*, 1975, 1977; *Kokubun et al.*, 1977; *Petrinec and Russell*, 1996], and it was suggested that the substorms might be triggered by IMF northward turnings or by solar wind pressure impulses. If the IMF B_z and/or the solar wind pressure have periodic variations with a period of 2-3 hours, such solar wind variations may be able to trigger corresponding periodic substorms. If this is the case, it means that the period of the substorms is determined by the solar wind. However, the IMF B_z was continuously southward and very stable and did not have any northward turnings in the cases on 18 April 2002 (Figure 2), 4 October 2000 (Figure 5), and 10-11 August 2000 (Figure 6). In the two cases shown in Figure 7, the IMF B_z was rapidly fluctuating. The periodic substorms always had periods of 2-3 hours. These observations show clearly that the period of the substorms was not determined by the IMF B_z. The solar wind pressure may play a role in the generation of the periodic substorms. The periodic substorms started to occur after a large solar wind pressure impulse impinged on the magnetosphere in the case of 18 April 2002 (Figure 2). In the cases shown in Figures 5 and 6, the periodic substorms started to occur when some large variations in the solar wind arrived at the magnetosphere. It appears that the solar wind pressure impulse (or variation) might have triggered the first cycle of the periodic substorms. However, in all the cases, the solar wind pressure became stable after the impulse or had only some small and irregular variations. The solar wind pressure did not show the same period as the substorms. The solar wind pressure and IMF B_z may oscillate with any periods from several minutes to many hours. However, the period of the substorms is always 2-3 hours. This implies that the periodic substorms occur with their own periods, no matter whether the solar wind is stable or fluctuating. The period of the substorms is not determined by variations in the solar wind.

Although it is unlikely for the solar wind to directly determine the period of substorms, is it possible that the energy transfer to the magnetosphere from the solar wind indirectly controls the periodic occurrence of substorms? Substorms have been explained as a process of energy loading and unloading (storage and release) in the magnetotail. It was suggested that the magnetosphere may have a maximum capacity of energy storage. When the energy transferred to the magnetosphere from the solar wind during the growth phase of a substorm reaches a critical value, some plasma instabilities in the magnetosphere may occur and result in the substorm onset. If the process of the energy storage and release has a time scale of 2-3 hours, it explains the periodic occurrence of substorms. This mechanism implies that the period of substorms depends on the magnitude of the southward IMF, because stronger southward IMF corresponds to larger energy transfer rate. The time for the magnetosphere to reach the maximum energy state should be shorter for stronger southward IMF than for weaker southward IMF. However, the observations are inconsistent with this scenario. In the cases of 18 April 2002 (Figure 2), 4 October 2000 (Figure 5), and 10-11 August 2000 (Figure 6), the IMF B_z was strongly negative with an amplitude of 10 nT or larger. In contrast, the IMF B_z fluctuated between -6 and 6 nT in the two cases shown in Figure 7. It is obvious that the energy transfer rate and the total energy transferred to the magnetosphere were much larger in the cases of continuously southward IMF than in the cases of small and fluctuating IMF B_z. However the periods of the substorms were always 2-3 hours under these very different IMF conditions and did not show a dependence on the energy transferred to the magnetosphere from the solar wind.

It is reasonable to conclude that the period of the substorms is determined by the magnetosphere but not by the solar wind. We suggest that magnetospheric substorms have an intrinsic period of 2-3 hours. The magnetosphere is highly dynamic during magnetic storms and becomes unstable to the generation of substorms. A solar wind pressure impulse may trigger the first cycle of periodic substorms. After being excited, the periodic substorms can last for many cycles without requirement of continuous external triggering. Each cycle of the substorms does not have to be triggered by an IMF northward turning or by a solar wind pressure impulse. The periodic substorms represent an intrinsic mode of the magnetosphere.

An important characteristic of periodic substorms is the correlation between periodic magnetic reconnection in the near tail and sawtooth-like energetic flux injections at geosynchronous orbit. There is a one-to-one correspondence between each cycle of the near-tail reconnection and each cycle of the flux injections at substorm onsets. It is generally believed that the near-tail reconnection is the mechanism responsible for substorm onsets. The observations show that the periodic substorms are indeed correlated with the near-tail reconnection. However, it is uncertain whether the reconnection onsets result from a plasma instability in the tail or in the inner magnetosphere. The periodic substorms can occur when the magnetosphere is continuously driven during persistent southward IMF. This implies that the onsets of periodic substorms may be triggered by some internal magnetospheric plasma instabilities. A plasma instability in the tail can cause a magnetic reconnection onset and result in the substorm onset in the NENL model [*Hones*, 1984; *Baker et al.*, 1996]. A plasma instability in the inner magnetosphere may cause a substorm onset there and give rise to the near-tail magnetic reconnection through a rarefaction wave in the current disruption model [*Lui*, 1991, 1996]. Our studies establish the correlation between the near-tail reconnection and periodic substorms but cannot provide a solution to the onset location. Further investigations are required to solve this controversy.

6. Summary

The Geotail satellite measured periodic variations in the magnetospheric magnetic field and plasma parameters in the near tail between X_{GSM} = -20 and -30 R_E during the magnetic storms on 18 April 2002 and on 24 March 2002. These magnetospheric variations lasted for nine cycles in each case, with a mean period of ~2.7 hours. For each cycle of the variations, an increase in the strength of the magnetic field and a southward turning of the magnetic field B_z component were detected when Geotail was in the lobe, and a sudden decrease in the ion energy and pressure and a tailward plasma flow were detected when Geotail was in the plasma sheet. The magnetospheric variations are interpreted as periodic reconnection onsets in a near-Earth neutral line and subsequent plasmoid formations. Geosynchronous satellites measured periodic variations of energetic plasma particle fluxes. The flux variations exhibit a well-defined sawtooth shape, with gradual flux decreases followed by rapid flux increases. There is a one-to-one correspondence between each cycle of the magnetic reconnection in the near tail and each increase of the energetic fluxes at geosynchronous orbit. The sudden increases of the energetic fluxes represent true plasma particle injections from the tail to the inner magnetosphere. The periodic magnetic reconnection and flux injections are identified as periodic substorms in the magnetosphere.

The periodic substorms can occur during stable and continuously southward IMF or during small and fluctuating IMF. The substorms last continuously for many cycles over an interval of up to 24 hours when the IMF is stable and does not show any northward turnings. When the IMF fluctuates rapidly between southward and northward, the substorms still occur every ~3 hours and do not follow the rapid northward turnings of the IMF. A solar wind pressure impulse may play a role in triggering the first cycle of the periodic substorms, while the substorms continue after the solar wind pressure has become stable. Each cycle of the substorms does not have to be triggered by a solar wind pressure impulse or by an IMF northward turning. The observations show that the period of substorms is not determined directly by variations in the solar wind. The merging rate between the IMF and magnetospheric magnetic field during strongly southward IMF is significantly different from that during small and fluctuating IMF, so the period of substorms is not determined indirectly by the energy transferred to the magnetosphere from the solar wind.

We suggest that magnetospheric substorms have an intrinsic period of 2-3 hours. The magnetosphere is highly dynamic during magnetic storms and can become unstable to the generation of substorms. After being excited, substorms can occur periodically and last for many cycles without requirement of continuous external triggering. Periodic substorms represent an intrinsic mode of the magnetosphere, and this eigenmode has a period of 2-3 hours.

Acknowledgements

Work at MIT Haystack Observatory was supported by a NSF cooperative agreement with Massachusetts Institute of Technology. The data of the AL index are provided by World Data Center for Geomagnetism at Kyoto University. The Geotail data are provided by S. Kokubun through DARTS at the Institute of Space and Astronautical Science (ISAS) in Japan. We acknowledge the CDAWeb for access to the Wind data.

References

Akasofu, S. I., *Polar and Magnetospheric Substorms*, Dordrecht: Reidel, 1968.
Baker, D. N., T. I. Pulkkinen, V. Angelopoulos, W. Baumjohann, and R. L. McPherron, Neutral line model of substorms: Past results and present view, *J. Geophys. Res., 101*, 12,975, 1996.
Baker, D. N., T. I. Pulkkinen, J. Buchner, and A. J. Klimas, Substorms: Aglobal instability of the magnetosphere-ionosphere system, *J. Geophys. Res., 104*, 14,601, 1999.
Borovsky, J. E., R. J. Nemzek, and R. D. Belian, The occurrence rate of magnetospheric substorm onsets: Random and periodic substorms, *J. Geophys. Res., 98*, 3807, 1993.
Cann, M. N., R. L. McPherron, and C. T. Russell, Substorm and interplanetary magnetic field effects on the geomagnetic tail lobes, *J. Geophys. Res., 80*, 191, 1975.
Cann, M. N., R. L. McPherron, and C. T. Russell, Characteristics of the association between the interplanetary magnetic field and substorms, *J. Geophys. Res., 82*, 4837, 1977.
Henderson, M. G., G. D. Reeves, R. D. Belian, and J. S. Murphree, Observations of magnetospheric substorms occurring with no apparent solar wind/IMF trigger, *J. Geophys. Res., 101*, 10,773, 1996.
Hones, E. W. Jr., Transient phenomena in the magnetotail and their relation to substorms, *Space Sci. Rev., 23*, 393, 1979.
Hones, E. W., Jr., Plasma sheet behavior during substorms, in *Magnetic Reconnection in Space and Laboratory Plasmas, Geophys. Monogr. Ser.*, vol. 30, edited by E. W. Horns, Jr., pp. 178-184, AGU, Washington DC, 1984.
Horton, W., and I. Doxas, A low-dimensional energy-conserving state space model for substorm

278

dynamics, *J. Geophys. Res., 101*, 27,223, 1996.

Horton, W., R. S. Weigel, D. Vassiliadis, and I. Doxas, Substorm classification with the WINDMI model, *Nonlinear Processes in Geophysics, 10*, 363, 2003.

Hsu, T.-S., and R. L. McPherron, An evaluation of the statistical significance of the apparent association between northward turnings of the IMF and substorm expansion onsets, *J. Geophys. Res., 107*(A11), 1398, doi:10.1029/2000JA000125, 2002.

Hsu, T.-S., and R. L. McPherron, The occurrence frequencies of IMF triggered and non-triggered substorms, *J. Geophys. Res., 108*(A7), 1307, doi:10.1029/2002JA009442, 2003.

Huang, C. S., Evidence of periodic (2-3 hour) near-tail magnetic reconnection and plasmoid formation: Geotail observations, *Geophys. Res. Lett., 29*(24), 2189, doi:10.1029/ 2002GL016162, 2002.

Huang, C. S., J. C. Foster, G. D. Reeves, G. Le, H. U. Frey, C. J. Pollock, and J.-M. Jahn, Periodic magnetospheric substorms: Multiple space-based and ground-based instrumental observations, *J. Geophys. Res., 108*(A11), 1411, doi:10.1029/2003JA009992, 2003a.

Huang, C. S., G. D. Reeves, J. E. Borovsky, R. M. Skoug, Z. Y. Pu, and G. Le, Periodic magnetospheric substorms and their relationship with solar wind variations, *J. Geophys. Res., 108*(A6), 1255, doi:10.1029/2002JA009704, 2003b.

Huang, C. Y., and L. A. Frank, A statistical survey of the central plasma sheet, *J. Geophys. Res., 99*, 83, 1994.

Kivelson, M. G., Pulsations and magnetohydrodynamic waves, in *Introduction to Space Physics*, edited by M. G. Kivelson and C. T. Russell, p. 330, Cambridge University Press, New York, 1995.

Klimas, A. J., D. N. Baker, D. Vassiliadis, and D. A. Roberts, Substorm recurrence during steady and variable solar wind driving: Evidence for a normal mode in the unloading dynamics of the magnetosphere, *J. Geophys. Res., 99*, 14,855, 1994.

Kokubun, S., R. L. McPherron, and C. T. Russell, Triggering of substorms by solar wind discontinuities, *J. Geophys. Res., 82*, 74, 1977.

Lui, A. T. Y., A synthesis of magnetospheric substorm models, *J. Geophys. Res., 96*, 1849, 1991.

Lui, A. T. Y., Current disruption in the Earth's magnetosphere: Observations and models, *J. Geophys. Res., 101*, 13,067, 1996.

McPherron, R. L., Magnetospheric dynamics, in *Introduction to Space Physics*, edited by M. G. Kivelson and C. T. Russell, p. 400, Cambridge University Press, New York, 1995.

McPherron, R. L., T. Terasawa, and A. Nishida, Solar wind triggering of substorm expansion onset, *J. Geomag. Geoelectr., 38*, 1089, 1986.

Nagai, T., M. Fujimoto, Y. Saito, S. Machida, T. Terasawa, R. Nakamura, T. Yamamoto, T. Mukai, A. Nishida, and S. Kokubun, Structure and dynamics of magnetic reconnection for substorm onsets with Geotail observations, *J. Geophys. Res., 103*, 4419, 1998.

Petrinec, S. M., and C. T. Russell, Near-Earth magnetotail shape and size as determined from the magnetopause flaring angle, *J. Geophys. Res., 101*, 137, 1996.

Reeves, G. D., et al., IMAGE, POLAR, and geosynchronous observations of substorm and ring current ion injection, in *Dsiturbances in Geospace: The Storm-Substorm Relationship, Geophys. Monogr. Ser.*, vol. 142, edited by A. S. Sharma, Y. Kamide, and G. S. Lakhina, pp. 91-101, AGU, Washington D. C., 2003.

Rostoker, G., Triggering of the expansive phase intensifications of magnetospheric substorms by northward turnings of the interplanetary magnetic field, *J. Geophys. Res., 88*, 6981, 1983.

Slavin, J. A., D. H. Fairfield, M. M. Kuznetsova, C. J. Owen, R. P. Lepping, S. Taguchi, T. Mukai, Y. Saito, T. Yamamoto, S. Kokubun, A. T. Y. Lui, and G. D. Reeves, ISTP observations of plasmoid ejection: IMP 8 and Geotail, *J. Geophys. Res., 103*, 119, 1998.

Slavin, J. A., E. J. Smith, B. T. Tsurutani, D. G. Sibeck, H. J. Singer, D. N. Baker, J. T. Gosling, E. W. Hones, and F. L. Scarf, Substorm associated traveling compression regions in the distant tail: ISEE-3 geotail observations, *Geophys. Res. Lett., 11*, 657, 1984.
Slavin, J. A., M. F. Smith, E. L. Mazur, D. N. Baker, E. W. Hones, Jr., T. Iyemori, and E. W. Greenstadt, ISEE 3 observations of traveling compression regions in the Earth's magnetotail, *J. Geophys. Res., 98*, 15,425, 1993.

CHAPTER FOUR

MAGNETIC RECONNECTION AND TURBULENCE

Multiscale Coupling of Sun-Earth Processes
A.T.Y. Lui, Y. Kamide and G. Consolini, editors.

MAGNETIC FIELD FLUCTUATIONS IN THE PLASMA SHEET REGION: A STATISTICAL APPROACH

G. Consolini and M. Kretzschmar

Istituto di Fisica dello Spazio Interplanetario, Consiglio Nazionale delle Ricerche,
Via del Fosso del Cavaliere, 100, 00133 Roma, Italy

Abstract. Recent observation of magnetic field fluctuations in the geotail regions evidenced the occurrence of sporadic, localized multiscale turbulent fluctuations. In this work, we approach these fluctuations from a statistical point of view, showing the existence of non-trivial scaling features and similarity of the observed scaling properties with those of a stochastic process generated by an Ornstein-Uhlembeck-Cauchy equation. Our results will be discussed in the framework of the recent stochastic view of the geotail dynamics in terms of a *forced and/or self-organized criticality.*

Keywords. magnetotail, magnetic fluctuations, scaling features, fractal noise.

1. Introduction

The effects of the continuous interaction between the solar wind and the Earth's magnetosphere manifest in a wide variety of phenomena and processes, which occur in the near-Earth space region. For example, magnetospheric substorms and storms are manifestations of such an interaction and are comprises of phenomena involving several magnetospheric regions.

In the last decade it was shown that the dynamics of the Earth's magnetosphere is equivalent to that of a complex system usually in a far-from-equilibrium configuration. Some evidences of this complex response to the solar wind changes are the impulsive and avalanching character of the magnetospheric dynamics as shown by the geomagnetic indices. Furthermore, scaling invariance and intermittent turbulence [*Consolini*, 1997; *Angelopoulos et al.*, 1999; *Lui et al.*, 2000, *Uritsky et al.*, 2002] were shown to play a relevant role in the dynamics of several magnetospheric regions. Although significant advances in the understanding of the magnetospheric dynamics were achieved from the early days to present, a complete description of the spotty, intermittent and turbulent dynamics displayed by the Earth's magnetosphere in response to solar wind changes is still far to be reached. In detail, the understanding of the role of the stochastic small-scale space-time fluctuations, and of the cross-coupling phenomena involving a wide range of scales in connection with the large-scale dynamics is still lacking. For instance, some substorm associated studies evidenced the relevance of the small-scale randomness of the spatial and temporal fluctuations in respect to the dynamical evolution of the magnetospheric substorms [*Consolini et al.*, 1996; *Chang*, 1998, 1999].

Historically, the study of the magnetospheric dynamics is based on the magneto-hydrodynamic (MHD) concepts. Although in the past this approach led to significant advances in the description of global and large-scale magnetospheric processes and phenomena, a detailed description of the small-scale randomness is still missing. This point is, indeed, particularly true when one refers to the large magnetic fluctuations and the rapid particle energization observed in the plasma and current sheet, both during substorm expansion phase [*Lui and Najimi*, 1997; *Lui*, 2002] or not [*Lui et al.*, 1998; *Angelopoulos et al.*, 1999]. Furthermore the observational evidence of some inconsistencies with some classical MHD-based substorm scenarios (e.g.: Near-Earth

286

Furthermore, Equation (2.1) involves the occurrence of a homogeneous cascading process, i.e. a uniform distribution of eddies at each scale. In the case of eddies, which do not uniformly fill the space, the scale-invariant features of the cascading process are local, i.e. we are in presence of *intermittency*.

A recent study by *Borowsky et al.* [1997], based on in-situ single-point measurements, clearly evidenced that the plasma transport in the PS region is strongly turbulent, and that the observed flow turbulence is of a different origin in comparison to the solar-wind turbulence. At the present stage, it is not yet clear whether PS turbulence can be better read in terms of incompressible MHD, compressible MHD, reduced MHD or of other theories.

A different scenario for the emergence of stochastic and turbulent fluctuations in space plasmas was proposed by *Tetrault* [1992a, 1992b] and *Chang* [1998, 1999] in terms of multiscale coherent magnetic and plasma structures. The emergence of such multiscale coherent structures follows from the existence of resonance sites, i.e. sites where propagation condition is not satisfied: e.g. $\mathbf{k} \cdot \mathbf{B}_0(\mathbf{x}) = 0$ for Alfvén waves (see *Tetrault* [1992a] and *Chang* [1998] for a detailed discussion). In other words, these coherent structures might be thought in terms of bundles of multiscale fluctuations, which take the form of cross-tail flux tubes in the PS region. Furthermore, these structures can twist, wander, deform, and coalesce under the influence of both the local and global magnetic field and of the plasma conditions. The evolution of coherent structures will involve the production of new fluctuations, which can generate new resonance sites and new structures. As clearly stated by *Chang* and his co-authors [1999, 2003, 2004] the emerging scenario is that of a "turbulent medium of coherent structures", where turbulence results on the evolution of coherent structures. We underline that such a scenario, recently substantiated by numerical reduced magnetohydro-dynamic simulations [*Wu and Chang*, 2000, 2001; *Chang et al.*, 2004], involves a highly disordered and stochastic topology of the magnetic field in the PS region. Furthermore, differently to the Alfvénic incompressible MHD wave turbulence the emerging turbulence is not due to a simple cascading process with local interaction in the Fourier k-space, but is the result of the coupling of multiscale coherent structures.

As a consequence of the external driving the dynamical complexity of *Chang'* scenario [*Chang*, 1999; *Consolini and Chang*, 2001; *Consolini et al.*, 2003] may evolve toward a dynamical critical configuration, although the emergence of scale-invariance in this system could be observed also out-of-criticality in a limited range of scales as a consequence of the randomness of the fluctuating fields. Here, the term *criticality* is taken with the same meaning of *Kadanoff et al.* [1967], i.e. refers to the absence of a characteristic scale and scale-invariance in the fluctuation fields.

Recently, *Chang et al.* [2004] using an approach based on the path integral formalism and the DRG showed that the above multiscale stochastic scenario implies scale-invariance, power-law correlations, criticality, non-Gaussian and scale-invariant distribution functions of turbulent magnetic field fluctuations at different scales. These authors underline that the observation of a mono-power scaling for the PDFs of the magnetic energy fluctuations $\delta B^2(\delta) = \mathbf{B}^2(x+\delta) - \mathbf{B}^2(x)$ (where δ is the scale) may be understood in terms of DRG arguments by assuming δB^2 to be a relevant parameter (eigen-operator) near a certain critical fixed point. Data collapsing of the PDFs at different scales thus indicates the emergence of a master curve for the statistics of the fluctuations of a relevant parameter near criticality where a scale-invariance property is expected. In detail, PDF data collapsing follows from the following transformation,

$$\begin{cases} \delta B^2 \rightarrow \delta^s \delta B^2 \\ P\!\left(\delta B^2, \delta\right) \rightarrow \delta^{-s} P\!\left(\delta B^2, \delta\right) \end{cases} \tag{2.3}$$

where s is the scaling exponent evaluated by studying the scaling features of the so-called *probability of return* $P(0,\delta)$ with δ. On the basis of reduced MHD simulations of the plasma sheet magnetic field structures they found $s \sim 0.35$. Their results suggest approaching to the randomness of the magnetic and plasma fluctuations in the PS region by investigating the probability distribution functions (PDFs). Similar approaches have been used by *Hnat et al.* [2002a, 2002b, 2003] in studying the scaling properties of magnetic and plasma parameters in solar wind.

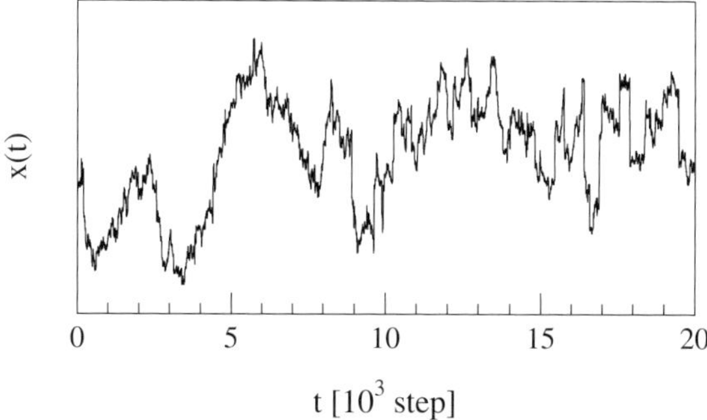

Figure 2. A sample of a 1d signal generated using the dissipative stochastic Langevin process described by Equation (2.4).

To clarify the aforementioned arguments on scaling features of the PDFs, let us consider the following stochastic process:

$$\frac{dx(t)}{dt} = -\lambda x(t) + \eta(t) \tag{2.4}$$

where $x(t)$ is a stochastic variable, λ is a positive constant describing dissipation, and $\eta(t)$ is a δ-correlated stochastic noise distributed according to a symmetric Lévy-stable distribution $L_\alpha(x, \gamma)$ with characteristic index α. Equation (2.4) is a dissipative stochastic Langevin equation describing an Ornstein-Uhlenbeck-Cauchy process.

Figure 2 shows a sample of 20000 points of a signal generated using Equation (2.4) with the following parameters: $\lambda = 2 \cdot 10^{-4}$, $\alpha = 1.35$ and $\gamma = 10$.

To study the scaling features of such a process, we have evaluated the probability distribution functions (PDFs) of the signal differences $\delta x(\tau) = x(t+\tau) - x(t)$ at different timescales τ and studied the dependence of the probability of return $P(0,\tau)$ on timescale τ. In Figure 3 we report the PDFs at different timescales (left panel) and the evolution of the probability of return $P(0,\tau)$ with τ (right panel). The probability of return $P(0,\tau)$ scales following a power-law with a scaling exponent $s = [0.74 \pm 0.01]$. This value agrees with the expected scaling exponent in the case of a simple Lévy process being $\alpha \sim 1/s$ [see *Sornette*, 2000 and references therein].

288

Using the scaling exponent s and transformation reported in Equation (2.3) it is possible to collapse all the PDFs in a single master curve, which is no-longer dependent on the timescale. Figure 4 shows the data collapsing relative to the PDFs of Figure 3. We may note that, although the observed data collapsing is a consequence of the existence of scaling-invariance in the fluctuations of the process described by Equation (2.4), it is also related with the existence of a fixed point in the RG description. As a matter of fact, in the case of multiscale complex and critical systems a link among the scaling features of the PDFs of the fluctuations and the nature of the RG fixed point exists [*Iona-Lasinio*, 1975]. This point will be discussed in the next Sections when this technique will be applied to magnetic field fluctuations in the plasma sheet.

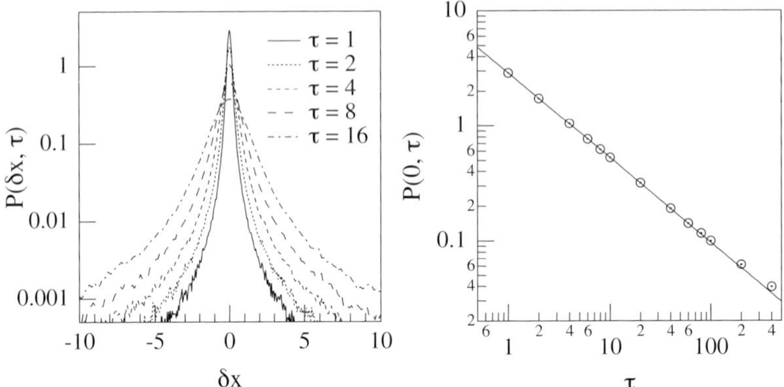

Figure 3. Evolution of the PDFs of the signal differences relative to the process described by Equation (2.4) [left panel] as a function of the timescale. Right panel shows the scaling features of the probability of return $P(0, \tau)$. The solid line is a power-law best fit, $P(0, \tau) \approx \tau^{s}$ with scaling exponent $s = [0.74 \pm 0.01]$.

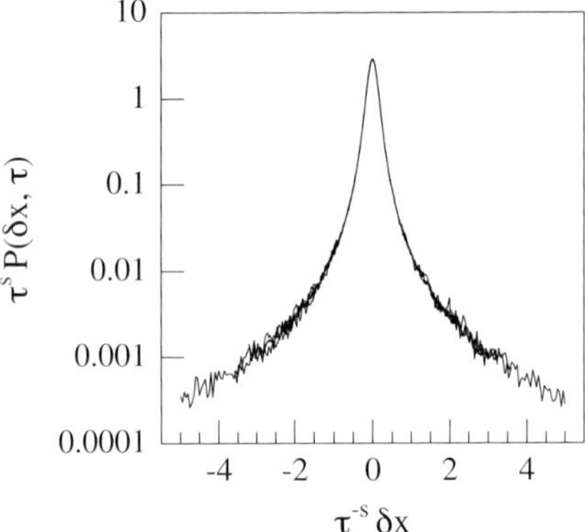

Figure 4. The master curve of the PDFs of signal fluctuations reported in Figure 3 (left panel) and obtained by the transformation described in Equation (2.3) with using the exponent s evaluated by the scaling of the probability of return (Figure 3 – right panel).

Let us now briefly discuss some aspects of the non-Gaussian features of the PDFs in turbulent plasmas. On the basis of the transformation reported in Equation (2.3) the information contained in the PDFs is separated into a master PDF curve and the scaling properties of the probability of return at different scales. Generally, the PDFs observed in several space plasma turbulent systems have a leptokurtic shape. This feature along with the occurrence of scaling invariance is assumed to be evidence of intermittency at the smaller scales; i.e. a non-homogeneous spatial distribution of the energy in the cascading process. In the past, several models were proposed to explain the non-Gaussianity of PDFs in turbulent systems [*Castaing et al.*, 1990, 1993; *Vassilicos*, 1995; *Frisch and Sornette*, 1997; *Beck*, 2000, 2002; *Treumann*, 1999a, 1999b, 2001; *Gell-Mann and Tsallis*, 2004]. These models are based on different ideas: i) a log-normal energy cascading process [*Castaing et al.*, 1990, 1993; *Vassilicos*, 1995]; ii) a general fragmentation process, based on the extreme deviations theory [*Frisch and Sornette*, 1997]; iii) the non-extensive thermostatistics [*Beck*, 2000, 2002], iv) a generalized-Lorentzian thermodynamics [*Treumann*, 1999a, 1999b, 2001] derived from the BBGKY hierarchy assuming a different collisional term. In all the aforementioned models a non-Gaussian statistics is recovered. Anyway, although in several cases it is possible to get good analytical expressions for PDFs that fit well with the results of experimental measurements, we believe that a deep understanding of the mechanisms producing leptokurtic PDFs has not been still reached. To clarify this statement let us concentrate our attention on two of the aforementioned models: the *energy cascade model* [*Castaing et al.*, 1990, 1993], and the model based on the nonextensive thermostatistics [*Beck*, 2000, 2002; *Tsallis*, 1988; *Gell-Mann and Tsallis*, 2004]. Both models provide good fit of non-Gaussian PDFs of velocity fluctuations in turbulent fluid flows, even if they start on different assumptions: the self-similar scenario for the energy cascade in turbulence [*Castaing et al.*, 1990, 1993], a new generalized shape for the entropy [*Tsallis*, 1988; Gell-*Mann and Tsallis*, 2004]. In detail, in the former model the non-Gaussianity of PDFs results on a superposition of Gaussian sets of fluctuations characterized by a log-normal distribution of variances, while in the latter the leptokurtic nature of PDFs follows from the assumption of a nonextensive entropy describing out-of-equilibrium systems [*Tsallis*, 1988; *Gell-Mann and Tsallis*, 2004]. In other words, the former model can be considered to be based on empirical considerations, while the latter is based on the assumption of a nonextensive entropy, whose shape is strongly debated [*Lavenda and Dunning-Davies*, 2003a, 2003b]. However, it is important to emphasize that the emergence of non-Gaussian distribution functions seems to be strongly related with the emergence of long-range correlations implying "ordering through correlations" [*Milovanov and Zelenyi*, 2000].

Here, we will restrict our attention only to the stable distributions [*Gnedenko and Kolmogorov*, 1954] that follow from a generalization of the well-known Central Limit Theorem (CLT) and that have been related to the fixed point features in the RG theories [*Iona-Lasinio*, 1975]. This choice has been done in order to study the features of the PDFs independently to every model assumption, and starting from the idea that the observed fluctuations come from a random system.

3. Data Description and Analysis

To study the scaling features of the magnetic field fluctuations in the plasma sheet region we have put our attention to a passage of GEOTAIL satellite in the mid-Earth's equatorial tail. The selected period refers to October, 1st, 1996 from 11 UT to 22 UT during which GEOTAIL satellite crosses the plasma sheet region and the current sheet at a distance from the Earth's in the range $13 \div 20\ R_E$.

Figure 5 shows the trajectory of GEOTAIL satellite for the selected period. In this period we have selected a time interval (from 19:30 UT to 22:00 UT) during which the satellite crosses the

290

central plasma sheet (CPS) region (average elevation angle $\theta = \tan^{-1}(B_z/B_{xy}) \sim 60°$, $<B_{xy}> \sim 5$ nT) and no-evident activity in this region is present (we have excluded time intervals during which bursty-bulk flows occur). The selection of the interval has been done according to the definition of the central plasma sheet given by *Angelopoulos et al.* [1992, 1996]. For the selected time interval the plasma beta β is about 11 and the average plasma velocity is $<v> \sim 50$ km/s.

Figure 6 shows the actual magnetic field magnitude $|B|$ for the selected time interval. The time resolution of the magnetic field measurements is $\Delta t = 3$ s. The average magnetic field is $<|B|> = [9.7 \pm 2.3]$ nT.

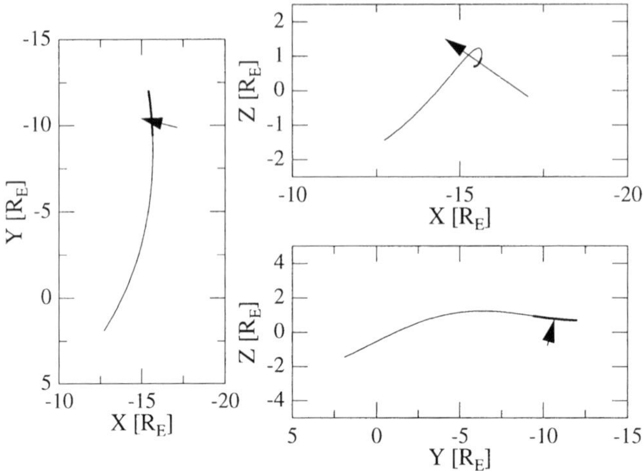

Figure 5. Trajectory of GEOTAIL satellite for the period here considered in GSM coordinate system. The solid black curves indicates the selected time interval. Arrows shows the average plasma flow direction, mainly in the plane XZ.

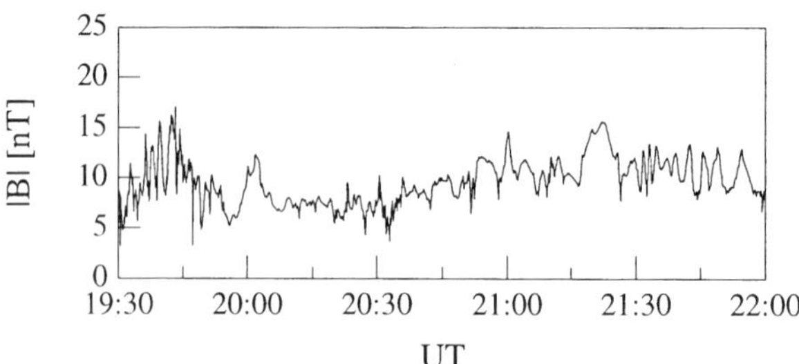

Figure 6. The actual magnetic field magnitude measurements for the selected time interval.

According to *Chang et al.* [2004], we have evaluated the fluctuations of the squared magnetic field $\delta B^2(\tau) = B^2(t+\tau) - B^2(t)$ at different timescales τ in the range 3 s $\div$ 300 s, and studied the evolution of the PDFs as a function of the timescale.

Figure 7 shows the PDFs for different timescales. PDFs look non-Gaussian at the smaller timescale showing a leptokurtic shape with enhanced tails. Furthermore, PDFs evolve with increasing timescale τ toward a Gaussian shape. This behavior resembles what is observed in the case of truncated Lévy process [*Sornette*, 2000] suggesting the occurrence of a sort of symmetry-breaking phenomenon.

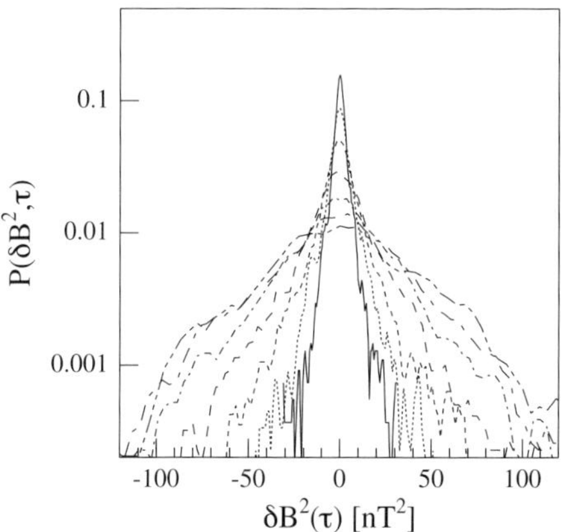

Figure 7. Evolution of the PDFs of B^2 fluctuations with the timescale τ. Here, $\tau = 3, 6, 12, 24, 48, 96,$ and 128 s.

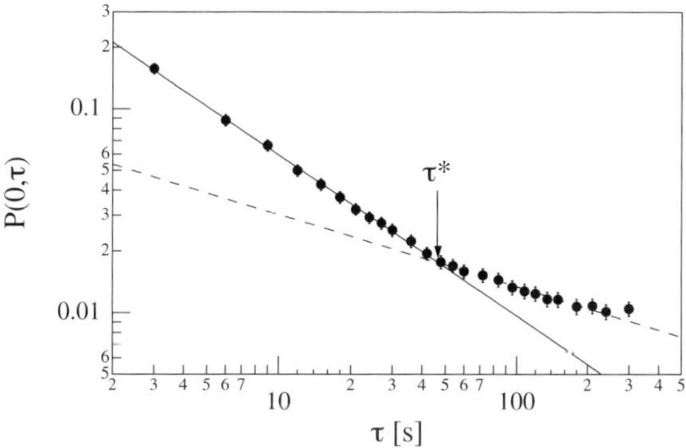

Figure 8. The probability of return $P(0,\tau)$ plotted versus the timescale τ. Solid and dashed lines are power-law best fit, $P(0,\tau) \sim \tau^{-s}$, with scaling exponents $s = [0.79\pm0.01]$ and $s = [0.35\pm0.04]$, respectively. $\tau^* \approx 46$ s indicates the crossover timescale between the two regimes.

To investigate the occurrence of scaling properties for the observed PDFs we analyzed the scaling features of the so-called *probability of return $P(0,\tau)$* with τ. The study of the probability of return is equivalent to the study of the dependence of the variance from the timescale.

292

Conversely to the conventional study of the variance from the timescale, this technique is applicable to all the cases in which the variance is not defined (see for example the case of Lévy distributions with characteristic index $\alpha < 2$).

Figure 8 shows the scaling features of $P(0,\tau)$. Two different scaling regimes characterized by different power-law scaling exponents (i.e. $P(0,\tau) \sim \tau^{-s}$) are found in the limit of small and large timescales, respectively. Scaling exponents are $s = [0.79 \pm 0.01]$ and $s = [0.35 \pm 0.04]$ for small ($\tau < \tau^*$) and large ($\tau > \tau^*$) timescales, respectively. The crossover timescale τ^* is found to be approximately 46 s.

The observed scaling regimes are both anomalous in respect to the usual scaling features of Gaussian process. In detail, the short timescale scaling exponent agrees with a Lévy domain for the basin of attraction of the PDFs at the smallest scales. As a matter of fact, in the case of a Lévy distribution function characterized by a Lévy index α, the probability of return is expected to scale with a scaling exponent $s = 1/\alpha$.

Figure 9 shows the data collapsing of the PDFs for small timescales $\tau < \tau^*$ obtained by the transformation of Equation (2.3), in comparison with a symmetric Lévy distribution with a characteristic index $\alpha = [1.27 \pm 0.02]$. The good agreement with the Lévy distribution and the master curve resulting from data collapsing seems to support the hypothesis of a Lévy basin of attraction at the smallest scales. This point will be discussed in the next section. Anyway, some deviations from the Lévy shape can be observed at very large fluctuations ($|\tau^{-s}\delta B^2| > 20$). These deviations might be the signature of multi-scaling properties in the observed fluctuations.

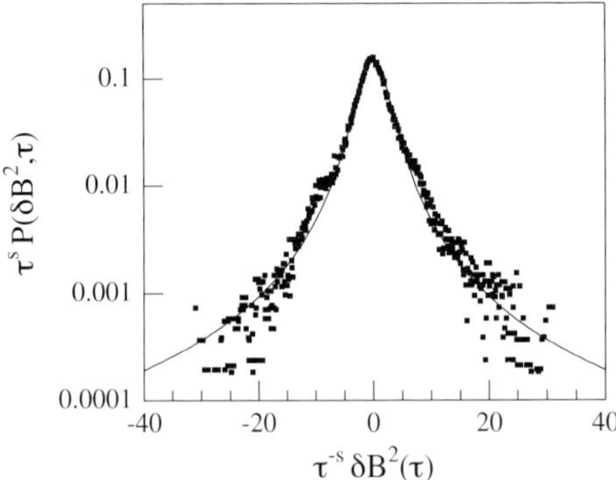

Figure 9. A sample of data collapsing of PDFs at the smaller timescales ($\tau < \tau^*$). Solid line is a nonlinear best fit by a symmetric Lévy distribution with characteristic index $\alpha = [1.27 \pm 0.02]$.

In Figure 10 we show a comparison for small scale PDF ($\tau = 3$ s) among Lévy-stable distribution fit, and fits by other two different models: the log-normal model [*Castaing et al.*, 1990, 1993]

$$P(x,\chi) = \frac{A}{2\pi\chi} \int_0^\infty \exp\left[-\frac{x^2}{2\sigma^2}\right] \exp\left[-\frac{\ln^2(\sigma/\sigma_0)}{2\chi^2}\right] \frac{1}{\sigma^2} d\sigma \tag{3.1}$$

where A is a normalization factor, χ represents the variance of the variances in the log-normal term, and σ_0 is the most probable variance, and the nonextensive thermostatistics model [*Tsallis*, 1988; *Beck*, 2000; *Gell-Mann and Tsallis*, 2004]

$$P(x) = \frac{1}{Z_q}\left(1 + \frac{1}{2}\beta(q-1)x^2\right)^{\frac{1}{q-1}} \tag{3.2}$$

where Z_q is a normalization factor, β is a parameter playing a role equivalent to an effective temperature, and q is a real parameter that provide a measure of the non-additivity in the Tsallis entropy [*Tsallis*, 1988; *Beck*, 2000, 2002; *Gell-Mann and Tsallis*, 2004].

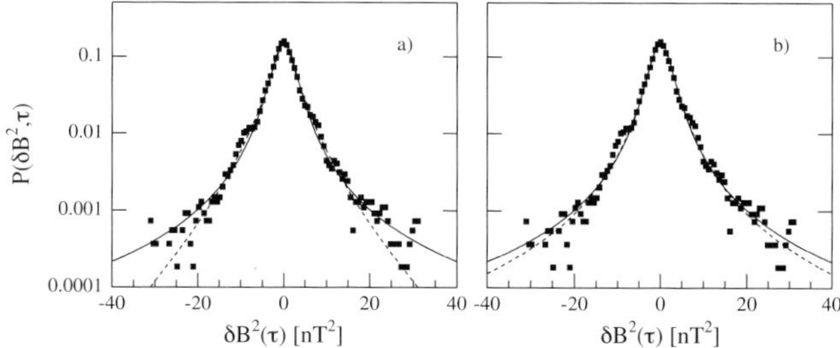

Figure 10. A comparison of the small scale PDF ($\tau = 3$ s) with different models: log-normal model [*Castaing et al.*, 1990] panel a); the nonextensive thermostatistics model [*Beck*, 2000, 2002] panel b). Solid and dashed lines are the Lévy-stable distribution reported in Figure 9, and fits by log-normal (a) and nonextensive thermostatistics model (b), respectively.

Although, all the models provide good fits in the central part of the master PDF, differences are observed in the tails. In detail, the Lévy-stable distribution and the non-extensive thermostatistics model provide better fits of the PDF tails in comparison with log-normal model. In any case, the Lévy-stable fit seems to be the best one when a comparison among the three models is made using an error function defined as follows:

$$X = \frac{1}{N-v}\sum_{i=1}^{N}\left[\frac{Y-y}{y}\right]^2 \tag{3.3}$$

where Y is the observed value, y is the estimated value, N in the number of points, and v is the number of parameters used in the fit. In our case we get, $X \sim .11$, $X \sim .16$, and $X \sim 1.75$ for Lévy-stable distribution, the non-extensive thermostatistics model, and the log-normal model, respectively.

At longer timescales ($\tau > \tau^*$) the scaling exponent is well in agreement with the mono-power scaling observed in RMHD simulations by *Chang et al.* [2004]. This result suggests the occurrence of a different basin of attraction (fixed point) in this domain. In Figure 11 we show the data collapsing of PDFs at timescales longer than the crossover time. In this regime, the data collapsing is worst than in the case of the shorter timescales. This is particularly evident in the tails of the PDFs. As a matter of fact, the tails of the scaled distributions do not fall onto one

294

single master curve. That suggests a non mono-power scaling structure of the PDFs in this regime. As already noted by *Chang et al.* [2004] this may be a consequence of the intrinsic nature of the strong intermittency (multiscaling features requiring a hierarchy of scaling indices). In other words, the scaling properties of the PDFs in this regime are subtler, and involve multi-parameters to be described. Another relevant aspect of PDFs in this timescale region is that from a very simple inspection they look similar to those observed away from the neutral sheet in *Chang et al.* RMHD simulations.

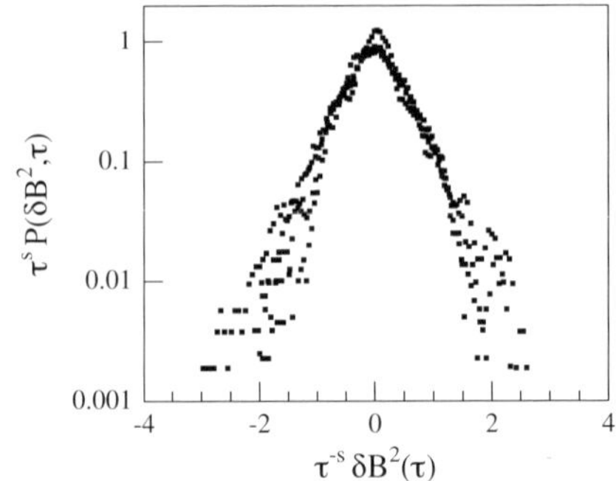

Figure 11. A sample of data collapsing of PDFs at the smaller timescales ($\tau > \tau^*$).

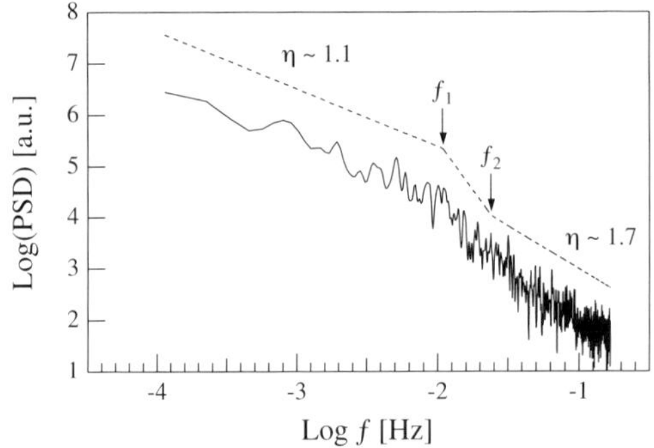

Figure 12. The PSD of the magnetic field magnitude for the selected time interval. Dashed line is a segmented fit (scaled of a factor 10) of the PSD.

To clarify the meaning of the crossover timescale τ^* we have evaluated the power spectral density (PSD) of the magnetic field magnitude (see Figure 12). PSD shows two different power-law domains below and above .01 Hz, respectively, and a transition interval in the range .01 ÷ .02 Hz. In detail, below .01 Hz the PSD scales following a power-law with scaling exponent $\eta = [1.12 \pm .02]$, while above .02 Hz the scaling exponent is $\eta = [1.69 \pm .05]$. We may note that the order of magnitude of the crossover timescale is well in agreement with the frequencies in the transition interval. As a matter of fact, the corresponding crossover frequency $f^* \sim 1/\tau^*$ is in the range $f_1 \sim 1/2\tau^* \leq f^* \leq f_2 \sim 1/\tau^*$. Furthermore, this crossover frequency f^* is well within the MHD domain being the corresponding proton-gyrofrequency f_c of the order of 0.15 Hz or higher, computed from the average magnetic field.

4. Discussion

In the previous section we have shown a case study of the scaling features of the PDF of the magnetic field fluctuations in the plasma sheet region. Our analysis evidenced the existence of two different scaling regimes both characterized by anomalous scaling exponents in comparison with what is expected for the PDF of a Gaussian stochastic signal. In detail, below a certain crossover timescale (i.e. at the shorter timescales) the scaling features of the probability of return agree with that of typical Lévy stochastic process. In this regime a nearly simple mono-power scaling domain is found and a good data collapsing is observed. Conversely at scale longer than the crossover the PDFs assume a more Gaussian shape and PDF data collapsing seems to fail in the tails, thus suggesting the occurrence of intermittency (multi-scaling features).

Another relevant aspect of the magnetic field fluctuations observed in the selected time interval is the kink-power spectrum where the small frequency regime has a nearly $1/f$ structure and the high frequency domain shows a power spectrum which is well in agreement with a Kolmogorov–5/3 spectrum. The transition interval between the two spectral regimes when the Taylor hypothesis (i.e. $f = vk$) is assumed to be valid, should be equivalent to a spatial scale in the range $2000 \div 5000$ km. We note how this characteristic spatial scale is in the typical range of tearing mode instability generating magnetic islands, and thus, it may play the role of a sort of correlation length for the fluctuations. In other words, the scaling features associated to the longer timescales seem to be related to the stochastic nature of magnetic islands. This point seems also to be corroborated by other observations in space plasmas [*Matthaeus and Goldstein*, 1986]. Furthermore, in analogy to what has been proposed in other different turbulent plasma scenario [*Matthaeus and Goldstein*, 1986; *Chang et al.*, 2004] the –1 spectral exponent, observed at small frequencies can be read as the result of a superposition of multiscale discrete magnetic structures. *Chang et al.* [2004] studying these large-scale (small frequency) fluctuations by the dynamic renormalization group (DRG) argued that the –1 spectrum exponent is mean-field-like.

Looking at the short timescale scaling features of the PDFs the emerging scenario is quite different. First of all we may note how such scaling properties are very well in agreement with the existence of a Lévy fixed point (basin of attraction) in the Renormalization Group (RG) transformation [*Iona-Lasinio*, 1975; *West et al.*, 2003]. This point can be understood when one considers the generalized Central Limit Theorem (CLT) in connection with Renormalization Group (RG) transformation [*West et al.*, 2003]. The study of the evolution of the PDFs at different timescales is equivalent to the coarse-graining procedure used in the RG transformation. Here, the existence of a scale transformation of the type of Equation (2.3) which is able to generate a data collapsing of the PDFs at different timescales is analogous to the existence of a fixed point in the RG transformation for the characteristic function $\phi(k)$ of the observed small scale PDF, i.e. the RG coarse-graining does not change the functional form of the characteristic function:

$$T_N\big[\phi(k)\big] = \phi(k) \qquad (4.1)$$

where T_N represents the RG transformation. The consequence of the transformation (4.1) is that the collective complex behavior of the system under investigation is parameterized and characterized by a single parameter/exponent related to the fractal dimension of the system. In our case, the complex collective behavior of the magnetic field fluctuations δB^2 is described by a

296

PDF that belongs to the class of stable Lévy distribution with exponent $\alpha = [1.27 \pm 0.02]$. The consequence of the PDF scaling invariance is that the topology of magnetic field in the plasma sheet region at these timescales (or spatial scales) is equivalent to self-similar patches of fluctuations in space, as already proposed by *Chang* [1999].

The change of the scaling features from short timescales to large timescale is equivalent to a symmetry breaking in the RG coarse-graining transformation. In other words, we are observing a transition in the RG-space flux from one fixed point to another as the scale approaches a characteristic correlation length or timescale. Figure 12 is a schematic representation of the observed behavior.

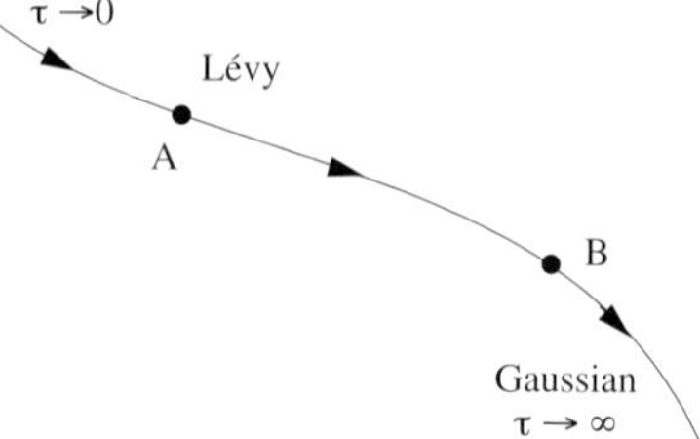

Figure 12. A schematic representation of the evolution of the PDFs with the scale in the RG space. Black dots represent the two fixed points at short (Lévy) and long (Gaussian) timescales, respectively.

The behavior showed in Figure 12 is analogous to the one generated by the Lévy dynamical process described by the stochastic Langevin equation reported in Section 2 (Equation 2.4). As a matter of fact, investigating the scaling features of the probability of return in a range of scales exceeding $\tau_c \sim 1/\lambda$ (see Equation (2.4)) it is possible to observe that the scaling features of $P(0,\tau)$ changes from one regime towards another going from timescales smaller than τ_c to timescales larger than τ_c. Here, τ_c plays the role of a correlation length. Figure 13 shows the behavior of $P(0,\tau)$ in the timescale interval $1 \div 50000$ as evaluated by a simulated signal of 10^6 points.

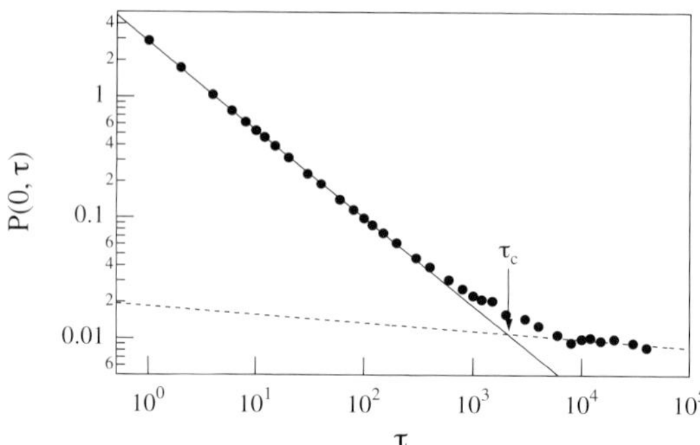

Figure 13. The behavior of $P(0,\tau)$ for a simulated signal using Equation (2.4).

Figure 13 clearly shows the existence of two different dynamical regime in this simple dynamical model: one at the smaller scales characterized by a Lévy fixed point, one other at the larger scales where the PDFs tend to assume a Gaussian shape. In any case, conversely to what is found for the magnetic field fluctuations, the larger scales do not seem to show scaling features in the probability of return and this could be due to the fact that in the simulated signal only one characteristic scale exists: the one related with the dissipation term in Equation (2.4). However, this point suggests that the scaling regime at the longer timescales in the case of the magnetic field fluctuations is the results of the interaction of discrete structures characterized by a wide range of characteristic scales.

The analogy between the case of the magnetic field fluctuations and the simulated Ornstein-Uhlembeck-Cauchy process seems to suggest that the symmetry breaking is due to the existence of a limited correlation length in the systems. We believe that this analogy establishes the relevance of the stochastic nature of the fluctuations observed in the plasma sheet. In other word, from our analysis it is evident that in order to understand the complex nature of the turbulent and intermittent fluctuations in the tail regions one has to consider the role that stochastic and random space-time fluctuations play in generating a complex hierarchy of coherent structures.

5. Summary and Conclusions

In this work we have presented a preliminary study of the scaling properties of the PDFs of the magnetic field fluctuations in the plasma sheet region as observed by GEOTAIL satellite. Although our study is limited only to a single case we found strong evidences for the existence of scaling features in the PDFs and for the occurrence of a symmetry-breaking phenomenon.

Our results support the point of view of *Chang et al.* [2004] for the existence of RG fixed point, although some differences have been found between our results and RMHD simulations [*Chang et al.*, 2004]. In detail, the short timescale PDFs seem to follow a different scaling regime characterized by a scaling exponent $s \sim 0.8$, while longer timescale PDFs scale with a scaling exponent $s \sim 0.35$ which is quite well in agreement with the one found by *Chang et al.* Conversely, the agreement seems to be better when one puts his attention to the spectral features, where two different spectral regimes are found with spectral exponents close to -1, and -1.7. Clearly, our results are by no means conclusive but can be a good starting point for future works in this direction.

Another potential implication of the study presented in this work is in underlying the relevance of such statistical approach in providing a different framework to understand the nature of the complex structure of the geotail plasma sheet region. Last but not least, let us emphasize that a relevant application of the present results might be the identification and modeling of the proper dynamical transport regime for plasma in the geotail PS. As a matter of fact, it can be demonstrated that the existence of scaling features in some relevant parameters may be modeled using a special class of transport equation: the fractional Fokker-Planck equation. This argument will be discussed in detail in a future work.

Acknowledgements

The Authors wish to thank their colleagues T. Chang, S.C. Chapman, P. De Michelis, A. Klimas, S.W.Y. Tam, and N.W. Watkins for useful discussion. GEOTAIL magnetic field data and plasma data, used in this work, were provided by S. Kokubun and T. Mukai through DARTS at the Institute of Space and Astronautical Science (ISAS) in Japan. This work is supported by European Commission contract number HPRN-CT-2001-00314.

References

Angelopoulos, V., W. Baumjohann, , C.F. Kennel, F.V. Coroniti, M.G. Kivelson, R. Pellat, R.J. Walker, H. Luher, and G. Paschmann, Bursty bulk flows in the inner central plasma sheet, *J. Geophys. Res.*, *97* (A4), 4027-4039, 1992.

Angelopoulos, V., F.V. Coroniti, C.F. Kennel, M.G. Kivelson, R.J. Walker, C.T. Russell, R.L. McPherron, E. Sanchez, C.-I. Meng, W. Baumjohann, G.D. Reeves, R.D. Belian, N. Sato, E. Friis-Christensen, P.R. Sutcliffe, et al., Multipoint analysis of a bursty bulk flow event on April 11, 1985, *J. Geophys. Res.*, *101* (A3), 4966-4990, 1996.

Angelopoulos, V., T. Mukai, and S. Kokubun, Evidence for intermittency in Earth's plasma sheet and implications for self-organized criticality, *Phys. Plasmas*, *6*, 4161-4168, 1999.

Bauer, T.M., W. Baumjohann, R.A. Treumann, and N. Sckopke, Low-frequency waves in the near-Earth plasma sheet, *J. Geophys. Res.*, *100*, 9605-9618, 1995.

Beck, C., Application of generalized thermostatistics to fully developed turbulence, *Physica A*, *277*, 115-123, 2000.

Beck, C., Non-extensive statistical mechanics approach to fully developed hydrodynamic turbulence, *Chaos, Solitons and Fractals*, *13*, 499-506, 2002.

Biskamp, D., *Nonlinear Magnetohydrodynamics*, Cambridge University Press, Cambridge, 1993.

Borowsky, J.E., R.C. Elphic, H.O. Funsten, and M. Thomsen, The Earth's plasma sheet as a laboratory for flow turbulence in high-β MHD, *J. Plasma Phys.*, *57*, 1-34, 1997.

Castaing, B., Y. Gagne, and E.J. Hopfinger, Probability density functions of high Reynolds number turbulence, *Physica D*, *46*, 177-200, 1990.

Castaing, B., Y. Gagne, and M. Marchaud, Log-similarity for turbulent flows ?, *Physica D*, *68*, 387-400, 1993

Chang, T., Low dimensional behavior and symmetry breaking of stochastic systems near criticality - can these effects be observed in space and in the laboratory ?, *IEEE Trans. Plasma Sci.*, *20*, 691-694, 1992.

Chang, T., Sporadic localized reconnections and multiscale intermittent turbulence in the magnetotail, in *Geospace Mass and Energy Flow*, Geophys. Monograph, *104*, ed. by J.L. Horwitz et al., AGU, Washington DC, 193-205, 1998

Chang, T., Self-organized criticality, multifractal spectra, sporadic localized reconnections and intermittent turbulence in magnetotail, *Phys. Plasmas*, *6*, 4137-4145, 1999

Chang, T., S.W.Y. Tam, C.C. Wu, and G. Consolini, Complexity, forced and/or self-organized criticality, and topological phase transitions in space plasmas, *Space Sci. Rev.*, *107*, 425-444, 2003

Chang, T., S.W.Y. Tam, and C.C. Wu, Complexity induced anisotropic bimodal intermittent turbulence in soace plasmas, *Phys. Plasmas*, *11*, 1287-1299, 2004.

Chapman, S. C., N. W. Watkins, R. O. Dendy, P. Helander, and G. Rowlands, A simple avalanche model as an analogue for magnetospheric activity, *Geophys. Res. Lett.*, *25*, 2397-2400, 1998.

Consolini, G., M. F. Marcucci, and M. Candidi, Multifractal structure of auroral electrojet index data, *Phys. Rev. Lett.*, *76*, 4082-4085, 1996

Consolini, G., Sandpile cellular automata and magnetospheric dynamics, *Proceedings of Cosmic Physics in the Year 2000*, *58*, ed. by S. Aiello et al. (eds.), Società Italiana di Fisica, Bologna, Italy, 123-126, 1997.

Consolini, G. and T. Chang, Magnetic field topology and criticality in geotail dynamics: relevance to substorm phenomena, *Space Sci. Rev.*, *95*, 309-321, 2001

Consolini, G., T. Chang, and A.T.Y. Lui, Complexity and topological disorder in the Earth's magnetoatil dynamics, in *Non-equilibrium Transitions in Plasmas*, ed. by A.S. Sharma and P. Kaw, Kluwer Academic Publishers, Dordrecht, The Netherlands, to be published.

Frisch, U., and Sornette, D., Extreme deviations and applications, *J. Phys. I France*, 7, 1155-1171, 1997.

Gell-Mann, M., and C. Tsallis, *Nonextensive Entropy – Interdisciplinary Applications*, Oxford Univerity Press, New York, 2004.

Gnedenko, B.V., and Kolmogorov, A.N., *Limit distributions for sums of independent random variables*, Addison-Wesley, 1954.

Hnat, B., S.C. Chapman, G. Rowlands, N.W. Watkins, and N.W. Farrell, Finite size scaling in the solar wind magnetic field energy density as seen by WIND, *Geophys. Res. Lett.*, *29*, 86-1, citeID 1446, doi 10.1029/2001GL014587, 2002a.

Hnat, B., S.C. Chapman, G. Rowlands, N.W. Watkins, and M.P.. Freeman, Scaling in long term data sets of geomagnetic indices and solar wind as seen WIND spacecraft, *Geophys. Res. Lett.*, *29*, 35-1, citeID 2078, doi 10.1029/2002GL016054, 2002b.

Hnat, B., S.C. Chapman, G. Rowlands, N.W. Watkins, and M.P.. Freeman, Scaling of solar wind ε, and the AU, AL and AE indices as seen WIND, *Geophys. Res. Lett.*, *30*, SSC 6-1, citeID 2174, doi 10.1029/2003GL018209, 2003.

Hoshino, M., A. Nishida, T. Yamamoto, and S. Kokubun, Turbulent magnetic field in the distant magnetotail: bottom-up process of plasmoid formation ?, *Geophys. Res. Lett.*, *21*, 2935-2938, 1994.

Iona-Lasinio, G., The renormalization group: a probabilistic view, *Il Nuovo Cimento*, *26B*, 99-119, 1975.

Kadanoff, L.P., W. Götze, D. Hamblen, R. Hecht, E.A. Lewis, V.V. palciauskas, M. Rayl, J. Swift, D. Aspnes, and J. Kane, Static phenomena near critical point: theory and experiment, *Rev. Mod. Phys.*, *39*, 345-431, 1997.

Kivelson, M., and C.T. Russell (eds.), *Introduction to Space Physics*, Cambridge University Press, New York, 1995

Klimas, A.J., D.V. Vassiliadis, D.N. Baker, and D.A. Roberts, The organized nonlinear dynamics of the magnetosphere, *J. Geophys. Res.*, *101* (A6), 13089-13114, 1996.

Lavenda, B.H., and J. Dunning-Davies, Qualms concerning Tsallis' use of the maximum entropy formalism, *eprint arXiV:* cond-mat/0312132v1, 2003a.

Lavenda, B.H., and J. Dunning-Davies, Qualms concerning Tsallis's condition of pseudo-additivity as a definition of non-extensivity, *eprint arXiV:* cond-mat/0311477v1, 2003b

Lui, A. T. Y., and A.-H Najimi, Time-frequency decomposition of signals in a current disruption event, *Geophys. Res. Lett.*, *24*, 3157-3160, 1997.

Lui, A. T. Y., K. Liou, P.T. Newell, C.I. Meng, S.-I Ohtani, T. Ogino, S. Kokubun, M. Brittnacher, and G. K. Parks, Plasma and magnetic flux transport associated with auroral breakups, *Geophys. Res. Lett.*, *25*, 4059-4062, 1998.

Lui, A. T. Y., S. C. Chapman, K. Liou, P. T. Newell, C.-I. Meng, M. Brittnacher, and G. K. Parks, Is the dynamic magnetosphere an avalanching system?, *Geophys. Res. Lett.*, *27*, 911-914, 2000.

Lui, A. T. Y., Multiscale phenomena in the near-Earth magnetosphere, *J. Atmos. Sol. Terr. Phys.*, *64*, 125-143, 2002.

Matthaeus, W.H., and M.L. Goldstein, Low-frequency $1/f$ noise in the interplanetary magnetic field, *Phys. Rev. Lett.*, *57*, 495-498, 1986.

Milovanov, A.V., and L.M. Zelenyi, Functional background of the Tsallis entropy: "coarse-grained" systems and "kappa" distribution functions, *Nonlin. Prog. Geophys.*, *7*, 211-221, 2000.

Montgomery, D., Remarks on the MHD problem of generic magnetospheres and magnetotails, in *Magnetotail Physics*, ed. by A.T.Y. Lui, Johns Hopkins Univ. Press, Baltimore, MD, 1987

300

Nagai, T., M. Fujimoto, R. Nakamura, Y. Saito, T. Mukai, T. Yamamoto, A. Nishida, S. Kokubun, G.D. Reeves, and R.P. Lepping, Geotail observations of a fast tailward flow at $X_{GSM} = -15R_E$, *J. Geophys. Res.*, *103* (A10), 23543-23550, 1998.

Parks, G. K., *Physics of Space Plasmas: An Introduction*, Westview Press, 2004

Sornette, D., *Critical Phenomena in Natural Sciences. Chaos, Fractals, Selforganization and Disorder: Concepts and Tools*, Springer-Verlag, Berlin Hedelberg, Germany, 2000

Tetrault, D., Turbulent relaxation of magnetic fields: 1. Coarse-grained dissipation and reconnection, *J. Geophys. Res.*, *97* (A6), 8531-8540, 1992a.

Tetrault, D., Turbulent relaxation of magnetic fields: 2. Self-organization and intermittency, *J. Geophys. Res.*, *97* (A6), 8541-8547, 1992b.

Treumann, R.A., Kinetic theoretical foundation of Lorentzian statistical mechanics, *Physica Scripta*, *59*, 19-26, 1999a.

Treumann, R. A., Generalized-Lorentzian thermodynamics, *Physica Scripta*, *59*, 204-214, 1999b.

Treumann, R. A., Statistical mechanics of stable states far from equilibrium: thermodynamics of turbulent plasmas, *Astrophys. & Space Sci.*, *277*, 81-95, 2001.

Tsallis, C., Possible generalization of Boltzmann-Gibbs statistics, *J. Stat. Phys.*, *52*, 479-487, 1988.

Uritsky, V. M., A. J. Klimas, D. Vassiliadis, D. Chua, and G. Parks, Scale-free statistics of spatiotemproal auroral emissions as depicted by Polar UVI images: The dynamic magnetosphere is an avalanching system, *J. Geophys. Res.*, *107* (A12), p. SMP 7-1, 1426, doi: 10.1029/2001JA000281, 2002.

Vassilicos, J.C., Turbulence and intermittency, *Nature*, *374*, 408-409, 1995.

West, B.J., M. Bologna, and P. Grigolini, *Physics of Fractal Operators*, Springer-Verlag, New York, 2003

Wu, C.C., and T. Chang, 2D MHD simulation of the emergence and merging of coherent structures, *Geophys. Res. Lett.*, *27*, 863-866, 2000.

Wu, C.C., and T. Chang, Further study of the dynamics of two-dimensional MHD coherent structures – a large scale simulation, *J. Atmos. Sol. Terr. Phys.*, *63*, 1447-1453, 2001.

Published by Elsevier B.V.
Multiscale Coupling of Sun-Earth Processes
A.T.Y. Lui, Y. Kamide and G. Consolini, editors.

NUMERICAL SIMULATIONS OF SOLAR WIND TURBULENCE

Melvyn L. Goldstein, D. Aaron Roberts, and Arcadi V. Usmanov
NASA Goddard Space Flight Center, Greenbelt, MD 20771, USA

Abstract. Alfvénic fluctuations are a ubiquitous component of the solar wind. Evidence from many spacecraft indicates that the fluctuations are convected out of the solar corona with relatively flat power spectra and constitute a source of free energy for a turbulent cascade of magnetic and kinetic energy to high wave numbers. Observations and simulations support the conclusion that the cascade evolves most rapidly in the vicinity of velocity shears and current sheets. Numerical solutions of the magnetohydrodynamic equations are clarifying not only how a turbulent cascade develops, but also the nature of the symmetries of the turbulence. Of particular interest is the origin of the *Maltese cross* two-dimensional correlation function of magnetic fluctuations that was deduced from ISEE 3 data. A central issue to be resolved is whether the correlation function indicates the existence of a quasi-two-dimensional component of the turbulence, or reflects another origin, such as pressure-balanced structures or small velocity shears. This paper reviews the current status of our simulations of turbulence in the solar wind. These simulations include a tilted rotating current sheet as well as variety of waves (*e.g.*, Alfvénic, quasi-two-dimensional, pressure balance structures) and "microstreams". Some aspects of our global models of heliospheric structure are also covered.

Keywords. Solar wind, magnetohydrodynamic turbulence, global heliospheric structure

1. Introduction: Observed Properties of the Solar Wind

The history of the solar wind dates back to Birkeland [*Birkeland*, 1896; *Birkeland*, 1908; *Birkeland*, 1913]. Later, *Parker* [1958] predicted the existence of the wind. Early experimental confirmation soon followed [*Gringauz*, 1961; *Gringauz et al.*, 1967; *Gringauz et al.*, 1960]. Measurements on Explorer 10 soon showed that the wind was continuous, filled the heliosphere, and, at least near solar minimum, existed in two states: fast, with velocities up to 800 km/s, and slow, with velocities more in the range of 200–350 km/s [*Bonetti et al.*, 1963; *Bridge et al.*, 1961; *Scherb*, 1964]. (For a review, see *Parker* [1997].) It was soon realized [*Barnes*, 1979; *Belcher and Davis*, 1971; *Coleman*, 1966; *Unti and Neugebauer*, 1968] that the fluctuations in the solar wind magnetic field and velocity were often so highly correlated that they appeared as nearly perfect Alfvén waves [*Alfvén*, 1950]. If one normalizes the fluctuating magnetic field, $\mathbf{b}$ (in cgs units), by dividing by $\sqrt{4\pi\rho}$ (where ρ is the mass density), Alfvén waves are defined by $\mathbf{v} = \pm\,\mathbf{b}$, where $\mathbf{v}$ is the fluctuating component of the velocity. The $\pm$ sign represents wave propagation antiparallel (+) and parallel (−) to the ambient magnetic field. *Belcher and Davis* [1971] demonstrated that the direction of propagation of Alfvén waves is predominantly outward from the Sun. If the Alfvénic wave flux is generated in the photosphere or lower corona where the radially outward flow of the atmosphere is both subsonic and sub-Alfvénic, only outward propagating fluctuations will be convected into the heliosphere. An example of an Alfvénic interval of solar wind flow is illustrated in Figure 1 [*Bruno et al.*, 1985].

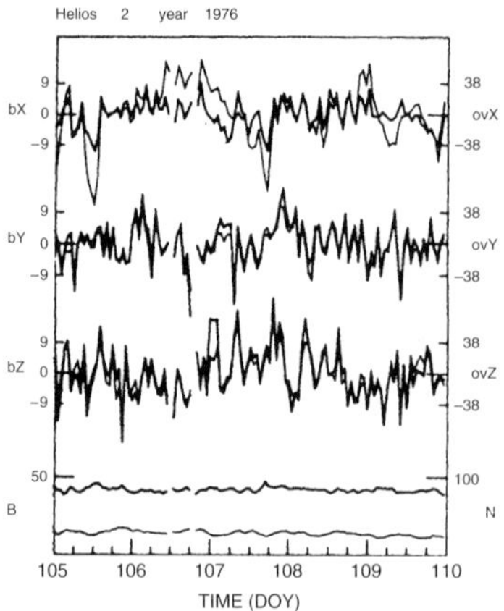

Figure 1. Five days of Helios 2 magnetic field and plasma data during 1976 at 0.29 AU. One hour averages of magnetic field (heavy line) and velocity. Data were detrended and rotated into the minimum variance direction. Adapted from *Bruno et al.*, [1985].

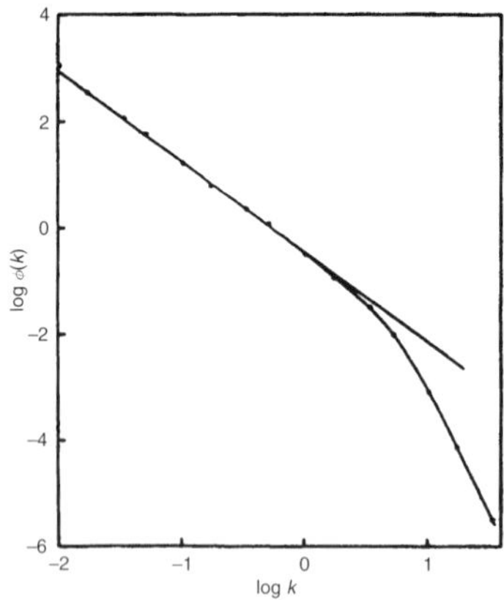

Figure 2. A plot of the one-dimensional spectrum of tidal channel data collected on March 10, 1959. The straight line has a slope of –5/3. From *Grant et al.* [1962].

The high Alfvénicity of solar wind fluctuations suggested that the medium was not actively evolving — pure Alfvén waves are an exact solution of the incompressible, ideal, equations of magnetohydrodynamics. However, the solar wind also exhibited a contrasting quality first noted by *Coleman* [1968], *viz.*, that the power spectra of the magnetic (and velocity) fields resembled those of fully developed fluid turbulence [*Kolmogorov*, 1941]. In particular, the spectral index of the power spectra tended to have a slope of –5/3. However, analyses of Helios data show that while, indeed, the power spectrum of fluctuations observed in slow wind and in fast wind at 1 AU and beyond typically has a spectral index of –5/3 in an extended "inertial" range, fast (or uniform slow) wind at 0.3 AU has a very limited inertial range, but a manifests an extended energy-containing scale where the spectral index is –1.

In Figure 2, we show the classic example of fully developed fluid turbulence [*Grant et al.*, 1962], and in Figure 3, an example computed from Voyager magnetometer data obtained near 1 AU [*Matthaeus and Goldstein*, 1982a]. Figure 4 shows the difference between fast and slow wind at 0.3 AU from [*Tu et al.*, 1989].

The Helios data has been plotted in terms of the Elsässer variables, which naturally project out fluctuations that are propagating outward (z^+) or inward (z^-) from the Sun.

As is clear from these spectra, when the turbulence has had the opportunity to develop an inertial range, the slope is essentially –5/3, the Kolmogorov value for fully developed, incompressible Navier-Stokes fluid turbulence. The remarkable result is still not fully understood, especially since the solar wind is neither a Navier-Stokes fluid nor is it incompressible. It is also significant that fast wind in the inner heliosphere is highly Alfvénic, *i.e.*, the fluctuations are essentially all propagating outward from the Sun. In contrast, slow wind, which is usually found in the vicinity of the heliospheric current sheet, where velocity gradients tend to be large and where the magnetic field changes sign as one crosses the current sheet. This association of Kolmogorov-like spectra with velocity gradients was used by *Coleman* [1978] as evidence that the solar wind was a turbulent magnetofluid.

That it is gradients, and not speed, that is important is shown by the very high Alfvénicity of slow, uniform solar wind regions [*Roberts et al.*, 1987a].

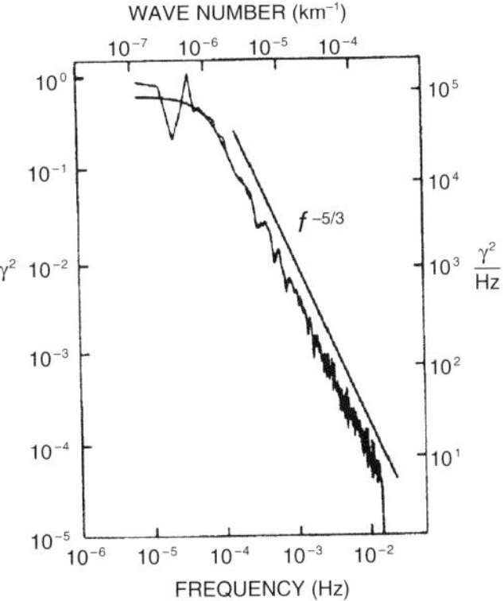

Figure 3. The magnetic energy spectrum of data collected at 1 AU in units of γ^2. The abscissa is given in both frequency and wave number. Spectra constructed using the Blackman-Tukey algorithm [*Blackman and Tukey*, 1958] and the fast Fourier transform technique are shown.

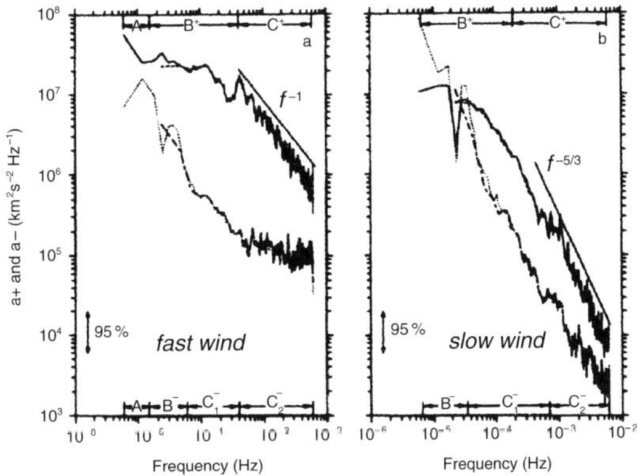

Figure 4. Power spectra of the Elsässer variables $\left(\delta z^\pm \equiv \delta v \pm \delta b\right)$ for fast wind (left panel) and slow wind (right panel) from data obtained by Helios at 0.3 AU. Adapted from *Tu et al.* [1989].

In a series of papers, *Roberts et al.* [1987a,b] examined the evolution of the Alfvénicity of solar wind fluctuations using data from Helios and Voyager. That analysis elaborated on the earlier results [*Matthaeus and Goldstein*, 1982a] that showed that in the outer heliosphere the Alfvénicity tended to decrease—higher wave numbers retained their Alfvénic nature to larger

306

urements of three-dimensional properties possible, but their maximum separation (~10,000 km) will still be less than the correlation length of the turbulence. It is possible, however, even with one spacecraft (and some additional assumptions) to explore at least two-dimensional symmetries. *Sari and Valley* [1976] analyzed the relatively few intervals during which the background magnetic field **B** was either parallel or transverse to solar wind velocity. Because of a paucity of examples they could not reach any definitive conclusions, but suggested that the magnetic fluctuations could not be isotropic, although for periods when **B** was perpendicular to the solar wind velocity, the isotropic model was a reasonable approximation and that *slab* waves (*i.e.*, parallel propagating Alfvén waves) offered a poor approximation that was only approximately valid during the radial **B** flow intervals. They suggested that a combination of magnetosonic and Alfvén waves with approximately one quarter as much power in magnetosonic waves as in Alfvén waves fit both data intervals.

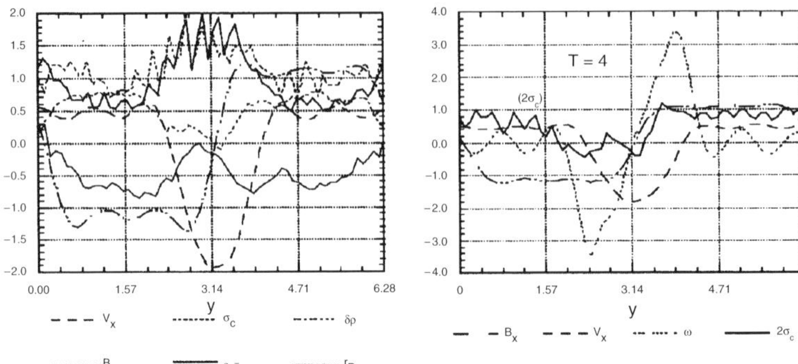

Figure 8. Left: Results at $T = 2$ eddy-turn-over times from a solution of the compressible equations. The panel includes density fluctuations, ρ, δz^-, and the correlation between density and magnetic field magnitude, r_B ($\delta\rho$ and δz^- are normalized to twice their maximum values of 0.042 and 0.075, respectively). Right: The solution at $T = 4$ eddy-turnover times from an incompressible simulation (ω is the vorticity). From *Roberts et al.* [1991].

The first comprehensive effort to deduce at least the two-dimensional structure of the magnetic fluctuations was carried out using approximately 16 months of nearly continuous data from ISEE 3 by *Matthaeus et al.* [1990]. The results suggested that the solar wind fluctuations were made of two components. The first were slab Alfvén waves; the second component comprised fluctuations with wave vectors nearly transverse to the mean magnetic field. The analysis was done on stationary data intervals [*Matthaeus and Goldstein*, 1982b; *Matthaeus et al.*, 1986; *Monin and Yaglom*, 1975; *Panchev*, 1971] and assumed that all the data intervals (463 15-min averages) were members of the same statistical ensemble. The two-dimensional correlation function from that analysis is shown in Figure 10.

Bieber et al. [1996] suggested that typically 80% of the magnetic fluctuation energy resided in the quasi-two-dimensional component and only 20% was in the "slab" component. Subsequently, *Smith* [2002], using Ulysses data, showed that between 1991 and 1995, a more typical value was 50% (see Figure 11). There was no obvious way to tell, however, whether the "quasi-2D component" reflected a nearly-incompressible population of fluctuations or merely one that contained nearly perpendicular k-vectors.

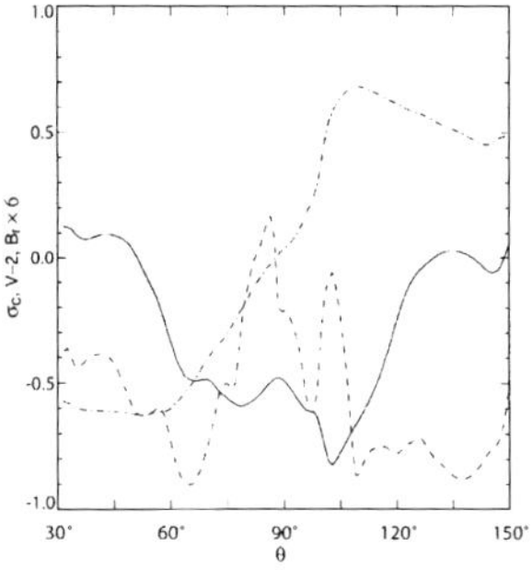

Figure 9. Effects of velocity shear from a three-dimensional spherical code. The current sheet is indicated by the change in sign of B_r (dot-dashed line); the velocity shear by V_r (solid line), and σ_c (dashed line). The values have been averaged over the radial distance of $r = 3$–4. All quantities are in dimensionless units. From *Goldstein et al.* [1999].

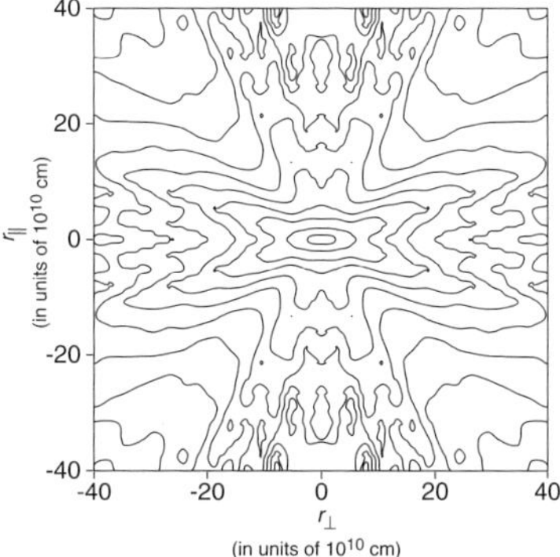

Figure 10. Contour plot of the two-dimensional correlation function of solar wind fluctuations as a function of distance parallel and perpendicular to the mean magnetic field. The four quadrant plot was produced by reflecting the data across the axes from the first quadrant. From *Matthaeus et al* [1990].

Ruderman et al. [1999] used a three-dimensional spherical simulation to demonstrate that small velocity shears would produce rapid "phase mixing" of planar Alfvén waves. Subsequently, *Goldstein et al.* [2002] generalized the results of *Ghosh et al.* [1998] to three dimensions, showing that small velocity shears superposed on an initial wave packet of planar Alfvén waves could evolve with distance to produce two-dimensional correlation functions reminiscent of the ISEE-3 analysis. Furthermore, *Roberts et al.* [1999] showed that nearly two-dimensional turbulence that was initially highly Alfvénic would evolve rapidly to a non-Alfvénic state dominated by magnetic fluctuations (also see, *Tu and Marsch* [1993]). Observationally, one would expect, then, that for ambient magnetic field directions close to 90°, $\sigma_c \approx 0$. In contrast, however, analysis of Voyager data near 8 AU indicated that, at least on one occasion, even with the field in the tangential direction, $\sigma_c \approx 0.8$ [*Goldstein et al.*, 1995b; *Goldstein et al.*, 1997]. The plot of σ_c is shown in Figure 12.

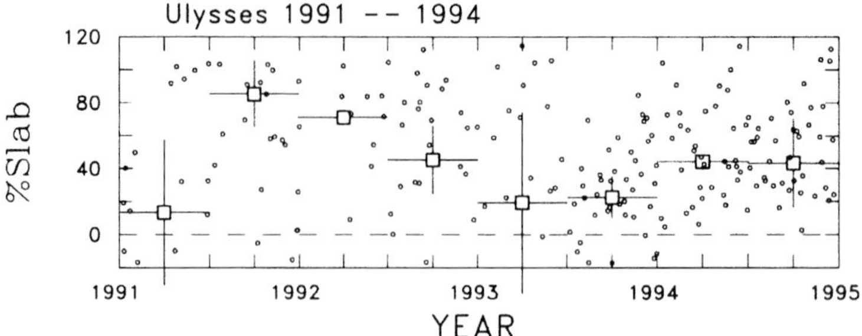

Figure 12. The %Slab. Square symbols represent averages over the time intervals represented by horizontal lines. Vertical lines give the variance of the underlying subset. Adapted from *Smith* [2002].

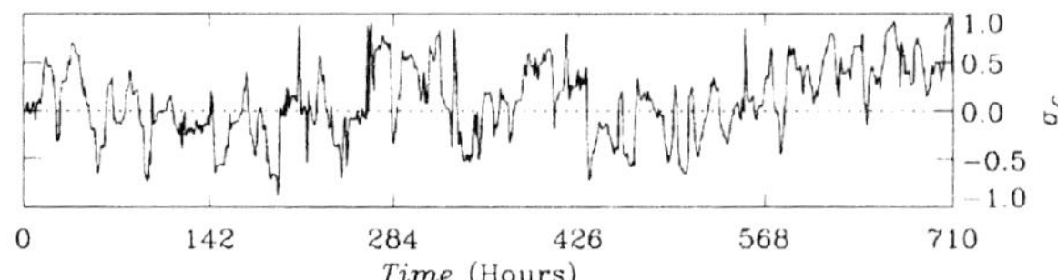

Figure 13. An example of observations taken by Voyager near 8 AU when **B** was nearly tangential to the solar wind flow velocity but the flow was highly Alfvénic as measured by σ_c (computed from hour-averaged magnetic field and plasma data). Adapted from *Goldstein et al.* [1997].

Consequently, it appears that the 80% of the fluctuation energy that is quasi-two-dimensional, does not reflect a nearly-incompressible component of the turbulence that arose in the corona, but rather indicates either local generation or phase mixing. Ascertaining the nature of these fluctuations is important because the distribution of power between parallel and perpendicular wave numbers determines the spatial diffusion of solar and galactic energetic particles since only power in $k_\parallel$ resonantly scatters energetic charged particles.

4. Simulations of Solar Wind Turbulence

To simulate the evolution of turbulence in a spherically expanding solar wind, we have solved the inviscid MHD equations using a fourth-order flux corrected transport (FCT) algorithm [*Boris and Book*, 1973; *DeVore*, 1991; *Zalesak*, 1979]. FCT is designed to capture shocks and preserve discontinuities. For a discussion of the application of this algorithm to the solar wind, see *Goldstein et al.* [1999]. The magnetic field is determined from the electric field using Faraday's law to ensure that $\nabla \cdot \mathbf{B} = 0$ is preserved to within numerical round-off errors. The code advances the density ρ, the momentum $\rho\mathbf{v}$, the total energy (internal (ρe) + kinetic ($\rho v^2/2$) + magnetic ($B^2/8\pi$)), and the magnetic field $\mathbf{B}$. The equation of state is that of an ideal gas ($p = \rho RT$), where R is the gas constant and T is temperature.

The MHD equations as solved in the code are

$$\frac{\partial \rho}{\partial t} + \nabla \cdot \rho \mathbf{u} = 0 \tag{4.1}$$

$$\frac{\partial \rho \mathbf{u}}{\partial t} + \nabla \cdot \left[\left(p + \frac{B^2}{8\pi} \right) \mathbf{I} + \rho \mathbf{u}\mathbf{u} - \frac{1}{4\pi} \mathbf{B}\mathbf{B} \right] = 0 \tag{4.2}$$

$$\frac{\partial}{\partial t} \left(\rho e + \tfrac{1}{2}\rho u^2 + \frac{B^2}{8\pi} \right) + \nabla \cdot \left[\left(\rho e + \tfrac{1}{2}\rho u^2 + p \right) \mathbf{u} - \frac{1}{4\pi} \left(\mathbf{u} \times \mathbf{B} \right) \times \mathbf{B} \right] = 0 \tag{4.3}$$

$$\frac{\partial \mathbf{B}}{\partial t} + \nabla \cdot \left[\mathbf{u}\mathbf{B} - \mathbf{B}\mathbf{u} \right] = 0 \tag{4.4}$$

with the relations $\nabla \cdot \mathbf{B} = 0$ and $p = (\gamma - 1)\rho e$.

$\mathbf{I}$ is the unit tensor and γ is the ratio of specific heats. Supersonic/super-Alfvénic time-dependent inflow boundary conditions are imposed at a radial distance above the critical point. Although gravity is included in the equations, it is not important for the cases discussed here, where the inner boundary is taken to be between $r = 0.2$–0.5 AU.

The code includes a tilted rotating current sheet, incoming Alfvénic or two-dimensional modes with finite transverse correlation lengths, microstreams, and pressure-balanced structures. The code can run in either Cartesian or spherical geometry. To illustrate some of the capabilities, Figure 13 shows an example of a simulated stream interaction region at about 20 AU. On the left, an r-θ cut of three-dimensional plane; the top right panel shows a radial cut through the ecliptic of the density (in cgs units). The bottom right panel shows the density from Voyager 1. The interaction between the fast and slow wind compresses the plasma and results in the alternating compression and rarefaction patterns visible in both simulation and observations.

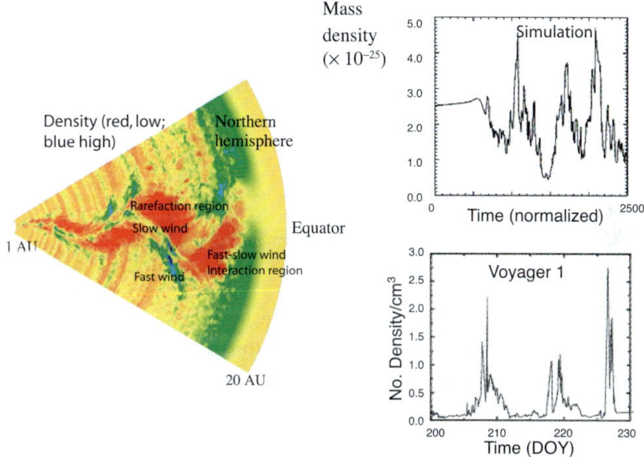

Figure 13. Simulation of a stream interaction region. The r-θ plane is shown with fast wind in the northern and southern hemispheres. Slow wind is found in the

310

warped plasma sheet region. As a result of the warping, fast wind overtakes slow wind and compresses it. Strong density enhancements (and rarefactions) result. The density from the simulation and Voyager is shown in the right-hand panels.

4.1 Simulations of a Turbulent Cascade.

The FCT algorithm is sufficiently dissipationless that one can study the development of a cascade of magnetic energy from the input energy-containing scales toward the dissipation range. Eventually, the residual numerical dissipation does, of course, steepen the spectrum and damp fluctuations. In Figure 14, a series of power spectra are plotted that were computed from a time series of magnetic field values taken at four sites in the simulation domain. The data were accumulated in slow flow near $\theta = 90°$ at four radial points ranging from 1.02–3.98 AU. This region near the current sheet is where the most rapid evolution takes place. The turbulent mixing generates about a decade of inertial range. The straight line on the plots has a slope of –5/3.

4.2 On the Origin of the Quasi-Two-Dimensional Component of Solar Wind Fluctuations

As discussed above, there are several possibilities for the origin of the power in $k_\perp$ component of solar wind magnetic fluctuations. The possibilities include:
- a two-dimensional component generated in the solar corona that is subsequently convected into the heliosphere
- compressive fast mode waves
- pressure balanced structures
- velocity shear-induced phase mixing

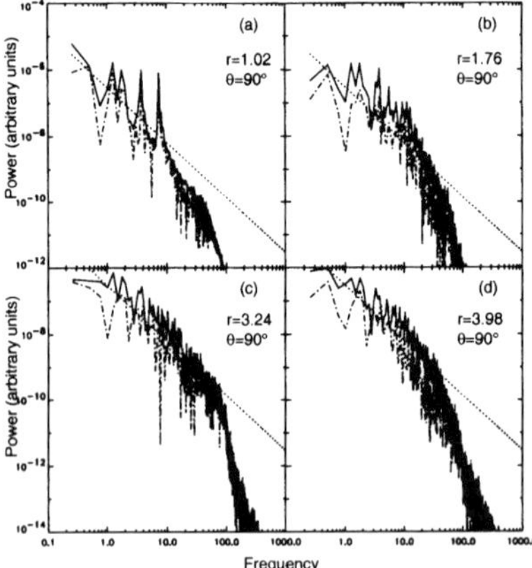

Figure 14. Power spectra deduced from the time series of the magnetic field components and |B| (dashed line) along a radius in slow speed flow at $\theta = 90°$ and four radii ranging from $r = 1.02$–3.98 AU. Also shown on each plot is a line with slope of –5/3. From *Goldstein et al.* [1999].

The solar wind does not appear to be sufficiently compressible for the fluctuations to be fast mode waves and fast modes are likely to damp relatively rapidly [*Barnes*, 1966]. Pressure-balanced structures are not likely to be highly Alfvénic, leaving only the first and last possibilities as the most likely. But, independent of any coronal source, the velocity shears that permeate all fast wind will mix the phase fronts of any initially planar waves so that a substantial amount of magnetic power in $k_\perp$ will result.

In a series of simulations we examined the effect of microstreams on the flow. A variety of initial conditions were run, some with no waves and others that contained an initial packet of plane-polarized Alfvén waves. The simulation domain was a three-dimensional wedge, $\pm 30°$ in both θ and ϕ. The inner boundary was placed at $r = 0.5$ AU and the outflow boundary was at 2.5 AU. The magnitude of the velocity shears was designed to approximate the ± 40 km/s observed in high-latitude fast wind by Ulysses. In the absence of waves, the velocity shears introduced some meandering of the magnetic field as illustrated in Figure 15. Although the field lines emanating from adjacent flow tubes mix and meander, they tend to remain together, maintaining their identity while staying aligned with the Parker flow pattern.

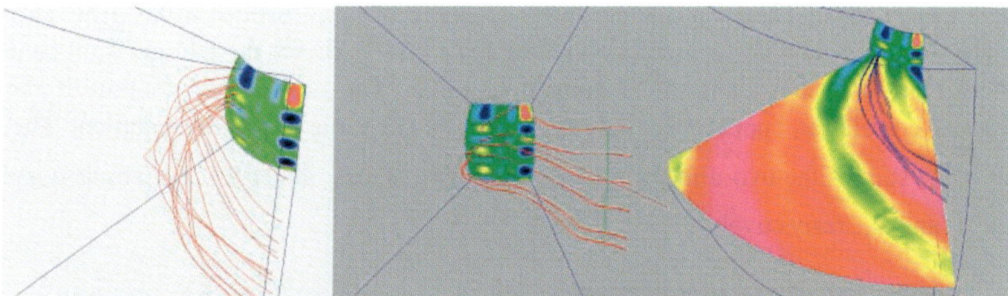

Figure 15. The inflow surface of this three-dimensional simulation is colored by the magnitude of the velocity. Fluctuations are shown about the mean, which is colored green. Lines of magnetic force (red) are traced out from adjoining flow tubes. The right-most panel adds a view of the flow in the r-φ plane showing that the meandering field lines tend to follow the average Parker spiral pattern.

In another set of runs, the effect of microstreams on planar Alfvén waves was investigated. Figure 16 shows a meridional slice from the simulation illustrating how the small velocity shears lead to a sharp reduction in the cross helicity. The left-hand panel shows σ_c in the absence of microstreams. In the right-hand panel, microstreams have been added.

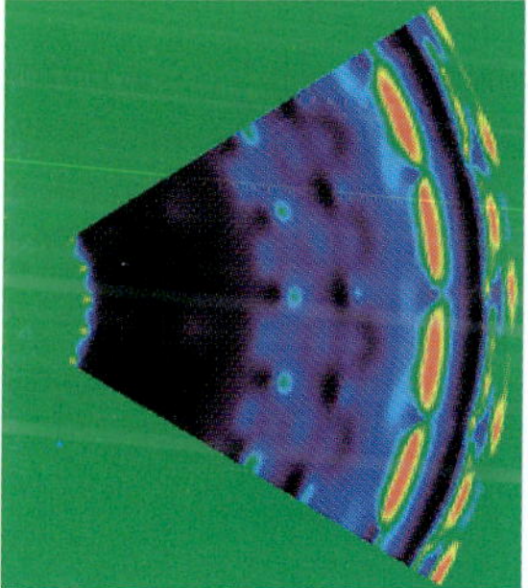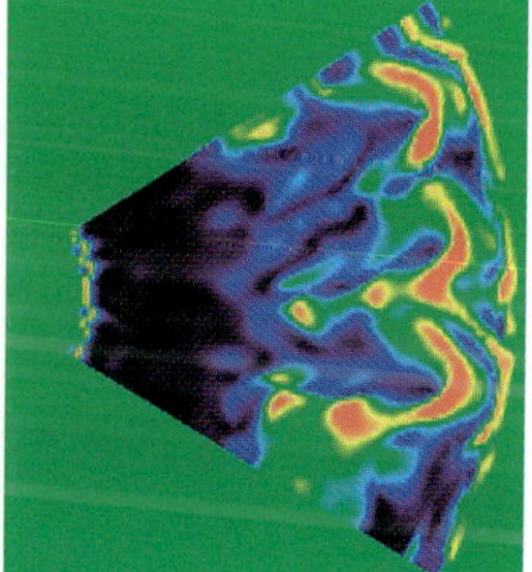

Figure 16. The effect of microstreams on the evolution of initially planar Alfvén waves. On the left, no microstreams are present, while on the right, small flow

312

tubes have been added as in Figure 15. Purple indicates pure outward propagating waves.

Figure 17 (left) from another run that also included planar Alfvén waves, shows the magnitude of the vorticity ω in the r–θ plane from a 304×154×154 run. The right-hand panel shows two cuts of the three-dimensional results. The in-flow boundary is the initial distribution of the velocity fluctuations, the shading in the r–θ plane is $\delta B_\perp = \sqrt{\delta B_\theta^2 + \delta B_\phi^2}$ in the r–ϕ plane, magnetic field lines are shown along with a color shading of $|v|$.

In Figure 18 we illustrate the evolution of the spectrum of the magnetic fluctuations from plane-polarized waves to something more reminiscent of the "Maltese cross" pattern [*Matthaeus et al.*, 1990]. Both panels contain three contours: In the lower left-hand corner is an image of $\delta B_\perp$; the alignment of these contours approximately follows that of the spiral field structure in that $\mathbf{B}_o$ is nearly parallel to r on the left and is closer to the ϕ-direction on the right. The radial ranges are $r = 0.1$–0.43 AU (in simulation units) and $r = 4$–5 AU for the left and right panels, respectively. The ordinate spans the ϕ domain of $\pm30°$. The next larger contour is the two-dimensional correlation function of δB_θ and δB_ϕ, in the same r-ϕ representation. The large contour is the sum of the two-dimensional power spectra of δB_θ and δB_ϕ. The abscissa and ordinate are k_r and k_ϕ, respectively.

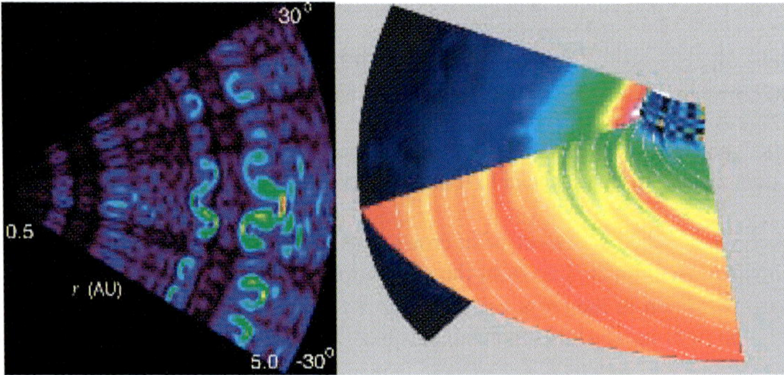

Figure 17. Left: Magnitude of ω in the r–θ plane from $r = 0.5$–5.0 AU from a run of size 304×154×154. Right: A three-dimensional view of the same simulation. The color contours at the in-flow boundary show the initial distribution of the velocity fluctuations; the contour in the r–θ plane is the fluctuation in $\delta B_\perp$ and, in the r-ϕ plane, magnetic field lines are shown in white along with color contours of $|v|$. From *Goldstein et al.* [2002].

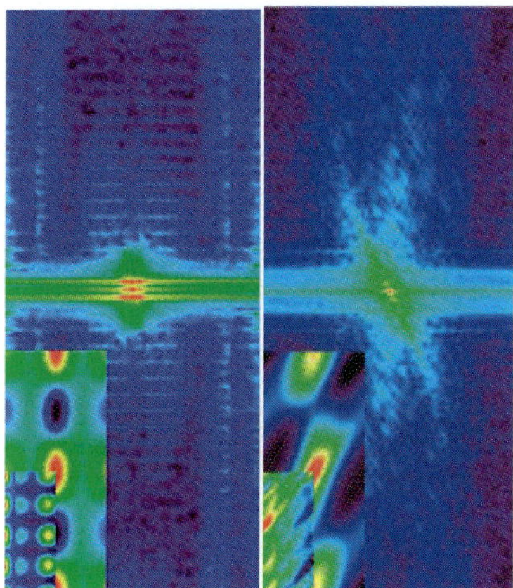

Figure 18. $\delta B_\perp$, two-dimensional correlation functions, and two-dimensional power spectra. Left: $r = 0.1$–0.43 AU with $\phi \pm 30°$. Right: $r = 4$–5 AU and the same range in ϕ. From *Goldstein et al.* [2002]

The two-dimensional correlation function at small radial distance is aligned nearly perpendicular to the wave vector of the fluctuations, as expected of radial wave vectors. Close to $r = 5$ AU the contours are nearly parallel to fluctuating magnetic field structures. Furthermore, the shape of the correlation function is reminiscent of that from ISEE 3 [*Matthaeus et al.*, 1990]. Although the two-dimensional power spectrum taken between ≈ 4–5 AU still contains power in wave vectors parallel to the spiral field direction, the power in the perpendicular direction dominates. The small velocity shears have deformed the distribution of initially parallel-propagating plane-polarized Alfvén waves so that by several AU the correlation function contains substantial power nearly perpendicular to the background magnetic field. The ISEE–3 interval analyzed by *Matthaeus et al.* [1990] was a mixture of fast and slow wind from near the ecliptic plane, while the simulation parameters are closer to those encountered by Ulysses at high latitudes. *Smith* [2002] has presented evidence that the two-component model of solar wind fluctuations also fits Ulysses data. Thus, velocity shear alone appears capable of generating nearly perpendicular wave vectors and correlation functions reminiscent of the observations.

4.3 Interplanetary Loops and Magnetic Holes

A curiosity that emerged from the three-dimensional simulations of the solar wind was the appearance of large-scale loops associated with the heliospheric current sheet. These loops appear in the simulations when the current sheet is both tilted and rotating. An example is shown in Figure 19. The formation of loops is not a result of reconnection, but rather results from an inconsistency of the imposed ideal field reversal conditions with real MHD flows [*Roberts et al.*, 2002]. In particular, the self-consistent fields cannot maintain the initial magnetic-kinetic pressure balance, leading to flows transverse to the current sheet. The natural configuration appears to be one with a small normal component and loops that come from the decoupled phasing of the zeros of the radial and tangential magnetic fields as a function of radial distance from the Sun.

314

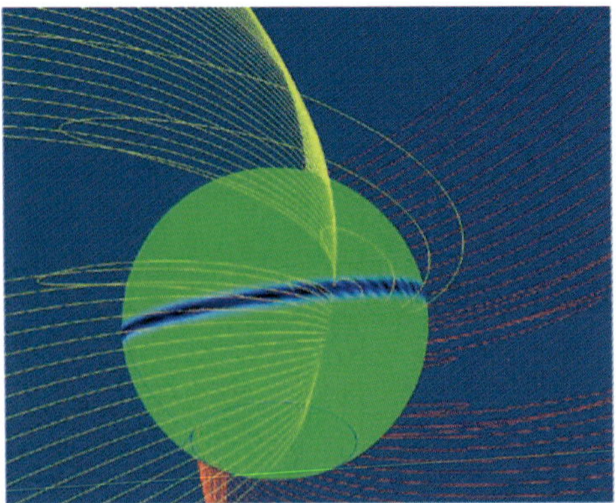

Figure 19. A view of a simple run that exhibits loops across the current sheet. The radial extent is from 0.5 to 2.5 AU. The inner boundary is colored by the inflow field magnitude. The current sheet, where the magnetic field is near zero, is the blue band at the equator of the simulation sphere. Field lines are yellow and red. From *Roberts et al.* [2002]

Figure 20 shows on the left, a typical interplanetary current sheet crossing observed by Helios 2 in which the field rotates without ever dropping to zero. The radial and tangential components of **B** go through zero at slightly out of phase with each other. This crossing is similar to many others [*Smith*, 2001]. On the right is a time series from the MHD simulation shown in Figure 19. There are neither waves nor velocity shear in that run. Note the general similarities between the Helios observations and the simulations.

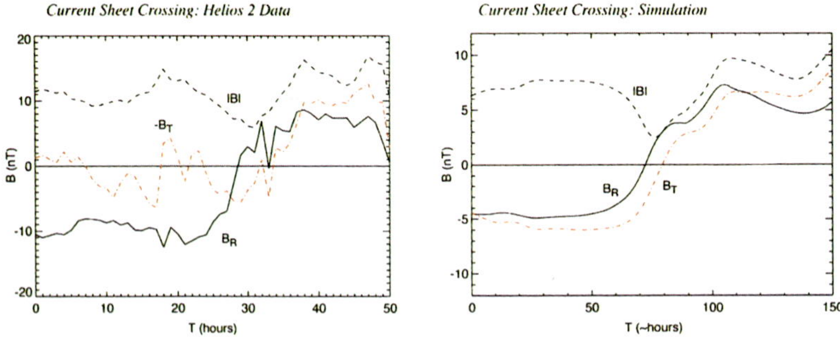

Figure 20. Left: Helios 2 hour-averaged observations of a sector crossing showing the common feature of the displacement in time of the zeros of the radial and tangential magnetic fields. Right: Time series from a point in the MHD simulation of a tilted, rotating current sheet. From *Roberts et al.* [2002].

Another way to look at these loops and magnetic holes is to ask how the field depressions and rotations would appear to one or more spacecraft moving past them. In left panel of Figure 21 we show a trace of magnetic field vectors (normalized in length, but colored by field magnitude) as a pair of virtual spacecraft are passed by a loops and magnetic holes. The software used to visualize this is the VisSBARD package developed by Aaron Roberts and available from http://nssdcftp.gsfc.nasa.gov/selected_software/visbard/. For comparison, the right panel of Figure 21 shows data from ACE and WIND as they encounter an interplanetary magnetic hole.

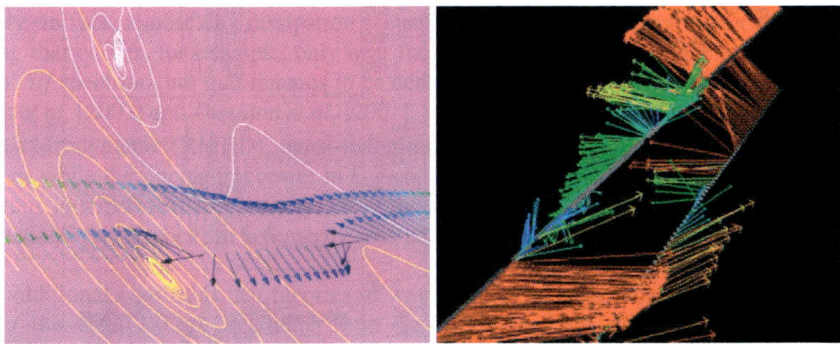

Figure 21. Left: Normalized magnetic field vectors colored by field magnitude from to virtual spacecraft trajectories in the vicinity of magnetic loops and holes from a three-dimensional simulation. Right: ACE and WIND spatial views of interplanetary magnetic holes visualized using *ViSBARD*.

5. Modeling the Global Structure of the Solar Corona and Solar Wind

In addition to studying the evolution of turbulence in the solar wind, we have also developed simulations of the large-scale structure of the expanding solar corona. Coronal structure is determined largely by the pattern of magnetic fields on the solar photosphere. As a test of these simulations, we have matched the solutions to Ulysses observations of an entire orbit around the polar regions of the heliosphere. The photospheric field serves as a boundary condition for the MHD equations [*Groth et al.*, 2000; *Mikic and Linker*, 1996; *Mikic et al.*, 1999; *Riley et al.*, 2001; *Sittler and Guhathakurta*, 1999; *Stewart and Bravo*, 1997; *Usmanov*, 1993]. To produce fast wind together with reasonable agreement with coronal densities and temperatures requires Alfvén waves are used as an additional source of momentum [*Alazraki and Couturie*, 1971; *Belcher and Davis*, 1971]. The Alfvén wave effects are incorporated into the governing equations in the WKB limit and it is assumed that the waves are damped by a mechanism that may be characterized by a constant dissipation length, L. The three-dimensional non-axisymmetric code [*Usmanov and Goldstein*, 2003] is a generalization of *Usmanov et al.* [2000].

Figure 22 compares the simulations with Ulysses data. The daily Ulysses data were scaled to 1 AU assuming that b_r and n fall off with radial distance as r^{-2}, the azimuthal magnetic field b_ϕ as r^{-1}, and the temperature T as $r^{-2(\gamma-1)}$ with $\gamma = 1.46$. The warping or the heliospheric current sheet and the latitudinal extension of the slow solar wind are modeled by inclining the solar dipole with respect to the solar rotation axis. Although the computed temperature is somewhat higher than that observed, this could be adjusted by varying L; larger L would provide less wave dissipation and ultimately lower temperatures. The electron temperature in the distant solar wind is larger than the proton temperature, so that the single-fluid temperature of the simulation can be higher than the observed proton temperature.

Birkeland, K., On the cause of magnetic storms and the origin of terrestrial magnetism, first section, in *The Norwegian Aurora Polaris Expedition 1902-1903*, edited by Cristiania, H. Aschehoug and Co., 1908.

Birkeland, K., On the cause of magnetic storms and the origin of terrestrial magnetism, second section, in *The Norwegian Aurora Polaris Expedition 1902-1903*, edited by Cristiania, H. Aschehoug and Co., 1913.

Blackman, R., and J. Tukey, *Measurement of Power Spectra*, Dover, New York, 1958.

Bonetti, A., H.S. Bridge, A.J. Lazarus, E.F. Lyon, B. Rossi, and F. Scherb, Explorer 10 plasma measurements, *J. Geophys. Res.*, *68*, 4017-4063, 1963.

Boris, J.P., and D.L. Book, Flux corrected transport, I, SHASTA A fluid transport algorithm that works, *J. Comput. Phys.*, *11*, 38, 1973.

Bridge, H., C. Dilworth, A.J. Lazarus, E.F. Lyon, B. Rossi, and F. Scherb, Explorer 10 plasma measurements, *J. Phys. Soc. Japan*, *17A*, 11, 1961.

Bruno, R., B. Bavassano, and U. Villante, Evidence for long period Alfvén waves in the inner solar system, *J. Geophys. Res.*, *90*, 4373, 1985.

Coleman, P.J., Hydromagnetic waves in the interplanetary medium, *Phys. Rev. Lett.*, *17*, 207, 1966.

Coleman, P.J., Turbulence, viscosity, and dissipation in the solar wind plasma, *Astrophysical Journal*, *153*, 371, 1968.

DeVore, C.R., Flux-corrected transport techniques for multidimensional compressible magnetohydrodynamics, *J. Comput. Phys.*, *92*, 142, 1991.

Dmitruk, P., W.H. Matthaeus, L.J. Milano, S. Oughton, G.P. Zank, and D.J. Mullan, Coronal heating distribution due to low-frequency, wave-driven turbulence, *Astrophysical Journal*, *575* (1), 571-577, 2002.

Ghosh, S., W.H. Matthaeus, D.A. Roberts, and M.L. Goldstein, Waves, structures, and the appearance of two-component turbulence in the solar wind, *J. Geophys. Res.*, *103* (A10), 23,705-23,716, 1998.

Goldstein, B.E., E.J. Smith, A. Balogh, T.S. Horbury, M.L. Goldstein, and D.A. Roberts, Properties of magnetohydrodynamic turbulence in the solar wind as observed by Ulysses at high heliographic latitudes, *Geophys. Res. Lett.*, *22* (23), 3393-3396, 1995a.

Goldstein, M.L., and D.A. Roberts, Magnetohydrodynamic turbulence in the solar wind, *Physics of Plasmas*, *6*, 4154, 1999.

Goldstein, M.L., D.A. Roberts, and A. Deane, The effect of microstreams on Alfvénic fluctuations in the solar wind, in *Solar Wind 10*, edited by M. Velli, R. Bruno, and F. Malara, pp. 405, AIP, Pisa, Italy, 2002.

Goldstein, M.L., D.A. Roberts, A.E. Deane, S. Ghosh, and H.K. Wong, Numerical simulation of Alfvénic turbulence in the solar wind, *J. Geophys. Res.*, *104* (A7), 14,437-14,452, 1999.

Goldstein, M.L., D.A. Roberts, and W.H. Matthaeus, Magnetohydrodynamic turbulence in the solar wind, *Ann. Rev. Astron. and Astrophys.*, *33*, 283, 1995b.

Goldstein, M.L., D.A. Roberts, and W.H. Matthaeus, Magnetohydrodynamic turbulence in cosmic winds, in *Cosmic Winds and the Heliosphere*, edited by J.R. Jokipii, C.P. Sonett, and M.S. Giampapa, pp. 521, Univ. of Arizona Press, Tucson, 1997.

Grant, H.L., R.W. Stewart, and A. Moilliet, Turbulence spectra from a tidal channel, *J. Fluid Mech.*, *12*, 241, 1962.

Gringauz, K.I., Some results of experiments in interplanetary space by means of charged particle traps on Soviet space probes, *Space Res.*, *2*, 539, 1961.

Gringauz, K.I., V.V. Bezrukikh, and L.S. Musatov, Solar wind observations with the Venus 3 probe, *Cosmic Res.*, *2*, 216, 1967.

Gringauz, K.I., V.V. Bezrukikh, V.D. Ozerov, and R.E. Ribchinsky, Study of the interplanetary high energy electrons and solar corpuscular radiation by means of three electrode traps for charged particles on the second Soviet cosmic rocket, *Soviet Phys. Doklady*, *53*, 61, 1960.

Groth, C.P.T., D.L.D. Zeeuw, and T.I. Gombosi, Global three-dimensional MHD simulation of a space weather event: CME formation, interplanetary propagation, and interaction with the magnetosphere, *J. Geophys. Res.*, *105*, 25,053, 2000.

Heinemann, M., and S. Olbert, Non-WKB Alfvén waves in the solar wind, *J. Geophys. Res.*, *85*, 1311, 1980.

Kohl, J.L., G. Noci, E. Antoucci, and e. al., First results from the SOHO ultraviolet coronagraph spectrometer, in *The First Results from SOHO*, edited by B. Fleck, and Z. Svestka, pp. 613-644, Kluwer Academic Publishers, Dordrecht, 1997.

Kolmogorov, A.N., The local structure of turbulence in incompressible viscous fluid for very large Reynolds numbers, *Dokl. Akad. Nauk SSSR*, *30*, 299, 1941.

Kudoh, T., and K. Shibata, Alfven wave model of spicules and coronal heating, *Astrophysical Journal*, *514* (1), 493-505, 1999.

Matthaeus, W.H., and M.L. Goldstein, Measurement of the rugged invariants of magnetohydrodynamic turbulence, *J. Geophys. Res.*, *87*, 6011, 1982a.

Matthaeus, W.H., and M.L. Goldstein, Stationarity of magnetohydrodynamic fluctuations in the solar wind, *J. Geophys. Res.*, *87*, 10347, 1982b.

Matthaeus, W.H., and M.L. Goldstein, Low-frequency 1/f noise in the interplanetary magnetic field, *Phys. Rev. Lett.*, *57*, 495, 1986.

Matthaeus, W.H., M.L. Goldstein, and J.H. King, An interplanetary field ensemble at 1 AU, *J. Geophys. Res.*, *91*, 59, 1986.

Matthaeus, W.H., M.L. Goldstein, and D.A. Roberts, Evidence for the presence of quasi-two-dimensional, nearly incompressible fluctuations in the solar wind, *J. Geophys. Res.*, *95*, 20,673, 1990.

Mikic, Z., and J.A. Linker, The large-scale structure of the solar corona and inner heliosphere, in *Solar Wind 8*, edited by D.W.e. al., pp. 104-107, AIP, Dana Point, CA, 1996.

Mikic, Z., J.A. Linker, D.D. Schnack, R. Lionello, and A. Tarditi, Magnetohydrodynamic modeling of the global solar corona, *Physics of Plasmas*, *6* (5), 2217-2224, 1999.

Monin, A.S., and A.M. Yaglom, *Statistical Fluid Mechanics: Mechanics of Turbulence*, MIT Press, Cambridge, Mass., 1975.

Odstrcil, D., J.A. Linker, R. Lionello, Z. Mikic, P. Riley, V.J. Pizzo, and J.G. Luhmann, Merging of coronal and heliospheric numerical two-dimensional MHD models, *J. Geophys. Res.*, *107* (A12), 2002.

Ofman, L., Three fluid model of the heating and acceleration of the fast solar wind, *J. Geophys. Res.*, *in press*, 2004.

Oughton, S., W.H. Matthaeus, P. Dmitruk, L.J. Milano, G.P. Zank, and D.J. Mullan, A reduced magnetohydrodynamic model of coronal heating in open magnetic regions driven by reflected low-frequency Alfven waves, *Astrophysical Journal*, *551* (1), 565-575, 2001.

Panchev, S., *Random Functions and Turbulence*, Pergammon, New York, 1971.

Parker, E.N., Dynamics of the interplanctary gas and magnetic fields, *Astrophysical Journal*, *128*, 664, 1958.

Parker, E.N., Mass ejection and a brief history of the solar wind concept, in *Cosmic Winds and the Heliosphere*, edited by J.R. Jokipii, C.P. Sonett, and M.S. Giampapa, pp. 3, Univ. of Arizona Press, Tucson, 1997.

Riley, P., J.A. Linker, and Z. Mikic, An empirically-driven global MHD model of the solar corona and inner heliosphere, *J. Geophys. Res.*, *106* (A8), 15889-15901, 2001.

Roberts, D.A., Interplanetary observational constraints on Alfvén wave acceleration of the solar wind, *J. Geophys. Res.*, *94*, 6899, 1989.

Roberts, D.A., S. Ghosh, M.L. Goldstein, and W.H. Matthaeus, Magnetohydrodynamic simulation of the radial evolution and stream structure of solar-wind turbulence, *Phys. Rev. Lett.*, *67*, 3741, 1991.

Roberts, D.A., M.L. Goldstein, and A. Deane, Three-Dimensional MHD Simulation of Solar Wind Structure, in *Solar Wind 10*, edited by M. Velli, R. Bruno, and F. Malara, pp. 133, AIP, Pisa, Italy, 2002.

Roberts, D.A., M.L. Goldstein, A.E. Deane, and S. Ghosh, Quasi-two-dimensional MHD turbulence in three-dimensional flows, *Phys. Rev. Lett.*, *82* (3), 548-551, 1999.

Roberts, D.A., M.L. Goldstein, L.W. Klein, and W.H. Matthaeus, Origin and evolution of fluctuations in the solar wind: Helios observations and Helios-Voyager comparisons, *J. Geophys. Res.*, *92*, 12,023, 1987a.

Roberts, D.A., M.L. Goldstein, W.H. Matthaeus, and S. Ghosh, Velocity shear generation of solar wind turbulence, *J. Geophys. Res.*, *97*, 17115, 1992.

Roberts, D.A., L.W. Klein, M.L. Goldstein, and W.H. Matthaeus, The nature and evolution of magnetohydrodynamic fluctuations in the solar wind: Voyager observations, *J. Geophys. Res.*, *92*, 11021, 1987b.

Ruderman, M.S., M.L. Goldstein, D.A. Roberts, A. Deane, and L. Ofman, Alfvén wave phase mixing driven by velocity shear in two-dimensional open magnetic configurations, *J. Geophys. Res.*, *104* (A8), 17,057-17,068, 1999.

Sari, J.W., and G.C. Valley, Interplanetary magnetic field power spectra: mean field radial or perpendicular to radial, *J. Geophys. Res.*, *81*, 5489, 1976.

Scherb, F., Velocity distributions of the interplanetary plasma detected by Explorer 10, *Space Res.*, *4*, 797, 1964.

Sittler, E.C., and M. Guhathakurta, Semiempirical two-dimensional magnetohydrodynamic model of the solar corona and interplanetary medium, *Astrophysical Journal*, *523* (2), 812-826, 1999.

Smith, C.W., The geometry of turbulent magnetic fluctuations at high heliographic latitudes, in *Solar Wind 10*, edited by M. Velli, R. Bruno, and F. Malara, pp. 413, AIP, Pisa, Italy, 2002.

Smith, E.J., The heliospheric current sheet, *J. Geophys. Res.*, *106* (A8), 15819-15831, 2001.

Spitzer, L., Jr., *Physics of fully ionized gases*, John Wiley Interscience, New York, 1962.

Stewart, G.A., and S. Bravo, Latitudinal solar wind velocity variations from polar coronal holes: A self-consistent MHD model, *Journal of Geophysical Research-Space Physics*, *102* (A6), 11263-11272, 1997.

Tu, C.-Y., and E. Marsch, A model of solar wind fluctuations with two components: Alfvén waves and convective structures, *J. Geophys. Res.*, *98*, 1257, 1993.

Tu, C.-Y., E. Marsch, and K.M. Thieme, Basic properties of solar wind MHD turbulence near 0.3 AU analyzed by means of Elsässer variables, *J. Geophys. Res.*, *94*, 739, 1989.

Unti, T.W., and M. Neugebauer, Alfvén waves in the solar wind, *Physics of Fluids*, *11*, 563, 1968.

Usmanov, A.V., A global numerical 3-D MHD model of the solar wind, *Solar Physics*, *146* (2), 377-396, 1993.

Usmanov, A.V., and M.L. Goldstein, Three-dimensional MHD modeling of the solar corona and solar wind, in *Solar Wind 10*, edited by M. Velli, R. Bruno, and F. Malara, AIP, Pisa, Italy, 2002.

Usmanov, A.V., and M.L. Goldstein, A tilted-dipole MHD model of the solar corona and solar wind, *J. Geophys. Res.*, *108* (A9), doi:10.1029/2002JA009777, 2003.

Usmanov, A.V., M.L. Goldstein, B.P. Besser, and J.M. Fritzer, A global MHD solar wind model with WKB Alfvén waves: Comparison with Ulysses data, *J. Geophys. Res.*, *105* (A6), 12675-12695, 2000.

Zalesak, S.T., Fully multidimensional flux-corrected transport algorithms for fluids, *J. Comp. Phys.*, *31*, 335, 1979.

Multiscale Coupling of Sun-Earth Processes
A.T.Y. Lui, Y. Kamide and G. Consolini, editors.

INTERMITTENT TURBULENCE
IN 2D MHD SIMULATION

C. C. Wu

Institute of Geophysics and Planetary Physics, University of California
Los Angeles, CA 90095, USA

Tom Chang

Kavli Institute for Astrophysics and Space Research
Massachusetts Institute of Technology, Cambridge, MA 02139, USA.

Abstract. In situ observations indicate that the dynamical processes in the space plasma environment generally entail anisotropic and localized intermittent fluctuations. Results from two-dimensional MHD simulations are presented to demonstrate the intermittency. Non-Gaussian probability distribution functions of the intermittent fluctuations are obtained and discussed. Power spectra and local intermittency measures using the wavelet analyses are presented to display the spottiness of the small-scale turbulent fluctuations.

Keywords. Intermittency, MHD simulation, Space plasmas, Turbulence

1. Introduction

Recent observations indicate that the dynamical processes in the space plasma environment generally entail anisotropic and localized intermittent fluctuations. It was suggested by *Chang* [1998a,b,c; 1999] that instead of considering this type of turbulence as an admixture of waves, such patchy intermittency could be more easily understood in terms of the development and interactions of coherent structures. We have carried out two-dimensional MHD simulations to study the dynamics of the coherent structures [*Wu and Chang*, 2000a,b; 2001]. Results, which include the calculated fluctuation probability distribution functions and local intermittency measures (LIM) based on the wavelet transforms, seem to validate the suggested characteristics of the intermittent stochastic processes [*Chang et al*, 2004]. Some results are presented briefly in this paper.

2. MHD Simulations

In our simulation, we solved compressible MHD equations (which, for example, are given in *Wu and Chang* [2000a]) on a computer as an initial and boundary value problem. The computer code is constructed in conservative form so that the total mass, total energy, total magnetic fluxes and total momenta are conserved. Also, we do not include explicit diffusion terms in the calculations. However, in the numerical scheme there is numerical diffusion, which concentrates at places where the gradient is high; it allows, for example, magnetic field reconnection to proceed. Reported here is a 2D calculation, first presented in *Wu and Chang* [2001], that was carried out with 512 by 512 grid points in a doubly periodic (x, y) domain of length 2π in both directions. The initial condition consists of constant density and pressure, and random magnetic field and velocity. The plasma β is high for the run.

Figure 1 shows the results at $t=300$; for reference, the sound wave and Alfvén wave traveling times through a distance of 2π are about 4.4 and 60, respectively. The left-hand panel shows the

322

magnetic field lines, which are the level curves of the magnetic flux and the right-hand panel plots the corresponding current distributions. The field lines point in the clockwise (counter-clockwise) direction if the current density is positive (negative). The figure shows magnetic islands of various sizes are formed. At the center of each island is an isolated current filament. The structures are very persistent and are the coherent structures of the system.

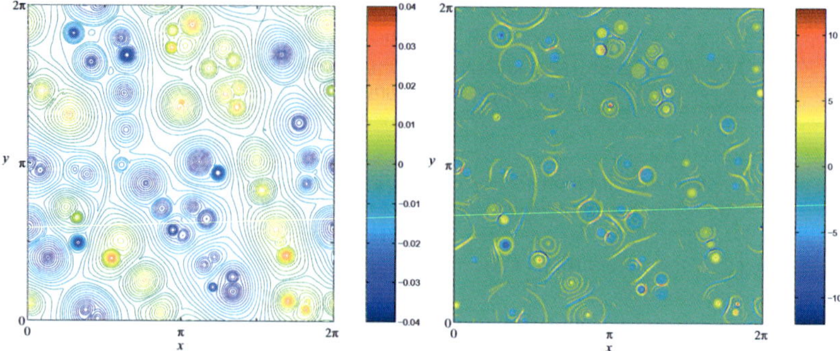

Figure 1.   2D MHD simulation of coherent structures (left panel) and current sheets (right panel) generated by initially randomly distributed current filaments after an elapsed time of $t = 300$. (From *Wu and Chang* [2001])

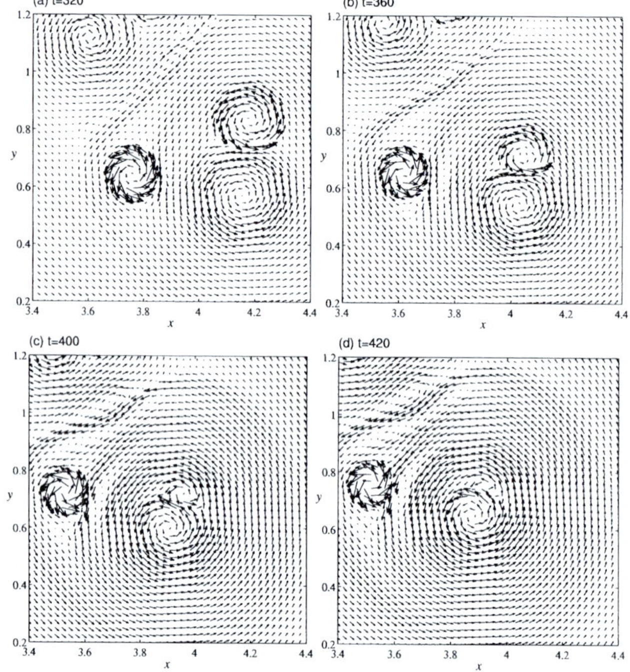

Figure 2.   Merging of two magnetic islands. (From *Wu and Chang* [2001])

These coherent structures can undergo different types of motions and interactions under the influence of the local plasma and magnetic topologies. When coherent magnetic flux tubes of the same polarity migrate toward each other, strong local magnetic shears are created. The coarse-grained dissipation will then initiate "fluctuation-induced nonlinear instabilities" [*Chang*, 1999; *Chang et al.*, 2003]; and, thereby reconfigure the topologies of the coherent structures into a combined lower local energetic state, eventually allowing the coherent structures to merge locally. Figure 2 shows an example of the magnetic merging process. The two islands near $(x, y)=(4.2, 0.7)$ at $t=300$ are shown to merge into one at t around 420. On the other hand, when coherent structures of opposite polarities approach each other due to the forcing of the surrounding plasma, they might repel each other, scatter, or induce magnetically quiescent localized regions. Under any of the conditions of the above interaction scenarios, new fluctuations will be generated. Figure 2 also provides an example that shows the persistence of the coherent structure: the one at $(x, y)=(3.75, 0.6)$ at $t=320$ lasts more than 100 units of time.

3. Intermittency

Nearly all fluctuations in space plasmas exhibit intermittency. For turbulent dynamical systems with intermittency, the transfer of energy (or other relevant scalars and tensors) due to fluctuations deviates significantly from uniformity. A technique of measuring the degree of intermittency is the study of the departure from Gaussianity the probability distribution functions of turbulent fluctuations at different scales. To demonstrate this point, results from the 2D numerical simulation are presented in this section. For example, we may generate the probability distribution function $P(\delta B^2,\delta)$ of $\delta B^2(x,\delta) \equiv B^2(x+\delta) - B^2(x)$ at a given time t for such simulations, where δ is the scale of separation in the x-direction. Figure 3 displays the calculated results of $P(\delta B^2,\delta)$ from the numerical simulation for several scales δ.

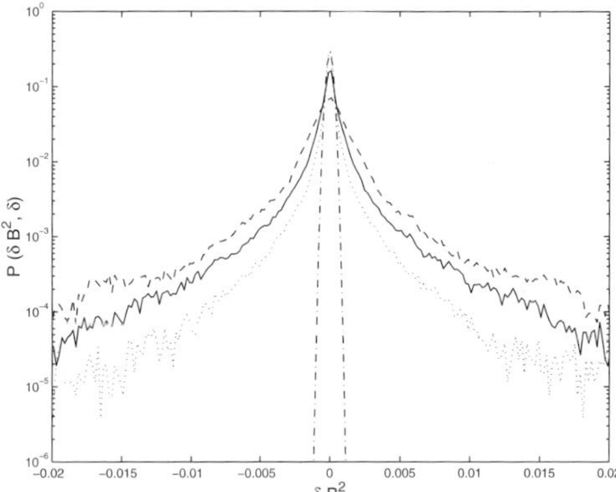

Figure 3. 2D MHD simulation result at $t=300$ of PDF's of B^2 at scales of 2 (dotted), 8 (solid), and 32 (dashed) units of grid spacing ε. The dash-dot curve is proportional to $\text{Exp}(-c\,(\delta B^2)^2)$ with c chosen to fit the PDF at scale of 2ε near $\delta B^2=0$. (Modified from *Chang et al.* [2004])

324

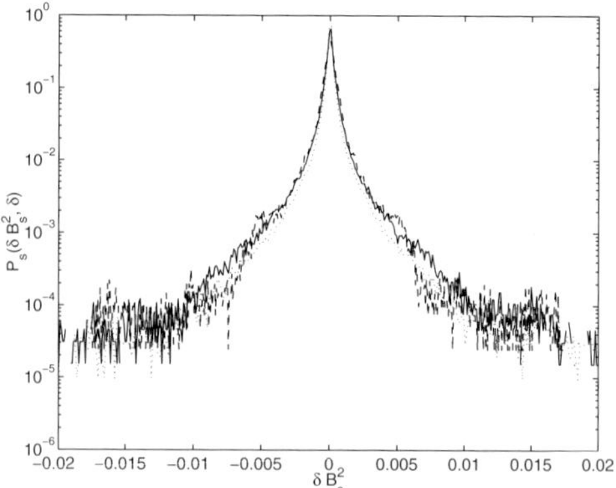

Figure 4. Scaled PDF's according to Eq. (3.1) with s=0.335; $\delta B^2_s \equiv \delta B^2 \delta^{-s}$. Line styles are the same as in Figure 3. (From *Chang et al.* [2004])

From this figure, we note that the deviation from Gaussianity becomes more and more pronounced at smaller and smaller scales. In an interesting paper by *Hnat et al.* [2002], they demonstrated that such probability distributions for solar wind fluctuations exhibit approximate mono-power scaling according to the following functional relation:

$$P(\delta B^2, \delta) = \delta^{-s} P_s(\delta B^2 \delta^{-s}, \delta) \qquad (3.1)$$

Where s is the mono-scaling power. We demonstrate that mono-power scaling also holds approximately for our simulated results with the value of s equal to approximately 0.335, Figure 4.

Actually the scaling relation (3.1) is approximate in that the tails of the distributions in Figure 4 do not exactly fall onto one curve. This is the intrinsic nature of the strong intermittency at small scales. Thus, representations of the probability density functions will involve multi-parameters in general and may sometimes, for example, be represented by the κ- or Castaing distributions [*Jurac,* 2003; *Forman,* 2003; *Weygand,* 2003; *Castaing et al.,* 1990].

Since the degree of intermittency generally increases inversely with scale, it will be interesting to study the degrees of intermittency locally at different scales. This can be accomplished by the method of Local Intermittency Measure (LIM) using the wavelet transforms. A wavelet transform generally is composed of modes which are square integrable localized functions that are capable of unfolding fluctuating fields into space and scale [*Farge,* 1992]. Figure 5b is the power spectrum of a complex Morlet wavelet transform of the current density for the 2D MHD simulations mentioned in Section 2. We notice that the intensity of the current density is sporadic and varies nonuniformly with scale.

We now define LIM(1) as the ratio of the squared wavelet amplitude $|\psi(x,\delta)|^2$ and its space averaged value $<|\psi(x,\delta)|^2>_x$. We note that LIM(1)=1 for the Fourier spectrum. To emphasize the variation of intensity with scale, we also consider the logarithm of LIM(1). It has been suggested by *Meneveau* [1991] that the space average of the square of LIM(1), which is a scale dependent measure of the kurtosis or flatness, is a convenient gauge of the deviation of intermittency from Gaussianity. We denote this measure by LIM(2). It is equal to 3 if the probability distribution is Gaussian. Figures 5d,e and 6 are graphical displays of the calculated results of LIM(1), logLIM(1) and LIM(2) for our 2D numerical simulations using the complex Morlet transform. We notice that the fluctuations are indeed scale dependent, localized and strongly intermittent at small scales. Similar experimental results using the wavelet transforms have been found, for example, by *Consolini et al.* [2004] for the magnetotail and *Bruno et al.* [2001] for the solar wind.

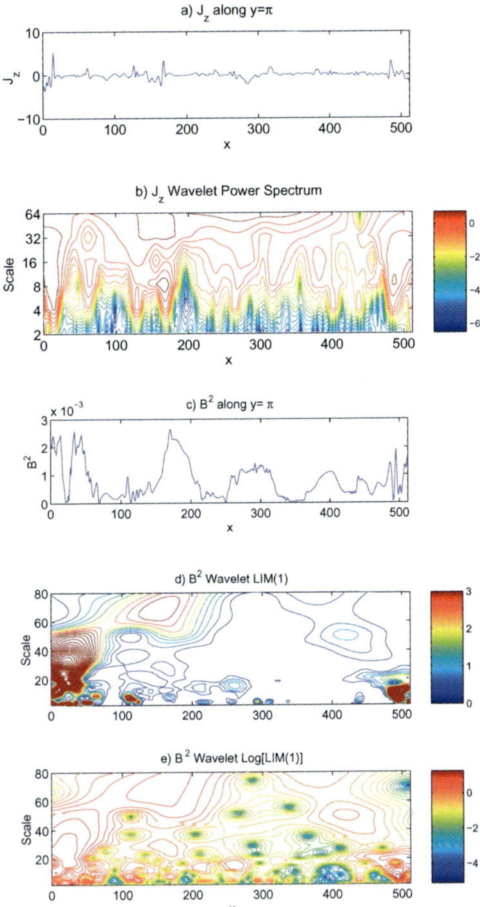

Figure 5. (a) 2D MHD simulation result of current density J_z along the z-axis at t=300. (b) Power spectrum of complex Morlet wavelet transform of J_z. (c) 2D MHD simulation result of B^2 distribution along the x-axis for y=π at t=300. (d and e) Contour plots of LIM(1) and LogLIM(1) of the B^2 distribution shown in (c). The x-axis and scale are in units of the grid spacing ε. (From *Chang et al.* [2004])

326

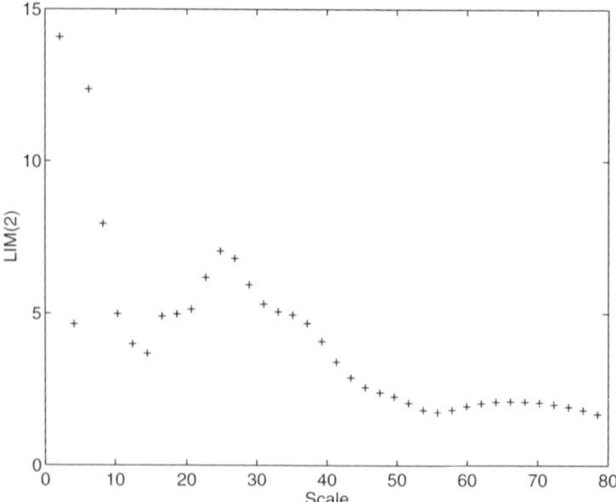

Figure 6. LIM(2) of B^2 for the same B^2 distribution of Figure 5c. (From *Chang et al.* [2004])

Acknowledgements

This research was partially supported by NASA and NSF.

References

Bruno, R., V. Carbone, P. Veltri, E. Pietropaolo, B. Bavassano, Identifying intermittency events in the solar wind, *Planetary and Space Science*, 49, 1201, 2001.

Castaing, B., Y. Cagne, and E.J. Hopfinger, Velocity probability density functions of high Reynolds number turbulence, *Physica D.*, 46, 177, 1990.

Chang, T., Sporadic, localized reconnections and multiscale intermittent turbulence in the magnetotail, *AGU Monograph on "Encounter between Global Observations and Models in the ISTP Era"*, vol. 104, p.193, Horwitz, J.L., D.L. Gallagher, and W.K. Peterson, Am. Geophys. Union, Washington, D.C., 1998a.

Chang, T., Multiscale intermittent turbulence in the magnetotail, *Proc. 4th Intern. Conf. on Substorms*, ed. Kamide, Y. et al., Kluwer Academic Publishers, Dordrecht and Terra Scientific Publishing Company, Tokyo, p. 431, 1998b.

Chang, T., Self-organized criticality, multi-fractal spectra, and intermittent merging of coherent structures in the magnetotail, *Astrophysics and Space Science*, ed. Büchner, J. et al., Kluwer Academic Publishers, Dordrecht, Netherlands, vol. 264, p. 303, 1998c.

Chang, T., Self-organized criticality, multi-fractal spectra, sporadic localized reconnections and intermittent turbulence in the magnetotail, *Physics of Plasmas*, 6, 4137, 1999.

Chang, T., S.W.Y. Tam, C.C. Wu, and G. Consolini, Complexity, forced and/or selforganized criticality, and topological phase transitions in space plasmas, *Space Science Reviews*, 107, 425, 2003.

Chang, T., S.W.Y. Tam, and C.C. Wu, Complexity induced bimodal intermittent turbulence in space plasmas, *Physics of Plasmas*, 11, 1287, 2004.

Consolini, G., T. Chang, and A.T.Y. Lui, Complexity and topological disorder in the Earth's magnetotail dynamics, to be published, 2003.

Farge, M., Wavelet transforms and their applications to turbulence, *Annual Reviews of Fluid Mechanics*, 24, 395, 1992.

Forman, M., private communication, 2003.

Hnat, B., S. C. Chapman, G. Rowlands, N. W. Watkins, and W. M. Farrell, Finite size scaling in the solar wind magnetic field energy density as seen by WIND, *Geophys. Res. Lett.*, *29*, 1446, 2002.

Jurac, S., private communication, 2003.

Meneveau, C., Analysis of turbulence in the orthogonal wavelet representation, *J. Fluid Mech.*, 232, 469, 1991.

Weygand, J., private communication, 2003.

Wu, C.C., and T. Chang, 2D MHD Simulation of the Emergence and Merging of Coherent Structures, *Geophys. Res. Lett.*, 27, 863, 2000a.

Wu, C.C., and T. Chang, Dynamical evolution of coherent structures in intermittent two dimensional MHD turbulence, *IEEE Trans. on Plasma Science*, 28, 1938, 2000b.

Wu, C.C., and T. Chang, "Further Study of the Dynamics of Two-Dimensional MHD Coherent Structures—A Large Scale Simulation", *Journal of Atmospheric Sciences and Terrestrial Physics*, 63, 1447, 2001.

Multiscale Coupling of Sun-Earth Processes

A.T.Y. Lui, Y. Kamide and G. Consolini, editors.

INTERMITTENCY AND SELF-SIMILARITY IN 'NATURAL PARAMETERS' IN SOLAR WIND TURBULENCE

S.C. Chapman[1,2], B. Hnat[1], and G. Rowlands[1]

[1]Physics Department, University of Warwick, Coventry, CV4 7AL, UK

[2]also at Radcliffe Institute for Advanced Studies, Harward University, USA

Abstract. The solar wind provides a natural laboratory for observations of MHD turbulence over extended temporal scales. We identify self-similarity in the probability density functions of fluctuations in certain solar wind bulk plasma parameters as seen by the WIND spacecraft. Whereas the fluctuations of speed v and IMF magnitude B are multi-fractal, we find that the fluctuations in the ion density ρ, energy densities B^2 and ρv^2 as well as MHD-approximated Poynting flux vB^2 are self-similar on the timescales up to ~26 hours. We argue that these are 'natural parameters' in which to cast the solar wind measurements which are taken in an Eulerian frame of reference. The intermittency of the system is then expressed in these parameters through the non-Gaussian nature of the single curve that describes the fluctuations PDF up to this timescale. Self-similarity implies that a Fokker-Planck model exists for the timeseries and we derive this here along with the associated Langevin equation.

1. Introduction

Statistical properties of velocity field fluctuations recorded in fluid experiments and in solar wind observations exhibit a range of similarities [*Carbone et al.*, 1995; *Veltri*, 1999]. A common feature found in these fluctuations is fractal or multi-fractal scaling, accompanied by a Probability Density Function (PDF) that shows a clear departure from the Normal distribution when we consider the difference in velocity on small spatial scales [*Frisch*, 1995; *Bohr et al.*, 1998]. On the large scale, these features appear to be uncorrelated and cross over to a Gaussian distribution. The approach is then to treat the solar wind fluctuations as arising in a manner similar to that of hydrodynamic turbulence [*e.g. Dobrowolny et al.*, 1980; *Tu and Marsch*, 1995; *Milovanov and Zelenyi*, 1998; *Goldstein and Roberts*, 1999]. In this context, empirical models have been widely used to approximate the fluctuation PDFs of data from wind tunnels [*Castaing et al.*, 1995] as well as the solar wind [*see e.g. Sorriso Valvo et al.*, 2001]. The picture of turbulence emerging from these models is more complex than the original Kolmogorov theory. It requires a multi-fractal phenomenology to be invoked as the self-similarity of the cascade is broken by the introduction of intermittency.

Recently, however, a new approach has emerged where the presence of intermittency in the system coincides with statistical self-similarity rather than multi-fractality in the fluctuations of certain quantities which simultaneously exhibit leptokurtic PDFs. An example of this *statistical intermittency* was discussed in *Mantegna and Stanley* [1995] (*see also Mandelbrot*, [1963]), and in the context of the solar wind in *Hnat et al.* [2002, 2003]. We will apply this model-independent and generic PDF rescaling technique to extract the scaling properties of the solar wind fluctuations as seen by WIND, directly from the data. The aim is to determine a set of plasma parameters that exhibit statistical self-similarity and to verify the nature of the PDF for their

332

A generic one parameter rescaling method [*Hnat et al.*, 2002, 2003] is applied to these PDFs. We seek a single scaling exponent α such that all unscaled PDFs $P(\delta x,\tau)$ collapse onto a single curve $P_s(\delta x_s)$ using the following change of variables:

$$P(\delta x,\tau) = \tau^{-\alpha} P_s(\delta x \tau^{-\alpha}) \tag{3.1}$$

We can thus extract the scaling index α with respect to τ, directly from the time series of the quantity δx by considering how any point on the curve $P(\delta x,\tau)$ varies with τ. Figure 2 shows the standard deviation $\sigma(\tau)$ of the unscaled PDFs plotted versus τ on log-log axes for the four bulk plasma parameters. We see that the $\sigma(\tau)$ of these PDFs are described by a power law (and thus also provide an estimate of the Hurst exponent H) so that $\sigma(\tau) \sim \tau^{H} \equiv \tau^{\alpha}$ for a range of τ up to $\sim$ 26 hours. In Figure 2, lines were fitted with R^2 goodness of fit for the range of τ between $\sim$ 2 minutes and $\sim$ 26 hours. The slopes of these lines yield the exponents α. We will next take α to be the scaling index in (3.1) and attempt to collapse the curves. For experimental data, an approximate collapse of the PDFs is an indicator of a dominant self-similar trend in the time series, i.e. this method may not be sensitive enough to detect multi-fractality that could be present only during short time intervals. One can treat the identification of the scaling exponent α and, as we will see, the non-Gaussian nature of the rescaled PDFs (P_s) as a method for quantifying the intermittent character of the time series. A consequence of the rescaling is that $P(\delta x,\tau)$ is the self-similar solution of the F-P equation describing the dynamics of the unscaled PDF in time and with respect to the coordinate δx [*e.g. van Kampen*, 1992].

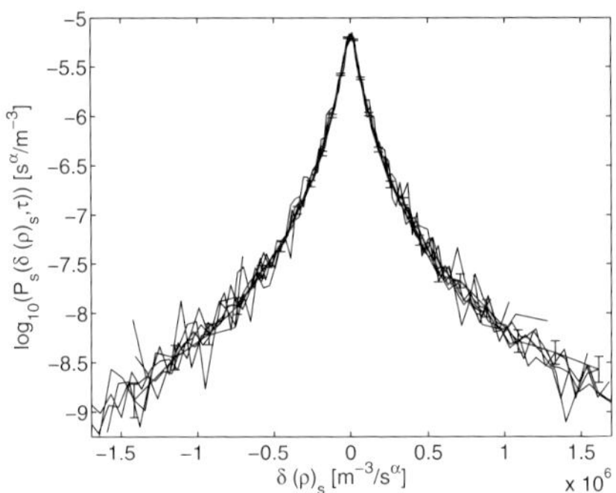

Figure 4. As in Figure 3 for ion density fluctuations $\delta\rho$.

Ideally, we would use the peaks of the PDFs to obtain the scaling exponent α, as the peaks are statistically the most accurate parts of the distributions. However for observations, the peaks may not be the optimal statistical measure for obtaining the scaling index. For example, the B_z component of the solar wind magnetic field is measured with an absolute accuracy of typically 0.1 nT. Such discreteness in the time series introduces large errors in the differences of B and B^2 as $B \to 0$ so that the peak values $P(0,\tau)$ may not be well determined. However, if the PDFs rescale, we can in principle obtain the scaling exponent from any point on the curve [*see Hnat et al.*, 2003].

4. PDF Rescaling Results

We now present results of the rescaling procedure as applied to the solar wind bulk plasma parameters. We now apply the result from Figure 2 to rescale the raw PDF (an example of which was shown in Figure 1). Figures 3 - 6 show the result of the one parameter rescaling applied to this unscaled PDF of fluctuations in ρ, ρv^2, B^2 and vB^2 respectively, for temporal scales up to ~26 hours. We see that the rescaling procedure (3.1) gives good collapse of each curve onto a single common functional form for the entire range of the data. These rescaled PDFs are leptokurtic rather than, Gaussian consistent with an underlying intermittent process.

The fluctuations PDFs for all self-similar quantities investigated here are nearly symmetric. This is in sharp contrast with the strong asymmetry of the PDF of uctuations in quantities known to be multifractal, such as the velocity and a comparison is made in *Hnat et al.* [2004].

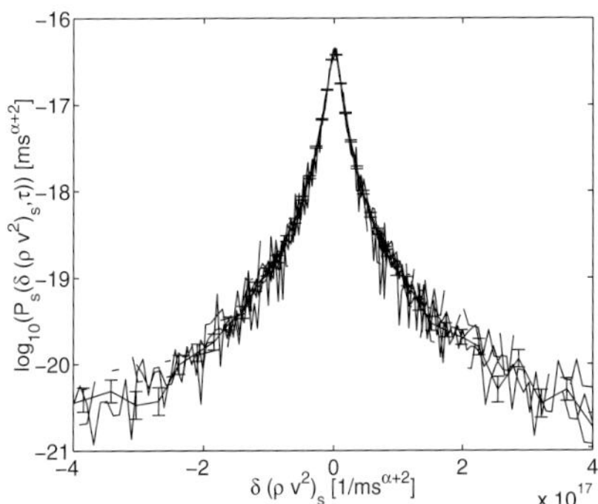

Figure 5. As in Figure 3 for kinetic energy density fluctuations $\delta(\rho v^2)$.

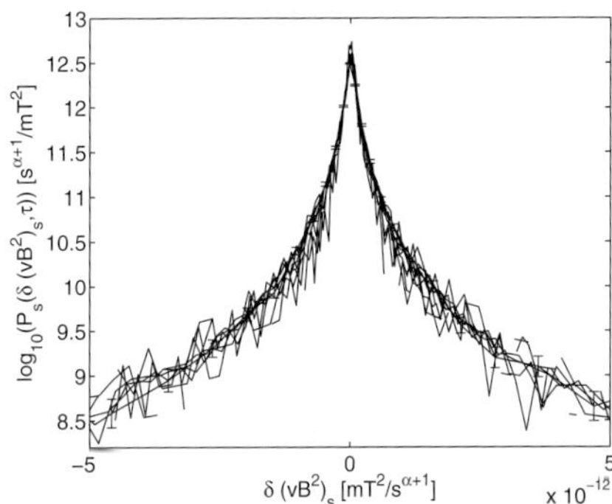

Figure 6. As in Figure 3 for Poynting flux $\delta(vB^2)$.

We can now directly compare the functional form of these rescaled PDFs by normalizing the curves to their σ_s and overlaying them on the single plot for a particular τ within the scaling range. Figure 7 shows these normalized PDFs $P_s(\delta x_s, \tau)$ for $\delta x_s = \delta(\rho)_s$, $\delta(B^2)_s$, $\delta(\rho v^2)_s$, $\delta(vB^2)_s$ and $\tau \approx 1$ hour overlaid on a single plot. These normalized PDFs have a remarkably similar functional form suggesting a shared process responsible for fluctuations in these four plasma parameters on temporal scales up to $\tau_{max} \sim 26$ hours.

It has been found previously [*Burlaga*, 2001] that the magnetic field magnitude fluctuations

334

are not self-similar but rather multi-fractal. Single parameter scaling would not then be expected to rescale the entire PDF. To illustrate this we applied the rescaling procedure for magnetic field magnitude differences $\delta B(t,\tau) = B(t+\tau) - B(t)$. Figure 8 shows the result of one parameter rescaling based on the exponent from the peak, $P(0,\tau)$ applied to the PDFs of the magnetic field magnitude fluctuations. We see that the scaling procedure is satisfactory only up to ~ 3 standard deviations of the original sample, despite the satisfactory scaling obtained for the peaks of the PDFs (see insert of the Figure 8). This confirms the results of *Sorriso-Valvo et al.* [2001] where a two parameter Castaing fit to values within 3 standard deviations of the original sample yields scaling in one parameter and weak variation in the other. We found similar lack of single exponent scaling in the fluctuations of the solar wind velocity magnitude and we show the rescaled PDF in Figure 9. We stress that the log-log plots of the PDF peaks $P(0,\tau)$ show a linear region for both velocity and magnetic field magnitude fluctuations (see insert in each figure). Their PDFs, however, do not collapse onto a single curve when the rescaling (3.1) is applied.

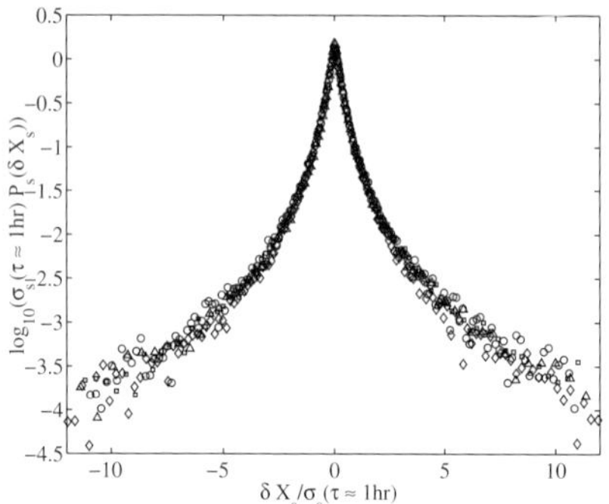

Figure 7. Direct comparison of the PDFs of fluctuations for all quantities. Circles correspond to $\delta(B^2)$, squares, to ion density $\delta(\rho)$, diamonds to $\delta(\rho v^2)$ and triangles, Poynting flux component $\delta(vB^2)$.

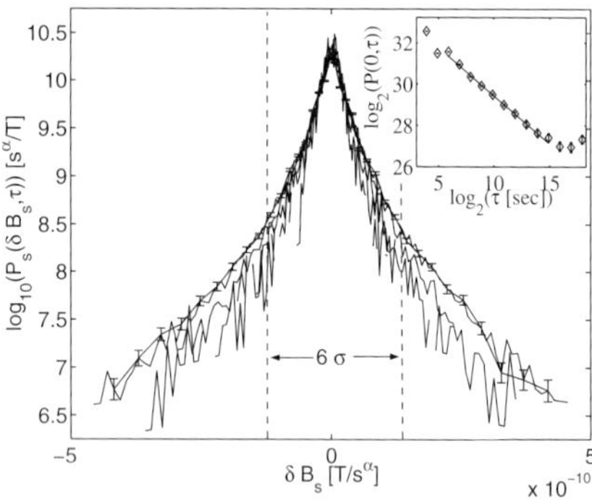

Figure 8. As in Figure 3 for the solar wind magnetic field magnitude.

Multiscale Coupling of Sun-Earth Processes
A.T.Y. Lui, Y. Kamide and G. Consolini, editors.

ACCELERATION OF PARTICLES BY LOWER-HYBRID WAVES IN SPACE PLASMAS

R. Bingham

*Rutherford Appleton Laboratory,Chilton, Didcot,Oxon, OX11 0QX, UK
and University of Strathclyde, Glasgow, G1 1XQ, Scotland, UK*

V. D. Shapiro

Physics Department, UCSD, San Diego, California, USA

P.K. Shukla

*Institut für Theoretische Physik IV, Ruhr Universit"at Bochum, D-44780
Bochum, Germany*

Abstract. Processes that control plasma behaviour at boundaries are important in our understanding of inter-acting plasmas, such as the interaction of the solar wind with planetary magnetospheres and comets. One of the important aspects of this interaction is its "collisionless" nature ie the lack of collisions between the plasma particles. In the absence of normal particle collisions wave particle interactions and plasma turbulence become dominant mechanisms for the transport of energy and momentum between different plasma populations. The interaction between the particles and turbulent fields leads to particle acceleration and heating producing nonthermal electron and ion distribution functions. Lower-hybrid waves are ideal at transferring energy between ions and electrons and the auroral region is an ideal laboratory to understand both linear and nonlinear interactions.

1. Introduction

Considerable attention has already been devoted to the interaction of lower-hybrid waves with ions in the auroral zones, they have previously been suggested as an explanation for ion acceleration in the plasma sheet boundary layer [*Chang et al.*, 1981] and ion heating in the auroral zone [*Shapiro et al.*, 1995]. In these models of ion heating by lower-hybrid waves the generation of lower-hybrid waves is by the precipitating electrons [*Chang et al.*, 1981; *Shapiro et al.*, 1995]. An interesting property of lower-hybrid waves is that they can resonate with electrons moving along the magnetic field and with ions moving across the field simultaneously. This allows the waves to be created by both parallel moving electrons and perpendicularly moving ions. Therefore if the free energy is in the electrons precipitating then the electrons generate the waves whereas if the free energy is in the ions the ions generate the waves. Since the process is reversible the waves can either accelerate electrons along field lines or ions across field lines. An example of waves generated by the precipitating electrons in the suprauroral region is auroral hiss. The frequency of these waves extends from the lower-hybrid resonance up to plasma frequency, falling within the VLF range. Simulations by *Retterer and Chang* [1988] show that the precipitating electrons can be responsible for wave generation in this frequency range.

The simulations by *Retterer and Chang* [1988] demonstrate that these waves could accelerate the ions transversely and the ambient electron parallel to the magnetic field forming counterstreaming electrons [*Liu et al.*, 1982].

Electron acceleration by lower-hybrid waves in the auroral zone has been discussed by *Bingham et al.* [1984, 1988], *Retterer et al.* [1988], *Dendy et al.* [1995] and *Bryant et al.* [1995]. Parallel electron acceleration arising from resonant interaction of lower-hybrid waves with the electrons are well known in the laboratory [*Fisch,* 1978] where they are used extensively in magnetic fusion devices to drive toroidal currents ie lower-hybrid current drive. Quasi-linear models are used to estimate the current produced by an extended tail in the electron distribution function as a result of the resonant interaction with lowerhybrid waves. Particle in cell simulations of electron acceleration by lower-hybrid waves have also been carried out to understand the complete self-consistent interaction process. In these simulations the lowerhybrid waves were intense enough to form localized field aligned filaments [*Bingham et al.,* 1999]. This property of localization of field aligned lower-hybrid wave packets is a common feature of these waves and will be discussed in Section 4.

The special role that lower-hybrid waves play in plasma physics is due to the following reasons. The frequency of the waves $\omega = \omega_{lh}(1+(m_i / m_e)(k_{\parallel}^2/k^2))^{1/2}$ where $\omega_{lh} = \omega_{pi}/(1+\omega_{pe}^2/\omega_{ce}^2)^{1/2}$ is the frequency of the lower-hybrid resonance, this frequency is always between the electron and ion gyro frequencies, this means that the electrons are strongly magnetized in these oscillations while the ions can be considered as unmagnetized. The waves are propagating almost perpendicular to the magnetic field with the dominant wavevector $k = (k_{\perp}^2+k_{\parallel}^2)^{1/2}$ component perpendicular to the magnetic field. For oblique propagation $k_{\parallel}/k > (m_i/m_e)^{1/2}$ the mode evolves into electrostatic whistler waves with the dispersive law $\omega = \omega_{pe}(k_{\parallel}/k)/(1+\omega_{pe}^2/\omega^2)^{1/2}$. The wave phase velocity perpendicular to the field $\omega/k_{\perp}$ is much smaller than the phase velocity along the magnetic field direction $\omega/k_{\parallel}$ resulting in the possibility of simultaneous Landau resonance between waves and fast field aligned electrons and relatively slow but unmagnetized ions moving perpendicular to the magnetic field. Because of the possibility of simultaneous Landau resonance lower-hybrid waves play a central role in the collisionless energy and momentum coupling between electrons and ions in different plasmas. They play the same role in many different space and astrophysical environments. At the earth's bow shock front creating a counterstreaming flow in the solar wind resulting in the excitation of lower-hybrid waves through the modified two stream instability [*McBride et al.,* 1972] in the upstream region of the bow shock. In the modified two stream instability the free energy is in the relative motion between counter streaming ions across the magnetic field. The most unstable branch is the lower-hybrid mode propagating almost perpendicular to the magnetic field. Landau absorption of these waves by energetic electrons results in electron acceleration along the magnetic field upstream of the earth's bow shock [*Anderson et al.,* 1981]. A similar coupling mechanism between different plasmas through lower-hybrid waves takes place between the solar wind and ionospheres of non-magnetic planets such as Mars [*Grard et al.,* 1989] and Venus [*Scarf et al.,* 1980]. In this case because of the absence of the magnetosphere the planetary ionosphere is directly exposed to the solar wind flow and the modified two stream instability develops because of the relative motion of the shocked solar wing and cold planetary ions. In the auroral ionosphere the precipitating field aligned electrons usually serves as a source of free energy exciting lower-hybrid waves by the bump on tail instability or fan instability also known as the anomalous doppler resonance instability [*Kadomtsev,* 1967]. Absorption of these waves by ionospheric ions leads to the formation of energetic ion conics [*Kintner et al.,* 1992].

Freja observations [*Eliasson et al.,* 1994] at 1700 km in the auroral oval show down-going beam-like ion distributions accompanied by a horseshoe shaped distribution, which has a parallel beam-like component and wings extending into the perpendicular domain. These were initially called ring distributions but we prefer to call them horseshoe type distributions. Ring distributions are normally used when referring to distribution functions, which have predominantly perpendicular components forming a torus in velocity space with its axis parallel to the magnetic field. Such ring distribution functions are commonly found at quasi-perpendicular shocks and also associated with transverse resonant acceleration. The down-going field aligned beam feature is peaked at about 400 km/sec [*Eliasson et al.*, 1994]. Due to the Freja orbit, which is below or in the lower part of the acceleration region mostly down-going beams are observed. These horseshoe shaped features in the ion distribution function are strongly correlated with broadband wave emission around 1 kHz close to the lower-hybrid frequency [*Eliasson et al.*, 1994].

Measurement of the low energy proton distribution function by Freja [*Eliasson et al.*, 1994] also reveal the presence of perpendicular acceleration which was estimated to have occured about a few hundred km below the satellite. The energy of the low energy protons is much lower than the energy of the ring or horseshoe protons. The horseshoe shaped distribution is composed of a parallel beam component and a loss-cone in the up-going direction. The horseshoe distribution function has a pronounced +*ve* slope in the perpendicular direction. This particular type of distribution is unstable to the generation of waves around the lower-hybrid waves which are observed at the same time as the observation of the horseshoe distribution [*Eliasson et al.*, 1994].

Our objective here is to present a stability analysis of the dispersion relation for plasma waves with the particular horseshoe distribution observed by Freja [*Eliasson et al.*, 1994]. We show that the horseshoe distribution is responsible for growth of broadband waves around the lower-hybrid frequency.

2. Generation of Lower-Hybrid Waves

It has been proposed by *Eliasson et al.* [1994] that the ring or horseshoe distribution generates lowerhybrid waves, which are then responsible for accelerating the low energy proton component producing transversely heated protons. The aim of this paper is to analyse the stability of the ring or horseshoe distribution to the generation of low frequency waves around the lower-hybrid frequency. The analysis we consider is closely related to the ion loss-cone instability [*Rosenbluth and Post*, 1965], which transfers the perpendicular energy of the gyrating ions to lower-hybrid waves through Landau resonance on the positive slope of the distribution function. This instability is also similar to the modified two-stream instability [*McBride et al.*, 1972] with a cross field ion beam. The modified two stream instability is a fluid type instability whereas the loss-cone is a kinetic instability. Both excite the same mode, the lower-hybrid mode, whose frequency and wavelength satisfy the following inequalities:

$$\left\{ \begin{array}{l} \omega_{ci} < \omega << \omega_{ce} \\ k_{\perp}\rho_e << 1 << k_{\perp}\rho_i \text{ and } k_{\parallel} << k_{\perp} \end{array} \right. \tag{2.1}$$

where $\rho_{e,i}$ is the electron, ion gyro-radius. These assumptions are appropriate for plasmas of moderate densities with $\omega_{pi} >> \omega_{ci}$.

When the above inequalities are satisfied we can neglect the influence of the magnetic field on the ion motion. The instability is then described by the kinetic equations for ions and electrons, together with Poisson's equation:

$$\left(\frac{\partial}{\partial t} + \mathbf{v} \cdot \nabla - \frac{e}{m_i} \nabla \phi \cdot \frac{\partial}{\partial \mathbf{v}} \right) f_i(\mathbf{x}, \mathbf{v}, t) = 0, \tag{2.2}$$

$$\left(\frac{\partial}{\partial t}+\mathbf{v}\cdot\nabla-\frac{e}{m_e}\left(-\nabla\phi+\mathbf{v}\times\mathbf{B}\right)\cdot\frac{\partial}{\partial \mathbf{v}}\right)f_e(\mathbf{x},\mathbf{v},t)=0, \tag{2.3}$$

$$\nabla\cdot\mathbf{E}=-\frac{1}{\varepsilon_0}\sum_{j=i,e}e_j\int f_j(\mathbf{x},\mathbf{v},t)d^3\mathbf{v}, \tag{2.4}$$

where $\mathbf{E}=-\nabla\phi$, ϕ is assumed to have the space-time dependence $\phi\sim\exp i\,(\omega t+k_{\parallel}z+k_{\perp}x)$. In the linear approximation, equations (2.2) to (2.4) reduce to the dispersion relation for the electrostatic lower-hybrid mode given by:

$$\varepsilon(\omega,\mathbf{k})=1+\omega_{pe}^2/\omega_{ce}^2>\omega_{pe}^2 k_{\parallel}^2/\omega^2 k^2+\frac{\omega_{pi}^2}{k^2}\int\frac{\mathbf{k}\cdot\partial f_i/\partial\mathbf{v}}{\omega-\mathbf{k}\cdot\mathbf{v}}d^3\mathbf{v} \tag{2.5}$$

In deriving equation (2.5) we have assumed that the perpendicular wavelength is larger than the electron gyroradius, $k_{\perp}\rho_e\gg 1$, and that the parallel phase velocity is larger than the electron thermal velocity, an assumption commonly referred to as the cold plasma approximation. The growth rate and frequency of the unstable modes can be obtained from the equation:

$$\gamma_g=\operatorname{Im}\varepsilon(\omega,\mathbf{k})\left(\frac{\partial\operatorname{Re}\varepsilon(\omega,\mathbf{k})}{\partial\omega_k}\right)^{-1} \tag{2.6}$$

To solve (2.6) we must carry out the ion integrals in equation (2.5).

Numerical solutions of equation (2.6) have been carried out for the particular horseshoe type distribution function observed by the Freja satellite with the total proton distribution function being represented by

$$f(v_{\parallel},v_{\perp})=\frac{1}{\left(2\pi v_{Ti}^2\right)^{3/2}}\left\{\exp\left[-\frac{v^2}{v_{Ti}^2}\right]+\alpha f_h(v_{\perp},v_{\parallel})\right\}. \tag{2.7}$$

The first term in (2.7) is a Maxwellian background ion distribution function and the second term represents the horseshoe distribution with

$$f_h(v_{\perp},v_{\parallel})=\exp\left[-\frac{\left(v(v_{\perp},v_{\parallel})-v_b\right)^2}{v_0^2}\right]\left(\mu(v_{\perp},v_{\parallel})+1\right)^2 \tag{2.8}$$

where $f_h(v_{\perp},v_{\parallel})$ represents the horseshoe component with v_0 the thermal width of this component normalized to the ion thermal speed v_{Ti} ($v_{Ti}=(2k_BT_i/m_i)^{1/2}$ where T_i is the ion temperature, m_i is the ion mass and k_B is Boltzmann's constant) and v_b is the displacement in parallel velocity of the peak of the horseshoe distribution. From the Freja data, v_b is typically 400 km/sec, again normalized with respect to v_{Ti}. Note that we have defined $\mu=v_{\parallel}/v$, $v=(v_{\perp}^2+v_{\parallel}^2)^{1/2}$ and α is the ratio of number of ions in a horseshoe distribution to the number in the Maxwellian background. The distribution function represented by Equation (2.7) consists of a Maxwellian component and a horseshoe component resembling Figure 3 given in the paper by *Eliasson et al.* [1994]. By integrating over the parallel velocity we can obtain the perpendicular distribution function $f_\perp(v_\perp^2)$ which has a positive slope on the perpendicular velocity, this is a source of free energy capable of generating waves.

Using the distribution given by equation (2.7) and assuming $k_\parallel \ll k_\perp$, allowing us to neglect the $k_\parallel \partial f/\partial v_\parallel$ component, we can obtain real and imaginary frequency components as a function of $k_\perp$. A broad range of frequencies just below ω_{pi} is found to be unstable. This broad range encompasses the lower-hybrid resonance frequency given by $\omega_{pi}/(1 + \omega_{pe}^2/\omega_{ce}^2)^{1/2}$ which is close to ω_{pi} in the auroral zone.

The physical nature of this instability is very similar to that of the modified two-stream fluid instability. The free energy comes from the gyro-motion of the ions across the magnetic field lines. The general effect of the instability would be to convey energy from perpendicular motion into waves, which would have the effect of filling in the region between the background Maxwellian and horseshoe distribution. The waves would also be responsible for heating the ions in the transverse direction from the background distribution, thus forming the distribution observed by *Eliasson et al.* [1994]. Heating in the transverse direction and filling in of the loss-cone would result.

The instability is convective in nature and therefore it is important to determine the exponentiation growth length along the field lines. Assuming $k_\parallel \ll k_\perp$ we can neglect $k_\parallel(\partial f_i/\partial v_\parallel)$ in equation (2.5) and solve for spatial growth at a given real ω, yielding:

$$ k_\parallel = k_\perp \left[\frac{\omega^2}{\omega_{LH}^2} - \frac{m_e}{m_i} \frac{\omega^2}{k_\perp^2} \int \frac{v_\perp \frac{\partial f}{\partial v_\perp}}{\omega - k_\perp v_\perp} dv_\perp \right]^{1/2} \tag{2.9} $$

Since for $k_\parallel \neq 0$ the lower-hybrid wave group velocity parallel to the magnetic field is greater than the group velocity perpendicular to **B** most of the wave energy is convected along the magnetic field. These waves can resonate with ions moving perpendicular to **B** and electrons moving parallel to B therefore acting as a mechanism for transferring ion energy to electrons or vice versa.

3. Quasi-Linear Diffusion

The quasi-linear theory of electron acceleration by lower-hybrid waves is based on diffusion in velocity space using the Fokker-Planck description [*Dendy et al.*, 1995]. This describes diffusion in velocity space in response to a given wave spectrum and allows for multiple wave packets interacting simultaneously with the electrons. The possibility of using multiple wave packets has been shown by *Dendy et al.* [1995] to have significant consequences for electron acceleration. electrons can be accelerated out of the cool bulk population by a wave packet with a relatively low phase velocity which takes them into a region where a higher phase velocity wave packet accelerates the electrons to much higher energies.

The collective evolution of the velocity distribution of the electrons in response to a spectrum of lower-hybrid waves is easily modelled as a diffusive process using the Fokker-Planck equation. This equation describes the time or space evolution of the electron velocity distribution due to diffusion driven by Landau resonant lower-hybrid waves described by

two terms describing excitation of monochromatic waves with $k = k_0$ by electrons and Landau damping by ions with the distribution function assumed to be Maxwellian:

$$\Gamma(r) = \frac{1}{(2\pi)^2} \int dk_\perp \Gamma_{k_\perp}^r \exp(ik_\perp r)$$

$$\Gamma_{k_\perp}^r = \Gamma_0 \delta(k_\perp - k_0) + \left(\frac{\pi}{8}\right)^{\frac{1}{2}} \frac{\omega_{LH}^4}{k_\perp^3 v_{T_i}^3} \exp\left\{-\frac{\omega_{LH}^2}{2k_\perp^2 v_{T_i}^2}\left(1 + k^2 R^2\right)\right\}$$

(4.4)

It should be noted that the system of equations investigated in *Shapiro et al.* [1993] is identical to Equations (4.1) and (4.2) except for the source term described in Equation (4.4) as well as the term $(m_i/m_e)(\omega_{LH}^2/c^2)(\omega_{pe}^2/\omega_{ce}^2)\phi$, which is omitted on the l.h.s. of equation (4.1). The latter term is responsible for the electromagnetic correction to the dispersion relation, and only becomes important in finite β plasmas. For the auroral ionosphere β is very small of order 10^{-6}, therefore the excited oscillations are definitely electrostatic.

There are two main differences between the system of equations (4.1) and (4.2) and those corresponding to the Zakharov system of equations for Langmuir waves. First, the nonlinear frequency shift in equation (4.1) is created by the adiabatic drift of electrons in the inhomogeneous density background $\delta n(t,r)$ appearing due to the slow mode. Since in LH oscillations electrons are magnetized in the plane perpendicular to the magnetic field their velocity of oscillation can be written in the form

$$v_e = -\frac{c}{B_0^2}\nabla\phi \times B_0 - \frac{i\omega}{\omega_{ce}}\frac{c}{B_0}\nabla\phi$$

(4.5)

The electron density perturbation in the LH oscillations can be represented as

$$n_e = \frac{n_0}{i\omega}\nabla \cdot v_e + \frac{1}{i\omega}v_e\nabla\delta n$$

(4.6)

Adiabatic drift corresponds to an incompressible fluid with $\nabla \cdot v_e = 0$ and it is important only in the presence of density perturbations $\delta n(t,r)$ coupled with the LH mode. The nonlinear dispersion relation for LH waves then has the form

$$\omega_k = \omega_{LH}\left(1 + \frac{1}{2}k^2 R^2\right) - \frac{\omega_{ce}}{2k^2 n_0}\left(k \times \nabla\delta n\right)_z$$

(4.7)

and the nonlinear frequency shift is significant even starting with small amplitude of the density perturbations

$$\frac{\delta n}{n_0} \sim k^2 R^2 \frac{\omega_{LH}}{\omega_{ce}} \sim \sqrt{\frac{m_e}{m_i}}k^2 R^2$$

(4.8)

The second difference between lower-hybrid and Langmuir waves is in equation (4.2) which describes how slow large scale density perturbations are driven via Reynolds stresses caused by short wavelength LH turbulence. Contrary to the Langmuir problem where Reynolds stresses generate an isotropic wave pressure here an anisotropy of magnetized electron motion results in mean shear flow along the magnetic field lines driven by the longitudinal Reynolds stress:

$$f_{\parallel} = -m_e \left\langle \left(\mathbf{v}_{eD} \underline{\nabla} \right) v_{e\parallel} \right\rangle \tag{4.9}$$

Brackets in this formula correspond to the averaging over the fast time scale $t \sim \omega_{LH}^{-1}$. As well as for the nonlinear frequency shift an adiabatic drift of electrons across magnetic field lines v_{eD} is dominant in the Reynolds stress term resulting in the appearance of the large parameter $\sim \omega_{ce}/\omega_{LH}$ on the r.h.s. of equation (4.2).

Shapiro et al. [1995] have obtained numerical solutions of equations (4.1) and (4.2) for ionospheric parameters and find that the waves undergo quasi-classical collapse resulting in density cavitons and trapped lower-hybrid waves.

The structure of these cavitons are cigar shaped in agreement with the observations. In the numerical solutions a randomisation of the collapsing structures are found. The trapped lower-hybrid waves have two possibilities for dissipating their energy. One is Landau resonance with magnetized electrons $\omega_{LH} = k_{\parallel} v_e$ due to cascade in their perpendicular wavenumber $k_{\parallel}$, resulting in longitudinal electron acceleration; the other one is Landau resonance with unmagnetized ions $\omega_{LH} = k_{\perp} \cdot v_i$ due to the cascade in $k_{\perp}$ leading to transverse ion heating. There are some indications of ion heating from observations which measure field strengths of $E \sim 200\text{-}300$ mV/m inside the cavitons. This is definitely above the threshold for the onset of the collapse process, which is of the order of 50-100 mV/m for typical ionospheric parameters and for perpendicular wavelengths of order $\lambda_{\perp} \sim 2$ m, as measured inside caviton structures. The trapping also results in a lowering of the eigen frequency, which for the quasiclassical case is estimated to be [*Shapiro et al.*, 1995] $\Delta\omega \simeq -0.3\omega_{\parallel}$ for $E \sim 300$mV/m which also agrees with observations. The papers by *Shapiro et al.* [1993, 1995] clearly demonstrate that lower-hybrid waves can be localized through the nonlinear coupling with density structures. The waves inside the cavitons are undergoing collapse that is accompanied by the cascade to larger wave numbers and absorption of these waves by either ions or electrons. In 3-D the main coupling in the equation for density perturbations is the vector type of nonlinearity. For such types of nonlinearity the cavities for lower-hybrid waves are not necessary in the form of density depletions the trapping of lower-hybrid waves is also possible inside density hills and can form dipolar cavitons of the vortex-antivortex type. So far no trapping of waves in density hills has been reported. The resonant wave particle interactions leading to particle acceleration within lowerhybrid cavitons structures is very efficient in producing high energy electrons or ions. The efficiency is higher than that obtained by quasi-linear diffusion. A new type of collisionless damping of cavitons exists [*Bingham et al.*, 1997; *Bingham and Tsytovich*, 2004] which can be described in terms of the usual Landau damping and transition damping [*Bingham et al.*, 1997]. The latter only exists in the presence of spatial inhomogencity of the particle distribution function. Lower-hybrid interaction with particles within the caviton results in inhomogeneous wavefields. The combination of inhomogeneous wavefields and inhomogeneous particle distribution functions gives rise to the new damping term for cavitons called transition damping.

5. Discussion

In summary we have presented a theoretical description of linear and nonlinear lower-hybrid wave processes that may be occurring in the auroral zone and other space environment. We have considered various wave generation mechanisms and possible wave acceleration processes involving lower-hybrid waves which have amplitudes sufficiently large for nonlinear effects to be important. Ion beams observed in the auroral ionosphere could be the result of a strong interaction

352

with kinetic Alfvén waves.The ion beams evolve into a horseshoe distribution which can generate lower-hybrid waves. Lower-hybrid waves are shown to be modulationally unstable producing filamentation of the wave and field aligned striations, which in some cases may explain the observed density structures.

A mechanism for producing lower-hybrid waves from the observed ion horseshoe distributions was presented. These waves are important in their ability to energise both electrons and ions and form in some cases filamentary density structures or cavitons. The structures have been shown through numerical simulations to be stochastic in nature forming at random in configuration space.

The auroral plasma provides an interesting laboratory for studying nonlinear wave phenomena leading to acceleration of electrons and ions.

Acknowledgements

This work was supported partially by a UK, PPARC grant and a European RTN grant.

References

Angelopoulos, V., W. Baumjohann, , C.F. Kennel, F.V. Coroniti, M.G. Kivelson, R. Pellat, R.J. Walker, H. Luher, and G. Paschmann, Bursty bulk flows in the inner central plasma sheet, *J. Geophys. Res.*, *97* (A4), 4027-4039, 1992.

Anderson, R.S., G. K. Parks, T. E. Eastman, D. A. Gurnett, and L. A. Frank, Plasma waves associated with energetic particles streaming into the solar wind from the earth's bow shock, *Geophys. Res.*, *86*, 4493-4510, 1981.

Bingham, R., D. A. Bryant, and D. S. Hall, A Wave model for the aurora, *Geophys. Res. Lett.*, *11*, 327-330, 1984.

Bingham, R., D. A. Bryant, and D. S. Hall, Auroral electron acceleration by lower-hybrid waves, *Annales Geophys.*, *6*, 159 - 168, 1988.

Bingham, R., U de Angelis and V.N. Tsytovich, Theory of Collisionless damping of cavitons and resonant particle acceleration, *J Plasma Physics.*, *Vol 58*, 41-59 (1997)

Bingham, R. and V.N. Tsytovich, Landau damping of lower-hybrid cavitons, *Physica Scripta* (in press)

Bingham, R., B. J. Kellett, R. A. Cairns, R. O. Dendy, and P. K. Shukla, Wave generation by ion horseshoe distributions on auroral field lines, *Geophys. Res. Lett.*, *26* (17), 2713-2716, 1999.

Bryant, D.A., and C. H. Perry, Velocity-space distributions of wave accelerated auroral electrons, *JGR.*, *100*, 23711 - 23725, 1995.

Bryant, D.A., *Electron Acceleration in the aurora and beyond*, IOP Publishing, Bristol, 1999.

Chang, T., and B. Coppi, Lower Hybrid Acceleration and Ion Evolution in the Supra auroral Region, *Geophys. Res. Lett.*, *8*, 1253-1256, 1981.

Dendy, R.O., R. Bingham, B. M. Harvey, and M. O'Brien, Fokker-Planck Modelling of Auroral Wave-Particle Interactions, *JGR, 100* (A11), 21973-21978, 1995.

Dovner, P. O., A. I. Eriksson, andR. Bostrom, Freja multiprobe observations of electrostatic solitary structures, *Geophys. Res. Lett.*, *21*, 1827-1830, 1994.

Dovner, P. O., A. I. Eriksson, R. Bostrom, B. Holback, J. Waldemark, L. Eliasson, and M. Bochm, The occurrence of lower-hybrid cavities in the upper ionosphere, *Geophys. Res. Lett.*, *24*, 619-622, 1997.

Eliasson, L., M. André, A. Eriksson, P. Norqvist, O. Norberg, R. Lundin, B. Holback, H. Koskinen, H. Borg, and M. Boehm. Freja observations of heating and precipitation of positive ions, *Geophys. Res. Lett.*, *21*, 1911-1914, 1994.

Fisch, N.J., *Phys. Rev. Lett.*,Confining a Tokamak Plasma with rf – driven currents *41*, 873-876, 1978.

Grard, R., A. Pederson, S. Klimov, S. Sovin, A. Skalsky, J. G. Trotignon, and C. Kennel, 1st Measurements of plasma waves near mars, *Nature, 341*, No 6243, 607 - 609, 1989.

Hall, D.S., Acceleration of auroral electrons by waves, RAL report 83-028, 1983.

Hultqvist, B., R. Lundin, K. Stasiewicz, L. Block, P.-A. Lindqvist, G. Gustafsson, H. Koskinen, A. Bahnsen, T. A. Potemra, and L. J. Zanetti, Simultaneous observation of upward moving field-aligned energetic electrons and ions on auroral zone field lines, *J. Geophys. Res., 93*, 9765 - 9776, 1988.

Kadomtsev, B.B., and O. P. Pogetse, Electric Conductivity of a plasma in a strong magnetic field. *Sov. Phys. JETP, 26*, 1146 - 1150, 1968.

Kjus, S. H., H. L. Pécseli, B. Lybekk, J. Holtet, J. Trulsen, H. L'uhr, and A. Eriksson, Statistics of the lower-hybrid wave cavities detected by the Freja satellite, *J. Geophys. Res., 103*, 26633 - 26647, 1998.

Kintner, P. M., J. Vago, S. Chesney, R. L. Arnoldy, K. A. Lynch, C. J. Pollock, and T. E. Moore, Localized lower-hybrid acceleration of ionospheric plasmas, *Phys. Rev. Lett., 68*, 2448-2451, 1992.

Liu, C. S., J. L. Burch, J. D. Winningham, and R. A. Hoffmans, De-1 observations of counterstreaming electrons at high altitudes. *Geophys. Res. Lett, 9*, 925-928, 1982.

McBride, J. B., E. Ott, B. Jay, and J. H. Orens, Theory and simulation of turbulent heating by the modified two stream instability, *Phys. Fluids, 15*, 2367-2383, 1972.

McClements, K. G., R. Bingham, J. J. Su, J. M. Dawson, and D. S. Spicer, Lower-Hybrid Resonance Assceleration of Electrons and Ions in Solar Flares and the Associated Microwave Emission, *Astrophys. J., 409*, 465 - 475, 1993.

Pécseli, H. L., K. Iranpour, O. Holter, B. Lybekk, J. Holltet, J. Trulsen, A. Eriksson, and B. Holback, Lower-hybrid wave cavities detected by the Freja satellite, *J. Geophys. Res., 101*, 5299 - 5316, 1996.

Retterer, J. M. and T. Chang, Plasma Simulation of Intense VLF Turbulence and Particle Acceleration in the Suprauroral Region, *Physics of Space Plasmas, SPI Conf. proc. 8*, 309 - 321, 1988.

Rosenbluth, M. N. and R. F. Post, High frequency electrostatic plasma instability inherent to "loss-cone" particle distributions, *Phys. Fluids, 8*, 547-550, 1965.

Scarf, F. L., W. W. L. Taylor, C. T. Russel, and R. C. Elphic, Pioneer Venus plasma wave observations: The solar wind – Venus interactions, *J. Geophys. Res., 85*, 7599 - 7612, 1980.

Shapiro, V. D., V. P. Kalinin, V. I. Shevchenko, G. I. Solovev, R. Bingham, R. Z. Sagdeev, and M. Ashour-Abdalla, Wave Collapse at the Lower-hybrid resonance, *Physics of Fluids, B5* (9), 3148-3162, 1993.

Shapiro, V. D., G. I. Solove'v, J. M. Dawson, and R. Bingham, Lower hybrid dissipative cavitons and ion heating in the auroral ionosphere. *Phys. Plasmas, 2*, 516-526, 1995.

Vago, J. L., P. M. Kintner, S. W. Chesney, R. L. Arnoldy, K. A. Lynch, T. E. Moore, and C. J. Pollock, Trancverse ion-acceledration by localized lower-hybrid waves in the topside auroral ionosphere, *J. Geophys. Res., 97* (A11), 16935 - 16957, 1992.

Zakharov, V. E., Collapse of Langmuir Waves, *Sov. Phys. JETP, 35*, 908 - 914, 1972.

CHAPTER FIVE

MODELING AND COUPLING OF SPACE PHENOMENA

Multiscale Coupling of Sun-Earth Processes
A.T.Y. Lui, Y. Kamide and G. Consolini, editors.

SUBSTORM DISTURBANCE PROPAGATION FROM A TWO-DIMENSIONAL CELLULAR AUTOMATON MODEL

A. T. Y. Lui

*Applied Physics Laboratory, The Johns Hopkins University,
Laurel, MD 20723-6099, USA*

G. Consolini

*1st Fiscia dello Spazio Interplanetario, Consiglio Nazionale delle Ricerche,
Via del Fosso del Cavaliere, 100, 00133 Roma, Italy*

Abstract. Systematic trends of westward motion of auroral surges and poleward expansion of the auroral bulge during substorms have been well established since the early days of auroral observations with all-sky cameras. Consensus on the physical interpretation of these movement trends in the magnetotail has yet to emerge. In this paper, we study a two-dimensional cellular automaton model with simple rules governing the disturbance propagation of avalanches. These rules are set up to mimic the substorm disturbance propagation in the magnetotail from the current disruption scenario. We show significant similarities between the direction of avalanche propagation in this two-dimensional cellular automata model with those observed in substorm auroral disturbance propagation. This preliminary result is interpreted to be consistent with, but not proof of, the view that substorm disturbance propagation in the magnetotail follows the current disruption scenario and it may be governed by the complexity behavior of the magnetosphere.

Keywords. substorm, auroral disturbance movement, avalanche, sandpile.

1. Introduction

Substorm problem is one of the grand challenges for magnetospheric physicists today. The substorm auroral morphology has been well established in the mid-1960's [*Akasofu*, 1964]. The expansion phase of an auroral substorm typically starts with a localized brightening of the most equatorward auroral arcs in the premidnight sector. Subsequently, the auroral disturbance spreads westward in the form of a westward traveling surge, poleward in the form of a bright boundary delineating the poleward extent of the auroral bulge, and eastward in the form of isolated auroral forms such as eastward drifting patches. This evolution of auroral disturbance was pieced together in the early days by a network of all-sky cameras.

Although it is well recognized that these motions of auroral disturbances reflect the propagation of substorm activity in the magnetosphere/magnetotail and presumably bear clues on the physical process responsible for onset of substorms, there is yet no general consensus on the interpretations of these movement trends. An extensive study on the statistical characteristics may help to gain a better insight to the nature of these movement trends. In addition, recent studies of auroral and substorm activities suggest that these disturbances may be manifestation of the magnetosphere behaving like a complex system [*Chang*, 1992; *Consolini*, 1997; *Lui et al.*, 2000;

360

This is followed by updating cells with even indices in the x-direction but not at the boundaries:

$$g(x,y) = g(x,y) - 5,$$
$$g(x+1,y) = g(x+1,y) + 1,$$
$$g(x-1,y) = g(x-1,y) + 1, \qquad (3.4)$$
$$g(x,y+1) = g(x,y+1) + 2,$$
$$g(x,y-1) = g(x,y-1) + 1.$$

This scheme of redistribution is motivated by the physics behind the propagation of current disruption in adjacent tail regions as described by *Lui et al.* [1991]. The current disruption scenario suggests a preference in disturbance propagation towards tailward and duskward. The tailward preference is due to a reduction of the B_z magnetic field component tailward of the current disruption site, providing tailward stretching of the magnetic field line and thus a larger cross-tail current density, which in turn enhances the chance of the current density exceeding the instability threshold to initiate current disruption in the tailward region. In a similar way, the duskward preference arises from the strong field-aligned current generated at the duskward portion of the current disruption site. The field-aligned current is stronger in the dusk side because the current density is stronger there from ions being accelerated by the cross-tail electric field in the duskward direction. This also reduces the B_z magnetic field component duskward of the current disruption site, eventually enhancing the chance of the current density to exceed the current disruption instability threshold. This scenario is illustrated schematically in Fig. 2.

It may be further noted that the substorm model based on multiple current disruptions has a basic ingredient of randomness. The randomness, as explained earlier, arises from thermal fluctuations in the current density of the plasma sheet. Observations clearly show that although the cross-tail current in the plasma sheet decreases progressively down the tail, there is some non-uniformity in the current density as revealed by non-systematic variations of the magnetic fields in the magnetotail. This is consistent with the random selection of a cell to be loaded with sand grains in this model.

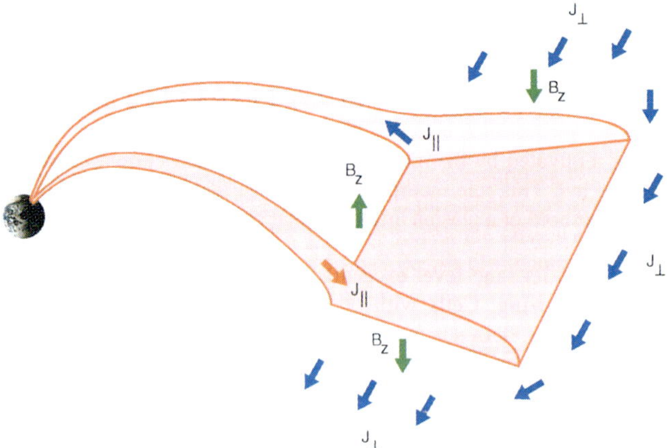

Figure 2. A schematic 3D diagram to illustrate the preference of initiating current disruption in the adjacent tailward and duskward locations. The currents are indicated by blue arrows and the magnetic perturbations associated with the current system are indicated by the green arrows.

In passing we remark that the simulation is performed in a running regime, i.e. the addition of grains of sand is not stopped when an avalanche is observed. This choice is justified by the fact that the magnetotail is continuously perturbed in a quasi-stationary regime.

4. Model Results

The movements of avalanches are followed in the time series created by running the model for 200,000 time steps. The results of the initial 5,000 steps are skipped to avoid effects from the initiation procedure. In other words, the simulation is performed in a nearly stationary regime.

A sample of avalanches found in this numerical sandpile model is shown in Fig. 3. In this time series, time progresses from left to right and top to bottom. For the purpose of clarity, only a quarter of the simulation domain, corresponding to the region closest to the Earth and midnight meridian (i.e., the lower portion of the xy plane), is shown at each time frame.

Cells with values below and at the critical threshold are indicated by blue and green, respectively. Cells with values exceeding the critical thresholds, broadened by a one-cell border for visibility, are given in red. These are the cells that will initiate avalanches to adjacent cells.

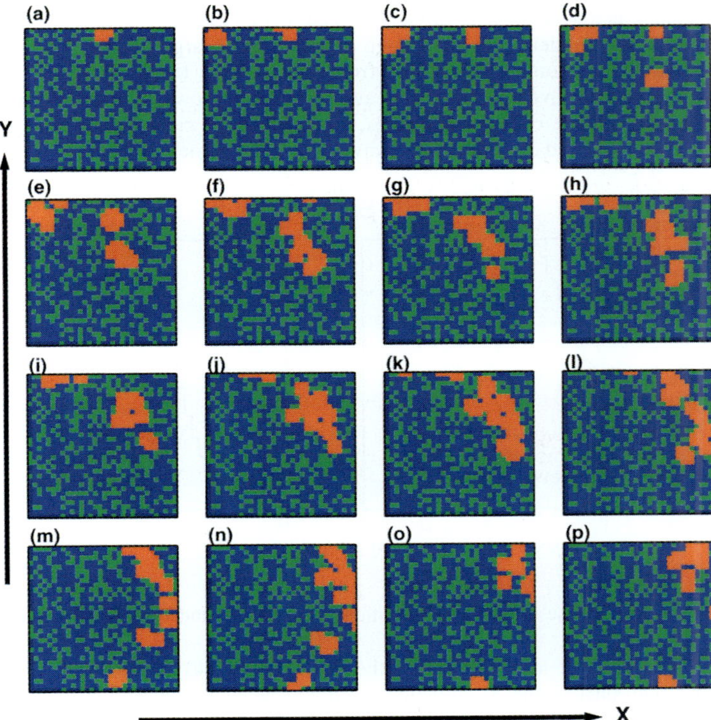

Figure 3. A time series of avalanches found in the sandpile model.

The continuity of these avalanches from frame to frame is tracked in a similar way as the continuity of auroral activity sites in *Lui et al.* [2003]. In other words, the continuation of an avalanche is based on the evolution of its location. The spatial extent of an avalanche is determined by extending each active cell by a border of one-cell thick. Active cells touching each other through this extension are considered to be part of a single avalanche. An avalanche is considered to be the subsequent evolution of one in the previous time frame if it overlaps in area with the previous one. When two or more avalanches merge into a single one in the subsequent time frame, the one in the previous time frame with the largest area is designated as the preceding one while the others are considered as being terminated. When a single avalanche splits into two or more, the subsequent avalanche with the largest area is designated as the continuation of the previous one and the others are assigned as new avalanches.

In this time series, several avalanches are shown. In Fig. 3a, a small avalanche site occurs near the middle top of the box. In the subsequent time frames, it moves mainly toward increasing x coordinate. It disappears in Fig. 3e. In Fig. 3d, a separate avalanche occurs near the middle of the box. It grows in size and splits into two activity sites in Fig. 3e. In the next time frame, these two activity sites merge. It then moves toward increasing x and y coordinates in subsequent time frames, occasionally splitting and merging. In Fig. 3m, a new activity site appears near the middle bottom of the box. It moves toward increasing x coordinate but at the same time toward decreasing y coordinate instead.

Overall, there is a general trend in the averaged motion of avalanches from visual inspection of avalanches in this model. They tend to go to locations of larger x and y coordinates as time progresses. There are notable deviations from this general trend, however. The statistical characteristics on the movements of these avalanches are shown in Fig. 4. The speeds of an avalanche V_x and V_y in the x and y coordinates, respectively, are calculated by the difference in its location at the start and end of its occurrence divided by its lifetime.

2D Sandpile Run

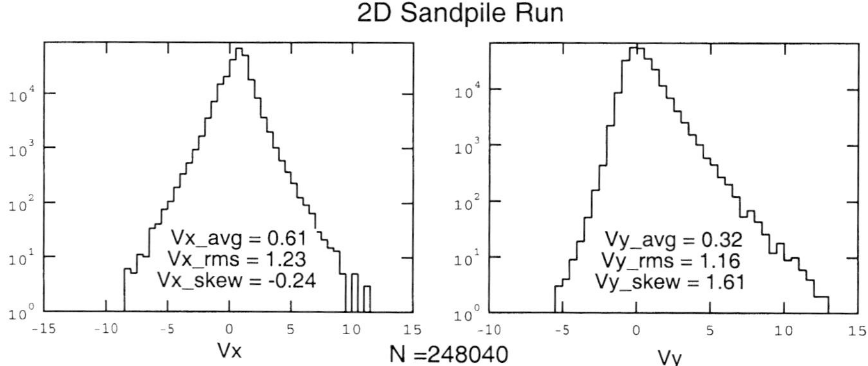

Figure 4. The statistical distributions of speeds V_x and V_y for avalanches

In this model run, we have determined the speed of 248,040 avalanches. The distribution of V_x shows a slight shift in the $+x$ direction, giving an average value of 0.61. Furthermore, the velocity distributions are both non-Gaussian, as also indicated by the value of the kurtosis κ, with the result of $\kappa_x = 8$ and $\kappa_y = 9.5$ for V_x and V_y, respectively. We remind that in the case of a Gaussian distribution the kurtosis κ is expected to be $\kappa = 3$. The non-Gaussian shape of the propagation velocity suggests the occurrence of intermittency in the turbulent evolution of the sandpile. This point will be discussed in detail in a future paper. There is a significant degree of randomness as indicated by the root-mean-square (Vx_rms) and a small amount of skew (Vx_skew). The distribution of V_y is slightly different. The median value of V_y is 0. However, there is an extended tail in the $+y$ direction, giving a positive average value and a large skew. There is also a significant degree of randomness just like that of the V_x value. Overall, there seems to be two components present in the propagation characteristics, namely, a general trend of a definite direction indicated by the positive average values in V_x and V_y and a random component indicated by the large root-mean-square values in the distribution. This is similar to the characteristics of propagation of auroral activity sites reported by *Carbary et al.* [2000]. The difference in V_x and V_y distributions is probably the result of different thresholds and anisotropic addition and evolution rules (Eqs. (3.1) - (3.4)) used in this model. This point requires further analyses.

5. Summary and Conclusions

The preliminary results of a two-dimensional cellular automation model are presented. This model is constructed to mimic the substorm disturbance propagation in the magnetotail based on the current disruption scenario. Simple rules in sand redistribution for avalanche and sand addition to the system are set up to portray the preferential excitation of a current-driven instability in the magnetotail. It is found that the disturbance propagation of activity sites in this simple sandpile model bears similarities to the auroral disturbance propagation observed previously. There seems to be two components in the distributions of propagation direction. One component shows a definite trend of motion and is governed possibly by the trend in the critical threshold of the system as a function of the two coordinates. Another component resembles a random walk that may be related to the complexity of the system. The similarity of these characteristics to those of auroral disturbance propagation in substorms supports the view that substorm disturbance propagation in the magnetotail follows the current disruption scenario for substorms and it may be governed to some extent by the complex behavior of the magnetosphere. However, these results are by no means conclusive evidence for the correctness of the current disruption model in describing the substorm disturbance propagation in the magnetotail since we have only demonstrated the similarities between these two models. It will be interesting to explore the association between these two models more definitively in the future.

One other potential implication of this work is that the statistical characteristics of this simple model may be used to explore the complexity nature of substorm disturbance propagation. Large statistics may be generated from this simple sandpile model for analysis to remedy the scarcity of global auroral data to build a large data set for this kind of study. Future work may also involve investigation on what features in avalanches may be used to gain reliable prediction of their motions, much like what has been done on the prediction of total energy dissipation in auroral activity sites by examining the statistical distributions of parameters associated with an individual auroral activity site at a given time [*Lui et al.*, 2003].

Acknowledgements

The work of A. T. Y. L. is supported by NASA Grant NAG5-10475 and NSF Grant ATM-0135667 to The Johns Hopkins University Applied Physics Laboratory. G. C. thanks the Italian National Research Council (CNR) for the financial support.

References

Akasofu, S.-I., The development of the auroral substorm, *Planet. Space Sci.*, 12, 273-282, 1964.

Carbary, J. F., K. Liou, A. T. Y. Lui, P. T. Newell, and C. I. Meng, "Blob" analysis of auroral substorm dynamics, *J. Geophys. Res., 105*, 16083-16091, 2000.

Chang, T., Low dimensional behavior and symmetry breaking of stochastic systems near criticality - can these effects be observed in space and in the laboratory? *IEEE Trans. Plasma Sci., 20*, 691-694, 1992.

Chapman, S. C., N. W. Watkins, R. O. Dendy, P. Helander, and G. Rowlands, A simple avalanche model as an analogue for magnetospheric activity, *Geophys. Res. Lett., 25*, 2397-2400, 1998.

Consolini, G., Sandpile cellular automata and magnetospheric dynamics, *Proceedings of Cosmic Physics in the Year 2000, 58*, ed. by S. Aiello et al. (eds.), Società Italiana di Fisica, Bologna, Italy, 123-126, 1997.

Lui, A. T. Y., C.-L. Chang, A. Mankofsky, H.-K. Wong, and D. Winske, A cross-field current instability for substorm expansions, *J. Geophys. Res., 96*, 11389-11401, 1991.

Lui, A. T. Y., S. C. Chapman, K. Liou, P. T. Newell, C.-I. Meng, M. Brittnacher, and G. K. Parks, Is the dynamic magnetosphere an avalanching system?, *Geophys. Res. Lett., 27*, 911-914, 2000.

Lui, A. T. Y., W. W. Lai, K. Liou, and C. I. Meng, A new technique for short-term forecast of auroral activity, *Geophys. Res. Lett.*, 30(5), 1258, doi:10.1029/2002GL016505, 2003.

Uritsky, V. M., A. J. Klimas, D. Vassiliadis, D. Chua, and G. Parks, Scale-free statistics of spatiotemproal auroral emissions as depicted by Polar UVI images: The dynamic magnetosphere is an avalanching system, *J. Geophys. Res., 107* (A12), p. SMP 7-1, 1426, doi: 10.1029/2001JA000281, 2002.

Multiscale Coupling of Sun-Earth Processes
A.T.Y. Lui, Y. Kamide and G. Consolini, editors.

A Perspective of E+VxB=0 from the Special Theory of Relativity

George K. Parks

Space Sciences Laboratory, University of California, Berkeley. CA 94720, USA

Abstract We demonstrate from observing how a solar or stellar wind interacts with electromagnetic fields of magnetized objects, like planets, that the ideal MHD equation $\mathbf{E} + \mathbf{V} \times \mathbf{B} = 0$ is not adequate to describe the dynamics of space plasmas. Consider a solar wind that is approaching a magnetized planet. Chapman and Ferraro (1931) showed that a boundary current will be formed as the solar wind interacting with the planetary magnetic field induces an electromotive force ($\mathcal{EMF}$). Ideal MHD fluids do not have this capability since the total magnetic flux is conserved, $d\phi/dt = 0$ and $\mathcal{EMF}$ is not generated. The small electric field in the plasma which is neglected in ideal MHD is necessary for the generation of the boundary current. The special theory of relativity shows the charge density responsible for this electric field is due to Lorentz contraction of the current source in the moving planet as seen in the solar wind frame. The small electric field in the solar wind must be retained if the physics is to be consistent with the relativity principle. Faraday's law provides a correct description of how plasmas interact with electromagnetic fields, relating changes of the magnetic flux to the electromotive force ($\mathcal{EMF}$) that includes both electric and magnetic fields in the plasma frame.

Keywords. Electric Field, Relativity principle, Ideal MHD Approximation

1. Introduction

Space plasmas are always in motion. Therefore, spacecraft measurements must specify the coordinate frames in which the measurements are made. Einstein postulated in 1905 that the laws of electromagnetism obey the principle of relativity which requires physics be the same in all inertial frames of reference. Maxwell's equations are covariant under Lorentz transformation. Electric $\mathbf{E}$ and magnetic $\mathbf{B}$ fields in two different inertial frames, one moving with a velocity $\mathbf{V}$ relative to a stationary frame, transform according to $\mathbf{E}'_{\parallel} = \mathbf{E}_{\parallel}$, $\mathbf{E}'_{\perp} = \gamma(\mathbf{E} + \mathbf{V} \times \mathbf{B})_{\perp}$, $\mathbf{B}'_{\parallel} = \mathbf{B}_{\parallel}$ and $\mathbf{B}'_{\perp} = \gamma(\mathbf{B} + \mathbf{V} \times \mathbf{E}/c^2)_{\perp}$, where $\gamma = 1/\sqrt{(1 - V^2/c^2)}$. Here, the prime ($'$) denotes quantities measured in the moving frame and the subindices $\parallel$ and $\perp$ refer to directions parallel and perpendicular to $\mathbf{V}$. For many classes of space plasma phenomena, the assumption $V \ll c$ is valid and the transformation equations simplify to

$$\begin{aligned} \mathbf{E}' &= \mathbf{E} + \mathbf{V} \times \mathbf{B} \\ \mathbf{B}' &= \mathbf{B} \end{aligned} \tag{1.1}$$

These equations are general and valid regardless of whether space is empty or has plasmas. However, when space is filled with plasmas, additional equations are needed to describe the dynamics. These additional equations are obtained by treating the plasma either as consisting of individual particles or as a continuum fluid. We investigate what special theory of relativity has to say about the behavior of electromagnetic fields associated with the motion of conducting magnetohydrodynamic (MHD) fluids. MHD theory ignores the individual particles and considers only the motion of a fluid element.

366

In space applications, the transformation equation for the electric field is often reduced further to

$$\mathbf{E} + \mathbf{V} \times \mathbf{B} = 0 \tag{1.2}$$

where $\mathbf{V}$ is now the flow velocity of the fluid, which is $\mathbf{V}_\perp = \mathbf{E} \times \mathbf{B}/B^2$ where $\mathbf{E}_\parallel = 0$. Equation (1.2) is also known as the ideal MHD approximation. This equation is the same as (1.1) except that $\mathbf{E}'=0$ and the moving frame is identified as the fluid frame. The justification for setting $\mathbf{E}' = 0$ is that the electric fields in high conductivity plasmas are assumed to be small and negligible. This article will examine what physics is missed by the premise that the electric field in the plasma is so small that it can be ignored from the *outset*. It is true electric fields are small in plasmas, but this article will show that ignoring $\mathbf{E}'$ will severely impact the physics.

It is important to note that the Lorentz transformation theory guarantees us that it is always possible to find a frame of reference in which the electromagnetic field is purely electric or magnetic. However, the physics here is incompatable with the physics of ideal MHD approximation. To obtain such a frame in the Lorentz theory, consider the Lorentz invariants of the transformation $B'^2 - E'^2/c^2 = B^2 - E^2/c^2$ and $\mathbf{E}' \cdot \mathbf{B}' = \mathbf{E} \cdot \mathbf{B}$. where as before $\mathbf{E}$ and $\mathbf{B}$ are fields measured in the stationary frame and $\mathbf{E}'$ and $\mathbf{B}'$ in the moving frame. Let us find the velocity $\mathbf{v}$ of the observer so in this moving frame, the electric field does not appear. Assume $\mathbf{v} = v_x$. Then, use of the relationships of electromagnetic fields $\mathbf{E}' = \mathbf{E} + \mathbf{v} \times \mathbf{B}$ and $\mathbf{B}' = \mathbf{B}$ will show the three components of $\mathbf{E}' = 0$ only if

$$v = \frac{E_y}{B_z} = -\frac{E_z}{B_y} \tag{1.3}$$

Note that since $\mathbf{v}$ is perpendicular to $\mathbf{E}$, $E'_x = E_x$. This pair of equations is satisfied only if

$$
\begin{aligned}
E_y B_y + E_z B_z &= 0 \\
E^2/c^2 &< (B_y^2 + B_z^2)
\end{aligned} \tag{1.4}
$$

These equations show that for $\mathbf{E}' = 0$, $\mathbf{E} \cdot \mathbf{B} = 0$ and the magnitude of $\mathbf{E}$ must be less than the magnitude of $\mathbf{B}/c$ in the frame moving with a velocity given by

$$\mathbf{v}_\perp = \frac{\mathbf{E} \times \mathbf{B}}{B^2} \tag{1.5}$$

In this moving plasma frame, $\mathbf{B}'$ is in the same direction as $\mathbf{B}$ and has the magnitude $\sqrt{B^2 - E^2/c^2}$. This moving frame velocity has exactly the same expression as the ideal fluid velocity given earlier. However, in special relativity theory, this frame is found by applying transformation properties and the Lorentz invariants. It is not necessary to assume electric fields ($\mathbf{E}'$) vanish in the plasma because of the high electrical conductivity as is done in ideal MHD theory. In general, $\mathbf{E} \cdot \mathbf{B} \neq 0$ since the induction process can produce electric fields parallel to the magnetic field direction (Note that a similar derivation can be made for the requirements of the moving frame so that in that frame the electromagnetic field is purely electric. See Parks, 2004.).

Solving a complicated problem in physics often requires that approximations be made to gain insight that would otherwise not be possible. However, one needs to ask what and how much physics is lost in the approximation, and in the case of ideal MHD theory, how important these losses are for explaining the dynamics of space plasmas. In simplifying the problem, it is important to ascertain that the assumptions made do not violate first principles. Our analysis shows the small electric field which the ideal MHD theory ignores is crucial; it is needed to move the particles to produce currents. Without this electric field, the particles experience no force and the particles are stuck on magnetic fields rendering them inactive. Ideal plasmas thus only include *preexisting* electromagnetic fields and currents that are *frozen* in the plasma. However, the concept of frozen-in magnetic fields has limited usefulness in space plasmas. Plasmas in stars and magnetospheres are dynamic and continually changing, forming new currents. We need to study what mechanisms form these new currents which are so central to our understanding of how auroras and flares work.

1.1 A Short History

Magnetohydrodynamic (MHD) theory is the brainchild of Hannes Alfvén for which he received a Nobel prize. MHD theory has been used widely to interpret observations from our solar system to the distant astrophysical regions. However, Alfvén grew increasingly concerned that the scientific community was applying MHD theory incorrectly. He cautioned against the indiscriminate use of the *ideal* MHD theory, which he derived as a special case to describe the hypothetical behavior of plasmas that are imagined to have infinite conductivity (zero resistance). He warned that the ideal theory does not apply to nature's collisionless plasmas (Alfvén, 1977; Fälthammar, 1989; Alfvén and Fälthammar, 1963). Later in his life when the ideal MHD theory had even become the centerpiece of many space models, he said in one of his lectures, "I wish I had not invented the *frozen-in-field* concept."

Eugene Parker is one of the strongest defenders of the ideal MHD theory. He has argued that because space plasmas have high electrical conductivity, electric fields in plasmas are negligible and can be ignored, and space plasma behavior can be described using the ideal theory with the approximation of infinite conductivity. He has used the ideal theory to interpret numerous magnetospheric, solar and astrophysical plasma observations. His well-known picture of the interplanetary magnetic field as being the frozen-in solar magnetic field carried out into space by the solar wind comes from application of the ideal MHD theory. Parker's justifications for his thinking are published in many journals (Parker, 1996; and references therein).

This "controversy" has gone on for over forty years without a clear resolution. The main difference between the two approaches is that Alfvén emphasizes solving for currents and electric fields to study the behavior of plasmas while Parker emphasizes plasma flows and magnetic fields. Alfvén's approach is cumbersome since it requires solving Maxwell's equations in a self-consistent way. Parker solves Maxwell's equations by using the infinite conductivity approximation without regard to self-consistency.

The simplicity of ideal MHD theory is clearly appealing and theorists and modelers have embraced Parker's simpler approach, not heeding Alfvén's warnings. While Parker's theory has been stimulating for space, solar and astrophysics, his results do not describe *real* physics of space plasma dynamics.

2. What Special Theory of Relativity Says

How space plasmas interact with electromagnetic fields can be seen by considering what happens when a solar wind approaches a magnetized planet, like Earth. Chapman and Ferraro (1931) studied this problem and concluded a boundary will be formed because the motion of the solar wind (a conductor) relative to the terrestrial magnetic field induces a current whose magnetic field opposes the planetary magnetic field (Faraday's law). This boundary prevents the planetary magnetic field from penetrating into the solar wind and vice versa. Magnetospheres are formed in this way (we realize that the details of the physics are more complicated because of the presence of ionospheric and magnetospheric particles, but the essential physics remains the same).

Let us analyze this problem from the perspective of the special theory of relativity to see how the solar wind interaction with the planetary magnetic field produces a current. Consider an observer who releases particles with a velocity $\mathbf{v}$ into a uniform magnetic field. This observer sees that positive ions and electrons move in opposite directions due to the Lorentz force $q\mathbf{v} \times \mathbf{B}$. Now what will the observer see who is travelling along with the particles with the same velocity $\mathbf{v}$? In this frame, the particles are stationary and so $q\mathbf{v} \times \mathbf{B} = 0$. But the physics must remain the same (relativity principle) and the observer will still see the positive ions and electrons pushed in opposite directions even though there are no magnetic forces. The observer therefore concludes the particles are pushed in opposite directions by the electric force $q\mathbf{E}$.

We carry the above observations to the solar wind problem. An observer in the planetary frame describes the boundary current formation in the solar wind interaction with the planetary magnetic field as being due to the deflection of solar wind ions and electrons caused by the magnetic force. However, while the observer in the solar wind frame detects a magnetic field, in this frame there is no magnetic force. Hence, this observer concludes there must be an electric force in the solar wind that pushes the particles to form currents. How is this electric field produced?

The source of the magnetic field is the magnetic field of the planet but the source of the electric field is not immediately obvious. Electric fields originate from charges. Where do the charges come from? The observer in the solar wind frame sees the magnetized planet coming toward the solar wind and concludes that somehow the motion of the magnetized planet must be responsible for the electric field. As explained below, special theory of relativity shows that the electric field the solar wind sees comes from the Lorentz contraction of the moving planetary current source observed in the solar wind frame (this is the same current source responsible for the planetary magnetic field).

3. Source of Electric Field in the Solar Wind Frame

Changing reference frame from an Earth-fixed to a solar wind-fixed frame results in a contribution to the electric field which can be interpreted as a space charge. Let us find out how this charge density is produced. The relationship of charge densities in the moving and rest frames for $V \ll c$ is (Jackson, 1975)

$$\rho' = (\rho - \frac{1}{c^2}\mathbf{V} \cdot \mathbf{J}) \tag{3.1}$$

where ρ and $\mathbf{J}$ are charge and current densities in the stationary frame, and the prime ($'$) as before denotes measurements made in the frame moving with a velocity $\mathbf{V}$. The current density $\mathbf{J} \neq 0$ in the planet because a current source is needed to generate the planetary magnetic field. However, we can assume that the total charge density vanishes because charge densities of electrons ρ_- and ions ρ_+ balance. Hence, $\rho = \rho_- + \rho_+ = 0$ and since the planet is not charged, there is no electric field outside the planet.

What charge will an observer see in the solar wind frame? This observer will detect $\rho' \neq 0$ because of the second term in (3.1). Assume now the current source behaves as a conductor and that ions are at rest and the conduction electrons are drifting with a velocity u. Then $\mathbf{J} = \rho_-\mathbf{u}_-$. The charge densities of electrons and ions are not equal in the solar wind frame, $\rho' \neq 0$ because the two sides of the volume in which the currents flow Lorentz contract differently as the planet moves toward the solar wind (one side moves toward and the other side away from the solar wind). Although the charge is invariant in the two frames, the volume experiences a Lorentz contraction due to motion with the consequence the charge density changes. The net charge density seen in the solar wind frame is approximately

$$\rho' = (uV/c^2)\rho_- \tag{3.2}$$

This charge density is a consequence of a relativistic effect and illustrates the difference in physics between stationary and moving conductors.

In the case where the magnetic dipole can be estimated in terms of a current loop, there will be a positive charge density on one side and negative charge density on the other side giving rise to an electric dipole moment, $\mathbf{p}' = \mathbf{m} \times \mathbf{V}/4\pi\epsilon_o$, where $\mathbf{p}$ and $\mathbf{m}$ are electric and magnetic dipole moments. Then, $\mathbf{E}' = (\mathbf{p}' \cdot \mathbf{r}/\mathbf{r}^3)$ and this electric field satisfies the Coulomb equation, $\nabla \cdot \mathbf{E}' = \rho'/\epsilon_o$.

To get an estimate of the charge density, we assume that the electron drift velocities in conductors are $\approx 10^{-2}$ cm/s. Let $V = 400$ km/s. This yields $vV/c^2 = (4/9) \times 10^{-15}$. Assume the conductor is iron and that each iron atom contributes an electron to the conduction band. The number of conduction electrons is $n = Ad/w$ where A is the Avagadro's number (6×10^{23}), w is the atomic weight (55.8 gms/mole) and d is the density ($7.86\,gm/cm^3$). Thus, $n \approx 8.5 \times 10^{22}$ electrons/cm^3. Multiplying this by the charge yields $\rho_- \approx (5/3) \times 10^{10}$ Coulomb/m^3 from which one obtains $\rho' \approx 8 \times 10^{-6}$ Coulomb/m^3. This ρ' is entirely produced by the moving magnetized planet and independent of the solar wind.

Our analysis verifies the conclusions of the observer in the solar wind frame that there is an electric force $q\mathbf{E}'$ that pushes the particles and produces a current in that frame. For the simple example considered here, Lorentz transformation equation shows that $q\mathbf{E}' = q\mathbf{V} \times \mathbf{B}$, which appears the same as the magnetic force the observer sees in the planetary rest frame. In the neighborhood of Earth, $V \approx 400$ km/s, and $B' \approx 100$ nT, yielding a value for $E' \approx 10^{-2}$ V/m.

4. Faraday's Law and Special Theory of Relativity

If we use the ideal approximation $\mathbf{E} + \mathbf{V} \times \mathbf{B} = 0$ in Maxwell equation, $\nabla \times \mathbf{E} = -\partial \mathbf{B}/\partial t$ and integrate this equation over an arbitrary surface area that encloses a magnetic flux $\Phi = \int \mathbf{B} \cdot d\mathbf{A}$ along the contour that is moving with a velocity $\mathbf{V}$, we obtain

$$\frac{\partial}{\partial t} \int \mathbf{B} \cdot d\mathbf{A} + \oint \mathbf{V} \times \mathbf{B} \cdot d\mathbf{l} = 0 \tag{4.1}$$

The first term represents the time rate of change of the magnetic flux through the surface and the second term the additional magnetic flux picked up by the moving contour. This equation states that the total change of the magnetic flux is conserved,

$$\frac{d\Phi}{dt} = 0 \tag{4.2}$$

The concept of frozen-in magnetic field comes from the fact that the total magnetic flux is conserved (Alfvén and Fälthammer, 1963).

If instead we use the correct equation $\mathbf{E}' = \mathbf{E} + \mathbf{V} \times \mathbf{B}$ in $\nabla \times \mathbf{E} = -\partial \mathbf{B}/\partial t$, and follow the same procedure as above, we obtain

$$\frac{d\Phi}{dt} = -\mathcal{EMF} \tag{4.3}$$

where the electromotive force $\mathcal{EMF}$ is defined as

$$\mathcal{EMF} = \oint (\mathbf{E} + \mathbf{V} \times \mathbf{B}) \cdot d\mathbf{l} \tag{4.4}$$

This is Faraday's well-known law of electromagnetic induction which states that the changing magnetic flux induces an $\mathcal{EMF}$. The negative sign in Faraday's law is in accordance with Lenz's law. The $\mathcal{EMF}$ is the line integral of the tangential electric field measured at the element $d\mathbf{l}$ of the moving contour in the plasma. Note that the time derivative is the *total* time derivative, $(d/dt = \partial/\partial t + \mathbf{V} \cdot \nabla)$ which means $\mathcal{EMF}$ is induced even in steady state ($\partial/\partial t = 0$) when the magnetic field is inhomogeneous. Magnetic fields in space are usually inhomogeneous, hence $\mathcal{EMF}$ will be usually associated with space plasma motions (Parks, 2004).

The main difference between (4.2) and (4.3) is that $\mathcal{EMF}$ which is induced in real plasmas is not in ideal plasmas. The $\mathcal{EMF}$ includes the total electromagnetic field, $\mathbf{E} + \mathbf{V} \times \mathbf{B}$, which the particles experience and which vanishes for ideal plasmas. When the force vanishes, the particles cannot move and therefore cannot generate currents. But we know they do. Electric fields in plasmas, while small, must be moving the particles so the plasma can produce currents. Without $\mathcal{EMF}$, the plasma can only push and pull particles existing at the time the "frozen-in-field" assumption is made. Note that Faraday's law is consistent with the theory of special relativity.

5. Momentum Equation

If $\mathcal{EMF}$ is not induced and currents are not generated in ideal MHD approximation, how should the momentum equation

$$\rho_m \frac{d\mathbf{U}}{dt} = -\nabla p + \mathbf{J} \times \mathbf{B} \tag{5.1}$$

be interpreted? MHD fluids differ from ordinary fluids because of the $\mathbf{J} \times \mathbf{B}$ term which produces electrodynamic interactions involving motion, $\mathcal{EMF}$, currents and magnetic fields. However, as shown above, when $\mathbf{E} + \mathbf{U} \times \mathbf{B} = 0$ is used, the fluid cannot induce $\mathcal{EMF}$ and thus no currents and magnetic field can be generated. Without $\mathcal{EMF}$, the $\mathbf{J} \times \mathbf{B}$ refers to preexisting currents and magnetic fields that are frozen in the fluid. Note also that $d\mathbf{U}/dt = 0$ if $\mathbf{U}_\perp = \mathbf{E} \times \mathbf{B}/B^2$ since otherwise there will be inertial terms from time dependence of $\mathbf{E}$ and $\mathbf{B}$ which is not allowed. Thus, $\nabla p = 0$ and pressure is uniform throughout when ideal MHD approximation is made. Recall that in the original derivation of the wave equation (Alfvén and Fälthammer, 1963), they started with σ finite and then later took the limit of infinite conductivity. If one starts from the outset that $\sigma = \infty$, it eliminates the capability of MHD fluids to be electrodynamically interactive.

6. Role of IMF

Faraday's law is general and it says nothing about the interplanetary magnetic field (IMF) or its direction and the induced current is always in such a direction as to oppose the cause that changed the magnetic flux. Thus, for Earth, the magnetopause boundary current always flows from dawn to dusk regardless of the IMF or its direction. The current that is actually formed at the magnetopause is more complicated because of the presence of ionospheric and magnetospheric particles, but Faraday's law correctly provides the essential physics.

Observations indicate that the direction of the interplanetary magnetic field (IMF) is important for the solar wind-geomagnetic field interactions. For example, if the IMF has a component whose direction is antiparallel to Earth's dipole magnetic field direction, more intense magnetic storms and substorms are observed. What is the role of the IMF? The IMF and its direction will affect the structure of the current since particle orbits depend on the magnetic field geometry. For example, particle orbits in the neighborhood of currents with reversals of the magnetic field direction are more complicated and richer than the geometry where the fields all point in the same direction. While the latter geometry has only simple Larmor orbits, the former will include complicated meandering orbits due to null magnetic regions. These different particle orbits contribute differently to the structure of the currents and boundaries. Hence, the IMF direction affects current structures which in turn can influence how the solar wind is coupled to the geomagnetic field across the magnetopause boundary. The suggestion here is that the structure of the current will affect how mass, momentum and energy are transported across boundaries and strongly influences the dynamics inside the magnetosphere.

7. How to Use the MHD Fluid Theory

An improved description of space plasmas in the context of MHD fluid theory is to use the one or two fluid MHD equations without restricting the conductivity (σ) to be infinite (∞). An important equation of the one-fluid MHD theory is the magnetic field equation

$$\frac{\partial \mathbf{B}}{\partial t} = \frac{1}{\mu_o \sigma} \nabla^2 \mathbf{B} + \nabla \times (\mathbf{V} \times \mathbf{B}) \tag{7.1}$$

which is obtained from $\nabla \times \mathbf{E} = -\partial \mathbf{B}/\partial t$ using $\mathbf{J} = \sigma(\mathbf{E} + \mathbf{V} \times \mathbf{B})$ and $\mu_o \mathbf{J} = \nabla \times \mathbf{B}$. The ideal assumption ignores the first term on the right because it vanishes when $\sigma = \infty$. Integration of this equation over an arbitrary surface $d\mathbf{A}$ moving with a velocity $\mathbf{V}$ as was done in (4.1) and (6.1) yields

$$\frac{d\Phi}{dt} = \int \frac{1}{\mu_o \sigma} \nabla^2 \mathbf{B} \cdot d\mathbf{A} \tag{7.2}$$

This is the analog of $\mathcal{EMF}$ in MHD theory and applicable to finite conductivity plasmas. In imperfect fluids, magnetic fields are no longer frozen and there is diffusion of the magnetic field. One might argue that the term on the right of (6.2) is small and neglible, but as shown above, it's not a question of how small this term is. We have shown the neglect of this small term will lead to inconsistent physics.

If a two fluid MHD theory is used, one obtains the generalized Ohm's law (Parks, 2004),

$$\mathbf{E} + \mathbf{V} \times \mathbf{B} = \eta \mathbf{J} + \frac{m_e}{q^2} \frac{\partial}{\partial t}(\mathbf{J}/n) + (\mathbf{J} \times \mathbf{B})/qn + \nabla p_e/qn \tag{7.3}$$

where $m_e \ll m_i$ is assumed and it is further assumed that pressure is isotropic. Here η is the resistivity of the fluid equal to $1/\sigma$, and the subindices (e) denote electrons. Quasi-neutrality condition ($n_e \approx n_i$) is required for this derivation (for pressure not constant, see Krall and Trivelpiece, 1973). The generalized Ohm's law replaces the simple Ohm's law $\mathbf{J} = \sigma(\mathbf{E} + \mathbf{V} \times \mathbf{B})$ used in one-fluid theory. Comparison of this equation with $\mathbf{E}' = \mathbf{E} + \mathbf{V} \times \mathbf{B}$ shows that all of the terms on the right contribute to $\mathbf{E}'$. The relative importance of each term in the generalized Ohm's law is discussed in Alfvén and Fälthammar (1963) Krall and Trivelpiece (1973).

In principle, one- and two-fluid MHD theories can describe space plasmas for phenomena with scale lengths much larger than the ion Larmor radius ($\gg \rho_i$) and time scales much longer than the ion gyrofrequency ($\gg \omega_i$). However, one needs a good model of σ for collisionless plasmas. Collisionless plasmas behave very differently from ordinary conductors and collision dominated plasmas, and the meaning of collisionless conductivity is not clear (Alfvén and Fälthammar, 1963). Thus far, there is *no* satisfactory conductivity model of collisionless plasmas and the numerous results derived from MHD theory and models must be interpreted cautiously as they can be erroneous.

MHD equations have other restrictions since they are derived as moments of the Boltzmann equation. For one- and two-fluid theories, the derivation assumes plasmas are in thermal equilibrium. For anisotropic fluid theory, the Chew-Goldberger-Low (CGL) prescription restricts application of the theory to phenomena when there is no heat conduction along the direction of the mangetic field. Collisionless space plasmas are rarely in thermal equilibrium and they do not generally satisfy the restrictive measures that are incorporated in these fluid theories (Alfvén and Fälthamamar, 1963; Fälthammar, 1989; Lui, 2001; Lui, 2002; Song and Yang, 2001; Heikilla, 1997).

8. Summary and Conclusion

Space plasma interactions require the total electromagnetic force q($\mathbf{E} + \mathbf{V} \times \mathbf{B}$) to act on the particles. This force drives the particles into motion resulting in currents. Physics is consistent only if the electric field in the solar wind frame is included. Faraday's law agrees with the special theory of relativity and provides a correct description of how space plasmas interact, since the $\mathcal{EMF}$ includes both electric and magnetic fields.

Space plasma theories must first formulate with $\mathbf{E}' \neq 0$ (Alfvén, 1977; Fälthammer, 1989) rather than to set $\mathbf{E}' = 0$ from the *outset* as is often done in ideal MHD formulation of space plasmas (Parker, 1996). At some later stage, this small electric field term may be dropped to study how plasma behave in such a limit. Hence, ideal plasmas must be looked at as the limiting case. If one begins formulating plasma theories with $\mathbf{E}' = 0$, the physics is severely limited. When there is no $\mathcal{EMF}$, the particles will not have the capability to move and interact and generate currents. Therefore, such ideal theories can describe only the dynamics of electromagnetic fields and currents already existing in the plasma when the frozen-in-field assumption is made. Ideal theory cannot produce new currents nor can it explain space plasma phenomena such as auroras and flares that require generation of currents and Joule heating ($\mathbf{J} \cdot \mathbf{E} \neq 0$).

MHD fluid theory would be useful for application to space if we had a good model of collisionless plasma conductivity and if space plasmas were in thermal equilibrium. Some modelers might assume that the numerical noise in MHD simulation can play the role of a number of processes outside the framework of ideal MHD and therefore their simulation results using ideal MHD formulation are meaningful. We disagree with this conclusion. The results of any simulation models must be consistent with the theory that defines the physics in the simulation. Numerical noise is artificial and numerical resistivities do not mimic resistivities of real plasmas. Thus, any result that arises as a consequence of numerical noise is meaningless and not real.

More and more observations are showing plasma features that are the result of kinetic physics in action. In the solar wind, multiple component plasmas are observed (Marsch 1982; Lin et al., 1997). Recently, plasmas at the magnetopause region were shown to violate the frozen-in-field condition (Mozer et al., 2002). Understanding these observations require kinetic theory. We recognize that at the present time analytical solutions to kinetic equations are available for only simple models. Kinetic simulation models are a powerful tool but here we are significantly constrained by extreme demands on computational resources and progress has been slow. However, we urge space theorists to develop kinetic models. Note that recent studies of the solar wind using kinetic models have been producing exciting physics not previously known (Scudder, 1994; Meyer-Vernet, 1999; Maksimovic et al., 2001; Issautier et al., 2001; Pierrard et al, 2001; Zouganelis et al, 2003).

We need to understand how space plasmas work using theories that do not violate first principles. The challenge to space physicists is to develop new understanding of space plasmas based on kinetic theory. Data from the current Cluster mission and future missions Themis and MMS interpreted in the context of kinetic physics will greatly enhance our knowledge of space plasmas. These data are needed to study the critical role currents play in dynamic situations. Only then will we understand the causes of the dynamic phenomena occurring in magnetospheres, stars and other astrophysical plasmas.

Acknowledgement

This research is supported by NASA grants to the Space Sciences Laboratory, UC Berkeley, NAG5-10131 and NAG5-11416. The author would like to thank Drs. Michael McCarthy, Carl Fälthammar and Mark Wilber for useful discussions.

References

Alfvén and Fälthammar, Cosmic Electrodynamics, Fundamental Processes, 2nd ed., *Oxford Press*, Oxford, England, 1963.

Alfvén, H., Electric currents in cosmic plasmas, *Rev. Geophys. Space Phys.*, *15*, 271, 1977.

Chapman, S., and V. C. A. Ferraro, A new theory of magnetic storms, *Terr. Magn. Atmos. Elec*, *36*, 171, 1931.

Fälthammer, C. G., Electric fields in the magnetosphere-A review, *Planet. Space Sci.*, *37*, 171, 1989.

Heikkila, W., Comment on alternative paradigm for magnetospheric physics by E. N. Parker, *J. Geophys. Res.*, *5*, 9809, 1997.

Issautier, K., N. Meyer-Vernet, V. Pierrard, and J. Lemaire, Electron temperature in the solar wind from a kinetic collisionless model: application to high latitude Ulysses observations, *Astrophys. Space Sci.*, *277*, 189, 2001.

Jackson, J. D., Classical Electrodynamics, *John Wiley and Sons*, New York, New York, 1975.

Krall, N. and A. Trivelpiece, Principles of Plasma Physics, *McGraw Hill Book Co.*, New York, NY, 1973.

Lin, R. P., D. Larson, R. Ergun, J. McFadden, C. Carlson, T. Phan, S. Ashford, K. Anderson, M. McCarthy, G. K. Parks, H. Réme, J. Bosqued, C. d'Uston, T. Sanderson, K. Wenzel, Observations of the solar wind, the bow shock and upstream particles with the Wind 3D plasma instrument, *Adv. Space Res.*, *20*, 645, 1997.

Lui, A. T. Y., Current controversies in magnetospheric physics, *Rev. of Geophysics*, *39*, 535, 2001.

Lui, A. T. Y.,The Correct Approach to Understanding Magntospheric Physics, *EOS, Trans. AGU*, *83*, October 8, 2002.

Maksimovic, M., V. Pierrard, and J. Lemaire, On the exospheric approach for the solar wind acceleration, *Astrophys. Space Sci.*, *277*, 181, 2001.

Marsch, E., Solar wind proton: Three dimensional velocity distributions and derived plasma parameters measured between 0.3 AU and 1 AU, *J. Geophys. Res.*, *87*, 181, 1982.

Meyer-Vernet, N., How does the solar wind blow? A simple kinetic model, *Euro. J. Phys.*, *20*, 167, 1999.

Mozer, F., S. Bale, and T. Phan, Evidence of diffusion regions at a subsolar magnetopause crossing, *Phys. Rev. Lett.*, *89*, 015002(4), 2002.

Parker, E. N., The alternative paradigm for magnetospheric physics, *J. Geophys. Res.*, *101*, 10,587, 1996.

Parks, George K., Physics of Space Plasmas, an Introduction, Second Edition, *Westview Press*, Boulder, Colorado, 2004.

Pierrard, V., K. Issautier, N. Meyer-Vernet, and J. Lemaire, Collisionless model of the solar wind in a spiral magnetic field, *Geophys. Res. Lett.*, *28*, 223, 2001.

Scudder, J., Ion and electron suprathermal tail strengths in the transition region: Support for the velocity filtration model on the corona, *Space Sci. Rev.*, *95*, 446, 1994.

Song, Y. and R. Lysak, Towards a new paradigm:From a quasi-steady description to a dynamical description of the magnetosphere,*Space Sci. Rev.*, *95*, 273, 2001

Zouganelis, I., M. Maksimovic, N. Meyer-Vernet, H. Lamy, and V. Pierrard, A new exospheric model of the solar wind acceleration: the transition solution, *Solar Wind 10*, 2003.

Multiscale Coupling of Sun-Earth Processes
A.T.Y. Lui, Y. Kamide and G. Consolini, editors.

ENERGIZATION OF IONS BY BIMODAL INTERMITTENT FLUCTUATIONS

Sunny W. Y. Tam and Tom Chang

Center for Space Research, Massachusetts Institute of Technology, Cambridge, MA 02139, USA

Abstract. Sporadic and localized interactions of coherent structures in magnetized plasmas can be the source for the observed broadband, low frequency electric field fluctuations composed of nonpropagating spatiotemporal oscillations and propagating modes. We demonstrate that the particle interactions with this type of bimodal intermittent turbulence can lead to the efficient energization of the auroral ions. For intermittent turbulence, the probability distributions of the fluctuations are non-Gaussian and the effects of the intermittency can manifest in the higher order correlations beyond the second order diffusion coefficient. We have performed global auroral conic simulations to study the effect of intermittency on particle energization processes.

Keywords. intermittent fluctuations, ion acceleration.

1. Introduction

It has long been recognized that the commonly observed broadband, low frequency electric field fluctuations are responsible for the acceleration of oxygen ions in the auroral zone. In order for the fluctuating electric field to resonantly accelerate the ions continuously as the ions evolve upward along the field lines, they must be in continuous resonance with the ions. There does not seem to exist a fully viable mechanism that can generate a spectrum of fluctuations broadband and incoherent enough to fulfill this stringent requirement.

Recently, it has been demonstrated that the sporadic and localized interactions of coherent structures arising from the plasma resonances in magnetized plasmas can be the source for the observed broadband, low frequency electric field fluctuations [*Chang*, 2001] as well as co-existing nonpropagating and propagating fluctuations [*Chang et al.*, 2004]. We demonstrate that the particle interactions with this type of bimodal intermittent turbulence can lead to the efficient energization of the auroral ions, which was observed, for example, in early missions such as the Dynamics Explorer and the Viking satellites [*Chang et al.*, 1986; *Retterer et al.*, 1987; *Hultqvist et al.*, 1988].

2. The Diffusion Approximation

Assuming the oxygen ions are test particles, they would respond to the transverse electric field fluctuations $\mathbf{E}_\perp$ near the oxygen gyrofrequency locally. As the resonant effect on non-relativistic particles due to the magnetic field fluctuations is small by comparison with the electric field counterpart, the ions respond according to the following Langevin equation:

$$d\mathbf{v}_\perp / dt = q_i \mathbf{E}_\perp / m_i .$$

$$(2.1)$$

To understand the stochastic nature of the Langevin equation, we visualize an ensemble of ions $f(\mathbf{v}_\perp)$ and study its stochastic properties. Assuming that the interaction times among the particles and the local electric field fluctuations are small compared to the global evolution time, we may write within the interaction time scale:

$$f(\mathbf{v}_\perp, t + \Delta t) = \int f(\mathbf{v}_\perp - \Delta\mathbf{v}_\perp, t) P_t(\mathbf{v}_\perp - \Delta\mathbf{v}_\perp, \Delta\mathbf{v}_\perp) d\Delta\mathbf{v}_\perp \, , \qquad (2.2)$$

where $P_t(\mathbf{v}_\perp - \Delta\mathbf{v}_\perp, \Delta\mathbf{v}_\perp)$ is the normalized transition probability of a particle whose velocity changes from $\mathbf{v}_\perp - \Delta\mathbf{v}_\perp$ to $\mathbf{v}_\perp$ in Δt, and $\Delta\mathbf{v}_\perp$ ranges over all possible magnitudes and transverse directions. The standard procedure at this point is to expand both sides of (2.2) in Taylor series expansions:

$$\frac{\partial f}{\partial t} \Delta t + O((\Delta t)^2) = -\frac{\partial}{\partial \mathbf{v}_\perp} \cdot \left[\langle \Delta\mathbf{v}_\perp \rangle f \right]$$

$$+ \frac{1}{2} \frac{\partial^2}{\partial \mathbf{v}_\perp \partial \mathbf{v}_\perp} : \left[\langle \Delta\mathbf{v}_\perp \Delta\mathbf{v}_\perp \rangle f \right] + O((\Delta\mathbf{v}_\perp)^3) \, , \qquad (2.3)$$

where

$$\langle \Delta\mathbf{v}_\perp \rangle = \int P_t(\mathbf{v}_\perp, \Delta\mathbf{v}_\perp) \Delta\mathbf{v}_\perp d(\Delta\mathbf{v}_\perp) \, , \qquad (2.4)$$

$$\langle \Delta\mathbf{v}_\perp \Delta\mathbf{v}_\perp \rangle = \int P_t(\mathbf{v}_\perp, \Delta\mathbf{v}_\perp) \Delta\mathbf{v}_\perp \Delta\mathbf{v}_\perp d(\Delta\mathbf{v}_\perp) \, . \qquad (2.5)$$

If we assume the $O((\Delta\mathbf{v}_\perp)^3)$ terms are of order $(\Delta t)^2$ or higher, then in the limit of $\Delta t \to 0$, we obtain a Fokker-Planck equation [Einstein, 1905; Chandrasekhar, 1943] where the drift and diffusion coefficients are defined as: $\mathbf{D}_1 = < \Delta\mathbf{v}_\perp > / \Delta t$ and $\mathbf{D}_2 = < \Delta\mathbf{v}_\perp \Delta\mathbf{v}_\perp > / 2\Delta t$ in the limit of $\Delta t \to 0$. These coefficients may be calculated straightforwardly using the Langevin equation. If the transition probability P_t is symmetric in $\Delta\mathbf{v}_\perp$, then $\mathbf{D}_1$ vanishes and (2.3) leads to a diffusion equation in the transverse direction. We note that if the electric field fluctuations are Gaussian, then the higher order correlations of the fluctuations are automatically equal to zero.

3. Global Evolution of Auroral Ions

We shall come back to the discussion of the effects of general intermittent fluctuations on particle energization processes. For the moment, let us assume the approach using the Fokker-Planck formulation is valid and proceed. Since we have assumed the time scale for the particle-fluctuation interactions is much smaller than the global evolution time of the ion populations, we may then write the steady global evolution equation along an auroral field line s under the guiding center approximation and neglecting the cross-field drift as [*Chang et al.*, 1986; *Retterer et al.*, 1987; *Crew and Chang*, 1988]:

$$\frac{\partial}{\partial s}\left[v_\parallel \frac{f}{B_z} \right] + \frac{\partial}{\partial v_\parallel}\left[-\frac{v_\perp^2}{2B_z} \frac{dB_z}{ds} \frac{f}{B_z} \right] + \frac{1}{v_\perp} \frac{\partial}{\partial v_\perp}\left[v_\perp \frac{v_\perp v_\parallel}{2B_z} \frac{dB_z}{ds} \frac{f}{B_z} \right] = \frac{1}{v_\perp} \frac{\partial}{\partial v_\perp}\left[v_\perp D_\perp \frac{\partial}{\partial v_\perp} \frac{f}{B_z} \right] \, , (3.1)$$

where $v_{\parallel}, v_{\perp}$ are the parallel and perpendicular components of the particle velocity with respect to the field-aligned direction. This expression may be interpreted as a convective-diffusion equation for the density of the guiding center ions per unit length of flux tube, which is proportional to f / B_z, in the coordinate space of $(s, v_{\parallel}, v_{\perp})$.

To evaluate $D_{\perp}$, the gyrotropic perpendicular diffusion coefficient, from $\mathbf{D}_2$, we recognize that the electrostatic fluctuations are broadband both in $k_{\perp}$ and ω. Therefore, at all times, some portion of the fluctuations will be in resonance with the ions. The resonance, however, is strongly dependent on the amplitude of the intermittent fluctuations, which varies at different scales and locations. We demonstrate below, as a simple illustrative example, how such resonance interactions may be accomplished by neglecting the Doppler shifts due to $\mathbf{k}$ such that only the intermittent fluctuations clustering around the instantaneous gyrofrequency of the ions provide the main contributions to the diffusion process. Standard arguments then lead to the following expression for the perpendicular diffusion coefficient:

$$D_{\perp} = (\pi q_i^2 / 2m_i^2)\left\langle \left|E^2\right|(\Omega_i)\right\rangle_r , \tag{3.2}$$

where $<\left|E^2\right|(\Omega_i)>_r$ is the resonant portion of the average of the square of the transverse electrostatic electric field fluctuations evaluated at the instantaneous gyrofrequency of the ions, Ω_i.

Measurements by polar orbiting satellites indicate that the electric field spectral density Σ follows an approximate power law $\Sigma^{-\alpha}$ in the range of the local oxygen gyrofrequencies, where α is a constant. If we make the additional approximations by assuming that the spectrum observed at the satellite is applicable to all altitudes and choosing the geomagnetic field to scale with the geocentric distance as s^{-3}, we would then expect $\Sigma(\Omega_i, s)$ to vary as $s^{3\alpha}$. Because we have made some rather restrictive resonance requirements for the fluctuations to interact with the ions, we expect the resonant portion of the average of the square of the transverse electrostatic electric field fluctuations to be only a fraction η of the total measured electric field spectral density. Therefore, we arrive at the following approximate expression for the diffusion coefficient:

$$D_{\perp} = (\eta \pi q_i^2 / 2m_i^2)\Sigma_0(s / s_o)^{3\alpha} . \tag{3.3}$$

We have performed global Monte Carlo simulations for Eqs. (3.1) and (3.3) for the conic event discussed by *Retterer et al.* [1987] with $\alpha = 1.7$ and $\Sigma_0 = 1.9 \times 10^{-7} (\text{V/m})^2 \sec/\text{rad}$. The top left panel of Figure 1 shows the measured oxygen velocity distribution contours and the top right panel shows the corresponding calculated contours for $\eta = 1/8$ at the satellite altitude of $s_o = 2R_E$. Thus, with one eighth of the measured electric field spectral density contributing, the broadband electrostatic fluctuations can adequately generate an oxygen distribution function with the energy and shape comparable to that obtained from observations. We have also calculated the oxygen ion distributions for a range of altitudes under the same conditions. Figure 2 is a plot of the average parallel energy versus the average perpendicular energy per oxygen ion as the ions evolve upward along the geomagnetic field line. We note that as the energies increase with altitudes, the ratio of the energies becomes nearly a constant.

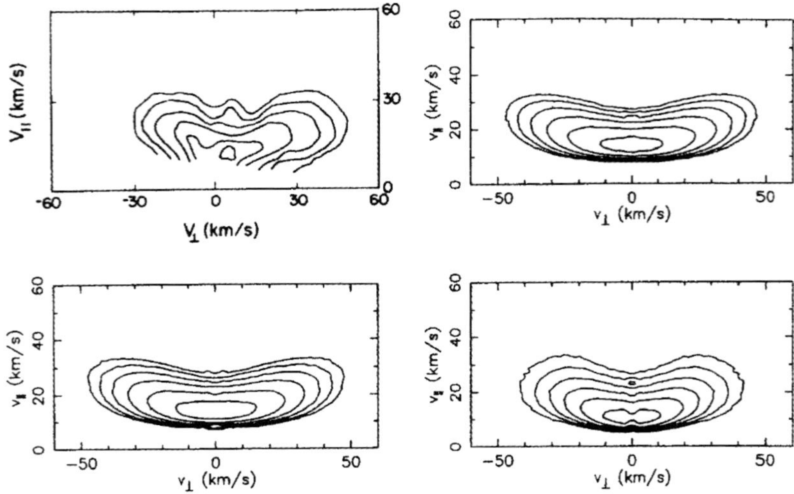

Figure 1. Observed and calculated velocity contour plots for conic event of *Retterer et al.* [1987]. Top left panel: observed contours. Top right panel: simulation results based on the diffusion approximation ($\lambda = 0$). Bottom left panel: simulation results for weak intermittency ($\lambda = 1$). Bottom right panel: simulation results for strong intermittency ($\lambda = 2$).

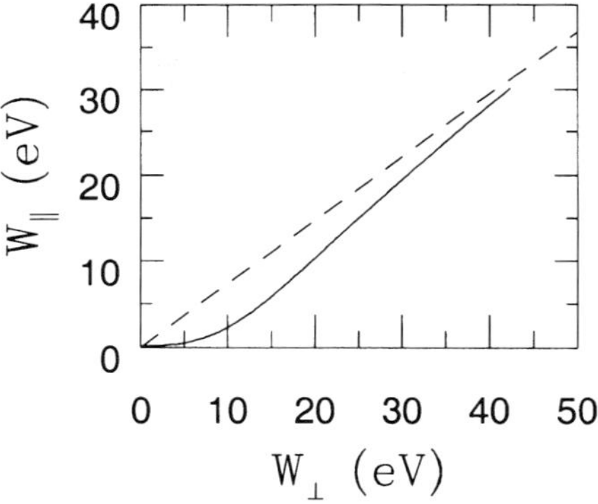

Figure 2. Solid line depicts $W_\parallel$ versus $W_\perp$ for simulated conic events. Dashed line is the asymptote predicted by *Chang et al.* [1986].

These results are comparable to our previous calculated results based on the assumption that the relevant fluctuations were field-aligned propagating electromagnetic ion cyclotron waves [*Chang et al.*, 1986; *Retterer et al.*, 1987; *Crew and Chang*, 1988]. We generally expect the coexistence of nonpropagating transverse electrostatic nonlinear fluctuations and a small fraction of field-aligned propagating waves in the auroral zone. Thus, the ion energization process in the auroral zone is probably due to a combination of both types of fluctuations. As it has been discussed in *Chang et al.* [1986], an asymptotic solution exhibiting such behavior may be obtained analytically in closed form. Therefore, in the asymptotic limit (i.e., at sufficiently high altitudes), it is expected that such an ion distribution will become entirely independent of its low altitude initial conditions. In fact, it has been shown by *Crew and Chang* [1988] that the ion distributions will become self-similar at sufficiently high altitudes and everything will scale with the altitude.

4. Effect of Intermittency

We now return to the discussion of the effect of intermittency on ion heating. Measurements of the electric field spectral density are generally limited by the response capabilities of the measuring instruments. The faster the instruments can collect data, the more refined the scales of the measurements. We expect the measured spectral density to exhibit small-scale intermittency behavior. In fact, it is known that fast response measurements generally exhibit strongly intermittent signatures of the fluctuations. In the diffusion approximation, the ion energization process is limited by the amplitude of the second moment of the probability distribution of the fluctuations. This amplitude may become smaller as the scale of measurements is reduced. Thus, in the limit of small scales, the amplitude of the measured spectrum may decrease and thereby requiring a larger value of η to accomplish the same level of energization.

But, the effects of the intermittency of the fluctuations on particle energization may be underestimated if we stay within the diffusion approximation. As it can be seen from the derivation of the diffusion approximation above, only the second order correlations of the fluctuations were included in the energization process. Since for intermittent turbulence, the probability distributions of the fluctuations are generally non-Gaussian, the effects of the intermittency can manifest in the higher order correlations beyond the second order diffusion coefficient. This implies that the higher order correlations of the velocity fluctuations may be of the order of Δt and therefore cannot be neglected in Eq. (2.3). Under such circumstances, the Fokker-Planck and diffusion approximations of the ion energization processes can become inadequate. A more appropriate approach to address such non-Gaussian stochastic processes is to refer directly to the functional equation (2.2) using the non-Gaussian transition probability or the Langevin equation (2.1) with the actual intermittent time series of the electric field fluctuations.

We have performed global simulations based on Eq. (2.2) for non-Gaussian intermittent fluctuations exhibiting the shape suggested by *Castaing et al.* [1990]:

$$\Pi_\lambda(\xi) = \frac{1}{2\pi\lambda} \int_0^\infty \exp\left(-\frac{\xi^2}{2\sigma^2}\right) \exp\left(-\frac{\ln^2(\sigma/\sigma_0)}{2\lambda^2}\right) \frac{d\sigma}{\sigma^2} , \qquad (4.1)$$

where ξ represents either the x- or y-component of the dimensionless transverse velocity fluctuations and $\lambda > 0$ is a parameter that characterizes the intermittency. We set $\ln \sigma_0 = -\lambda^2$, to ensure the variance equal to unity. For $\lambda = 0$, Eq. (4.1) reduces to a Gaussian distribution. As

380

λ increases, the degree of intermittency increases. The bottom left panel of Fig. 1 shows the contours calculated for $\lambda = 1$ with $\eta = 1/8$. For this case, the degree of intermittency is not strong enough to significantly affect the value of η. But with strong intermittent fluctuations ($\lambda = 2$, bottom right panel of Fig. 1), a value of η equal to 1/5 is required to adequately generate the ion conic to observed energies.

The above sample calculations did not include the self-consistent electric field that must be determined in conjunction with the energization of the ions as well as the electrons. This is particularly relevant in the downward auroral current region where the electric field can provide a significant pressure cooker effect such as that suggested by *Gorney et al.* [1985] and demonstrated convincingly by *Jasperse* [1998] and *Jasperse and Grossbard* [2000] based on global evolutional calculations similar to those considered by *Tam and Chang* [1999a,b; 2001; 2002] for the solar wind and *Tam et al.* [1995; 1998] for the polar wind. We shall consider these ideas in a separate treatise.

Acknowledgements

The authors are indebted to G.B. Crew, J. R. Jasperse, P.M. Kintner, and J.M. Retterer for discussions. This research was partially supported by grants from NASA and NSF.

References

Castaing, B., Y. Cagne, and E.J. Hopfinger, Velocity probability density functions of high Reynolds number turbulence, *Physica D.*, *46*, 177, 1990.

Chandrasekhar, S., Stochastic problems in physics and astronomy, *Rev. Mod. Phys.*, *15*, 1, 1943.

Chang, T., G.B. Crew, N. Hershkowitz, J.R. Jasperse, J.M. Retterer, and J.D. Winningham, Transverse acceleration of oxygen ions by electromagnetic ion cyclotron resonance with broadband left-hand polarized waves, *Geophys. Res. Lett.*, *13*, 636, 1986.

Chang, T., Colloid-like behavior and topological phase transitions in space plasmas: intermittent low frequency turbulence in the auroral zone, *Physica Scripta*, *T89*, 80, 2001.

Chang, T., S.W.Y. Tam, and C.C. Wu, Complexity induced bimodal intermittent turbulence in space plasmas, *Physics of Plasmas*, to be published, 2004.

Crew, G.B., and T. Chang, Path-integral formulation of ion heating, *Physics of Fluids*, *31*, 3425, 1988.

Einstein, A., Ann. d. Physik, *17*, 549, 1905.

Gorney, D.J., Y.T. Chiu, and D.R. Croley, Trapping of ion conics by downward parallel electric fields, *J. Geophys. Res.*, *90*, 4205, 1985.

Hultqvist, B., R. Lundin, K. Stasiewicz, L. Block, P.-A. Lindqvist, G. Gustafsson, H. Koskinen, A. Bahnsen, T.A. Potemra, and L. J. Zanetti, Simultaneous observation of upward moving field-aligned energetic electrons and ions on auroral zone field lines, *J. Geophys. Res.*, *93*, 9765, 1988.

Jasperse, J.R., Ion heating, electron acceleration, and self-consistent E field in downward auroral current regions, *Geophys. Res. Lett.*, *25*, 3485, 1998.

Jasperse, J.R., and N.J. Grossbard, The Alfvén-Fälthammar formula for the parallel E-field and its analogue in downward auroral-current region, *IEEE Trans. Plasma Science*, *28*, 1874, 2000.

Retterer, J.M., T. Chang, G.B. Crew, J.R. Jasperse and J.D. Winningham, Monte Carlo modeling of ionospheric oxygen acceleration by cyclotron resonance with broadband electromagnetic turbulence, *Phys. Rev. Lett.*, *59*, 148, 1987.

Tam, S.W.Y. and T. Chang, Kinetic evolution and acceleration of the solar wind, *Geophys. Res. Lett.*, *26*, 3189, 1999a.

Tam, S.W.Y. and T. Chang, Solar wind acceleration, heating, and evolution with wave-particle interactions, *Comments on Modern Physics,* Vol. I, Part C, 141, 1999b.

Tam, S.W.Y. and T. Chang, Effect of electron resonant heating on the kinetic evolution of the solar wind, *Geophys. Res. Lett., 28,* 11351, 2001.

Tam, S.W.Y. and T. Chang, Comparison of the effects of wave-particle interactions and the kinetic suprathermal electron population on the acceleration of the solar wind, *Astronomy & Astrophysics, 395,* 1001, 2002.

Tam, S.W.Y., F. Yasseen, T. Chang, and S.B. Ganguli, Self-consistent kinetic photoelectron effects on the polar wind, *Geophys. Res. Lett., 22,* 2107, 1995.

Tam, S.W.Y., F. Yasseen, and T. Chang, Further development in theory/data closure of the photoelectron-driven polar wind and day-night transition of the outflow, *Ann. Geophys., 16,* 948, 1998.

Multiscale Coupling of Sun-Earth Processes
A.T.Y. Lui, Y. Kamide and G. Consolini, editors.

A PHYSICS-BASED SOFTWARE FRAMEWORK FOR SUN-EARTH CONNECTION MODELING

Gabor Toth[1,2], Ovsei Volberg[1], Aaron J. Ridley[1], Tamas I. Gombosi[1],
Darren L. De Zeeuw[1], Kenneth C. Hansen[1], David R. Chesney[1], Quentin
F. Stout[1], Kenneth G. Powell[1], Kevin J. Kane[1], Robert C. Oehmke[1]

[1]*Center for Space Environment Modeling, The University of Michigan, Ann Arbor, MI
48109-2143, USA*
[2]*Department of Atomic Physics, Eotvos University, Budapest 1117, Hungary*

Abstract. The Space Weather Modeling Framework (SWMF) has been developed to provide NASA and the modeling community with a high-performance computational tool with "plug-and-play" capabilities to model the physics from the surface of the Sun to the upper atmosphere of the Earth. Its recently released working prototype includes five components for the following physics domains: Inner Heliosphere, Global Magnetosphere, Inner Magnetosphere, Ionosphere Electrodynamics and Upper Atmosphere. The SWMF is a structured collection of software building blocks to develop components for Sun-Earth system modeling, to couple them, and to assemble them into applications. A component is created from the user-supplied physics module by adding a wrapper, which provides the control functions, and coupling interface to perform the data exchange with other components. In its fully developed form, the SWMF will incorporate several more components (for example Solar Energetic Particles and Radiation Belt). It can also incorporate different versions – developed by the Sun-Earth modeling community – for each of the components. The SWMF Control Module is responsible for component registration, processor layout for each component, execution, and coupling schedules. We discuss the SWMF architecture, physics and implementation of component coupling, and results of some preliminary simulations that involve all five components.

Keywords. Space weather, software framework, numerical modeling, model coupling, component based software.

1. Introduction

The Sun-Earth system is a complex natural system of many different interconnecting elements. The solar wind transfers significant mass, momentum and energy to the magnetosphere, ionosphere, and upper atmosphere, and dramatically affects the physical processes in each of these physical domains. One example of such interaction effects is the population of highly energetic particles, which are produced in the inner magnetosphere. These particles precipitate into the upper atmosphere causing the aurora. A variety of complex electric currents in the magnetosphere, ionosphere, and upper atmosphere is another example of these interaction effects.

The various domains of the Sun-Earth system can be simulated with stand-alone models if simplifying assumptions are made about the interaction of the particular domain with the rest of the system. Sometimes the effects of the other domains can be taken into account by the use of satellite and ground based measurements. In other cases statistical and/or phenomenological models can be used. For the prediction of extreme space weather events, however, one must use first principles based physics models for all the involved domains and these models must be run and coupled in an efficient manner, so that the simulation can run faster than real-time. The ability to simulate and predict space weather phenomena is important for many applications, for instance the success of spacecraft missions and the reliability of satellite communication equipment. In extreme cases, the magnetic storms may have significant effects on the power grids used by millions of households.

As an illustrative example of modeling multiple domains of the Sun-Earth system with a highly integrated numerical code, we describe the evolution of the space plasma simulation program BATS-R-US developed at the University of Michigan. Originally BATSRUS was designed as a very efficient, massively parallel MHD code for space physics applications [*Powell et al*, 1999; *Gombosi et al*, 2001]. It is based on a block adaptive Cartesian grid with block based domain decomposition, and it employs the Message Passing Interface (MPI) standard for parallel execution. Later the code was coupled to an ionosphere model [*Ridley et al*, 2001], various upper atmosphere models [*Ridley et al*, 2003] and the inner magnetosphere model [*De Zeeuw et al*, 2003]. These couplings were done in a highly integrated manner resulting in a monolithic code, which makes it rather difficult to select an arbitrary subset of the various models, to replace one model with another one, to change the coupling schedules of the interacting models, and to run these models concurrently on parallel computers. Thus, although BATS-R-US is successfully used for the global MHD simulation of space weather [*Gombosi et al*, 2004], monolithic programs like it have their limitations.

The Center for Space Environment Modeling (CSEM) at the University of Michigan and its collaborators are building a Space Weather Modeling Framework (SWMF). The SWMF is designed to couple the models of the various physics domains in a flexible yet efficient manner, which makes the prediction of space weather feasible on massively parallel computers. Each model has its own dependent variables, a mathematical model in the form of equations of evolution, and a numerical scheme with an appropriate grid structure and temporal discretization. The physics domains may overlap each other or they can interact with each other through a boundary surface. The SWMF should be able to incorporate models from the community and couple them with minimal changes in the software of an individual model. In this paper we present the design and implementation of the first working prototype of the SWMF.

2. Architecture of the SWMF

The SWMF aims at providing a flexible and extensible software architecture for multi-component physics-based space weather simulations, as well as for various space physics applications [*G. Toth et al*, 2003; *Volberg et al*, 2003]. The main SWMF design goals were defined in *Volberg et al* [2002] as:

- Incorporate computational physics modules with only modest modification.
- Achieve good parallel performance in the coupling of the physics components.
- Allow physics components to interact with the SWMF as efficiently as possible.

One of the most important features of the SWMF is that it can incorporate different computational physics modules to model different domains of Sun-Earth system. Each module for a particular domain can be replaced with alternatives, and one can use only a subset of the modules if desired.

There are several a priori known problems, which need to be solved so that the heterogeneous computational models of the different domains of the Sun-Earth system can properly inter-operate. An incomplete list of these problems:

1. There are serial and parallel models.
2. An individual model is usually developed for stand-alone execution.
3. Input/output operations do not take into account potential conflicts with other models.
4. Models often do not have checkpoint and restart capabilities.
5. The majority of models are written in non-object oriented languages (e.g. Fortran 77 and Fortran 90), which means that data and procedure name conflicts can easily occur.
6. The efficient coupling of any arbitrary pair of parallel applications, each of them having its own grid structure and data decomposition, is not easily achieved.

There are several potential solutions that provide the necessary interoperability mechanism between parallel modules [*Edjlali et al*, 1997]. The most promising approach is to define a standard set of interface functions that every physics component must provide. In the SWMF a component is created from a physics module, for example BATSRUS, by making some minimal required changes in the module and by adding two relatively small units of code:

1. A wrapper, which provides the standard interface to control the physics module;
2. A coupling interface, to perform the data exchange with other components.

From a component software technology perspective, both the wrapper and coupling interface are component interfaces: the wrapper is an interface with the high-level Control Module (CON) of the framework, and the coupling interface is an interface with another component. As shown in Figure 1, the wrapper interface functions have standard names, which makes swapping between various versions of a component possible. Both the wrapper and the coupling interface are constructed from the building blocks provided by the framework.

The physics modules must comply with a minimum set of requirements before they are transformed into a component:

1. The parallelization mechanism (if any) should employ the MPI standard;
2. The module needs to be compiled as a library that could be linked to another executable;
3. The module should be portable to a specific combination of platforms and compilers, which include Linux workstations and NASA super-computers [*Volberg et al*, 2002];
4. The stand-alone module must successfully run a model test suite provided by the model developer on all the required platform/compiler combinations;
5. The module should read input from and write output to files which are in a subdirectory unique for the component;
6. A module should be implemented in Fortran 77 and/or Fortran 90;
7. A module should be supplied with appropriate documentation.

The SWMF requirements for a component are defined in terms of a set of methods to be implemented in the component wrapper. The methods enable the component to perform the

386

following tasks:

1. being registered by the Control Module;
2. being initialized in parallel configuration;
3. accept and check input parameters obtained from the Control Module;
4. provide grid description to Control Module;
5. initialize for session execution;
6. execute one time step which cannot exceed a specified simulation time;
7. receive and provide data to another component via an appropriate coupler;
8. write its state into a restart file when requested;
9. finalize at the end of the execution.

The structure of a component and its interaction with the Control Module (CON) and another component are illustrated in Figure 1.

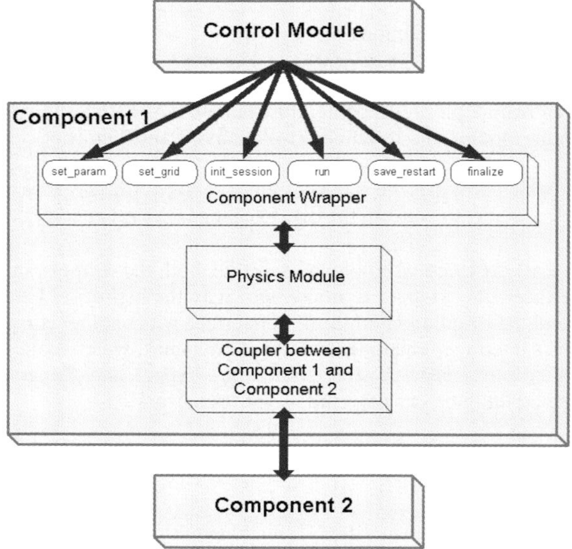

Figure 1. Integration of physics modules into the SWMF architecture

The framework's layered architecture is shown in Figure 2. The Framework Services consist of software units (classes) which implement component registration, session control, and input/output operations of initial parameters. The Infrastructure consists of utilities, which define physics constants and different coordinate systems, time and data conversion routines, time profiling routines and other lower level routines. The Superstructure Layer, Physics Module Layer, and Infrastructure Layer constitute the "sandwich-like" architecture similar to the Earth System Modeling Framework (ESMF) [*Hill et al*, 2004]. The SWMF will also contain a web-based Graphical User Interface, which is not part of the ESMF design.

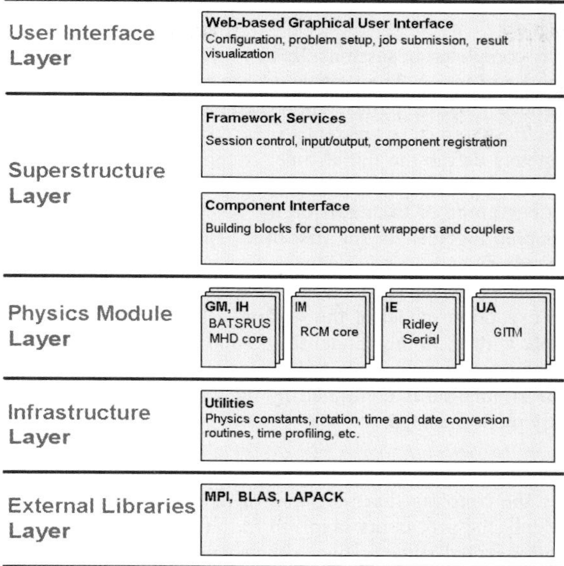

Figure 2. The layered hierarchy of the SWMF

```
ID    first last stride
#COMPONENTMAP
GM     0  999   2
IE     0   1    1
IH     1  999   2
IM    11  11    1
UA    12  15    1
#END
```

Figure 3. An example of the LAYOUT.in file

3. Execution of the SWMF

At the beginning of a run, the SWMF registers the components. The registration provides information about the component to CON, such as the name and the version of the component. In return, the control module assigns an MPI communication group to the component based on the processor layout defined in the LAYOUT.in file. An example of this file is shown in Figure 3. The first column identifies the components by their abbreviated names, while the rest of the columns define the ranks of the first and last processing elements (PE), and the PE stride. In the example shown in Figure 3, the Global Magnetosphere (GM) component runs on every even processor, the Ionospheric Electrodynamics (IE) component runs on the first two processors, the Inner Heliosphere (IH) component runs on every odd processor, the Inner Magnetosphere (IM) component runs on processor 11, and the Upper Atmpsphere (UA) component runs on 4 processors from ranks 12 to 15. Only registered components can be used in the run.

The execution is completed in sessions. In each session the parameters of the framework and the components can be changed. The parameters are read from the PARAM.in file, which may contain further included parameter files. These parameters are read and broadcast by CON and the component specific parameters are sent to the components for reading and checking. The CON related parameters define the initial time, coupling schedules, frequency of saving restart files, final simulation time of the session, and other information which is not restricted to a single component. At the beginning of each session the components are initialized and the interacting components are coupled together for the first time. The SWMF currently contains two different session execution models.

The sequential execution model of the session is based on the BATS-R-US module, and is backward compatible with it. In this model the components are synchronized at every time step, so typically only one component is executing (possibly on many processors) at any given time. The progress of the simulation is controlled by the GM module, and the coupling patterns and schedules are mostly predetermined.

In the concurrent execution model, the components communicate only when necessary. This is possible because the coupling times are known in advance. The components advance to the coupling time and only the processors involved in the coupling need to communicate with each other. In this execution model all components are 'equal', any physically meaningful subset can be selected for the run, and their coupling schedules are determined by the parameters given in the PARAM.in file. The concurrent execution model allows execution of more than one component on the same PE, in which case the components advance in a time-shared manner. The possibility of dead-locks is carefully avoided.

The coupling of the components is realized either with plain MPI calls, which are specifically designed for each pair of interacting components, or via the general SWMF coupling toolkit. The toolkit can couple components based on the following types of distributed grids:

- 2D or 3D block adaptive grid
- 2D or 3D structured grid

Structured grids include uniform and non-uniform spherical and Cartesian grids.

The toolkit obtains the grid descriptors from the components at the beginning of the run. The grid descriptor defines the geometry and parallel distribution of the grid. At the time of coupling the receiving component requests a number of data values at specified locations of the receiving grid (for example all grid points at one of the boundaries). The geometric locations are transformed, sometimes mapped, to the grid of the provider component. Based on the grid descriptor of the provider component, the data values are interpolated to the requested locations and sent back to the requesting component. The interpolation weights and the MPI communication patterns are calculated in advance and saved into a 'router' for sake of efficiency. The router is reused as much as possible in subsequent couplings. The routers are updated only if one of the grids has changed (e.g. due to grid adaptation) or when the mapping between the two components has changed (e.g. due to the rotation of one grid relative to the other). In certain cases the coupling is achieved via an intermediate grid, which is stored by the receiving component, but its structure is based on the providing component. The intermediate grid can be described to CON the same way as the base grid of the receiving component.

As specified in the PARAM.in file, CON instructs the components to save their current state into restart files periodically. This makes possible the seamless continuation of a run from a given point of the simulation. Check-point restart is an essential feature of a robust, user-friendly, and fault-tolerant software design.

At the end of the last session each component finalizes its state. This involves writing out final plot files, closing log files, and printing performance and error reports. After the components have finalized, CON also finalizes and stops the execution.

4. The SWMF Prototype

The recently released working prototype of the SWMF includes components for five physics domains: Global Magnetosphere (GM), Inner Heliosphere (IH), Ionosphere Electrodynamics (IE), Upper Atmosphere (UA) and Inner Magnetosphere (IM):

1. The GM and IH components are based on the University of Michigan's BATS-R-US MHD module. The highly parallel BATS-R-US code uses a 3D block-adaptive Cartesian grid.
2. The IM component is the Rice Convection Model (RCM) developed at Rice University. This serial module uses a 2D non-uniform spherical grid.
3. The IE component is a 2D spherical electric potential solver developed at the University of Michigan. It can run on 1 or 2 processors since the northern and southern hemispheres can be solved in parallel.
4. The UA component is the Global Ionosphere Thermosphere Model (GITM) recently developed at the University of Michigan as a fully parallel 3D spherical model.

A detailed description of the 5 components and their coupling with each other exceeds the limitations of this paper. Below we briefly outline the coupling mechanisms so that the complexity and variability of the couplings may be better appreciated.

4.1 Coupling the Inner Heliosphere to the Global Magnetosphere

The GM is driven by the solar wind flow. One method of accomplishing this is to use an upstream solar wind monitor to feed the upstream boundary conditions of the GM. Alternatively, the GM can be embedded into a full MHD IH model that provides all relevant flow conditions to the GM at the necessary time and location. The GM obtains a temporally and spatially varying set of data at its inflow boundary, which consists of the solar wind density, pressure, velocity, and magnetic field.

4.2 Coupling the Global Magnetosphere and the Inner Magnetosphere

This coupling is challenging, as the IM needs the magnetic field line flux tube volumes and the average density and pressure in the flux tubes connected to its 2D spherical grid points. This requires the integration along many magnetic field lines in GM. In turn GM needs to know where each of its 3D grid points are mapped onto the ionosphere along the magnetic field lines, so that the total pressure calculated by IM can be applied in GM. A parallel method has been developed to accomplish these steps in an efficient manner.

390

4.3 Coupling the Global Magnetosphere and the Ionosphere Electrodynamics

The IE requires the field aligned currents on its 2D spherical grid. The currents are calculated in GM at an appropriate radial distance (e.g. 3.5 Earth radii), and mapped along the dipole field lines of the planet to the ionosphere. In return, IE provides the electric potential on its grid. This potential is mapped back to the inner boundary of GM (typically at 2.5 Earth radii). The potential is used to calculate the electric field and the corresponding plasma velocities, which are used as the inner boundary condition for GM.

4.4 Coupling the Inner Magnetosphere and the Ionosphere Electrodynamics

This is a one-way coupling. IE provides the radial current and the electric potential for the 2D spherical grid of IM. The coupling is complicated by the non-uniformity of the IM grid.

4.5 Coupling the Ionosphere Electrodynamics and the Upper Atmosphere

The upper atmosphere model provides the Pederson and Hall conductivities and field aligned currents (in addition to the field aligned currents obtained from GM) to the IE. On the other hand, the electric potential calculated by IE is mapped to the upper atmosphere model in each grid cell along the magnetic field lines. The gradient of the potential provides the electric field which is used to drive the ion motion.

5. The First Simulation Results

We present some preliminary simulations which involve all five components of the SWMF prototype. This simulation is mainly used as a test bed for the development of the framework, thus the grid resolution is relatively coarse and the setup of the simulation is relatively simple. The first part of the simulation involves the IH, GM and IE components only. Both the IH and GM grids contain about a quarter million cells. For GM the cell sizes range from 0.5 to 4 Earth radii, while for IH they range from 0.125 to 16 Solar radii. The IE component has a 2D spherical grid with an approximately 1.5 degree resolution. All 3 components are run in a steady-state mode, and there is no coupling between IH and GM. The purpose of this start up is to obtain a physically reasonable steady-state solution in a relatively short time, but the steady state solutions do not match at the IH-GM boundary (for a real production run this would not be the case). The steady state solution obtained by the IH component is shown in Figure 4.

The simulation is continued in a time-accurate mode for an hour of physical time. All five components are involved and they are fully coupled in this part of the simulation. Since the steady state obtained with GM and IH do not match at the inflow boundary of GM, the magnetosphere undergoes a severe compression due to the incoming density jump in the solar wind. Figure 5 also shows how the orientation of the upwind magnetic field changes. Meanwhile the interaction of the Global Magnetosphere with the Inner Magnetosphere model causes a gradual increase of the plasma pressure on the night side of the Earth as shown in Figure 6. The compression of the magnetotail also increases the pressure, but the increase confined to the closed field lines is characteristic to the influence of the IM component. Another consequence of the IM-GM coupling is the strengthening of region 2 currents. These currents are used by the Ionosphere Electrodynamics model to solve for the electric potential (Figure 7). Finally we show the electron density and velocity field obtained by the Upper Atmosphere model in Figure 8. The

UA component is strongly dependent on the conductance and electric potential calculated by the IE component. The IM component uses a non-uniform 78 by 48 2D spherical grid, while the UA component runs on a 3D spherical grid with 5 degrees angular resolution and 25 altitude levels.

In the future we plan to apply the SWMF to more realistic problems, which involve the initiation and propagation of Coronal Mass Ejection from the Sun to the Earth. Such a simulation will require finer grid resolution and much longer time evolution.

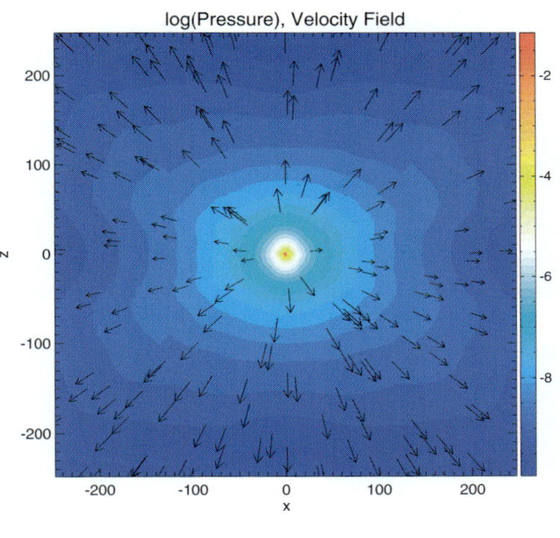

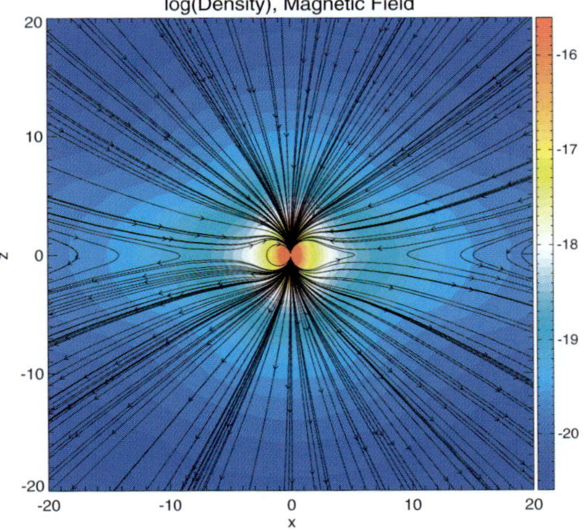

Figure 4. The steady state solution obtained by the Inner Heliosphere model. The top panel depicts the overall pressure and velocity fields out to 256 Solar radii, while the bottom panel

shows the magnetic field and density in the vicinity of the Sun.

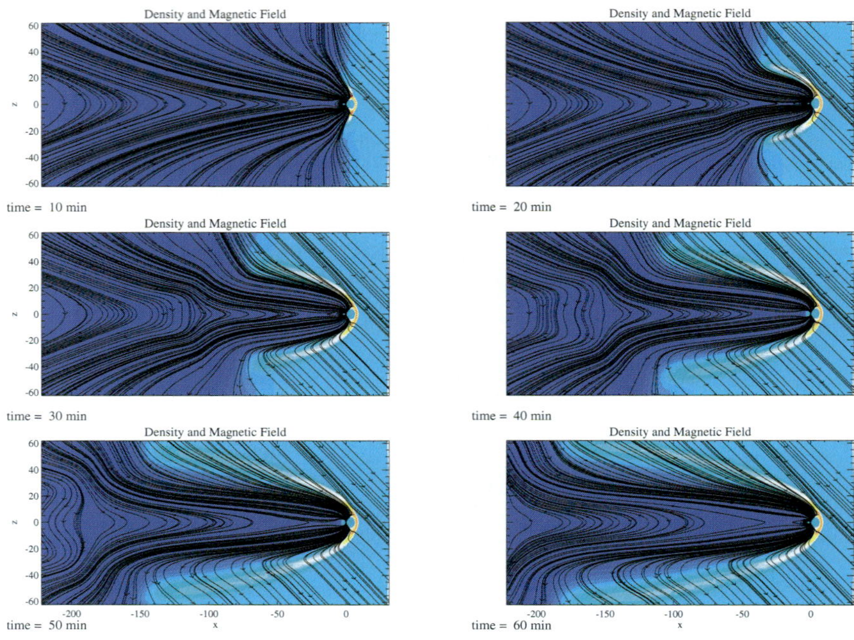

Figure 5. The evolution of the Global Magnetosphere due to the sudden changes in the solar wind density and magnetic field. The units on the axes are in Earth radii.

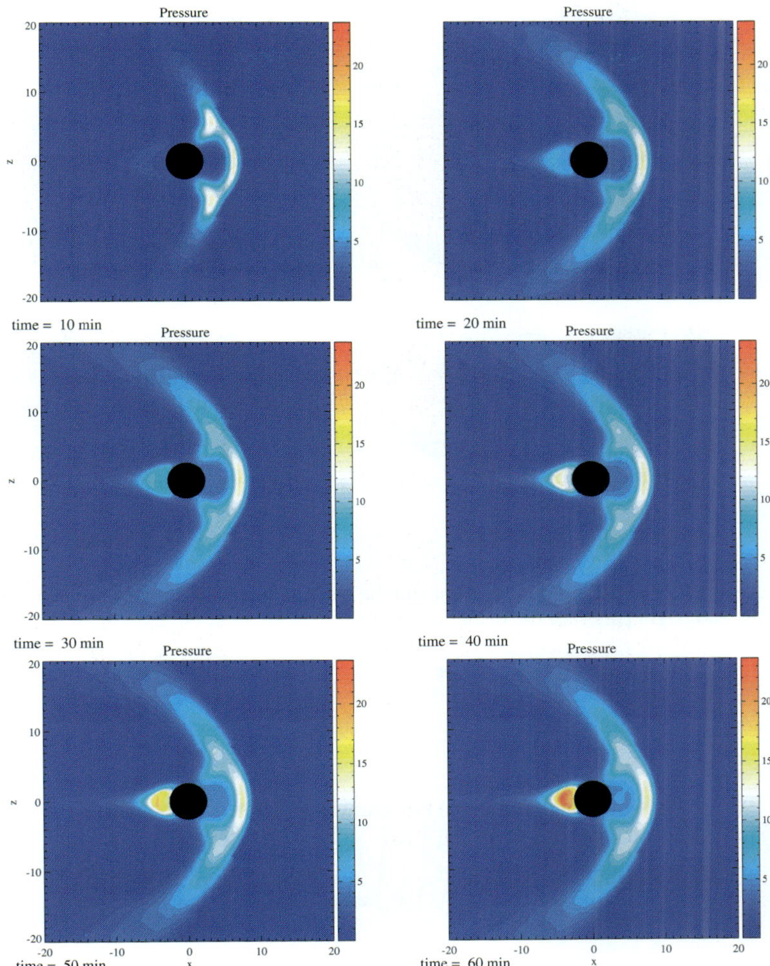

Figure 6. The interaction of GM with IM increases the pressure on the night side of the Earth.

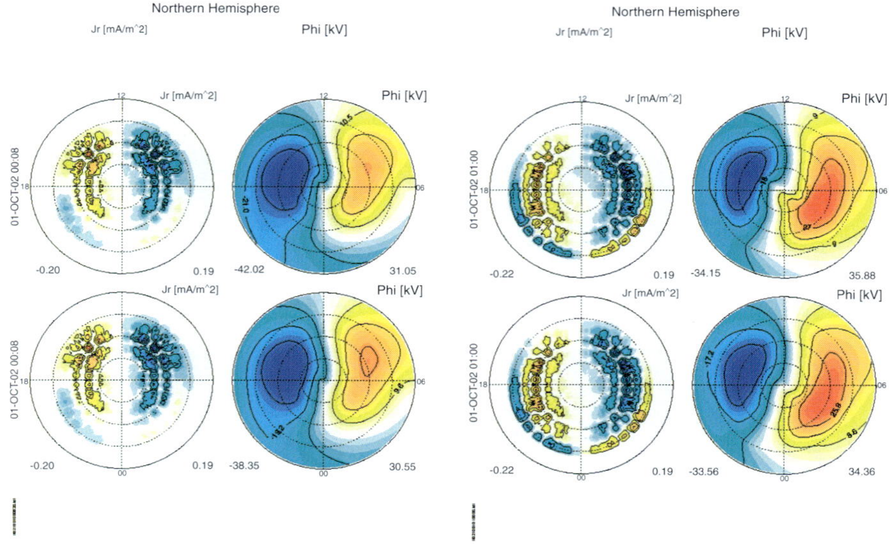

Figure 7. The field aligned current and the electric potential at the beginning (left panel) and at the end (right panel) of the one hour simulation. Note the strengthening of region 2 currents.

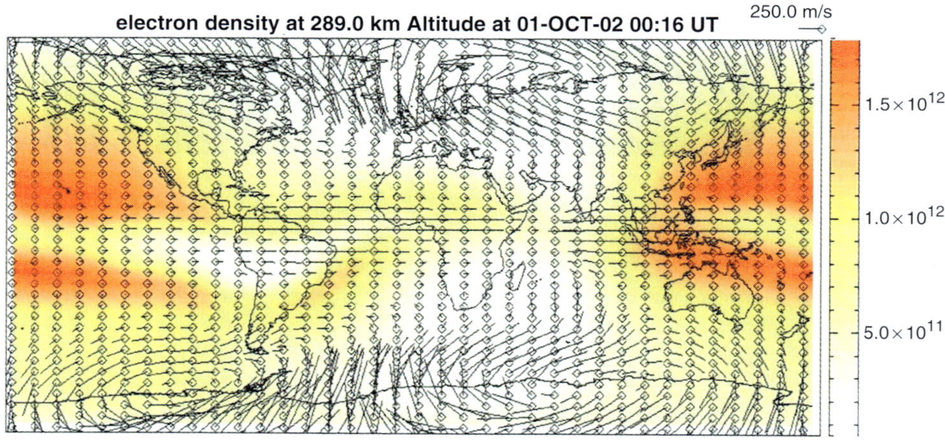

Figure 8. The electron density and velocity field obtained by the Upper Atmosphere model.

6. Conclusions and Future Work

The first results of our work are summarized as follows:

1. In its fully developed form, the SWMF comprises a series of interoperating models of physics domains, ranging from the surface of the Sun to the upper atmosphere of the Earth. In its current form the SWMF links together five physics models: Inner Heliosphere, Global Magnetosphere, Inner Magnetosphere, Ionosphere Electro-dynamics and Upper Atmosphere.
2. The SWMF contains utilities and data structures for creating model components and coupling them.
3. A component is created from the user-supplied physics code by adding a wrapper, which provides the control functions, and a coupling interface which performs the data exchange with other components.
4. The SWMF contains a Control Module, which controls initialization and execution of the components and is responsible for component registration, processor layout for each component, and coupling schedules.
5. The framework allows a subset of the physics components to execute and can incorporate several different models for the same physics domain.
6. The SWMF parallel communications are based on the MPI standard. In its current implementation the SWMF builds a single executable.

The SWMF is not only more flexible, but in some cases more efficient, than a highly integrated code. For example, when only the GM and IE components are used, the performance depends to a large extent on the execution model. The SWMF allows the concurrent execution of the two components. While IE solves for the electric potential on 2 processors, the GM code can proceed with the MHD simulation on the rest of the processors. As a result, the IE solution is applied to the inner boundary of GM with a small time shift, but this is an acceptable approximation. In this concurrent execution model, the SWMF can run almost twice as fast as the serial model with less flexible coupling schedules.

We will add two additional components for new physics domains in the next scheduled release: the Radiation Belt (RB) model developed at Rice University by A. Chan, D. Wolf and Bin Yu) and the Solar Energetic Particle (SP) model developed at the University of Arizona by J. Kota. The coupled GM and IM components will provide the time variations of the electric and magnetic fields in the region occupied by the outer radiation belt. The RB component will use those fields to estimate the time evolution of the outer belts, with particular emphasis on the relativistic electrons that are crucial for space-weather modeling.

The SP component will calculate both the acceleration of charged particles to high energies and the subsequent transport of these particles along the interplanetary magnetic field lines connecting the Earth to the site of acceleration, i.e. the strong shock driven by the CME. The code is an implicit scheme that provides numerical solutions to the Fokker-Planck equation including acceleration, focusing, convection, and random scattering in pitch angle. The code is designed so that the geometry of the magnetic field and the position and strength of the shock are constantly updated as the shock evolves. Multiple shocks are also allowed. The code will compute the predicted time profiles, energy spectra, and pitch-angle distribution of solar energetic particles events at the Earth.

396

Acknowledgments

We would like to express our gratitude to the governors of the Conference on Sun Earth Connection, A. T. Y. Lui and Y. Kamide. We also wish to thank the National Center for Atmospheric Research (NCAR), the NASA Goddard Space Flight Center (GSFC), Rice University and the University of Arizona for collaboration on this project. Our special thanks to J. Kota from the University of Arizona, A. Chan, D. Wolf, S. Sazykin and Bin Yu from Rice University, Cecelia Deluca from NCAR, J. Fisher from NASA GSFC, J. W. Larson and R. Jacob from Argonne National Laboratory. The SWMF project is funded by the NASA Earth Science Technology Office (the NASA CAN NCC5-614 grant). G. Toth is partially supported by the Hungarian Science Foundation (OTKA grant T047042).

References

De Zeeuw, D. L., S. Sazykin, A. Ridley, G. Toth, T. I. Gombosi, K. G. Powell, D. Wolf, Inner Magnetosphere Simulations - Coupling the Michigan MHD Model with the Rice Convection Model, *Fall AGU Meeting*, San Francisco, 2003.

Edjlali, G., A. Sussman, and J. Saltz, Interoperability of Data Parallel Runtime Libraries. In *Proceedings of the Eleventh International Parallel Processing Symposium*, IEEE Computer Society Press, 1997.

Gombosi, T. I., D. L. De Zeeuw, C. P. Groth, K. G. Powell, C. R. Clauer, and P. Song, From Sun to Earth: Multiscale MHD Simulations of Space Weather, in *Space Weather*, edited by P. Song, H. J. Singer, and G. L. Siscoe, vol. 125, pp. 169–176, AGU, 2001.

Gombosi T. I., K. G. Powell, D. L. De Zeeuw, R. Clauer, K. C. Hansen, W. B. Manchester, A. Ridley, I. I. Roussev, I. V. Sokolov, Q. F. Stout, G. Toth, Solution-Adaptive Magnetohydro-dynamics for Space Plasmas: Sun-to-Earth Simulations, *Computing in Science and Engineering, Frontiers of Simulation*, March/April, p. 14-35, 2004.

Hill, C., C. DeLuca, V. Balaji, M. Suarez, A. da Silva, and the ESMF Joint Specification Team, The Architecture of the Earth System Modeling Framework, *Computing in Science and Engineering*, Volume 6, Number 1, 2004.

Powell, K.G., P.L. Roe, T.J. Linde, T.I. Gombosi, and D.L. De Zeeuw, A solution-adaptive upwind scheme for ideal magnetohydrodynamics, *J. Comp. Phys.*, 154, 284, 1999.

Ridley, A. J., D.L. De Zeeuw, T. I. Gombosi, and K.G. Powell, Using steady-state MHD results to predict the global state of the magnetosphere-ionosphere system, *J. Geophys. Res.*, 106, 30,067, 2001.

Ridley, A. J., T. I. Gombosi, D. L. De Zeeuw, C. R. Clauer, and A. D. Richmond, Ionospheric control of the magnetospheric configuration: Neutral winds, *J. Geophys. Res.*, 108(A8), 1328, 2003.

Toth, G., O. Volberg, A. Ridley, Space Weather Modeling Framework Manual, Code version 1.0, *Center for Space Environment Modeling, the University of Michigan*, Ann Arbor, Michigan, 2003.

Volberg, O., D. R. Chesney, D. L. De Zeeuw, K. C. Hansen, K. Kane, R. Oehmke, A.J. Ridley, I.V. Sokolov, G. Toth, T. Weymouth, Space Weather Modeling Framework: Design Policy for Interoperability, http://csem.engin.umich.edu, *Center for Space Environment Modeling, The University of Michigan*, Ann Arbor, Michigan, 2002.

Volberg, O., G. Toth, I. V. Sokolov, A. J. Ridley, T. I. Gombosi, D. L. De Zeeuw, K. C. Hansen, D. R. Chesney, Q. F. Stout, K. G. Powell, K. Kane, R. C. Oehmke, Doing it the SWMF Way: From Separate Space Physics Simulation Programs to The Framework for Space Weather Simulation, *Fall AGU Meeting*, San Francisco, 2003.

Multiscale Coupling of Sun-Earth Processes
A.T.Y. Lui, Y. Kamide and G. Consolini, editors.

A COMPARATIVE STUDY OF PROBABILITY DISTRIBUTION FUNCTIONS AND BURST LIFETIME DISTRIBUTIONS OF B_S AND AE AT SOLAR MAXIMUM AND MINIMUM

R. D'Amicis[1,2], R. Bruno[1] and U. Villante[2]

[1]Istituto di Fisica dello Spazio Interplanetario (CNR), Via Fosso del Cavaliere 100,
00133, Rome, Italy
[2]Dipartimento di Fisica, Università di L'Aquila, Via Vetoio, 67100, L'Aquila, Italy.

Abstract. It is commonly accepted that the AE index reflects in some sense the state of the solar wind-magnetosphere-ionosphere dynamical system. Actually, the AE index is related to the solar wind driving function $|VB_s|$. Since the solar wind parameters have a different behavior for different phases of the solar cycle, a comparative study of the AE index and $|VB_s|$, including studies of probability distribution functions (PDF) and burst lifetime distributions, is needed in order to investigate the differences between solar maximum and solar minimum. In this work we report some of our preliminary results and possible implications.

Keywords. Solar wind-magnetosphere coupling, scaling laws

1. Introduction

The Earth's magnetosphere is a highly dynamical system, which continuously exchanges energy, mass and momentum with the solar wind and the Earth's ionosphere. Although, it is well known that the magnetospheric system is closely coupled to the solar wind [*Tsurutani et al.*, 1990; *Hoshino et al.*, 1994; *Klimas et al.*, 1996; *Borovsky et al.*, 1998], the understanding of the magnetospheric response to the solar wind variations is still an open problem as it involves different mechanisms of energy release and multi-scale coupling phenomena.

Evidences of intermittent and scale-invariant dynamics for the magnetosphere have been found in analyzing the impulsive character of the magnetotail dynamics in terms of the study of the auroral electrojet (AE) indeces [*Davis and Sugiura*, 1966] that are a rough estimate of the energy release by the magnetosphere into the ionosphere during the magnetospheric activity. As a matter of fact, *Consolini et al.* [1996; 1998] have shown that the AE index has a multifractal structure and its fluctuations are not distributed in accordance with a gaussian both in quiet and disturbed periods. Moreover, *Vörös et al.* [2002] have studied the dependence of AE intermittency on the solar cycle finding a more intermittent behavior at solar maximum than at minimum.

These measures of scaling and non-gaussian fluctuations in magnetospheric output need to be understood in the context of the system's driver, the solar wind, which shows an intermittent behavior [*Marsch and Lui*, 1993]. Recent works have focused on the comparison between some properties of the input parameters, such as ε and $|VB_s|$, that are the transferred energy flux and the transverse component of the induced electric field, respectively, and the auroral indexes to establish whether they are directly related. In particular, *Hnat et al.* [2002] have characterized and compared the fluctuations of the input parameters and the AE indeces finding similar scaling laws. Moreover, *Freeman et al.* [2000] have shown that the burst lifetime probability distributions of AL and AU have the same power law components as those of the input parameters.

400

However, the preceding studies were carried out not considering that the solar wind has different intermittent behaviors during different phases of the solar cycle. As a matter of fact, the probability of finding extreme events, that cause an anomalous scaling of the fluctuations, is higher at solar maximum than at minimum, given the higher variability of the parameters that characterize the solar wind. This implies a more intermittent behavior at solar maximum than at minimum. This conclusion is supported by an unpublished study we performed on several parameters of the solar wind as proton number density, bulk velocity and magnetic field.

Since the efficiency of the reconnection process depends also, but not only, on the duration for which B_z is southward (B_s) [*Tsurutani and Gonzalez,* 1995], we are interested in the study of the scaling of the burst lifetimes and of the waiting times between a burst and the next one, in order to emphasize possible correlations between events and to compare their behavior to that of AE index. We focus on this comparative study for different phases of the solar cycle.

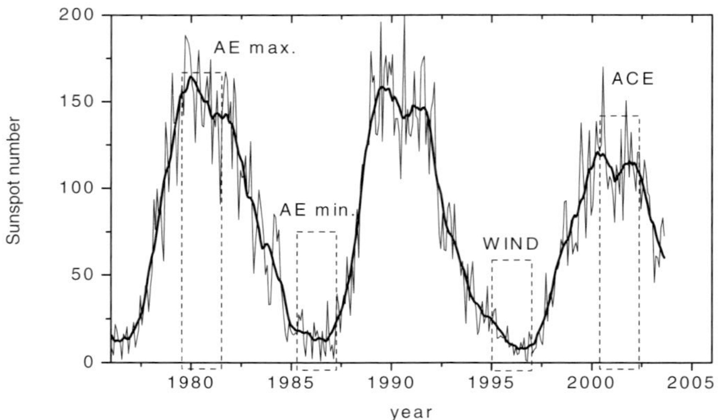

Figure 1. Daily sunspot number as a function of time and its monthly average (smoothed curve). The dashed boxes show the selected time intervals.

2. Data Selection

The present analysis was performed using data from the magnetic field instruments (MFI) in GSE coordinates, onboard WIND and ACE satellites, available at a time resolution of 16 s and 3 s, respectively. While the ACE satellite is continuously monitoring the solar wind at the L_1 point, WIND has a more complicated trajectory that crosses the magnetosphere from time to time. For our analysis we chose only time periods when WIND was in the solar wind. Even if there are gaps in the data, they are negligible, as they are only about 0.5% of the total data length. In order to study the response to the solar wind energy input and consequently of the magnetospheric dynamics, we used the AE index data available at the World Data Center C-2 (Kyoto, Japan), at a time resolution of 1 min.

In order to have the same time resolution for the two magnetic field datasets, we performed averages at 48 s, a value that is a multiple of both datasets and it is close to the resolution of AE data. We selected a period of 16 months for both minimum (from January 1995 to April 1996 inclusive) and maximum (from May 2000 to August 2001 inclusive) for B_z, with an amount of nearly 900000 points for each period. However, there are no contemporary AE data, as the only revised available time period is comprised between January 1978 and June 1988 inclusive. Therefore we selected 16 months at solar minimum, namely from June 1985 to September 1986 inclusive, and at solar maximum as well, namely from September 1979 to December 1980. Thus we have not been working with contemporary datasets. Nevertheless, *Freeman et al.* [2000] have already used non-contemporary data, showing that it is justified to make such comparison between the solar wind and the magnetospheric datasets. They compared lifetime probability distributions of auroral indexes, AU and |AL|, for two datasets, one comprised between January 1978 and June 1988 inclusive, and the other comprised between 1984 and 1987. They found no significant differences between the power law regimes in both datasets. Consequently, we think it is reasonable to choose our datasets as shown in Figure 1.

3. Characterization of B_s Versus Solar Cycle

In Figure 2 we show the flatness of B_z, that is an index of the intermittent behavior of the fluctuations of a given data sample, as a function of the scale [*Bruno et al.*, 2003]:

$$F(\tau) = \frac{<S_\tau^4>}{<S_\tau^2>^2} , \qquad (3.1)$$

where τ is the scale of interest and $S_\tau^p = <|V(t+\tau) - V(t)|^p>$ is the structure function of order p of the generic function $V(t)$. Actually fluctuations are said to be intermittent if the flatness increases when considering smaller and smaller scales [*Frisch*, 1995].

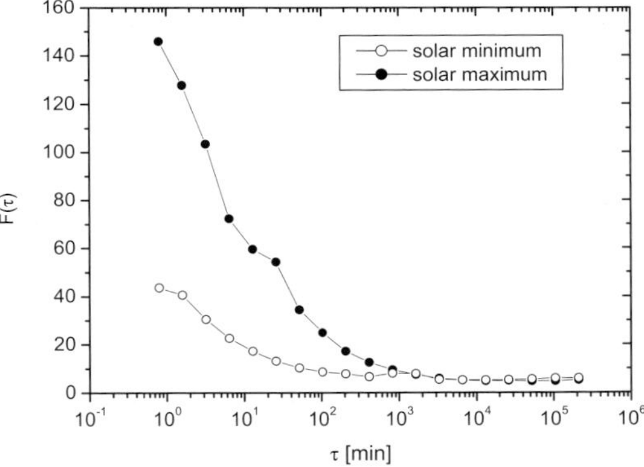

Figure 2. Flatness of B_z at minimum and maximum.

We notice that B_z is more intermittent at solar maximum than at solar minimum in agreement with what we stated in the introduction. As a matter of fact, at solar minimum the distribution evolves towards a Gaussian for time lags greater than 100 minutes, while at solar maximum we observe that the same behavior occurs for time lags greater than 1000 minutes. Therefore, as the flatness at solar minimum and at maximum show different intermittent behavior, we are interested in studying whether or not the magnetospheric response changes accordingly.

Moreover, the probability distribution functions (PDF) of B_z at solar minimum and maximum show a different behavior, as well. In Figure 3, we notice that the two distributions, apart from their cores, are not Gaussian. In both distributions, we have used standardized variables, that is B_z values normalized to their standard deviation value. The two distributions have different shape. In fact, at solar minimum it is nearly symmetric, while at solar maximum it has a tail towards negative values. As a matter of fact, the computation of the skewness returns a value of -0.18 for solar minimum and a value of -0.29 for solar maximum. The negative tail at solar maximum is caused by events that are mainly shocks and coronal mass ejections.

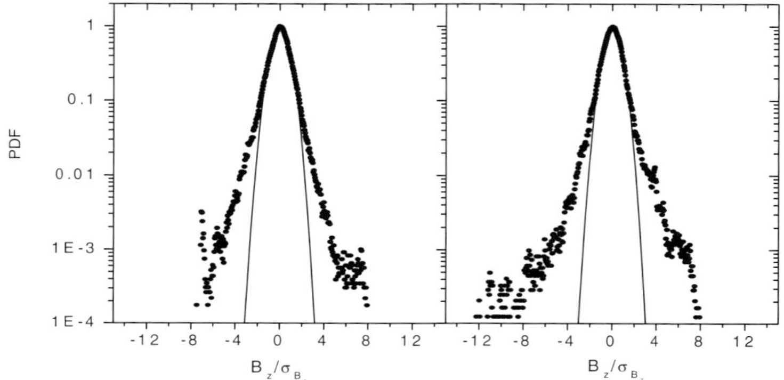

Figure 3. Probability Distributions Functions of B_z at solar minimum (on the left) and at solar maximum (on the right). The solid lines are gaussian fits. In the above diagrams we have used standardized variables (B_z/σ).

3.1 Burst Lifetime and Waiting Time Distributions

In order to compute the burst lifetime distributions at solar minimum and maximum, it is necessary to introduce the definition of burst lifetime and waiting time. A burst lifetime is defined as the duration for which a generic measurement $x(t)$ exceeds a given threshold value. If we refer to B_s, which is different from zero only when B_z is southward, the burst lifetime coincides with the duration of time for which B_z is negative. A waiting time is defined as the duration for which a generic measurement $x(t)$ is under a given threshold value. For B_s it coincides with the duration of time for which B_z is positive. Figure 4 displays the burst lifetime distributions of B_s at solar minimum and maximum, in which the points are derived from an average of the PDF over logarithmically equally spaced intervals. Power law scaling with an exponential cut-off

$$y(T) = \frac{A}{T^\alpha} \exp(-T/T_c),$$

(3.2)

seems to be appropriate to describe the behavior of such distributions. In (3.2) the presence of a power law takes into account the contribution of long term correlations, while the exponential function describes events that follow a Poissonian distribution. The parameters of these fits are shown in Table 1.

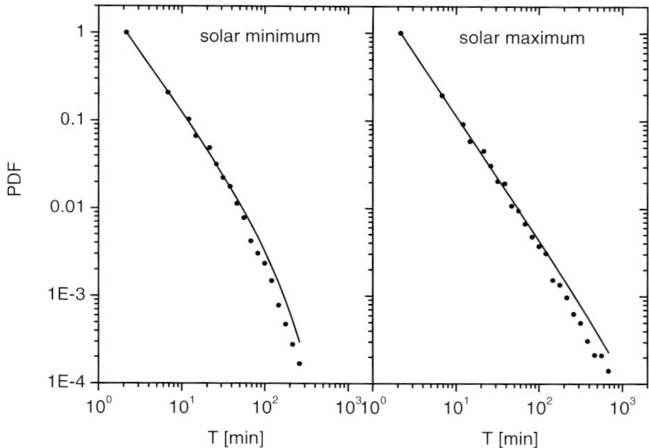

Figure 4. Burst lifetime probability distributions of B_s at solar minimum and maximum. The solid lines refer to expression (3.2). The points are derived from an average of the PDF over logarithmically equally spaced intervals.

Table 1. Comparing fits for burst lifetime distributions at solar minimum and maximum.

	Solar minimum	Solar maximum
A	2.80 ± 0.03	2.96 ± 0.03
α	1.32 ± 0.02	1.41 ± 0.02
T_c	146 ± 70	--

By comparing the parameters shown in Table 1, we notice that at solar minimum both components play a role, suggesting that long term correlations are partially lost. On the contrary, at solar maximum, we found that a simple power law is sufficient to describe the trend of the distribution since the exponential tends to 1. Moreover, at solar maximum the power law is steeper, meaning that long duration events have a less probability than at solar minimum.

Waiting time distributions are shown in Figure 5 and Table 2. In Figure 5 as well, the fit is expressed by (3.2). Even for the waiting time distributions, we find similar results of those already pointed out in the burst lifetime distributions, namely, at solar minimum, the presence of a cut-off, even if greater than that for the burst lifetimes, and a less steep power law.

As a consequence, we find differences between minimum and maximum. At solar minimum both burst lifetime and waiting time distributions can be described by a power law with an exponential cut-off. On the contrary, at solar maximum, the same distributions are well fitted by a simple power law. Moreover, the power laws are steeper at solar maximum than at minimum. In conclusion, the statistics is essentially similar for waiting times and for burst lifetimes.

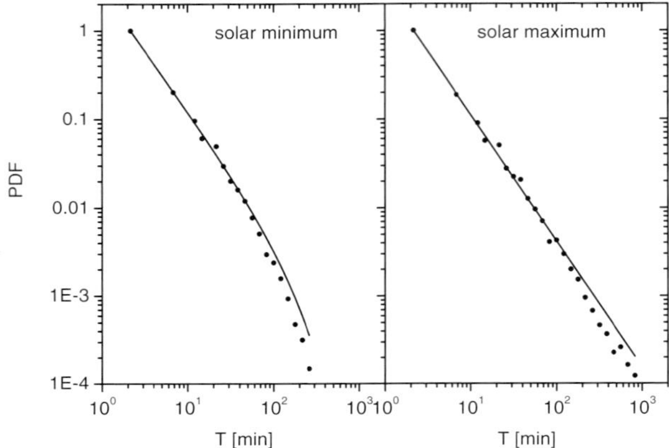

Figure 5. Waiting time distributions of B_s at solar minimum and maximum. The solid lines refer to the fits computed by (3.2). The points are derived from an average of the PDF over logarithmically equally spaced intervals.

Table 2. Comparing fits for waiting time distributions at solar minimum and maximum.

	Solar minimum	Solar maximum
A	2.88 ± 0.04	2.99 ± 0.05
α	1.36 ± 0.03	1.43 ± 0.01
T_c	184 ± 127	--

4. Data Analysis for AE

Kamide et al. [1999] suggested that the AE index is a compound index which is representative for two current systems, DP1 (disturbance polar of the first type) and DP2 (disturbance polar of the second type), related to distinct dissipative processes: an unloading process and a directly-driven process that act in solar wind-magnetosphere interaction. While the latter is linked to an enhancement of convection and is a typical signature of the substorm growth phase, the former appears in the form of bursts of activity and is generally associated with the rapid energy releases occurring in the tail regions. Evidences of this impulsive dissipative events have been found by *Angelopoulos et al.* [1999] who identified the scaling as a key property of magnetospheric energy release in the form of bursty bulk flows in the magnetotail.

In Figure 6, we report the probability distribution function of the AE index as evaluated on the basis of the datasets considered in standardized variables. *Consolini* [2002] and *Consolini and De Michelis* [2002] have already shown the bimodal character of the PDF of AE using a double lognormal distribution, defined as:

$$y(x) = \frac{A_1}{\sqrt{2\pi}w_1 x}\exp\left[\frac{-\ln(x/x_1)^2}{2w_1^2}\right] + \frac{A_2}{\sqrt{2\pi}w_2 x}\exp\left[\frac{-\ln(x/x_2)^2}{2w_2^2}\right]. \tag{4.1}$$

These two components represent periods in which the magnetospheric activity is low (quiet) and periods in which it is enhanced by the presence of storms and substorms (disturbed). In Table 3, we show the parameters obtained by the fits for (4.1).

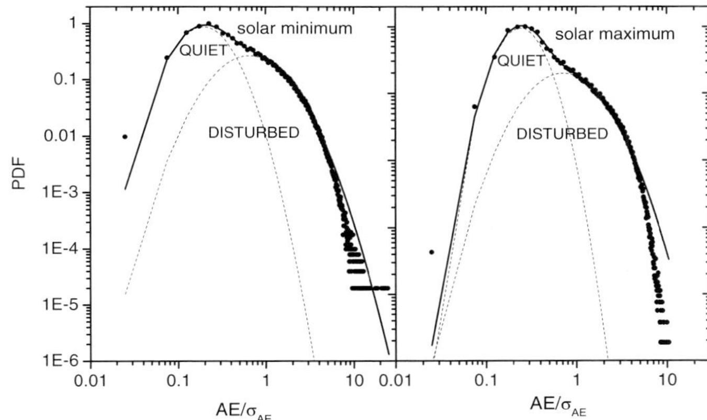

Figure 6. Probability distribution functions of AE values at solar minimum and at maximum in standardized variables. The fits are obtained as the sum of two lognormal distributions as in (4.1) (solid line) whose parameters are shown in Table 4. The dot lines refer to individual lognormal distributions.

Table 3. Comparing results between maximum and minimum obtained from the fits.

Parameters	Solar minimum	Solar maximum
A_1	0.276 ± 0.008	0.30 ± 0.01
w_1	0.556 ± 0.005	0.464 ± 0.007
x_1	0.260 ± 0.003	0.297 ± 0.003
A_2	0.42 ± 0.01	0.35 ± 0.02
w_2	0.74 ± 0.02	0.75 ± 0.04
x_2	1.13 ± 0.02	1.25 ± 0.04

Note that both peaks move towards higher values at solar maximum. If we compare the values of A_1 and A_2, that have the meaning of probabilities, the contribution of the first component (indicated by subscript 1) is smaller than the second component (indicated by subscript 2) at solar minimum while at solar maximum, they have the same importance with respect to the entire PDF. This suggests that even the quiet activity of the magnetosphere become more intense at solar maximum as we expect. Note that in both panels, the fit does not reproduce the trend of the tails of the distributions. We have also used the entire dataset available, namely from 1978 to 1988 but we have not found different results. It would be worth making a careful study of this topic that does not seem to be related to the poor statistics.

4.1 Burst Lifetime and Waiting Time Distributions

Recalling the definition of the burst lifetime and the waiting time given in the preceding section, we need to define the threshold for the AE index. We computed this threshold as suggested in *Consolini et al.* [2002] finding a value very similar for both solar minimum and maximum, 65 nT and 70 nT respectively, values that are in accordance with that found in *Consolini at al.* [2002]. As shown in *Freeman et al.* [2000], one can fit the AE burst lifetime PDF with the sum of a power law with an exponential cut-off plus a lognormal, attributing the power law component to the DP2 convection electrojet and the lognormal component to the so-called unloading component associated with the DP1 electrojet [*Nishida*, 1971; *Kamide and Baumjohann*, 1993]. Nevertheless, we could not fit the entire distribution as in Freeman et al [2000], even if we can identify two components. We were able to fit them only separately as shown in Figure 7. The parameters obtained by the fits are shown in Table 4.

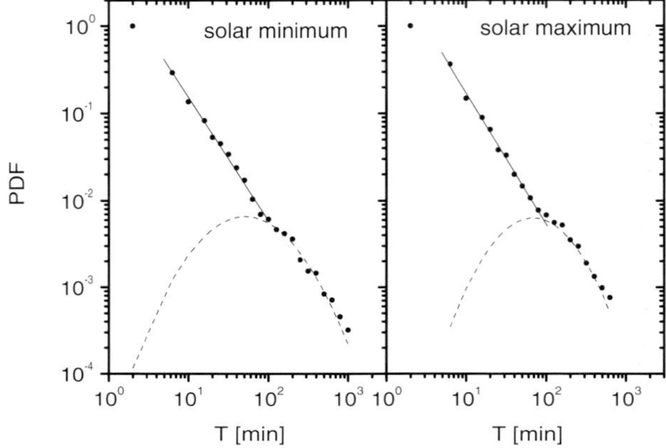

Figure 7. Lifetime distributions of AE at solar minimum and at maximum. The solid lines are power laws while the dot lines are lognormal distributions. The points are derived from an average of the PDF over logarithmically equally spaced intervals.

Table 4. Comparing results between maximum and minimum obtained from the fits. α refers to the slope of the power law, while the other parameters refer to the lognormal distribution.

Parameters	Solar minimum	Solar maximum
α	1.42 ± 0.04	1.51 ± 0.04
A	1.8 ± 0.4	1.8 ± 0.2
w	1.1 ± 0.3	1.0 ± 0.1
x_c	187 ± 28	182 ± 13

We notice that the power law regime is steeper at solar maximum than at minimum, in analogy with the results found in B_s waiting time distribution. Moreover, the second component can be fitted using the same parameters both at maximum and at minimum. This would suggest that statistically the same amount of energy is released in the magnetosphere during different phases of the solar cycle. Unfortunately, the statistics does not allow us to go further.

Completely different results are found for the waiting time distributions (see Figure 8). From the analysis of the trend of the probability distribution functions, it is evident that they follow power laws whose slopes, 1.54 ± 0.02 and 1.40 ± 0.03 for solar minimum and maximum respectively, are very similar to the one found by *Consolini and De Michelis* [2002]. These results show clearly that there is a time correlation among the bursts. We observe that the slopes of the two distributions are slightly different. In fact, at solar minimum, it is steeper indicating that longer waiting times are less probable than at solar maximum.

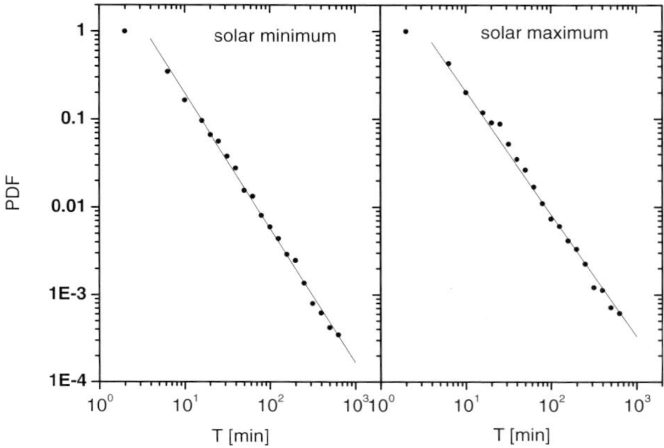

Figure 8. Waiting time distributions of AE at solar minimum and maximum. The solid lines are linear fits with slope 1.54 ± 0.02 for solar minimum and 1.40 ± 0.03 for solar maximum. The points are derived from an average of the PDF over logarithmically equally spaced intervals.

5. Summary and Conclusion

We report a comparative study of the burst lifetime and waiting time distributions for B_s and AE index, at minimum and at maximum of the solar activity. We show that, even though there are differences between maximum and minimum in B_s, the magnetospheric output does not change accordingly.

By comparing the results obtained for different periods of the solar cycle, we notice that the PDFs of B_z, even if they have both gaussian cores, have different shapes. While at solar minimum the distribution is nearly symmetric, the one at solar maximum clearly shows a negative skewness. Moreover, we find differences in the burst lifetime distributions. Even though, we can fit both distributions with a power-law plus an exponential cut-off, the power law behavior lasts longer at solar maximum than at minimum. This suggests that at solar maximum the correlations last for a longer time than at solar minimum. As a matter of fact, we showed that B_z fluctuations are less intermittent during solar minimum and we attributed this behavior to an enhanced presence of stochastic Alfvén waves [see also *Bruno et al.* 2003]. This conclusion is supported also by the fact that we find a cut-off of the order of the typical time period of these waves, namely few hours [*Belcher and Davis*, 1971; *Bruno et al.*, 1985]. During solar minimum, the presence of fast wind is larger than during solar maximum. On the other hand, the fast wind is more alfvénic than the slow wind. Consequently, during solar minimum, we expect that the fluctuations are less intermittent with a consequent less duration of correlations. The same argument can be used to understand the waiting time distributions.

As far as AE index is concerned, the PDFs show the typical bimodal character already emphasized in previous works. Comparing the distributions during different phases of the solar cycle, we notice that both peaks of the distribution move towards higher values at solar maximum. Moreover, while at solar minimum the probability of finding quiet events is smaller, at solar maximum, the two components have nearly the same probability. This suggests that at solar maximum even the quiet activity of the magnetosphere becomes more intense. The comparison between burst lifetimes shows that there are two components, namely a power law and a lognormal, that can be identified as the response of the DP2 and DP1 electrojets. In analogy with the results shown for B_s, the power law regime is steeper at solar minimum, supporting the theory according to which this distribution is due to the directly-driven process. On the contrary, waiting time distributions show a single component, namely a power law, suggesting long term correlations. However, the distribution is steeper at solar maximum, indicating that longer waiting times are less probable.

Acknowledgements

We are indebted to Giuseppe Consolini for useful discussions. We are grateful to the following people and organizations for the provision of data used in this study: R. P. Lepping and NASA Goddard Space Flight Center for WIND MFI data, N. Ness and the Bartol Research Institute for ACE MFI data and the World Data Center C2, Kyoto, Japan, for AE data. This work has been supported by the Italian National Research Council (CNR).

References

Angelopoulos, V., T. Mukai, and S. Kokubun, Evidence for intermittency in Earth's plasma sheet and implications for self-organized criticality, *Phys. Plasma*, 6(11), 4161-4168, 1999.

Belcher , J. W. and L. Davis, Jr., Large-amplitude Alfvén waves in the interplanetary medium, *J. Geophys. Res.*, 76(16), 3534-3563, 1971.

Borovsky, J. E., M. F. Thomsen, H. O. Funsten, and R. C. Elphic, The driving of the plasma sheet by the solar wind, *J. Geophys. Res.*, 103(A8), 17617-17639, 1998.

Bruno, R., B. Bavassano, and U. Villante, Evidence for long period Alfvén waves in the inner solar system, *J. Geophys. Res.*, 90(A5), 4373-4377, 1985.

Bruno, R., V. Carbone, L. Sorriso-Valvo, and B. Bavassano, Radial evolution of solar wind intermittency in the inner heliosphere, *J. Geophys. Res.*, 108(A3), SSH 8-1, 2003.

Consolini, G., M. F. Marcucci, and M. Candidi, Multifractal structure of auroral electrojet index data, *Phys. Rev. Lett.*, 76(21), 4082-4085, 1996.

Consolini, G. and P. De Michelis, Non-gaussian distribution function of AE index fluctuations: evidence for time intermittency, *Geoph. Res. Lett.*, 25(21), 4087-4090, 1998.

Consolini, G. and P. De Michelis, Fractal time statistics of AE-index burst waiting times: evidence of metastability, *Nonlinear Processes in Geophysics*, 9(5), 419-423, 2002.

Davis, T. and M. J. Sugiura, Auroral electrojet activity index AE and its universal time variations, *J. Geophys. Res*, 71(3), 785-791, 1966.

Frisch, U., *Turbulence: the legacy of A. N. Kolmogorov*, Cambridge University Press, 1995.

Hnat, B., S. C. Chapman, G. Rowlands, N. W. Watkins, and M. P. Freeman, Scaling in solar wind epsilon and AU, AL and AE indeces as seen by WIND, *Geophys. Res. Lett.*, 29(22), 35-1, 10.1029/2002GL016054, 2002.

Hoshino, M., A. Nishida, T. Yamamoto, and S., Kokubun, Turbulent magnetic-field in the distant magnetotail: bottom-up processes of plasmoid formation?, *Geophys. Res. Lett.*, 21(25), 2935-2938, 1994.

Kamide, Y. and W. Baumjohann, *Magnetosphere-Ionosphere Coupling*, Springer-Verlag, New York, 1993.

Kamide, Y., S., L. F. Kokubun, L. F. Bargatze, and L. A. Frank, The size of the polar cap as an indicator of substorm energy, *Phys. Chem. Earth (C)*, 24 (1-3), 119-127, 1999.

Klimas, A. J., D. Vassiliadis, D. N. Baker, and D. A. Roberts, The organized nonlinear dynamics of the magnetopshere, *J. Geophys. Res.*, 101(A6), 13089-13113, 1996.

Marsch, E. and S. Lui, Structure functions and intermittency of velocity fluctuations in the inner solar wind, *Ann. Geophysicae*, 11(4), 227-238, 1993.

Nishida, A., DP 2 and polar substorm, *Planet. Science Sci.*, 19(2), 205-221, 1971.

Tsurutani, B. T., M. Sugiura,, T. Iyemori, B. E. Goldstein, W. D. Gonzalez, S.-I. Akasofu, and E. J. Smith, The nonlinear response of AE to the IMF B_s driver - A spectral break at 5 hours, *Geophys. Res. Lett.*, 17(3), 279-282, 1990.

Tsurutani, B. T., and W. D. Gonzalez, The future of geomagnetic storm predictions: implications from recent solar and interplanetary observations, *J. of Atmospheric and Solar-Terrestrial Physics*, 57(12), 1369-1384, 1995.

Vörös, Z., D. Jankovičová, and P. Kovács, Scaling and singularity characteristics of solar wind and magnetospheric fluctuations, *Nonlinear Processes in Geophysics*, 9(2), 149-162, 2002.

Multiscale Coupling of Sun-Earth Processes
A.T.Y. Lui, Y. Kamide and G. Consolini, editors.

411

THE DAYTIME CUSP AURORA IN THE O(^{1}D) EMISSION OBSERVED BY WINDII ON UARS

S. P. Zhang and G. G. Shepherd
Centre for Research in Earth and Space Science
York University, Toronto, M3J 1P3 Canada

Abstract.   The Wind Imaging Interferometer (WINDII) provides an unprecedented database of the O(^{1}D) airglow emission rate measured during 1991-1995. Results from the database show a strong correlation between the daytime O(^{1}D) emission rate and the solar irradiation, in which the emission rate for a given day increases with increasing $cos^{1/e}\chi$ almost linearly, where χ is the solar zenith angle. The daytime aurora therefore can be separated from the background airglow. The daytime cusp (or cleft) aurora is clearly identified in both hemispheres at geomagnetic latitudes 70°-80° even during geomagnetic quiet times. The emission rate in the cusp region is enhanced by 20-120%, i.e. an increase of electron energy deposition by several hundred to 3000 Rayleighs.

Keywords. airglow, aurora, atomic oxygen.

1. Introduction

The O(^{1}D) atomic oxygen emission at 630 nm, also called the red line, is located in the thermosphere at altitudes between 100 and 400 km. The history of the discovery, measurements, and research of the red line were described in detail by *Shepherd* [1979] and *Chakrabarti* [1998]. The dayside aurora of both the O(^{1}D) and O(^{1}S) (557.7 nm) in the polar region and near the local winter solstice, when the sun is below the horizon, has been studied intensively by ground-based observations, showing different types of aurora related to the orientation of the interplanetary magnetic field (IMF) [*Sandholt et al.*, 2002]. But the sunlit airglow and aurora are extremely difficult to measure from the ground due to Rayleigh scattering in the atmosphere. There are just a few ground-based measurements of the integrated red line emission rates and rocket measurements of the emission profiles [*Chakrabarti*, 1998; *Pallamraju et al.*, 2001]. Only in the past three decades was a global view of the airglow layer achieved with observations from spacecraft. The early mapping of the integrated dayside O(^{1}D) emission was produced by *Shepherd and Thirkettle* [1973] seen from the ISIS-II spacecraft. In the polar region, ISIS measurements showed emission rate enhancement at geomagnetic latitudes (MLAT) 70°-80° as evidence of the cusp (also called by some, the cleft) effect. Measurements of the red line by the Atmospheric Explorer E and the Dynamic Explorer mission increased our understanding of global thermospheric dynamics [e.g. *Hays et al.*, 1981; *Killeen and Roble*, 1988]. Most recently, the Wind Imaging Interferometer (WINDII) on the Upper Atmospheric Research Satellite (UARS) provides an unprecedented and unique red line emission rate database with more than 130,000 daytime red line profiles (solar zenith angles less than 90°) measured during the 1990's [*Zhang and Shepherd*, 2004]. Details about the WINDII instrument were described by *Shepherd et al.* [1993].

The red line airglow is produced by reactions between the neutral atmosphere, the ionosphere, and the magnetosphere. The production and loss reactions are complex, which involve the densities of neutral atomic oxygen [O], molecular oxygen [O_2], ionized molecular oxygen [O_2^+], electrons and other species, solar UV and EUV irradiation, and neutral and plasma temperatures. The most important three production mechanisms during daytime are (1) dissociative recombination: $O_2^+ + e \rightarrow O + O(^1D)$; (2) photoelectron excitation: $e^* + O \rightarrow e^* + O(^1D)$; and (3) photodissociation: $O_2 + hv \rightarrow O(^3P) + O(^1D)$. For detailed discussion of the mechanisms of the $O(^1D)$ excitation, readers are referred to papers by *Solomon and Abreu* [1989] and *Witasse et al.* [1999].

In the polar region between 70° and 80° MLAT on the dayside, both electrons and protons from the magnetosheath are able to penetrate down to the thermosphere, where they produce dayside auroras. These regions are referred to as the dayside magnetospheric cusps, or polar cusps, originally predicted by *Chapman and Ferraro* in 1931 [e.g. *Shepherd*, 1979]. The dominant dayside cusp auroras were observed in the $O(^1D)$ (630.0 nm) and $O(^1S)$ (557.7 nm) emissions during polar winter darkness [e.g. *Shepherd*, 1979; *Carlson and Egeland*, 1995]. Further studies of the dayside cusp aurora observed in the two emissions during the northern polar winter were described by *Sandholt et al.* [1986, 2002] and *Meng and Lundin* [1986].

This article describes the response of the sunlit red line emission rate to the solar zenith angle and the electron energy deposition enhancement in the cusp regions of both hemispheres using WINDII data measured during 1991-1995. It should be noted that the solar-illuminated cusp is different from the dayside cusp in darkness according to observations by the Polar ultraviolet imager particle data, in that the cusp ion (electron) precipitating energy flux was larger in the summer hemisphere (the cusp in sunlight) than in the winter hemisphere (the cusp in darkness) by 61% (51%) [*Newell et al.*, 1988].

2. WINDII Measurements of the Red Line Emission Rate

The vertical range of WINDII red line measurements is usually between 100 and 300 km, which does not cover the entire layer but mostly cover the peak of the $O(^1D)$ layer. Figure 1 shows two profiles of the daytime red line volume emission rate (V) in photon cm^{-3} s^{-1} as examples. The entire profile can be reconstructed using a Gaussian fitting by

$$V(z) = V_p \, exp[-(z\text{-}H_p)^2/(2W^2)], \qquad (2.1)$$

where z is altitude in km, V_p is the peak of V, H_p is the height of V_p in km, and W is the profile width of the Gaussian function in km. The three unknown variables, V_p, H_p, and W, can be derived using a curve fitting procedure provided by IDL software [IDL Reference Guide, 2003]. The fitting results are plotted as the solid curves in Fig. 1. Using the fitting curve the integrated emission rate (I) in Rayleigh (R, 10^6 photon cm^{-2} s^{-1}) is easily calculated.

Totally there are about 130,000 profiles measured during 121 days from December 1991 to July 1995. The daily latitudinal coverage of WINDII data is 42°N-72°S or 42°S-72°N geographic latitude (GLAT) according to the orientation of the UARS spacecraft towards south or north. On most days, the measurements reach 80° MLAT north or south, covering the cusp region.

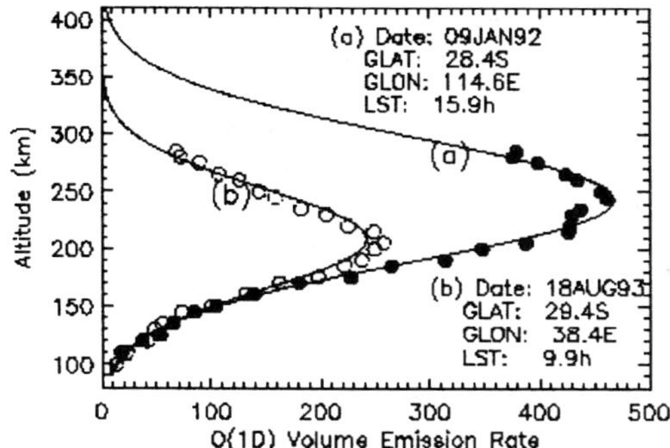

Figure 1. Two samples of the red line volume emission rate (photon cm^{-3} s^{-1}) profile measured by WINDII along with the corresponding Gaussian fitting curves.

By the nature of space-based measurements for an inclined orbit, all profiles for a given day are taken at different times and locations. Specifically, data at a given latitude correspond to similar local times, although at different universal times and different longitudes. WINDII employs two orthogonal fields of view, at 45° and 135° from the velocity vector on the anti-sun side of the spacecraft. There are 15 orbits per day, and the number of daytime O(^{1}D) profiles in each orbit is about 200. The measurements have fine vertical resolutions, about 2 km and 3-5 km below and above 120 km, respectively; but the horizontal separation between profiles along an orbit is about 100 km, and the distance between two consequent orbits is about 1000 km at 70° GLAT, and 3000 km at the equator. Thus horizontal variations of small scale cannot be detected.

Figure 2 shows the daytime integrated volume emission rate on February 17th, 1992 (17FEB92). The data are measured at 42°S-70°N GLAT. The dots (most of which overlap) indicate the footprints of the WINDII data, but do not represent the exact area of the measurements. Totally, there are about 2600 profiles with SZA< 90°. The contours of magnetic latitude, which are calculated using a computer program provided by *D. Lummerzheim* and *A. Richmond* [*Richmond*, 1995], are also plotted. It should be kept in mind that these data are not simultaneous measurements but are collected over a day. There is a local time difference for different latitudes. For example, on this day, the data taken at 40°S GLAT are for local times between 7.7-8.0 hr, and those at 40°N GLAT are for local times 12.6-12.7 hr.

The solar zenith angle, the tidal variation, and magnetic storms are all local time dependent and affect the airglow emission rate in different ways. The direct effect of the sun on each measurement, regardless of location and universal time, can be displayed using the zenith angle (X) of the incident solar radiation. By the fitting method described above, each profile provides its V_p, I, and H_p. The fitting results on 17FEB92 are depicted in Fig. 3 showing the responses of the three variables to $cosX$. Both V_p and I increase (Fig. 3 (a1) and (b1)) but H_p decreases (Fig. 3 (c1)) with increasing $cosX$. A linear relationship between the emission rate and $cos^{1/e}X$ is found such that the V_p and I can be expressed as follows:

$$V_p = 398 \, cos^{1/e} \, X + 45 \quad \text{and} \quad I = 5573 \, cos^{1/e} \, X + 101. \qquad (2.2)$$

The slopes of the fitting lines (solid lines in Fig. 3 (a2) and (b2)) vary from day to day and are found to be dependent on the solar flux [*Zhang and Shepherd*, 2004]. Among the measurements on that day the minimum SZA occurs at 3°S, which corresponds to the maximum emission rate (5.6 kR) related to the direct solar radiation. The departure from the baselines at high latitudes is the subject of Section 3.

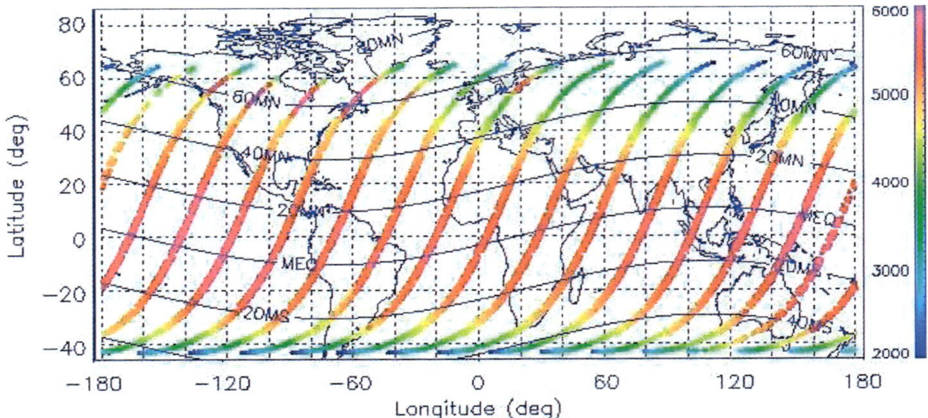

Figure 2. The O(^{1}D) integrated emission rates (R) measured on 17FEB92. The daily F107 and *Kp* are 206 and 2.7, respectively. The measurements at different GLAT correspond to different local solar times. The minimum SZA of the entire measurements is at 3°S GLAT. Note that the contours of the magnetic latitude (MN and MS) are also plotted.

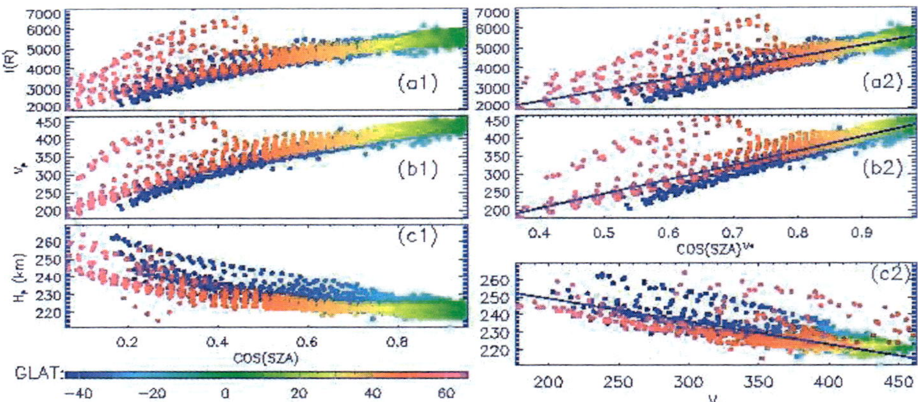

Figure 3. (a1) The integrated emission rate (I), (b1) the peak volume emission rate (V_p), and (c1) the height of V_p (H_p) with $cos\chi$; (a2) I and (b2) V_p with $cos^{1/e}\chi$, and (c2) H_p with V_p measured on 17FEB92. The colours are for GLAT.

3. Sun-illuminated Aurora

One prominent feature in Figs. 2 and 3 is the enhancement of the emission rate above the normal baseline in the northern polar region. The daytime airglow directly related to the SZA given by Eq. (2.2) may be regarded as the 'undisturbed' airglow or the background emission, and can be removed from the measurements using the fitting line. The residual of the emission rate in the polar region, therefore, is dominantly the daytime aurora produced by the penetration of the plasma from the magnetosheath.

Fig. 4 depicts the daytime aurora after removing the undisturbed emission rate on 17FEB92. Because the maximum GLAT of the WINDII data is about 72°N, the daily measurements only cover a sector of the northern aurora oval at geographic longitudes 10°W-130°W. The aurora emission rates of 6-7 kR at 50°N-70°N GLAT corresponding to 58°N-75°N MLAT are significantly enhanced above the background emission rates of 3-4 kR, indicating the daytime cusp aurora produced by electron energy deposition. The geomagnetic local time (MLT) of the cusp is around 1330-1530; the maximum cusp emission is at about 70°N MLAT; and there is a clear equatorward boundary at about 57°N MLAT. The cusp emission rate of 4-7 kR on this day is larger than the typical dayside O(^{1}D) cusp emission rate of 2-5 kR in darkness, and the equatorward boundary is much lower compared to about 70°N MLAT reported by *Sandholt et al.* [1998].

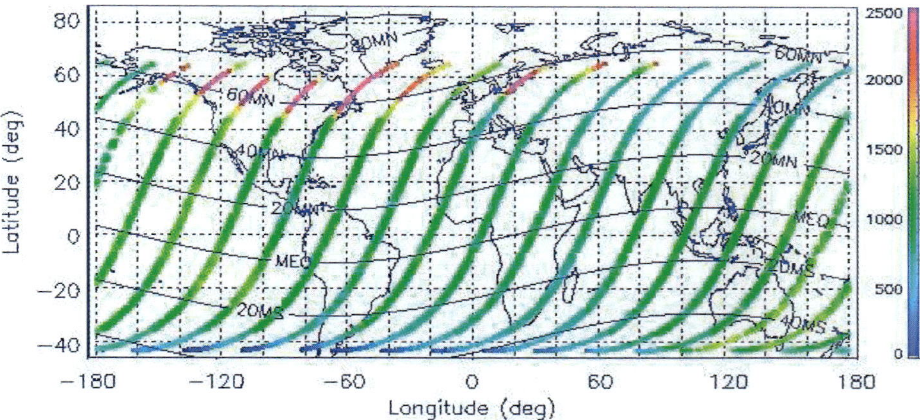

Figure 4. The red line emission rate on 17FEB92 after removing the undisturbed dayglow.

In the southern hemisphere, similar to the northern hemisphere, the daytime cusp aurora occurs at 60°S-80°S MLAT. Because WINDII data only reach 72°S GLAT, the observed sector of the southern aurora oval is located between geographic longitudes 60°E and 200°E as shown in Fig. 5 for January 4th, 1995 (04JAN95). The emission rate is much lower on this day compared to 17FEB92, mainly due to the much lower solar flux, which can be represented by the F10.7 cm flux (10^{-22} Wm^{-2} Hz^{-1}) as a proxy. The daily F10.7 cm flux is only 76 on 17FEB92 compared to 206 on 17FEB92. The fitting functions for V_p and I are respectively:

$$V_p = 189\ cos^{1/e}\ \chi\ + 47 \quad and \quad I = 2934\ cos^{1/e}\chi + 120. \tag{3.1}$$

The slopes of the two fitting lines depend on the solar irradiance, so are much smaller than those on 17FEB92 (Eq. 2.2). Although the airglow emission rate is lower, the emission rates of 3-4 kR in the aurora oval are significantly enhanced from a background of 1.5-2 kR, which are mostly larger than those at 38°S GLAT where the minimum SZA (about 23°) occures among the measurements during that day. The net maximum enhancement of V_p is about 120 photon cm^{-3} s^{-1},

416

or 2300 R for the integrated emission rate, which is about the same as the undisturbed emission rate. After removing the undisturbed emission rate the cusp aurora is clearly displayed in the polar region between 70°S and 80°S MLAT with an equatorward boundary at about 70°S MLAT (Fig. 5 (b)). The maximum emission rate of 4.5 kR is at about 78°S MLAT around 1400 MLT. The energy flux originating in the cusp region extends equatorwards reaching as far as 25°S MLAT or 20°S GLAT (Fig. 5 (a)). The broad illumination of the daytime aurora may be related to a geomagnetic storm of medium size with Kp 2.7-4.0 on that day. This daytime cusp region in the southern polar region during the summer season is wider than that of the cusps observed in emissions of wavelength 400-1100 nm between 75°S and 79°S MLAT by the Defense Meteorological Satellite Program spacecraft in the southern polar region during a winter season [*Meng and Lundin*, 1986].

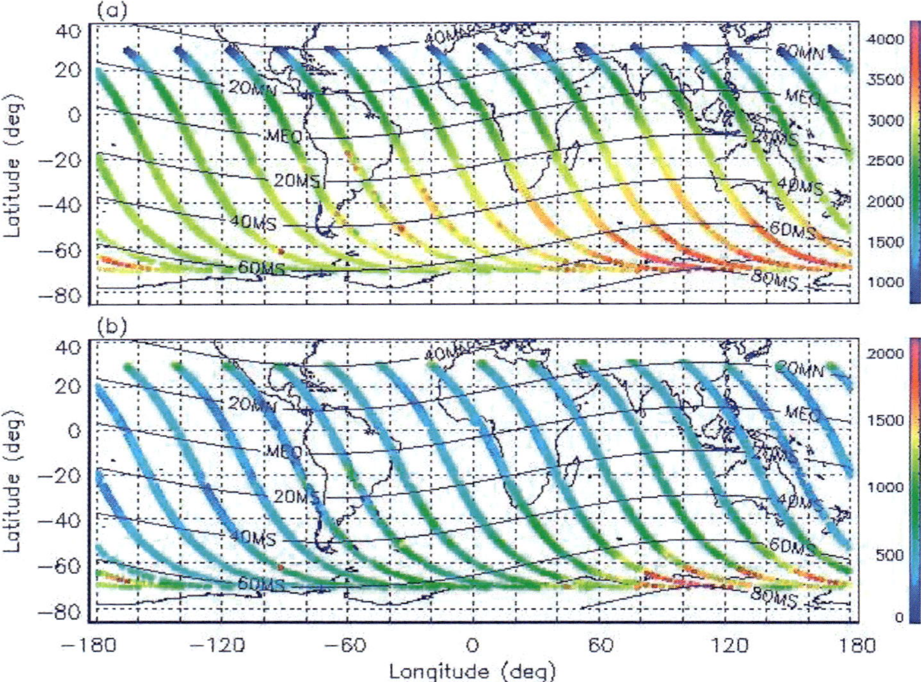

Figure 5. (a) The integrated emission rates measured on 04JAN95. The daily F10.7 and Kp are 76 and 3.3, respectively. The measurements at 20°N and 40°S GLAT are at local solar times about 7.8 hr and 10.6 hr, respectively, and the minimum solar zenith angle of the measurements is at 38°S GLAT. (b) After removing the dayglow, the cusp aurora is clearly displayed in the south-polar region.

Daytime aurora can also be presented using the measured vertical profiles. As mentioned above, for a given day WINDII data for the same GLAT are from different longitudes and different MLAT but at similar local times. Therefore, the local time related effects on those measurements are similar. Because these profiles are from different MLAT, the effect of the geomagnetic field on the emission profiles can be identified by comparing these profiles. Fig. 6

depicts the measured vertical profiles at 60°N (Fig. 6 (a1)) and 40°N (Fig.6 (b1)) GLAT on 17FEB92 for example. Each pair of profiles are from the two fields of view of WINDII, and the time separation of the two is about 7 min. Using the profiles in Fig. 6 (a1) the average of the profiles with MLAT less than 56°N (the solid curve in Fig.6 (a2)) can be treated as the undisturbed airglow volume emission rate with V_p equal 250 photons cm^{-3} s^{-1}, and the individual profiles (thin curves) as the disturbed ones between 56°N-72°N MLAT with V_p 230-400 photons cm^{-3} s^{-1}. In contrast with these high latitude emission rates, at 40°N GLAT the difference of the average profiles with MLAT less and greater than 36°N (Fig.6 (b2), the solid curve and the dashed curve, respectively) is less than 10 photon cm^{-3} s^{-1}, which means the geomagnetic effect at 40° GLAT is negligible on that day.

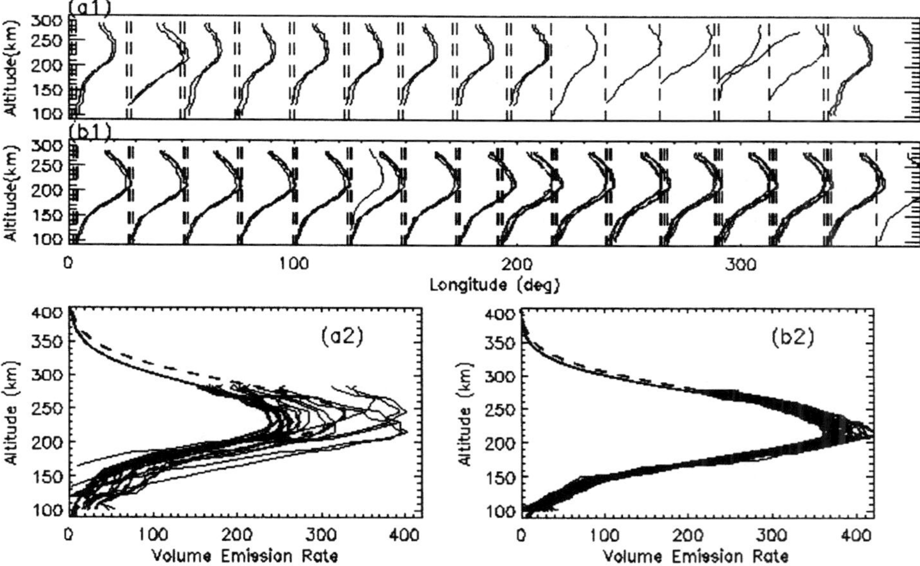

Figure 6. Measured O(^{1}D) volume emission rate profiles at (a1) 60°N and (b1) 40°N GLAT on 17FEB92 at different longitudes. (a2) The solid curve and the dashed curve are for the average profiles with MLAT less and greater than 56°N, respectively, and the solid thin curves are the individual profiles at 56°N-72°N MLAT. (b2) Same as (a2) but the dividing MLAT is 36° N and the maximum MLAT is 51°N.

The daytime cusp aurora is observed almost every day in the WINDII data base, even on quiet days. The geomagnetic disturbance can be estimated using the relative enhancement of the emission rate. As described above, for a given GLAT, the ratio of the measurement at higher MLAT and the undisturbed emission rate can be calculated as a function of MLAT. Results of the ratios from 112 days separated by $Kp <= 2$ and $Kp > 2$ are plotted in Figure 7 for 60°N-81°N MLAT, and in Figure 8 for 60°S-82°S MLAT. In the northern hemisphere, the ratios are 1-1.7 and 1-2.3 for Kp less and greater than 2, respectively; and the daily peak enhancements are mostly in the cusp region at 70°N-80°N MLAT. In the southern hemisphere, the ratios, 1-1.5 for $Kp <= 2$ and 1-2.0 for $Kp > 2$, are not as large as those in the northern hemisphere, but the daily peaks are also at 70°S-80°S MLAT. The cusp enhancement of 1-3 kR in the integrated emission

418

rate may be converted to an equivalent electron flux using the formulation of *Shepherd et al.* [1980], yielding values in the range 0.65 to 2.0 erg cm^{-2} s^{-1}, assuming electrons in the energy range 60 to 300 eV. The different features of the cusp auroras in the two polar regions requires further investigation, since the measurement days used in this study are not evenly distributed by seasons, and seasonal variabilities of auroral particle acceleration and precipitation should not be ignored [*Newell et al.*, 1996; *Liou et al.* , 2001].

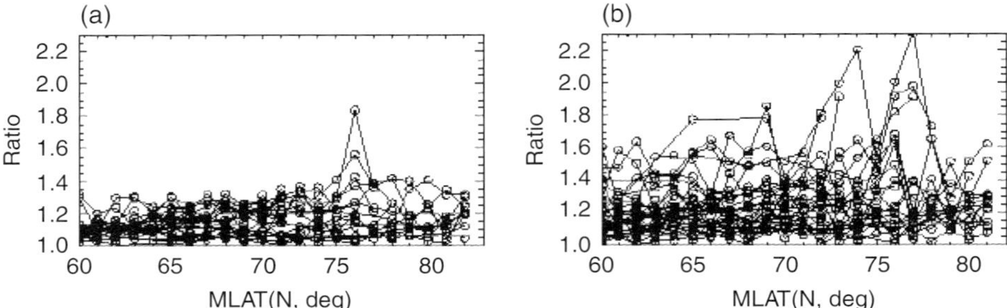

Figure 7. (a) The ratios of the integrated emission rate, *I*, at 60°N-81°N MLAT to the undisturbed *I* for the Norther hemisphere (a) *Kp* <=2 and (b) *Kp* >2.

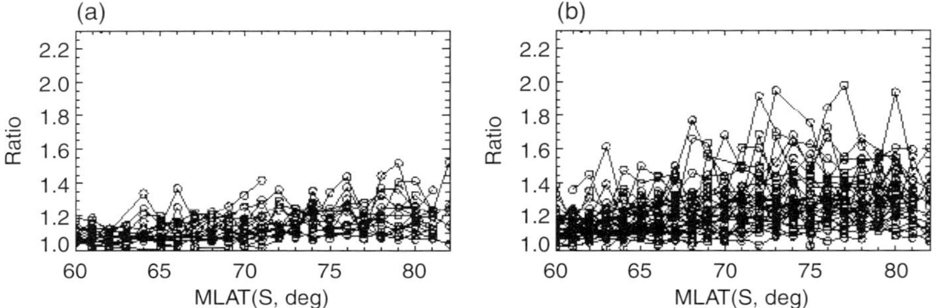

Figure 8. Same as Figure 7, but for the southern hemisphere.

4. Summary

WINDII provides a massive and unique database for daytime (sunlit) O(^{1}D) emission rates measured during the 1990's. One of the applications of the data is for studying the daytime aurora precipitation in the cusp regions and the effects of the disturbance of the geomagnetic field. Using entire measurements in a day, the undisturbed airglow emission rate directly related to the solar zenith angle can be extracted and removed to yield the daytime aurora. The daytime aurora is seen almost every day at 60°-80° geomagnetic latitudes, and conspicuous cusp auroras are observed primarily at geomagnetic latitudes 70°-80°. The O(^{1}D) emission rate in the cusp regions is 3-8 kR, which is about 1.2-2.3 times the undisturbed airglow emission rate, depending on the solar irradiance and geomagnetic disturbance. The enhancements of the emission rate on days with *Kp* greater than 2 are significantly larger than those with *Kp* less than 2, in particular, in the northern hemisphere. The equatorward boundary of the daytime cusp aurora is generally lower than that of the dayside cusp in darkness by a few to 10 degrees. The cusp enhancement of 1-3 kR in the integrated emission rate is equivalent to the electron flux in the range 0.65 to 2.0 erg cm^{-2} s^{-1}, assuming electrons in the energy range 60 to 300 eV. The findings reported in the article are preliminary, and detailed studies on the daytime aurora are in progress, which include the relationship of the $O(^1D)$ daytime aurora with the interplanetary magnetic field, the $O(^1S)$ daytime aurora from a much larger WINDII $O(^1S)$ database, and the correlation of dayglow and nightglow using WINDII nighttime measurements.

Acknowledgments

The WINDII project was jointly sponsored by the Canadian Space Agency and the Centre National d'Etudes Spatiales. The authors are grateful to the WINDII team for providing excellent measurements. The scientific analysis was supported by the Canadian Foundation for Climate and Atmospheric Sciences.

References

Chakrabarti, S., Ground based spectroscopic studies of sunlit airglow and aurora, *J. of Atmos. Sol. Terr Phys., 60*, 1403-1423, 1998.

Carlson, H.C., and A. Egeland, *Introduction to Space Physics,* Chapter 14, Edited by M.G. Kivelson   and C.T. Russell, Cambridge University Press, 1995.

Hays, P.B., T.L. Killeen, and B.C. Kennedy, The Fabry-Perot interferometer on Dynamics Explorer, *Space Sci. Instrum., 5*, 395, 1981.

IDL Reference Guide, Research System Inc. 2003.

Killeen, T.L., and R.G. Roble, Thermospheric dynamics: contributions from the first 5 years of the dynamics explorer program, *Rev. Geophys., 25*, 329-367, 1988.

Liou, K., P.T. Newell, and C.-I. Meng, Seasonal effects on auroral particle acceleration and precipitation, *J. Geophys. Res., 106*, 5531-5545, 2001.

Meng, C.-I. And R. Lundin, Auroral Morphology of the Midday Oval, *J. Geophys. Res., 91*, 1572-1584, 1986.

Newell, P.T. and C.-I. Meng, Hemispherical asymmetry in cusp precipitation near solstice, *J. Geophys. Res., 93*, 2643-2648, 1988.

Newell, P.T., K.M. Lyons, and C.-I. Meng, A large survey of electron acceleration events, *J. Geophys. Res., 101*, 2599-2614, 1996.

Pallamraju, D., J. Baumgardner, and S. Chakrabarti, Simultaneous ground based observations of an auroral arc in daytime/twilighttime OI 630.0 nm emission and by incoherent scatter radar, *J. Geophys. Res., 106*, 5543-5549, 2001.

Richmond, A.D., Ionospheric electrodymanics using magnetic apex coordinates, *J. Geomag. Geoelectr., 47, 191-212, 1995.*

Sandholt, P.E., C.S. Deehr, A. Egeland, B. Lybekk, R. Viereck, and G.J. Romick, Signatures in the Dayside Aurora of Plasma Transfer From the Magnetosheath, *J. Geophys. Res., 91*, 10,063-10,079, 1986.

Sandholt, P.E., H.C. Carlson, and A. Egeland, Dayside and Polar Cap Aurora, Kluwer Academic Publishers, Boston, 2002.

Shepherd, G.G., and F.W. Thirkettle, Magnetospheric dayside cusp: A topside view of its 6300 A emission, *Science, 180,* 737, 1973.

Shepherd, G.G., Dayside cleft aurora and its ionospheric effects, *Revs. Geophys. and Space Physics, 17,* 2017-2033, 1979.

Shepherd, G.G., J.D. Winningham, F.E. Bunn and F.W. Thirkettle, An empirical determination of the production efficiency for auroral 6300-Å emission by energetic electrons, *J. Geophys. Res. 85,* 715-721, 1980.

420

Shepherd, G.G., G. Thuillier, W. Gault, B. Solheim, C. Hersom, J. Alunni, J. Brun, S.Brune, P. Charlot, L. Cogger, D. Desaulniers, W. Evans, R. Gattinger, F. Girod, D. Harvie, R. Hum, D. Kendall, E. Llewellyn, R. Lowe, J. Ohrt, F. Pasternak, O. Peillet, I. Powell, Y. Rochon, W. Ward, R. Wiens, and J. Wimperis, WINDII: The Wind Imaging Interferometer on the Upper Atmosphere Research Satellite, *J. Geophys. Res.*, *98*, 10,725-10,750, 1993.

Solomon, S.C. and V.J. Abreu, The 630 nm Dayglow, *J. Geophys. Res.*, *94*, 6817-6824, 1989.

Witasse, O., J. Lilensten, and C. Lathuillère, Modeling the OI 630.0 and 557.7 nm thermospheric dayglow during EISCAT-WINDII coordinated measurements, *J. Geophys. Res.*, *104*, 24,636-24,655, 1999.

Zhang, S.P., and G.G. Shepherd, Solar influence on the O(^{1}D) dayglow emission rate: Global-scale measurements by WINDII on UARS, *Geophys. Res. Lett.*, 31, L07804, doi:10.1029/2004GL019447, 2004.

Multiscale Coupling of Sun-Earth Processes
A.T.Y. Lui, Y. Kamide and G. Consolini, editors.

THE WAVELENGTH OF SLOW MHD WAVES OBSERVED IN THE NIGHT-SIDE PLASMA SHEET

A. Nakamizo

*Solar-Terrestrial Physics Laboratory, Department of Earth and Planetary Sciences,
Kyushu University, Fukuoka 812-8581, Japan*

Abstract. Applying the linear plane wave theory to parallel velocity perturbations and plasma pressure perturbations observed by the Geotail satellite the parallel wavelength of slow MHD waves was estimated. We focused on the frequency range of 40-150 seconds that corresponds to that of Pi2 geomagnetic pulsations. It is found that the parallel wavelength is about several earth radii that may reflect the spatial scale of the region where the ∇P force is comparable to the $\mathbf{J} \times \mathbf{B}$ force.

Keywords. plasma sheet, slow MHD waves, magnetotail Pi2

1. Introduction

The night-side plasma sheet is one of the key regions of the earth's magnetosphere. In the convection system, the plasma sheet is responsible for the earthward transport of the mass, momentum and energy densities after the tail reconnection. Associated with this fundamental process, plasma sheet motions accompany many plasma processes occurring sporadically. These global and local aspects of plasma sheet dynamics are dominant during substorms.

The plasma sheet was thought earlier to undergo the way of steady, slow and adiabatic earthward convection. An issue was raised against such an ideal picture as 'Pressure balance inconsistency' by *Erickson and Wolf* [1980]. As the flux tube is transported earthward, the plasma pressure inside the flux tube becomes too large, if the flux tube continues to hold the plasma inside it without a loss. This loss-less assumption is equivalently mentioned that the motion inside the high-beta plasma sheet is dominated by the fast mode motion. *Erickson and Wolf* [1980] originally proposed that the substorm is the most promising process to resolve the inconsistency.

In the 1990s the ideal picture of the plasma sheet convection was entirely revised. On the basis of AMPTE/IRM and ISEE 2 observations, *Angelopoulos et al.* [1994] suggested that intermittent fast flows are responsible for most of the transport in the near-earth plasma sheet in spite of their relatively short time durations. As the generation mechanism of the fast flows, a patch of magnetic reconnection is generally accepted among many researchers. Another candidate of their generation mechanism is the 'plasma bubble.' The idea of the 'plasma bubble,' originally suggested to resolve the pressure balance inconsistency [*Pontius and Wolf*, 1990] and evolved by subsequent studies [*Chen and Wolf*, 1993, 1999], has been accepted as one of the possible mechanisms of the fast earthward flows. In the bubble model fast flows are naturally explained by the interchange motion of the under populated flux tube (bubble) interacting with the surrounding regions. In these ideas, considerations for the plasma sheet convection depend on the local process rather than the global process.

While the occurrence frequency of fast flows is correlated with geomagnetic activity, the causality between fast flows and substorms is not fully understood (see *Ohtani* [2001] and references therein). On the other hand, fast flows accompany a variety of phenomena, such as magnetic field fluctuations, ionospheric activity, etc. When the fast flows are discussed in relation with these phenomena, the problem shifts to the local process rather than the global convection process. We are missing the linkage between the bursty nature of the plasma sheet and its consequence for the magnetospheric convection system. The investigation of this linkage will lead to understanding the plasma sheet dynamics during substorms.

The present study is the extension of *Nakamizo and Iijima* [2003], in which it is proposed that the compressional component of magnetotail disturbances is primarily caused by the slow MHD mode and that coupling occurs between the transverse Alfven mode and the slow mode. In the present paper, we concentrate only on the Pi2 range of the disturbances. We hope, however, that our investigation would be useful for understanding plasma sheet dynamics. Section 2 briefly describes the data set we use and runs over the basic feature of the magnetotail disturbances. In section 3, applying the MHD linear wave theory, we estimate the parallel wavelength of slow MHD waves. Discussion is given in the last section.

2. Overview of Magnetotail Disturbances

We use the magnetic field and the plasma data acquired with the Geotail satellite for a full year of 1997. The magnetic field data were obtained by the MGF (magnetic field) instrument [*Kokubun et al.*, 1994] and were averaged every 12 seconds. The 12-second resolution plasma moment data were provided by the LEP (low-energy particle) instrument [*Mukai et al.*, 1994]. The GSM coordinate system is adopted throughout this paper.

Figure 1(a) shows an overview of magnetotail disturbances observed on March 19, 1997. The Geotail location was $(X, Y, Z)_{GSM} = (-20.13, -0.06, -0.26)$ R_E at 12:00 UT and $(X, Y, Z)_{GSM} = (-18.40, -2.52, -0.77)$ R_E at 15:00 UT. The magnetic field **B** (nT), plasma number density N (/cc), plasma temperature T (keV), plasma β (the ratio of the plasma pressure to the magnetic pressure) and bulk flow velocity **V** (km/s) are shown from top. The bulk flow velocity is decomposed into components parallel and perpendicular to the instantaneous magnetic field. The bottom panel shows the concurrent magnetic variation δBx obtained by band-pass filtering Bx in the period range of 40-150 seconds. This period range corresponds to that of the geomagnetic Pi2 pulsations. At 13:00 UT, the magnetic field and the plasma show drastic changes and the fast flow appears to coincide with the activation of δBx. This Pi2-range variation observed at the location of Geotail is termed as the 'magnetotail Pi2' and the start of the disturbances, designated as a vertical dashed line, is defined as the 'local onset' below.

Figure 1(b) focuses on the pressure changes and the magnetotail Pi2. The upper panel shows the magnetic pressure Pm (nPa) and the plasma pressure P (nPa) for the same interval as in Figure 1(a). The details of the magnetotail Pi2 during the expanded interval around the local onset are shown in the lower panel. The magnetic field perturbation $\delta\mathbf{B}$ (nT) and the velocity perturbation $\delta\mathbf{V}$ (km/s) are decomposed into components parallel and perpendicular to the background magnetic field given by $\mathbf{B} - \delta\mathbf{B}$. The magnetic pressure Pm and the plasma pressure P are in out-of-phase throughout the disturbances. It should be noted that the magnetic pressure perturbation δPm and the plasma pressure perturbation δP are also in out-of-phase as clearly shown in the lowest part of the magnetotail Pi2.

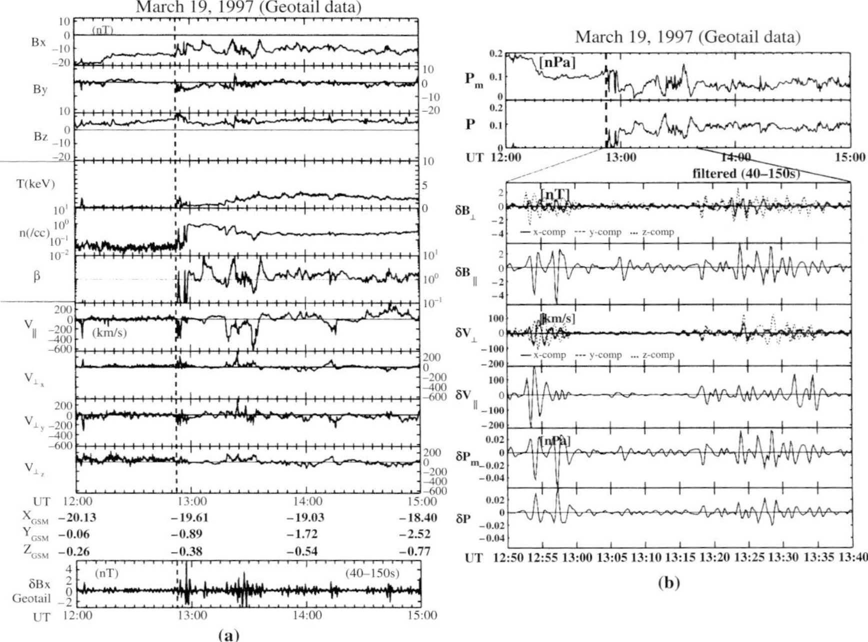

Figure 1. (a) Overview of magnetotail disturbances observed by the Geotail satellite, represented in GSM coordinates. Bulk flow velocity **V** is decomposed into components parallel and perpendicular to the instantaneous magnetic field **B**. Bottom panel: Concurrent Pi2-range oscillations at the same place. (b) Upper panel: Signatures for the plasma pressure P and the magnetic pressure Pm. Lower panel: Details of the magnetotail Pi2. Vector quantities are decomposed into components parallel and perpendicular to the background magnetic field (**B** − δ**B**). Out-of-phase variations between P and Pm and between δP and δPm are noteworthy.

3. Wavelength Estimates

Applying the MHD linear wave theory, we estimate the wavelength of the magnetotail Pi2 from observations by one satellite. In an usual linear plane wave theory, any quantity Q is expressed as $Q = Qo + \delta Q$, where Qo is the zero-order quantity (constant in space and time) and δQ is the first-order quantity represented as $\delta Q \exp\{i(\omega t - \mathbf{k}\cdot\mathbf{r})\}$ (ω is the angular frequency and **k** is the propagation vector). Characteristics of three MHD waves are summarized in Table 1. While the Alfven mode is incompressible, the fast and slow modes are compressible. These two compressible modes are distinguished by the phase relationship between the plasma pressure perturbation δP and the magnetic pressure perturbation δPm. δP and δPm are in-phase for the fast mode, whereas they are out-of-phase for the slow mode as noted in Table 1.

To discuss the magnetotail Pi2 in the context of the linear MHD wave theory, we regard the filtered quantity as the first-order quantity and adopt the following simplifying assumption: Q is the measured quantity, δQ is the band-pass filtered quantity and Qo is the background quantity given by $Q - \delta Q$. The out-of-phase variation between δP and δPm, clearly seen in Figure 1(b),

Table 1. Characteristics of MHD waves. The Alfven mode accompanies transverse perturbations but no parallel perturbations in the magnetic field and the bulk flow velocity, and does not cause pressure perturbations (indicated by (O) and (-) marks). The fast and slow modes accompany both of perpendicular and parallel perturbations and entail pressure perturbations (indicated by (O) marks). The plasma pressure perturbation and the magnetic pressure perturbation are in-phase for the fast mode but out-of-phase for the slow mode.

Wave mode	$\delta B_\perp$	$\delta B_\parallel$	$\delta V_\perp$	$\delta V_\parallel$	δP	δPm
Alfven	O	-	O	-	-	-
Fast	O	O	O	O	O	O (in-phase)
Slow	O	O	O	O	O	O (out-of-phase)

strongly suggests that the magnetotail Pi2 shows a slow MHD wave signature. In addition, the characteristics of wave energy fluxes provide the supporting evidence that the compressional component of the magnetotail Pi2 originates from the slow MHD wave. We also infer the co-existence and/or possible coupling of the slow and the transverse Alfven mode (see section 2.1.2 of *Nakamizo and Iijima* [2003]). Thus, $\delta B_\parallel$, $\delta V_\parallel$, δP and δPm are thought to be ascribable to the slow MHD wave, and $\delta B_\perp$ and $\delta V_\perp$ are thought to result from both the slow MHD wave and the concurrent Alfven wave.

Basic relationships between $\delta\rho$ and $\delta V_\parallel$ and between δB and $\delta V_\perp$, derived from the continuity equation and Faraday's Law, are as follows [e.g., *Siscoe*, 1983]:

$$Cs^2 \frac{(\delta\rho)}{\rho_0} = \frac{\omega_{F,S}}{k_\parallel}(\delta V_\parallel)_{F,S}, \tag{1}$$

$$\omega_{F,S} \frac{(\delta B)}{B_0} = k_\perp (\delta V_\perp)_{F,S}, \tag{2}$$

where Cs is the sound speed and the subscripts 'F' and 'S' refer to the fast and slow modes respectively. Replacing $\delta\rho$ with δP by the use of the polytropic law (P is proportional to ρ^γ, where γ is the adiabatic index) and multiplying equation (2) by $B_0^2/2\mu_0$, relationships between δP and $\delta V_\parallel$ and between δPm and $\delta V_\perp$ can be obtained as

$$\delta P = \rho_0 \frac{\omega_{F,S}}{k_\parallel}(\delta V_\parallel)_{F,S}, \tag{3}$$

$$\delta Pm = \rho_0 V_A^2 \frac{k_\perp}{\omega_{F,S}}(\delta V_\perp)_{F,S}, \tag{4}$$

where V_A is the Alfven speed. By plotting the observed values of ($\delta V_\parallel$, δP) on the XY plane, and by least-squares fitting a line to it, one can estimate $k_\parallel$ with equation (3) as the slope of the line (where ω is the dominant frequency determined by FFT and ρ_0 is the mean of background quantity). In a similar manner, the observed ($\delta V_\perp$, δPm) and equation (4) yield an estimate of $k_\perp$.

Figure 2(a) shows the estimate of $k_\parallel$ by applying the above-stated method to $\delta V_\parallel$ and δP during the interval 13:16-13:22 UT of Figure 1(b). The horizontal axis and the vertical axis represent $\delta V_\parallel$ and δP, respectively. The slope of the regression line yields the wavelength parallel to the ambient magnetic field of 2.30 R_E. Since the observed $\delta V_\perp$ contains both $\delta V_{\perp,S}$ and $\delta V_{\perp,A}$, where $\delta V_{\perp,A}$ is the contribution from the Alfven wave, $\delta V_{\perp,A}$ must be first identified and removed from $\delta V_\perp$ to obtain the perpendicular wavelength of the slow MHD wave.

Figure 2(b) shows other examples of the parallel wavelength estimation plotted against the GSM X position. Points are distinguished according to the β value. First, the events were selected from the magnetotail disturbance events used in *Nakamizo and Iijima* [2003] by scanning the correlation between $\delta V_\|$ and δP visually. The examples where the correlation coefficient (CC) between $\delta V_\|$ and δP exceeds 0.7 were then used. The substorm phases for these examples were checked by referring to ground magnetic records provided by the CANOPUS network, the Alaska network, CPMN and the IMAGE network. It was found that these examples correspond to widespread geomagnetic activities including the substorm growth phase, expansion phase and recovery phase.

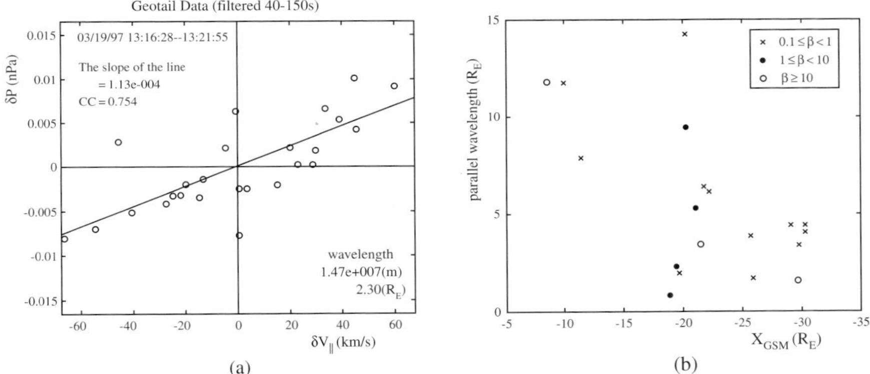

Figure 2. (a) Scatter plots of $\delta V_\|$ versus δP. Solid line represents the least-squares fit. The slope of the line corresponds to the coefficient in equation (4). The correlation coefficient (CC) is 0.754. (b) Plots of the estimated parallel wavelengths versus the GSM X position.

4. Discussion

By using the data from the Geotail satellite, we estimated the parallel wavelength of slow MHD waves that occurred in the night-side plasma sheet. We discuss some implications of the findings.

The out-of-phase signature between the plasma pressure and the magnetic pressure associated with fast flows was reported by earlier works using the AMPTE/IRM satellite [*Bauer et al.*, 1995] and the Cluster satellite [*Volwerk et al.*, 2003]. From the detailed analysis, *Bauer et al.* [1995] interpreted the signature as the neutral sheet oscillation associated with the substorm current wedge formation. In our observations, the parallel velocity perturbations are remarkable. Thus, we regard the observed signature as a disturbance or a wave that is generated at or passing through the Geotail location rather than the spacecraft's passage of the boundary between the low-β region and the high-β region due to the tail flapping or neutral sheet oscillations.

A linkage between fast earthward flows and ground Pi2 pulsations has been argued by many workers. *Kepko et al.* [2001] examined the timing relationship between earthward flows observed by Geotail and ground Pi2 pulsations during substorms. They proposed that the low-latitude Pi2 is directly driven by compressional pulses caused by braking fast flows at the inner magnetosphere. Here our interest lies in the connection between the Pi2-range variations observed in the vicinity of the plasma sheet and the ground Pi2 pulsations. Remembering the plasma bubble model, the analytic equations of the bubble yield two eigenmode solutions, the intermediate Alfven and the slow MHD mode [*Chen and Wolf*, 1999]. In the bubble simulation the slow MHD and the Alfven waves, initially generated by the magnetic tension near the equatorial plane, propagate toward the earth along the field line of the bubble-type flux tube as parallel and perpendicular velocity perturbations, respectively. This characteristic is consistent with our observations. We are uncertain, however, about whether slow MHD waves can exist everywhere in the magnetotail. As well as the $\mathbf{J} \times \mathbf{B}$ force, the ∇P force plays a role in driving slow MHD waves. Therefore, the observed slow MHD waves may be confined to the region where the ∇P force is comparable to the $\mathbf{J} \times \mathbf{B}$ force. The parallel wavelengths obtained here may reflect the spatial scale of such a favorable region. Even if this is the case, the Alfven waves that are coupled to the slow MHD wave would be able to propagate along field lines to the ionospheric altitude, contributing to the high-latitude Pi2 magnetic pulsations. Further study based on multipoint observations is needed for this subject.

The slow mode seems to be the common signature in the near-earth magnetotail not only in the region of the Geotail observations but also in other regions. *Saka et al.* [1997, 1999] showed that the magnetic field and plasma signatures at geosynchronous altitude are related to the Pi2 pulsations at the dip-equator. From the anti-correlation between the particle flux and magnetic field excitation/modulation, they invoked the slow mode as a possible source of the observed signatures.

Finally, we emphasize that the slow mode is the primary process as the compressional motion in the earth's magnetotail. In this study, our discussion is mainly concentrated on the time scale of Pi2 pulsations. However, the slow mode signature is applicable to variations of all time scales. The plasma pressure and the magnetic pressure always undergo the diamagnetic process. During disturbances, the enhanced parallel bulk flow appears together with the drastic but exact out-of-phase change of the plasma pressure and the magnetic pressure. In such a slow mode motion, which is different from the convective motion, the plasma fluid inside the flux tube moves along field lines without causing a local enhancement in the total pressure. The transport in the parallel direction is not included in the framework of the convective or the fast mode motion. To search for spatial scales of the background slow mode disturbances might be a key to understand the overall configuration change of the near-earth magnetotail. Considering the transport of the momentum and energy densities, the role of the slow mode in terms of the plasma sheet dynamics will be discussed in a future study.

Acknowledgements

The author would like to gratefully thank Drs. T. Mukai, S. Kokubun, and T. Nagai for providing Geotail LEP and MGF data. The author also thanks T. Tanaka, H. Kawano, and T. Uozumi for useful discussions.

References

Angelopoulos, V., C. F. Kennel, F. V. Coroniti, R. Pellat, M. G. Kivelson, R. J. Walker, C. T. Russell, W. Baumjohann, W. C. Feldman, and J. T. Gosling, Statistical characteristics of bursty bulk flow events, *J. Geophys. Res.*, *99*(A11), 21,257-21,280, 1994.

Bauer, T. M., W. Baumjohann, and R. A. Treumann, Neutral sheet oscillations at substorm onset, *J. Geophys. Res.*, *100*(A12), 23,737-23,742, 1995.

Chen, C. X., and R. A. Wolf, Interpretation of high-speed flows in the plasma sheet, *J. Geophys. Res.*, *98*(A12), 21,409-21,419, 1993.

Chen, C. X., and R. A. Wolf, Theory of thin-filament motion in Earth's magnetotail and its appreciation to bursty bulk flows, *J. Geophys. Res.*, *104*(A7), 14,613-14,626, 1999.

Erickson, G. M., and R. A., Wolf, Is steady convection possible in the Earth's magnetotail?, *Geophys. Res. Lett.*, *7*(11), 897-900, 1980.

Kepko, L., M. G. Kivelson, and K. Yumoto, Flow bursts, braking, and Pi2 pulsations, *J. Geophys. Res.*, *106*(A2), 1903-1915, 2001.

Kokubun, S., T. Yamamoto, M. H. Acuna, K. Hayashi, K. Shiokawa, and H. Kawano, The Geotail magnetic field experiment, *J. Geomag. Geoelectr.*, *46*(1), 7-21, 1994.

Mukai, T., S. Machida, Y. Saito, M. Hirahara, T. Terasawa, N. Kaya, T. Obara, M. Ejiri, and A. Nishida, The low energy particle (LEP) experiment onboard the Geotail satellite, *J. Geomag. Geoelectr.*, *46*(8), 669-692, 1994.

Nakamizo, A., and T. Iijima, A new perspective on magnetotail disturbances in terms of inherent diamagnetic processes, *J. Geophys. Res.*, *108*(A7), 1286, doi:10.1029/2002JA009400, 2003.

Ohtani, S., Substorm trigger processes in the magnetotail: Recent observations and outstanding issues, *Space Sci. Rev.*, *95*(1-2), 347-359, 2001.

Pontius, D. H., and R. A. Wolf, Transient flux tubes on the terrestrial magnetosphere, *Geophys. Res. Lett.*, *17*(1), 49-52, 1990.

Saka, O., K. Okada, and O. Watanabe, Pi2 associated particle flux and magnetic field modulations in geosynchronous altitudes, *J. Geophys. Res.*, *102*(A6), 11,363-11,373, 1997.

Saka, O., O. Watanabe, K. Okada, and D. N. Baker, A slow mode wave as a possible source of Pi2 and associated particle precipitation: A case study, *Ann. Geophys.*, *17*(5), 674-681, 1999.

Siscoe, G. L., Solar system magnetohydrodynamics, in *Solar-Terrestrial Physics, Principles and Theoretical Foundations*, edited by R. L. Carovillano and J. M. Forbes, pp. 11-100, D. Reidel, Hingham, Mass., 1983.

Volwerk, M., R. Nakamura, W. Baumjohann, R. A. Treumann, A. Runov, Z. Voros, T. L. Zhang, Y. Asano, B. Klecker, I. Richter, A. Balogh, and H. Reme, A statistical study of compressional waves in the tail current sheet, *J. Geophys. Res.*, *108*(A12), 1429, doi:10.1029/2003JA010155, 2003.

CHAPTER SIX

TECHNIQUES FOR MULTISCALE SPACE PLASMA PROBLEMS

Multiscale Coupling of Sun-Earth Processes
A.T.Y. Lui, Y. Kamide and G. Consolini, editors.

WINDMI: A FAMILY OF PHYSICS NETWORK MODELS FOR STORMS AND SUBSTORMS

W. Horton, M. J. Mithaiwala, and E. A. Spencer
Institute for Fusion Studies, The University of Texas, Austin, TX 78712

I. Doxas
*Center for Integrated Plasma Studies, University of Colorado,
Boulder, CO 80309*

Abstract. An important problem in magnetospheric physics is to develop integrated dynamical systems capable of modeling storm and substorm databases with the long term aim of developing space weather forecasting tools. WINDMI is a family of physics-based models that range in dimensionality, d, from low-order models which model the flow of energy between the dominant global energy components to high-order models with $d \sim 10^2 - 10^3$, for models that resolve the nonlinear dynamics of the system into different latitudes. The models are intrinsically three-dimensional in configuration space, and use the basic geometry of the Tsyganenko magnetic field model to define the geometrical quantities. Optimal values of the model parameters are found using the genetic algorithm on given storm and substorm databases. The model satisfy the constraints of the conservations laws of energy and electrical charge in their network of nodes and branches that follow the topology of the ambient magnetic field. The new WINDMI-RC model, which builds on the WINDMI model, includes the energy coupling of injected plasma from the plasma sheet across the Alfvén layer into the ring current.

1. Introduction

The WINDMI model is a physics based low-dimensional model for the coupled Solar Wind-Magnetosphere-Ionosphere system that couples the four basic energy components of the nightside magnetosphere: (i) the lobe magnetic energy, (ii) plasma thermal energy, (iii) streaming kinetic energy, and (iv) cross-tail perpendicular kinetic energy to the ionosphere via the nightside region-1 currents (*Horton and Doxas*, 1996 and 1998a). This model, which was originally used to model substorms, is extended to include the energy of the ring current driven by plasma injection across the Alfvén layer.

In Sec. 2 we review the results of WINDMI-SR, which is the high order spatially resolved model of substorms. The WINDMI-SR vector model includes effects such as compressional dipolarization pulses and Alfvénic pulses. In Sec. 3, the energy coupling of the WINDMI model to the ring current plasma is shown to result in the WINDMI-RC model. After a brief review of the results that the WINDMI model has given for substorms, we describe how the genetic algorithm is used to optimize the parameter values of the model and indicate the current work being performed to optimize the WINDMI-RC model for storms. We conclude in Sec. 4 with comparisons to existing storm models.

2. Multi-Mode Dynamical Model for the Synthesis Model of Substorms

In this section we describe the progress made in constructing a multimode model of the substorm physics, called WINDMI-SR, that is largely motivated by the synthesis model advanced by *Lui* (1991a). The synthesis model is described qualitatively in the chapter entitled "Extended Consideration of a Synthesizing for Substorms" by *Lui* (1991b) and "Synthesizing a Global Model" by *Kan* (1991) in the Geophysical Monograph 64 on Magnetospheric Substorms. The scientific val-

432

idation and the practical importance of the model requires a quantitative realization of the model. It is of particular importance to clarify the relation between auroral field line substorm onset versus the near-Earth neutral line (NENL) substorm onset, or more specifically to find the timing and the interaction of the two processes spatially localized in these two distinct regions of the nightside magnetosphere. Finally, the model must integrate both the microscopic physics of the stability of collective modes, including the near-Earth ballooning instability, the current diversion instability (*Yoon and Lui*, 1996), and the tearing mode instability. The current diversion instability and the tearing mode instability are themselves complicated, and occur in various forms, depending on the details of the system parameters and background models used. Thus, building a quantitative synthesis model is a long-term project suitable for modular massively-parallel simulation code. Here we describe a few of the results we have obtained in building the framework for such a simulation code.

There are six energy components of the WINDMI model. The largest reservoir of energy is the total magnetic energy in the system which is given by the inductance matrices and current loops. We define the cross-tail loops by I and the field-aligned current loops by I_1 and I_2, where 1 and 2 denote the field-aligned currents in the region 1 and 2 sense respectively. The total system energy, W, is the sum of magnetic energy, $\mathbf{E} \times \mathbf{B}$ flow energy, the plasma thermal and parallel flow energy components. The $\mathbf{E} \times \mathbf{B}$ flow energy equals the geotail convection flow plus the magnetosphere-ionosphere coupled driven flows.

The basic idea of dividing the nightside geotail region into connected spatial cells to investigate the propagation and timing of signals and shocks between the auroral field-line crossing points and the near-Earth neutral line (NENL) region is given in *Horton et al.* (2002). Figure 1 shows a schematic of the spatial cells and the network of cross-field and field-aligned current loops used in the model. Since the model clearly will have considerable kinetic theory physics, it is useful to define the spatial cells based on a classical magnetosphere field model, rather than an MHD code. Due to its wide-spread use, it is convenient to use the Tsyganenko field models to define and calculate various quantities as required for the model (*Tsyganenko*, 1987). Even with the limitation that the Tsyganenko model represents average magnetospheric conditions, we feel that the universality of the model and relative simplicity of its use makes it the best choice for formulating a synthesis or multimode model of the driven nightside magnetosphere-ionosphere system. The accuracy of the mapping from the ionosphere to the geotail by the Tsyganenko models (*Maynard et al.*, 2001) is probably as high as can be currently achieved with any standard model.

The cellular network model in Fig. 1 is a projection of the global currents into current-voltage vectors $\{I_j, I_j^1\}$ and $\{V_j, V_j^1\}$, the intensive pressure field a vector field $\{p_j\}$ onto the cells ω_j that form the central plasma sheet volume $\Omega = \cup \omega_i$ is the union of spatial cells ω_i. Each cell then is coupled to neighboring cells by the flux of plasma and the Poynting flux. Current diversion occurs by a reduction of the cross-field currents in one or more cells, and from drops in the cell's conductance Σ_i. Magnetic reconnection readily takes place with $B_i(t)$ becoming negative when the current distributions between cells are strongly disturbed from the Tsyganenko values.

The dynamical equations are obtained by applying the integral forms of the pre-Maxwell equations to each cell and current loop. In this way the derivation parallels that of the finite element method, where a partial differential equation is integrated over some suitable zone and not forced to be satisfied at particular spatial points or nodes. Making the projections of the partial differential equation of plasma physics on the magnetized cells then yields a system of vector dynamical equations that conserve energy and charge.

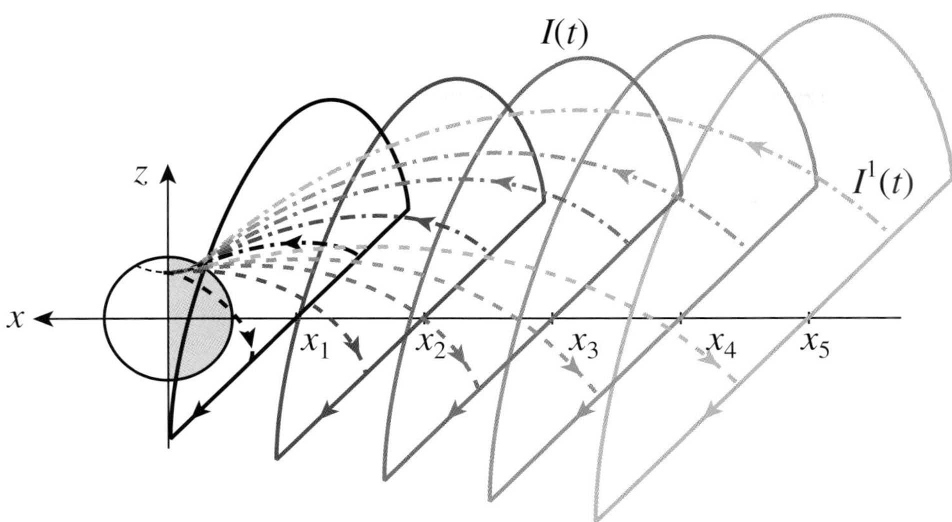

Figure 1. A schematic of the spatial cells and the network of cross-field and field-aligned current loops of the vector model. In Sec. 3.1 we add the ring current $I_{\text{rc}}(t)$ and the key plasma energy reservoir $W_{\text{rc}}(t)$ for the ring current.

The dynamics of the spatially-resolved WINDMI system shown in Figure. 1 is given by the discretized current-voltage-pressure vector equations:

$$L_{ij}\frac{dI_j}{dt} = V_i^{\text{SW}} - V_j(t) + M_{ij}\frac{dI_j^1}{dt} \tag{2.1}$$

$$C_j\frac{dV_j}{dt} = I_j - I_j^1 - \frac{(B_j - B_{j-1})\Delta z}{\mu_0} - I^{\text{MHD}}(p_j) - \Sigma_j V_j \tag{2.2}$$

$$\frac{\partial B_j}{\partial t} = -\frac{V_{j+1} - V_j}{\Delta x_j L_y} \tag{2.3}$$

$$\frac{3}{2}\frac{dp_i}{dt} = \frac{\Sigma_j}{\Omega_j}V_j^2 - u_{0j}\left(\frac{K_{\|j}}{\rho_j \Omega_j}\right)^{1/2}\Theta(I_j - I_{j,c})p_j - \frac{3}{2}\frac{p_i}{\tau_E} \tag{2.4}$$

$$\frac{dK_j}{dt} = I^{\text{MHD}}(p_j)V_j - \frac{K_j}{\tau_\|} \tag{2.5}$$

$$L_{ij}^1\frac{dI_j^1}{dt} = V_i - V_i^1 + M_{ij}\frac{dI_j}{dt} \tag{2.6}$$

$$C_j^1\frac{dV_j^1}{dt} = I_j^1 - \Sigma_j^1 V_j^1. \tag{2.7}$$

The current loops $I_i(t)$ now interact through the magnetic flux from the time-dependent magnetic dipole loop $A_i I_i$ at x_i through dipole loop $A_j I_j$ at x_j. Here A_i and A_j use the areas of the lobe field regions at x_i and x_j. This magnetic interaction is described by the geotail inductance

434

matrix L_{ij} with

$$L_{ij} = \frac{\mu_0 A_i A_j}{((x_i - x_j)^2 + R^2)^{3/2}} \tag{2.8}$$

and the total lobe magnetic energy in the geotail (gt) is given by

$$W_M^{gt} = \frac{1}{2} \sum_{i,j}^{N} L_{ij} I_i I_j \tag{2.9}$$

which is of order 5×10^{15} J when computed from the Tsyganenko model (*Horton et al.* 1998b). The plasma capacitance C_j is

$$C_j = \frac{\Delta x}{L_y} \left(\int_{\Delta z_j} dz \frac{\rho}{B^2} + \Delta z_j \epsilon_0 \right) \tag{2.10}$$

where $\epsilon_0 = 8.895 \times 10^{-12}$ F/m and ρ is the mass density. The collisionless kinetic plasma sheet conductance is $\Sigma = \Delta x \Delta z \sigma / L_y$ with $\sigma \simeq 0.1(ne/B_n)(\rho_i/\Delta z)^{1/2}$ (*Horton and Tajima* 1991a, 1991b).

The key property of this network system is the conservation of energy W. The system energy is

$$
\begin{aligned}
W &= \frac{1}{2} \sum_{i,j}^{N} \left[L_{ij} I_i I_j - M_{ij} I_i I_j^1 + L_{ij}^1 I_i^1 I_j^1 \right] \\
&\quad + \frac{1}{2} \sum_{j}^{N} \left[C_j V_j^2 + C_j^1 (V_j^1)^2 \right] + \sum_{j}^{N} \left[\frac{3}{2} p_j \Omega_j + K_j \right].
\end{aligned} \tag{2.11}
$$

The rate of change of the total energy in W associated with Eqs. (2.1)-(2.7) is determined by

$$\frac{dW}{dt} = \sum_{j}^{N} \left[I_j V_j^{\text{SW}} - \Sigma_j^1 (V_j^1)^2 - u_{0j} \left(\frac{K_{\|j}}{\rho_j \Omega_j} \right)^{1/2} \Theta(I_j - I_{j,c}) p_j \right] - \frac{K_\|}{\tau_\|}. \tag{2.12}$$

The nine physical quantities, $L, L_I, C, C_I, \Sigma, \Sigma_I, M, u_0$ and Ω_{cps} are the magnetospheric and ionospheric inductance, capacitance, and conductance respectively. M is the mutual inductance between the geotail lobe flux and the field-aligned current loop closing in the nightside auroral ionosphere. The plasma sheet pressure gradient drives the current given by $I_j^{\text{MHD}} = \alpha_j p_j^{1/2}(t)$ as derived from force balance and Ampere's law. The parameter α_j is an average over the pressure profile in cell ω_j of the current sheet with an approximate value $(\alpha_j = \Delta x_j \mu_0^{-1/2})$ computed from the lobe magnetic pressure in cell ω_j balanced with the plasma sheet pressure.

The plasma sheet capacitance C_j is determined by the mass density in cell ω_j so that Eq. (2.2) is the projection of the perpendicular component of Newton's second law in the jth cell.

The solar wind driving voltage in Eq. (2.1) is the input time series for this nonlinear driven-dissipative system. The central plasma sheet unloading term $u_0 K^{1/2} \Theta(I - I_c) p$ in Eq. (2.3) represents the rapid unloading of the stored plasma energy when the current exceeds a critical value, I_c.

The trigger value I_c can be taken from the basic instability criteria of magnetic reconnection or of the current diversion instability (*Chang, Lui and Yoon*, 1994) and (*Yoon and Lui*, 1996). The parallel heat flux limit that is neglected in the MHD closure (*Horton and Doxas*, 1996) determines the loss rate parameter u_0. In the absence of driving ($V_{\mathrm{SW}} = 0$) and damping ($\Sigma_1 = 0$ and $\tau_{\parallel} \sim \tau_E \to \infty$), and below the unloading limit, $I < I_c$, the total energy is conserved. All parameters of the model can be calculated explicitly from their integral definitions by the plasma conditions in the tail. Examples of these calculations for the Tsyganenko models are in *Horton et al.* (1998b).

The collisionless large ion gyroradius transport for the geomagnetic tail, which appears in the conductivity, Σ, and transfers energy between Eqs. (2.2) and (2.4). The plasma sheet conductivity, Σ, arises from the large ion gyroradius momentum stress tensor with the cross-tail voltage $V(t)$ driving the sheared Earth flow. The viscous heating of the ions is given $\Sigma_j V_j^2(t)$ in cell ω_j. This heating vanishes in the MHD limit where the gyroradius is taken as infinitesimal. The conductivity was derived from theory and particle simulations (*Horton and Tajima* 1991a, 1991b) and contains the *Lyons and Speiser* (1982) energization mechanism for the transient and reflected ions. Alternatively, the cross-current diversion instability of *Yoon and Liu* (1996) gives a drop in the value of Σ_j to 0.1 mho during the period $j_y(x_j)/en_j \gtrsim 0.5(T_i/m_i)^{1/2}$ for their reference case.

In the absence of solar wind driving and ionospheric loss terms, the total energy is constant of the complex dynamical system. The energy transfer coupling elements cancel in terms of six pairs for each spatial cell ω_j. For example, the energy transfer to the plasma pressure from the $\mathbf{E} \times \mathbf{B}$ kinetic energy $K_\perp(t)$ is given by $P_{p|K_\perp} = \Sigma_j V_j^2(t)$ as found by adding the energies formed from Eqs. (2.2) and (2.4). A typical value for $P_{p|K_\perp}$ is a few gigawatts. The six power transfer terms are $P_{I|V} = I \cdot V = \Sigma_j I_j V_j, P_{V|\mathrm{MHD}} = V \cdot I^{\mathrm{MHD}}, P_{V|p} = V \cdot \Sigma \cdot V, P_{I_1|V_1} = I_1 \cdot V_1,$ $P_{I|I_1} = \partial_t(I \cdot M \cdot I_1)$, and $P_{I_2|V_1} = I_2 \cdot V_1$. In the no driving-no damping limit the system is Hamiltonian in structure. The conservation of energy is a key feature of many systems with multiple energy reservoirs. In the closed system (Hamiltonian) limit with no driving and dissipation there is a wide spectrum of eigenmodes to the system.

In the lowest frequency modes the currents oscillate synchronously so that there are two global WINDMI modes of approximately one to two hours and of 15 minutes. We have a parallel Fortran 90 code that solves the system of Eqs. (2.1)-(2.7) by first inverting the 2×2 block matrices of L, M, and L_1 and then using an adaptive integrator to advance the system of $8N$ ordinary differential equations. We are investigating models with $N = 20$ and $N = 100$.

2.1 Dynamics of Compressional Dipolarization Pulses

In the present Multi-Mode Dynamical Model we break up the large geotail system with total current $I(t)$ of the global WINDMI model into subcells ω_i that cover the lobe region from the near-Earth to the distant tail. For example, the Tsyganenko model (*Tsyganenko*, 1987) has been used to find partitions $\{x_i\}^N$ along the geotail that contain comparable amounts of current of 1 MA per cell to give the full $I \simeq 20\,\mathrm{MA}$ geotail current could be represented by $N = 20$ cells.

In the spatially-resolved model, each geotail cell contains dynamical equations for the current $I_i(t)$, cross-tail voltage $V_i(t)$ and the magnetic field component, $B_i(t)$, of the normal ($\hat{z}$) to the equatorial plane. The net cross-tail current is $I_j^{\mathrm{gt}} = I_j - I_j^1$ and is equivalent to that cell's contribution of the lobe magnetic field $B_{xj} = \mu_0 I_j^{\mathrm{gt}}(t)/\Delta x_j$. The total northward magnetic field is

$$B_z(x_j, t) = B_{\mathrm{dp}}(x_j) + B_j(x_j, t) \tag{2.13}$$

where B_{dp} is from the Earth's magnetic dipole. The cross-tail is a convection electric field,

$E_y(x_j, t) = V_j(t)/L_y$, where L_y could be taken as $j-$dependent, if required. Thus, there is an Earthward directed Poynting flux of

$$S_j = \frac{1}{\mu_0} E_y(x_j, t) B_z(x_j, t). \tag{2.14}$$

In the midtail region $x \approx -20 R_E$, the dynamics easily gives regions of negative $B_z(x_j, t)$ due to current disturbances. Cross-tail current diversions are created by decreases in I_j, increases in I_j^1 and abrupt changes in Σ_j and $I^{\text{MHD}}(p_j)$ in Eq. (2.2).

2.2 Dipolarization Alfvénic Pulse Propagation

The simulations with Eqs. (2.1)-(2.7) show magnetic pulses propagating both Earthward and tailward with speeds slower than the local Alfvén speed. The pulses are complex structures and have a nonlinear steepening when the capacitance in Eq. (2.2) uses the local magnetic field in Eq. (2.10). The full description of the pulses requires a discussion of the eigenmodes of the vector-matrix equations, Eqs. (2.1)-(2.7). Here we simplify the discussion of the eigenvalue problem by using a local continuum approximation in Eqs. (2.2) and (2.3). For the continuum approximation we replace $(f_j - f_{j-1})/\Delta x$ with df/dx in Eqs. (2.3) and (2.3).

A simple numerical response to a current diversion event is shown in Figure. 2 where the steady state is perturbed by decreasing by 25% the currents I_j in the region between $-6 R_E$ and $-18 R_E$. The resulting dynamics is shown for 10 traces of the cross-tail voltage $V(x, t)$ after applying the 25% cross-tail current reduction and 25% increase V_i to keep the power, $I_i V_i$, constant in the perturbation. There are two large scale pulses emitted. The smaller, slower, pulse travels Earthward from $x = -15 R_E$ to $-10 R_E$ in 120s indicating a velocity of $2.5 R_E/$ min. The larger pulse propagates tailward from $x = -16 R_E$ to $-25 R_E$ with velocity of $-5 R_E/$ min. After 5-10 minutes, the system returns to its original state. From other numerical experiments, we find the perturbations in the auroral field-line crossing zone creates large, fast tailward propagating pulses and slower, smaller and partially reflecting Earthward propagating pulses. We now analyze the equations for these pulses.

To understand pulse propagation, we take the time derivative of Eq. (2.2) to form $\partial_t(C_j \partial V_j/\partial t)$ and eliminate $\partial_t B_j$ with Eq. (2.3) and then using the continuum limit of $(V_{j+1} - V_j)/\Delta x_j$ and $(B_{j+1} - B_j)/\Delta x_j$. We thus derive the local, approximate equation valid for the propagation of long waves $k_x \Delta x_j \ll 1$. The lower-order eigenmodes of the full system, Eqs. (2.1)-(2.7), follow these continuum limit modes when $\Delta x = L_x/N$ is small. The nonlinear, nonlocal pulse equation is

$$\frac{\partial}{\partial t}\left(C \frac{\partial V}{\partial t}\right) = -\mathcal{L}^{-1}\left(\begin{array}{c} V \quad -V^{\text{SW}} \\ V \quad -V_1 \end{array}\right) + \frac{\Delta z \Delta x}{\mu_0 L_y} \frac{\partial^2 V}{\partial x^2} \tag{2.15}$$

$$- \Sigma \frac{\partial V}{\partial t} - \frac{dI^{\text{MHD}}}{dp}\left(\frac{2\Sigma V^2}{3\Omega_{\text{cps}}} - u_0 \Theta P\right) \tag{2.16}$$

where $\mathcal{L}$ is the 2×2 symmetric block matrix composed of the $N \times N$ matrices $L, M,$ and $L1$. The local waves, $\delta V(t)\cos(k_x x)$, are of frequency $\omega(k_x)$ where

$$\omega^2 = C^{-1}\mathcal{L}^{-1} + \frac{\Delta z}{\mu_0 L_y} C^{-1} k_x^2 = \omega_0^2 + k_x^2 \overline{V}_A^2 \tag{2.17}$$

with $\omega_0^2 = (C\mathcal{L})^{-1}$ — the lowest global eigenvalue and $\overline{V}_A^2 = \Delta z \Delta x/\mu_0 L_y C$ the composite Alfvén velocity made up of both the lobe magnetic field, $B_l = \Delta z B_x'$ and the north-south

$B_n(t)$ equatorial plane magnetic field through $\overline{V}_A^2 = \Delta z B_x' B_n / \pi \rho_0 \mu_0$ with ambient mass density, $\rho_0 = \sum_i m_i n_i (z = 0)$ along the geotail axis. *Moore et al.* (1981) suggest that the mass density, ρ_0, may contain a substantial contribution of O^+ ions from ionospheric outflows if there has been earlier substorm activity.

The important feature of the compressional waves given by Eq. (2.15) is that there is a cut-off frequency ω_0 below which all fluctuations are evanescent and thus, non-propagating. Large-scale initial disturbances from a current disruption or magnetic reconnection event will then have only the smaller space scale spectral components, $|k_x| > \omega_0 / \overline{V}_A$ in the propagating pulse. Near the cut-off frequency, ω_0, the group velocity, $v_g = d\omega / dk_x$, vanishes and the phase velocity, v_p, goes to infinity with the product constant $v_g v_p = \overline{V}_A^2$ as is standard for TE (and TM) waves in arbitrarily-shaped geometry (*Jackson*, 1999). Here TE and TM are the transverse electric ($E_y \neq 0$) and transverse magnetic ($B_y \neq 0$) cavity modes of the lobe-plasma sheet cavity.

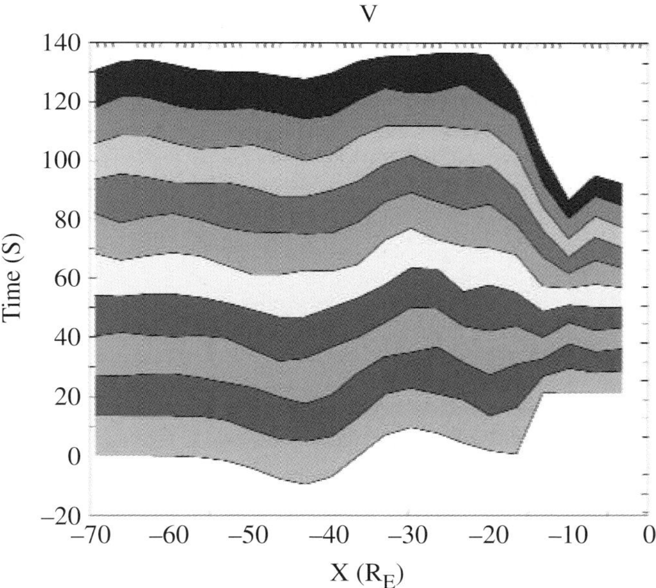

Figure 2. Propagation of the cross-tail voltage, $V(x_i, t)$, disturbance for 2 min after a localized current diversion in the region $-6.6R_E$ to $-12.2R_E$ where a 25% reduction in the cross-tail current occurs.

Thus, in addition to the oxygen loading used by *Moore et al.* (1981) to explain the sub-Alfvénic speeds of the dipolarization pulse, the arbitrary large space structures such as those hypothesized by *Li et al.* (1993) will propagate with sub-Alfvénic speeds, $v_A(1 - \omega_0^2/\omega^2)^{1/2}$, owing to their low-frequency-long axial wavelengths.

The value of the longest cut-off period, $T_0 = 2\pi(\mathcal{L}C)^{1/2}$ has not been well determined yet.

438

Rough estimates with $\mathcal{L} = 50H$ and $C = 5 \times 10^4$ F yield $T_0 \approx 10^4$ s$\simeq$ 2.8 hr. However, we find, using the genetic algorithm to optimize the performance of the global $d = 8$ model given in the next section, that $L = 100$ H, $L_1 = 10$ H and $M = 3$ H. Clearly, the value of L and C are dependent on the shape and size of the magnetospheric tail cavity. The product LC is proportional to the mean height (north-south) of the magnetolobe cavity, but independent of the dimensions of L_x and L_y. The shape and size of the cavity are known to increase with strong southward IMF. The relevant estimate of T_0 requires knowledge of the solar wind conditions over the period of time T_0 to $2T_0$ starting somewhat prior to the interval of concern.

3. Ring Current Energy $\mathbf{W}_{rc}$ and the $\mathbf{D}_{st}$ Prediction of WINDMI-RC

Consider the magnetosphere as the union, $\cup$, of the volumes of the lobe Ω_l, the central plasma sheet Ω_{cps}, the small transition (Alfvén) volume surrounding the separatrix (Alfvén layer) between the ring current region and the outer magnetospheric plasma Ω_{tr}, and the volume of the ring current itself Ω_{rc}. The volume of the magnetosphere, Ω_{mag}, is then $\Omega_{\mathrm{rc}} \cup \Omega_{\mathrm{tr}} \cup \Omega_{\mathrm{cps}} \cup \Omega_l$. The total plasma energy is decomposed as $W_p = \sum_a \int_{\Omega_s} d^3x \frac{3}{2}p = \frac{3}{2}p_{\mathrm{cps}}\Omega_{\mathrm{cps}} + \frac{3}{2}p_{\mathrm{rc}}\Omega_{\mathrm{rc}}$, since the particle pressure is negligible in the lobe plasma and the transition layer volume is a small fraction of Ω_{cps} and Ω_{rc}. The Dessler-Parker relation (*Dessler and Parker*, 1959; *Sckopke*, 1966) gives the D_{st} index through the total ring current energy component $W_{\mathrm{rc}}(t) = \frac{3}{2}\Omega_{\mathrm{rc}}p_{\mathrm{rc}}$. The principal source of ring current energy, W_{rc}, is the fraction of the central plasma sheet thermal flux, $p_{\mathrm{cps}}v_x$, that crosses the Alfvén layer. Through test particle integration we define the effective area, A_{eff}, for the scattering or admittance of central plasma sheet (CPS) plasma across the Alfvén layer. This gives a source of power, $P_{\mathrm{rc}} = v_x p_{\mathrm{cps}} A_{\mathrm{eff}}$, for the ring current energization.

The Dessler-Parker relation

$$\frac{\Delta B_{\mathrm{particles}}}{B_E} = -\frac{\mu_0}{2\pi}\frac{W_{\mathrm{rc}}(t)}{B_E^2 R_E^3} \tag{3.1}$$

is a robust relation that gives the change in magnetic field, $\Delta B_{\mathrm{particles}}$, from the ring current energy, $W_{\mathrm{rc}}(t)$. The change in field, $\Delta B_{\mathrm{particles}}$, is closely related to the magnetometer measurements used for the D_{st} magnetic index. We note that there are other contributions to D_{st} as given by *Liemohn* (2003) from the partial ring current and *Turner et al.* (2000) from the geotail current. Thus the ring current $I_{\mathrm{rc}}(t)$ and the associated induction electric field from the ring current are conjugate variables for the inner magnetosphere electrodynamics. In Eq. (3.1), $B_E = 3.1 \times 10^{-5}$ T is the magnetic field strength of the Earth's internal dipole field at the Earth's surface on the equator.

The fractional diversion of the incoming central plasma sheet particles across the separatrix formed by the corotation electrostatic field and the storm time varying convection electric field can be considered as a scattering and trapping problem for ensembles of electron and ion in plasma sheet kappa distributions. Such a process can then be modeled by an effective admission cross-section, A_{eff} for the fraction of the particles that cross the separatrix. Of course this fraction depends on the nature of the storm time electric field and the state of the transitional region of the transitional magnetic field so only a rough characterization of the trapping fraction is possible from theory. The parameter A_{eff} is then one of the critical parameters of the model, and can either be estimated from theory, estimated by Monte Carlo simulations, or viewed as a parameter to be optimized within a specified physical bounds. A schematic of the ring current model is shown in Figure 3.

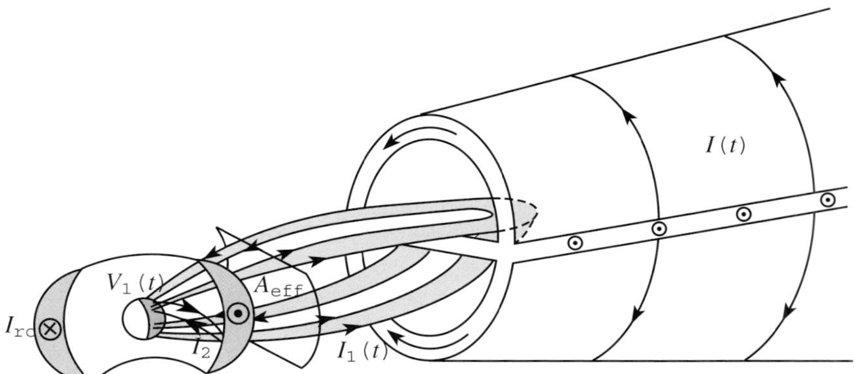

Figure 3. This figure gives the overall geometry of the model, including the ring current and the shielding region 2 current closing in the ring current.

The effective aperture, A_{eff}, is shown schematically as the cross-section that defines the fraction of the plasma sheet particles that cross into the ring current plasma.

Table 1 lists the parameters of the model, with estimates of typical values. Figure 4 shows the results of the WINDMI-RC model for the storm of May 15, 1997. The values of the model parameters used in that particular simulation are given in parenthesis in Table 1. We see that the model gives good agreement with observations for reasonable values of the parameters.

3.1 Global WINDMI-RC Model

The base WINDMI-RC model is an eight-dimensional $(d = 8)$ 21-parameter system, given by:

$$L\frac{dI}{dt} = V_{\text{sw}}(t) - V + M\frac{dI_1}{dt} \tag{3.2}$$

$$C\frac{dV}{dt} = I - I_1 - I_{\text{ps}} - \Sigma V \tag{3.3}$$

$$\frac{3}{2}\frac{dp}{dt} = \frac{\Sigma V^2}{\Omega_{\text{cps}}} - u_0 K_{\|}^{1/2}\Theta(I - I_c) - \frac{pV A_{\text{eff}}}{\Omega_{\text{cps}} B_{\text{tr}} L_y} - \frac{3p}{2\tau_E} \tag{3.4}$$

$$\frac{dK_{\|}}{dt} = I_{ps}V - \frac{K_{\|}}{\tau_{\|}} \tag{3.5}$$

$$L_1\frac{dI_1}{dt} = V - V_1 + M\frac{dI}{dt} \tag{3.6}$$

$$C_1\frac{dV_1}{dt} = I_1 - I_2 - \Sigma_1 V_1 \tag{3.7}$$

$$L_2\frac{dI_2}{dt} = V_1 - (R_{\text{prc}} + R_{A2})I_2 \tag{3.8}$$

$$\frac{dW_{\text{rc}}}{dt} = R_{\text{prc}}I_2^2 + \frac{pV A_{\text{eff}}}{B_{\text{tr}} L_y} - \frac{W_{\text{rc}}}{\tau_{\text{RC}}} \tag{3.9}$$

where the D_{st} signal is given by

$$D_{\text{st}} = -\frac{\mu_0}{2\pi}\frac{W_{\text{rc}}(t)}{B_E R_E^3}.$$

(3.10)

L_2 is the inductance of the region 2 current loop, and R_{prc} and R_{A2} are the resistances of the partial ring current and the region 2 foot point in the auroral region. The system of Eqs. (3.2)-(3.9) is the WINDMI-RC model, a $d = 8$ system of equations for the solar wind driven coupled Inner Magnetosphere, Ionosphere, and Tail. Together with Eq. (3.10), the model gives a prediction of the D_{st} index while conserving energy in the coupled Inner-Magnetosphere-Tail system. We conclude this section with a discussion of the shielding current and the ring current.

From the *Siscoe* (1982) model of the R1-R2 auroral coupling region, we infer nominal-to-maximum currents and voltages of $I_1 \sim 1\,\text{MA}$, $I_2 \sim 500\,\text{kA}$, $V_1 \sim 100\,\text{kV}$, and $R_{\text{rpc}}I_2 \simeq 50\,\text{kV}$ - $100\,\text{kV}$, so that the partial ring current energization power is of order 25-50 GW. *Stern* (1990) gives an estimate of the combined power of the region 1 and region 2 currents of 300 GW so that 30 GW would only be 10% of Stern's power estimate (*Stern*, 1990,1991).

The coupling of the region 1, region 2, and ring current contained in the model is basically the *Siscoe* (1982) model. Some of the details of the north-south Hall currents and Hall conductivities have been abridged in writing Eqs. (3.7)-(3.9). The Siscoe model allows for the compression to a smaller radius $R(t)$ of the ring current during the energization phase.

Equations (3.8) and (3.9) contain new power transfer terms $P_{p|r_c} = pV A_{\text{eff}}/B_{\text{tr}}L_y$, $P_{J_2|r_c} = R_{\text{prc}}I_2^2$ and power loss terms $P_{A2}^{\text{loss}} = R_{A2}I_2^2$ and $P_{\text{rc}}^{\text{loss}} = W_{\text{rc}}/\tau_{\text{RC}}$. The ring current characteristics including estimates of the decay time τ_{RC} are taken from *Chen and Schultz* (1996).

The WINDMI-RC network in Eqs. (3.2)-(3.9) has numerous physical parameters that are known only roughly. Thus, we specify the acceptable physical range of the parameters and use the genetic algorithm to determine the optimal values. Table 1 shows one example.

Table 1: WINDMI-RC Model and Genetic Algorithm Optimization GEM Storm October 21, 2001

Parameter	Physics Estimate	Range	GA Value
L (H)	40	10-100	85
M (H)	1	0.1-3	1.5
C (F)	10^4	1500-15000	6500
Σ (mho)	10	2-20	18
$\Omega_{\text{cps}}(m^3)$	4×10^{24}	$(1 - 10) \times 10^{24}$	1.4×10^{24}
u_0	1.5×10^{-9}	$(0.5 - 2.5) \times 10^{-9}$	1.5×10^{-9}
I_c [A]	2×10^7	$(0.5 - 5) \times 10^7$	1.5×10^7
$\alpha[A \cdot \text{Pa}^{-1/2}]$	8×10^{11}	5×10^{10} to 9×10^{11}	8.9×10^{11}
$\tau_{\|}$ [s]	600	300-900	600
τ_E [s]	2000	500-4000	2300
L_I [M]	5	1-10	7.6
C_I [F]	800	100-1500	850
Σ_I [mho]	5	0.5-10	8.5
β	0.7	0.05-0.5	0.38
R_{prc} [ohm]	0.1	0.05-0.5	0.06
τ_{rc} [hr]	12	1-24	5.55
L_2 [H]	8	1-20	1.2
R_{A2} [ohm]	0.3	0.05-2	1.7
B_{tr} [nT]	5	1-10	7.4×10^{-9}
A_{eff} $[m^2]$	$2R_E^2$	$2 \times 10^{13} - 2 \times 10^{14}$	4.1×10^{13}
L_y [m]	$5R_E$	$6 \times 10^6 - 6 \times 10^7$	1.5×10^7

The genetic algorithm is an algorithm to search the parameter space in Table 1 for a global optimum by combining, sharing and/or changing parameter values using inheritance, crossover and mutation based on evolutionary biological principles.

3.2 Connection of WINDMI-RC with the Burton Equation and the Temerin-Li Model

Equations (3.9) should be compared with the Burton equation (*Burton et al.*, 1975), and the new, more complex (*Temerin and Li*, 2002) model. The *Burton et al.* (1975) formulation gives the D_{st} as

$$D_{\text{st}} = D_{\text{st}}^{\text{rc}} + b\sqrt{P_{\text{dyn}}} - c. \tag{3.11}$$

The D_{st} is a sum of three terms: the contribution from the ring current, the contribution from the magnetopause current, which is dependent on the dynamic pressure, p_{dyn}, and the quiet condition represented by c. The ring current contribution to the D_{st}, $D_{\text{st}}^{\text{rc}}$, is

$$\frac{d}{dt}D_{\text{st}}^{\text{rc}} = F(E_{\text{sw}}) - aD_{\text{st}}^{\text{rc}} \tag{3.12}$$

where $F(E_{\text{sw}})$ is the injection rate of energy into the ring current, which depends on the solar wind electric field E_{sw}, and a is the inverse decay time of the ring current. $F(E_{\text{sw}})$ is modeled as the bifurcation event with bifurcation parameter E_{sw}. For $E_{\text{sw}} < 0.5\,\text{mV/m}$, $F(E_{\text{sw}}) = 0$, and for $E_{\text{sw}} > 0.5\,\text{mV/m}$, $F(E_{\text{sw}}) = -1.5 \cdot 10^{-3}(E_{\text{sw}} - 0.5)\,\text{nT/(s} \cdot \text{mV/m)}$.

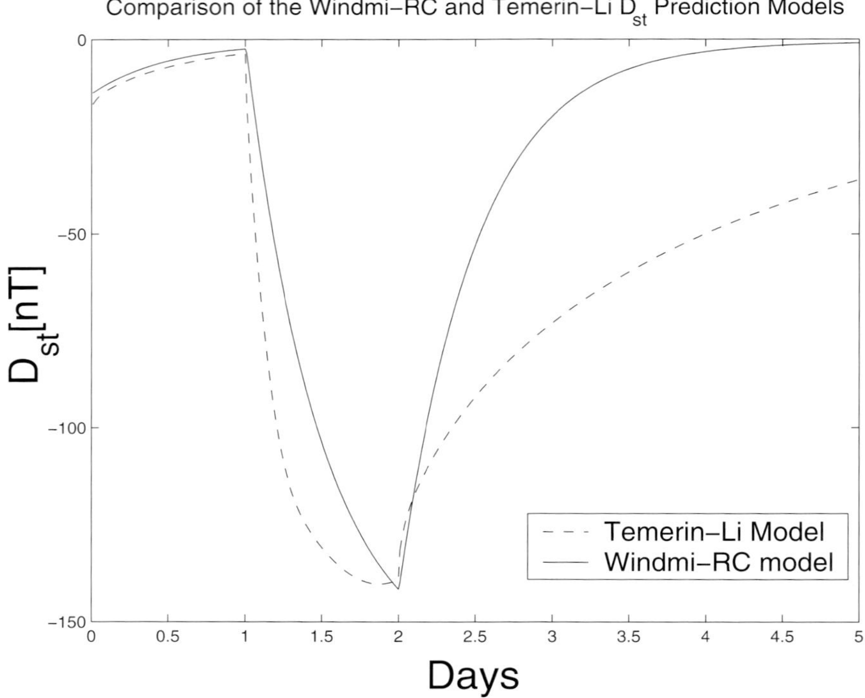

Figure 4. Comparison of the WINDMI-RC model (solid line) to the Temerin-Li D_{st} model for a sample solar wind input. The solar wind input here is for constant a solar wind speed of 500 km/s and a z-component of the magnetic field of 3 nT which turns south for one day to -3 nT and returns to its initial value. For a suitable choice of parameters, the WINDMI-RC model quantitatively agrees with the results of the Temerin-Li model. The parameters of the Temerin-Li model have been obtained by minimizing the root-mean-square between prediction and data by hand.

The Temerin-Li formulation represents the D_{st} as a sum of several terms:

$$D_{st} = d_{st}^1 + d_{st}^2 + d_{st}^3 + \text{Pressure term} + f(B_z^{IMF}). \tag{3.13}$$

The *Pressure Term* is not given only by, $\sqrt{p_{dyn}}$, as in the *Burton et al.* (1975) formulation, but includes a dependence on the solar wind density. The three terms, d_{st}^i, are introduced empirically where d_{st}^1 represents the ring current contribution as done by Burton and $d_{st}^2 + d_{st}^3$ "probably represents some combination of the so-called partial ring current and the tail currents" (*Temerin and Li*, 2002). The key new physics in the WINDMI-RC model is the effective aperture, A_{eff}, which will need to be parameterized in terms of B_z^{IMF} similar to the $f(B_z^{IMF})$ term in Eq. (3.13).

In Fig. 4, we make a comparison of the D_{st} as predicted by WINDMI-RC and by the Temerin-Li model for a simple input described in the figure caption. The WINDMI-RC model remains to be optimized by means of the Genetic Algorithm employed on the model but shows the potential to compete with emperical models. More details are given in *Doxas et al.* (2004).

4. Conclusion

The low-dimensional WINDMI model describes the energy storage and coupling in large cells of the solar wind-magnetosphere-ionosphere system. A spatially resolved, high dimensional version of the WINDMI model is shown to be capable of determining the timing and causal connection issues involving the auroral field line onset of substorms versus the near Earth neutral line model. We call this the Multimode Model of substorms which is a quantitative synthesis model. This realization of the of the synthesis model of *Lui* (1991) has two important results: the fast tailward propagation of a current diversion at the position of the auroral oval field-line crossing of the magnetic equatorial plane in dusk-to-midnight sector, and that the 3D geometry of the lobe-magnetotail cavity reduces the speed of the electromagnetic pulse propagation for large scale disturbances. The dipolarization pulse speeds of 50-100 km/s below the Alfvén speed are predicted by the vector model.

The vector WINDMI model may have the potential to resolve the controversy over the substorm onset mechanism by giving a quantitative description of the time for a near-Earth neutral line formation to trigger an auroral field line substorm signature and the inverse process of an auroral field line current diversion to trigger. To realize this potential actual solar wind input signals of density, velocity and IMF conditions may be needed for driving the model.

Furthermore, the global model has been extended to have the shielding currents, I_2, and the ring current, I_{rc}. A dynamical network model of the coupled inner magnetosphere to the solar wind-driven M-I system has been constructed by considering the coupling of the ring current energy to the central plasma sheet. The Dessler-Parker relation is used to compute the D_{st} from the ring current energy. The effective aperture A_{eff} for transferring energy from the central plasma sheet (CPS) through the Alfvén separatrix layer to the ring current can be either estimated or used as a parameter to be optimized.

The major advantage of this proposed network model is that the separate components of the magnetospheric current system, such as the ring current, tail current, and partial ring current, are introduced explicitly and not in an empirical manner. The terms and coefficients have physical meaning in contrast to those in digital signal processing filters. It is important to determine the relative contributions of the current systems to the total D_{st}. For example, *Turner et al.* (2000) determine the contribution from the tail current to be approximately one quarter of the total D_{st}. *Liemohn et al.* (2003) gives the contribution from the partial ring current I_2 to the D_{st}.

It may well be that the network model will give a useful intermediate level tool for interpreting and ordering storm data. Contracting a large storm observational database into a smaller physically-ordered database would be of considerable benefit to space science and MHD modelers. In any event, we claim that the network method of describing the complex, three-dimensional magnetosphere is a mathematically and physically sound approach that needs to be developed.

Acknowledgments

The University of Texas acknowledges support from the National Science Foundation through Grant ATM-0229863.

References

Burton, R. K., R. L. McPherron, and C. T. Russel, An empirical relationship between interplanetary conditions and D_{st}, *J. Geophys. Phys.*, *80*, 4204, 1975.

Chang, C. L., A. T. Y. Lui, and P. H. Yoon, Nonlocal stability analysis of the current driven instabilities in magnetotail, *EOS Trans. AGU*, *75*, no. 16 supplement, 308, 1994.

Chen, M. W., and M. Schulz, Ring current formation and decay: a review of modeling work, *Advances in Space Research*, *17*(10), 7-16, 1996 Elsevier, UK.

Dessler, A. J., and E. N. Parker, Hydromagnetic theory of geomagnetic storms, *J. Geophys. Res.*, *64*, 2239, 1959.

I. Doxas, W. Horton, W. Lin, S. Seibert, and M. Mithaiwala, A dynamical model for the coupled inner magnetosphere and tail, *IEEE Transactions on Plasma Science*, *32*, 1443, Aug. 2004.

Horton, W., R. S. Weigel, D. Vassiliadis, and I. Doxas, Substorm classification with the WINDMI model, *Nonl. Proc. Geophys.*, *10*, 363, 2003.

Horton, W., C. Crabtree, I. Doxas, and R. S. Weigel, Geomagnetic transport in the solar wind driven nightside magnetosphere-ionosphere system, *Phys. Plasmas*, *9*, 3712, 2002.

Horton, W., H. V. Wong, J. W. Van Dam, and C. Crabtree, Stability properties of high-pressure geotail flux tubes, *J. Geophys. Res.*, *106*(A9), 18803-18822, 10.1029/2000JA000415, 2001.

Horton, W., and I. Doxas, A low-dimensional dynamical model for the solar wind driven geotail-ionosphere system, *J. Geophys. Res.*, *103A*, 4561, 1998a.

Horton, W., M. Pekker, and I. Doxas, Magnetic energy storage and the nightside magnetosphere-ionosphere coupling, *Geophys. Res. Lett.*, *25*(21), 4083-4086, 1998b.

Horton, W., and T. Tajima, Collisionless conductivity and stochastic heating of the plasma sheet in the geomagnetic tail, *J. Geophys. Res.*, *96*, 15,811, 1991a.

Horton, W., and T. Tajima, Transport from chaotic orbits in the geomagnetic tail, *Geophys. Res. Lett.*, *18*, 1583, 1991b.

Jackson, J. D., Classical Electrodynamics, John Wiley and Sons, Inc., 1999. p. 364.

Kan, J. R., Synthesizing a Global Model of Substorms, in *Magnetospheric Substorms, Geophysical Monograph 64*, p. 73, 1991.

Li, X., I. Roth, M. Temerin, J. Wygant, M. K. Hudson, and J. B. Blake, Simulation of the prompt energization and transport of radiation particles during the March 23, 1991 SSC, *Geophys. Res. Lett.*, *20*, 2423, 1993.

Liemohn, M. W., Yet another caveat to using the Dessler-Parker Sckopke relation, *J. Geophys. Res.*, *108*(A6), 1251, doi:10.1029/2003JA009839, 2003.

Lui, A. T. Y., A synthesis of magnetospheric substorm models, *J. Geophys. Res.*, *96*, 1849, 1991a.

Lyons, L. R., and T. W. Speiser, Evidence for current sheet acceleration in the geomagnetic tail, *J. Geophys. Res.*, *87*, 2276, 1982.

Maynard, N. C., W. J. Burke, J. Moen, D. M. Ober, J. D. Scudder, J. B. Sigwarth, G. L. Siscoe, B. U. Ö Sonnerup, W. W. White, K. D. Siebert, D. R. Weimer, G. M. Erickson, L. A. Frank, M. Lester, W. K. Peterson, C. T. Russell, G. R. Wilson, A. Egeland, Responses of the open-closed field line boundary in the evening sector to IMF changes: A source mechanism for Sun-aligned arcs, *J. Geophys. Res.*, *108*(A1), 10.1029/2001JA000174, 2001

Moore, T. E., R. L. Arnoldy, J. Feynmann, and D. A. Hardy, Propagating substorm injection fronts, *J. Geophys. Res.*, *86*, 6713, 1981.

Sckopke, N., A general relation between the energy of trapped particles and the disturbance field near the Earth, *J. Geophys. Res.*, *71*, 3125, 1966.

Siscoe, G. L., Energy coupling between regions 1 and 2 Birkeland current systems, *J. Geophys. Res.*, *87*(A7), 5124, 1982.

Stern, D. P., Substorm electrodynamics, *J. Geophys. Res.*, *95*, 12057, 1990.

Stern, D. P., The beginning of substorm research, p. 11 in Magnetospheric Substorms, eds. J. R. Kan, T. A. Potemra, S. Kokubun, and T. Iijima, American Geophysical Union, 1991.

Temerin, M., and X. Li, A new model for the prediction of D_{st} on the basis of the solar wind, *J. Geophys. Phys.*, *07*, 1472, 2002.

Tsyganenko, N. A., Global quantitative models of the geomagnetic field in the cislunar magnetosphere for different disturbance levels, *Planet. Space Sci.*, *35*, 1347, 1987.

Tsyganenko, N. A., Global quantitative models of the geomagnetic field in the cislunar magnetosphere for different disturbance levels, *Planet. Space Sci.*, *35*, 1347, 1987.

Turner, N. E., D. N. Baker, T. I. Pulkkinen, and R. L. McPherron, Evaluation of the Tail Current Contribution to D_{st}, *J. Geophys. Res.*, *105*, 5431-5439, 2000.

Yoon, Peter H., and A. T. Y. Lui, Nonlocal ion-Weibel instability in the geomagnetic tail, *J. Geophys. Res.*, *101*, 4899-4906, 1996.

Multiscale Coupling of Sun-Earth Processes   447
A.T.Y. Lui, Y. Kamide and G. Consolini, editors.

A THREE-FLUID MODEL OF SOLAR WIND-MAGNETOSPHERE-IONOSPHERE-THERMOSPHERE COUPLING

P. Song[1], V. M. Vasyliūnas[2], and L. Ma[1]

1. Center for Atmospheric Research and Department of Environmental, Earth & Atmospheric Sciences, University of Massachusetts Lowell, Lowell MA 01854 USA

2. Max-Planck-Institut für Sonnensystemforschung, 37191 Katlenburg-Lindau, Germany

Abstract. We develop a systematic method for studying the coupling from the solar wind and magnetosphere to the ionosphere and thermosphere, on the basis of a three-fluid treatment: electrons, ions, and neutrals, with collisions among the three species and electromagnetic coupling. The momentum equations of the three species are transformed into the generalized Ohm's law and the plasma momentum and neutral momentum equations. By allowing the collision frequencies and the densities to vary as functions of height, this formalism continuously and smoothly describes the plasma conditions from the collisionless solar wind and magnetospheric plasma to partially ionized ionosphere/thermosphere gases, thus bridging a significant gap between the magnetospheric and ionospheric communities. Our approach treats a structured ionosphere with a finite thickness and a completely coupled neutral wind. We solve this equation set under the simplifying assumptions of a steady state and one-dimensional (vertical) spatial dependence, to model the magnetosphere-ionosphere/thermosphere coupling at high latitudes   near the poles where Birkeland (field-aligned) currents and precipitating particles are not important. Since the solution is self-consistent, the neutral wind speed is a function of height and thus cannot be taken as specifying a single frame of reference as treated in conventional methods. The results show that the F- and E-layers of the ionosphere may behave differently, particularly under dynamic conditions when the IMF changes its direction after a long period of steady state.

1. Introduction

This work is the continuation of our recent investigations on solar wind-magnetosphere-ionosphere coupling. Song et al. [2001] derived a three-fluid Ohm's law to describe the steady state relationship in a structured ionosphere that is coupled to both the magnetosphere and the thermosphere. They showed that for a given magnetospheric condition, the ionospheric plasma velocity varies continuously in both direction and magnitude from the magnetosphere to the ionosphere as collisions become more and more important. The present study provides a quantitative description. Vasyliūnas [2001] treated the deterministic relation between plasma velocity and electric field in ideal MHD. One question not touched by Song et al. [2001] and Vasyliūnas [2001] is how the physical processes at the magnetopause (steady or variable) driven by the solar wind are coupled to the ionosphere. To address this question, Song and Vasyliūnas [2002] discussed the propagation of perturbations from the magnetosphere to ionosphere when the magnetosphere is driven by reconnection at the magnetopause or in the magnetotail. They assumed, however, a collisionless plasma, and thus their results can be applied only at and above the top boundary of the ionosphere. Including the collision effects in this description leads to time-dependent coupled plasma and neutral equations, which are derived in this paper. Applying them to propagating perturbations is a very complicated mathematical problem, studies on which

448

will be reported elsewhere. Most recently, Vasyliūnas and Song [2004] treated the energy equations for a three-fluid system including the collisions among the three species and electromagnetic coupling. They showed that in such a system, the electromagnetic energy transferred from the magnetosphere is mostly dissipated in the form of frictional heating, distributed nearly equally between plasma and neutrals and associated with relative flow of the two; contrary to common belief, true resistive (Joule or Ohmic) heating is negligibly small.

In this paper, we present a general treatment of a three-fluid system for a partially ionized gas to describe the response of the ionosphere-thermosphere system to reconnection at the magnetopause or in the tail. The three fluids are electrons, ions, and neutrals, and inter-species collisions among all the three are included. This system can be used to describe the collocated ionosphere and thermosphere, as well as the magnetospheric and solar wind collisionless plasmas where the collision frequencies are negligible. This is a very complicated physical/mathematical system, and it will take a few steps before it can be treated realistically. In this paper, we establish the basic theory and derive the solution under highly idealized assumptions. In section 2, we derive the fundamental equations, including in particular: three-fluid generalized Ohm's Law, plasma momentum equation, and neutral momentum equation. In section 3, as the first step solving this equation set for the magnetosphere-ionosphere system, we solve it under the assumption of a 1-D steady state problem.

2. Three-Fluid Treatment

A complete treatment of three-species, i.e., electrons, ions, and neutrals, can be obtained from the kinetic equations for each species by integrating these equations over the phase space, defining macroscopic quantities, and deriving the various moment equations for each species [e.g., Sen and Wyller, 1960; Akasofu and Chapman, 1972; Gombosi, 1994; Schunk and Nagy, 2000]. These moment equations and macroscopic quantities describe each species as a fluid, without treating the motion of each individual particle. In this study, we focus on the behavior of the momentum equation for each of the three species. These equations are, in an inertial frame of reference, for ions,

$$m_i \frac{\partial N_i \mathbf{u}_i}{\partial t} + \nabla \bullet \mathbf{K}_i = q_i N_i (\mathbf{E} + \mathbf{u}_i \times \mathbf{B}) + \mathbf{F}_i - N_i m_i \nu_{in} (\mathbf{u}_i - \mathbf{u}_n) - N_i m_i \nu_{ie} (\mathbf{u}_i - \mathbf{u}_e), \tag{1}$$

for electrons,

$$m_e \frac{\partial N_e \mathbf{u}_e}{\partial t} + \nabla \bullet \mathbf{K}_e = -e N_e (\mathbf{E} + \mathbf{u}_e \times \mathbf{B}) + \mathbf{F}_e - N_e m_e \nu_{en} (\mathbf{u}_e - \mathbf{u}_n) - N_e m_e \nu_{ei} (\mathbf{u}_e - \mathbf{u}_i), \tag{2}$$

and for neutrals,

$$m_n \frac{\partial N_n \mathbf{u}_n}{\partial t} + \nabla \bullet \mathbf{K}_n = \mathbf{F}_n - N_n m_n \nu_{ni} (\mathbf{u}_n - \mathbf{u}_i) - N_n m_n \nu_{ne} (\mathbf{u}_n - \mathbf{u}_e), \tag{3}$$

where q_i, e, N_ζ, $\mathbf{u}_\zeta$, m_ζ, $\mathbf{E}$, $\mathbf{B}$, and $\nu_{\zeta\xi}$, are the absolute value of the electric charge of an ion and an electron, the number density, velocity, and average particle mass of species ζ, the electric field, the magnetic field, and the collision frequency between particles of species ζ and ξ, respectively. The subscripts i, e, and n denote ions, electrons, and neutrals, respectively. We assume, following the approximations discussed by Song and Vasyliūnas [2002], that the ions are singly charged, $q_i = e$, and that charge quasi-neutrality holds in the plasma, $N_e = N_i$. The tensor $\mathbf{K}_\zeta$, the kinetic tensor of particle species ζ defined as the integral of moment $m_\zeta \mathbf{v}_\zeta \mathbf{v}_\zeta$ over the velocity distribution, may be written as $\mathbf{K}_\zeta = N_\zeta m_\zeta \mathbf{u}_\zeta \mathbf{u}_\zeta + \mathbf{P}_\zeta$, where $\mathbf{P}_\zeta$ is the pressure tensor of species ζ, provided $\mathbf{P}_\zeta$ is defined in the frame of reference of bulk flow of each ζ separately. The kinetic tensor of the plasma as a whole $\mathbf{K} = \mathbf{K}_i + \mathbf{K}_e$ may be written as $N_e(m_i + m_e)\mathbf{UU} + \mathbf{P}$ where $\mathbf{P}$ is the pressure tensor of the plasma defined in the frame of reference of bulk flow $\mathbf{U}$ of the plasma. Note that $\mathbf{P}$ is <u>not</u> equal to $\mathbf{P}_i + \mathbf{P}_e$ if $\mathbf{P}_i$ and $\mathbf{P}_e$ are each defined in their frame; for further discussion, see Rossi and Olbert [1970, their equations 9.30 (p.252) and 10.69 (p.281) and associated discussion]. The vector $\mathbf{F}_\zeta$ denotes all other forces exerted on species ζ, such as gravity, viscous force, and (if the equations are written in a rotating frame) centrifugal and Coriolis forces. We have neglected possible wave-particle interactions and photochemical processes. The effects of the neutrals on charged particles are contained in the collision terms with neutrals in equations (1) and (2).

2.1 Momentum Equations and Generalized Ohm's Law

The bulk velocities of ions and electrons can be combined to give the electric current density

$$\mathbf{j} = eN_e(\mathbf{u}_i - \mathbf{u}_e),\tag{4}$$

and the bulk flow velocity of the plasma

$$\mathbf{U} = \frac{m_i \mathbf{u}_i + m_e \mathbf{u}_e}{m_i + m_e}.\tag{5}$$

The inverse relations, expressing the velocities of ions and electrons in terms of the plasma flow and the electric current, are

$$\mathbf{u}_i = \mathbf{U} + \frac{m_e}{m_e + m_i}\frac{1}{eN_e}\mathbf{j}\tag{6}$$

$$\mathbf{u}_e = \mathbf{U} - \frac{m_i}{m_e + m_i}\frac{1}{eN_e}\mathbf{j}.\tag{7}$$

The equations for the time derivatives of the plasma velocity and the current density, $\partial U/\partial t$ and $\partial j/\partial t$, can be derived by inserting (6) and (7) into (1) and (2) and rearranging. It is simpler, however, to sum over species the appropriate moments of the kinetic equations to obtain directly the two equations (detailed derivation is similar to that discussed by Song et al. [2001])

$$m_i \frac{\partial N_e \mathbf{U}}{\partial t} = -\nabla \bullet \mathbf{K} + \mathbf{j} \times \mathbf{B} + \mathbf{F} - N_e(m_i \nu_{in} + m_e \nu_{en})(\mathbf{U} - \mathbf{u}_n) + \frac{m_e}{e}(\nu_{en} - \nu_{in})\mathbf{j},\tag{8}$$

and

$$-\frac{m_e}{e}\frac{\partial \mathbf{j}}{\partial t} = -\nabla \bullet \mathbf{K}_e + \mathbf{j} \times \mathbf{B} - N_e e(\mathbf{E} + \mathbf{U} \times \mathbf{B}) + \mathbf{F}_e - N_e m_e(\nu_{en} - \nu_{in})(\mathbf{U} - \mathbf{u}_n)$$

$$+ \frac{m_e}{e}(\frac{m_e}{m_i}\nu_{in} + \nu_{en} + \nu_{ei})\mathbf{j} + \frac{m_e}{m_i}\nabla \bullet \mathbf{K}_i - \frac{m_e}{m_i}\mathbf{F}_i\tag{9}$$

450

where $\mathbf{K} = \mathbf{K}_i + \mathbf{K}_e$ and $\mathbf{F} = \mathbf{F}_i + \mathbf{F}_e$. In deriving (8) and (9), we have dropped m_e/m_i compared to 1 and have used $m_i N_e \nu_{ie} = m_e N_e \nu_{ei}$ implied by momentum conservation in collisions between two species.

Equation (8) is the plasma momentum equation. Because the difference between the plasma flow and the ion flow is smaller by a factor m_e/m_i than either of the flow differences plasma-electron or ion-electron, this equation can also be viewed for many purposes as describing the ion motion. The last two terms of (8) are associated with collisions. The first of the two is a drag on plasma by the neutrals, and the last is a differential neutral drag on ions and electrons. The electron-ion collision frequency does not appear because electron-ion collisions conserve momentum within the plasma system.

Equation (9) is the three-fluid generalized Ohm's law, as distinct from the two-fluid generalized Ohm's law [e.g. Gombosi, 1994; Schunk and Nagy, 2000]. It contains additional terms involving neutral collisions and neutral wind velocity. The forces that act solely on ions, such as ion pressure gradient, ion inertial force, and $\mathbf{F}_i$, are reduced by a factor of m_e/m_i and act in the direction opposite to those acting on the electrons. To some extent, this equation can be viewed as describing the electron motion in the frame of reference moving with the ions. An additional constraint, however, is provided by the fact that another relation between $\partial \mathbf{j}/\partial t$ and $\mathbf{E}$, derived from Maxwell's equations, can be used to show [Vasyliūnas, 1996] that $\partial \mathbf{j}/\partial t$ is negligible compared to other terms in the generalized Ohm's law whenever time scales are much longer than the electron plasma period and length scales much larger than the electron inertial length. The generalized Ohm's law then becomes, not an equation for the time evolution of the current but rather a constraint on the relative forces acting on electrons and ions, to ensure that they do not violate charge quasi-neutrality.

Momentum conservation during neutral-plasma collisions implies $N_n m_n \nu_{ni} = N_e m_i \nu_{in}$ and $N_n m_n \nu_{ne} = N_e m_e \nu_{en}$, whence equation (3) becomes

$$m_n \frac{\partial N_n \mathbf{u}_n}{\partial t} = -\nabla \bullet \mathbf{K}_n + \mathbf{F}_n + N_e (m_i \nu_{in} + m_e \nu_{en})(\mathbf{U} - \mathbf{u}_n) - \frac{m_e}{e}(\nu_{en} - \nu_{in})\mathbf{j}. \qquad (10)$$

Combining equations (8) and (10) yields

$$m_n \frac{\partial N_n \mathbf{u}_n}{\partial t} = -\nabla \bullet \mathbf{K}_n + \mathbf{F}_n + \left(-\nabla \bullet \mathbf{K} + \mathbf{j} \times \mathbf{B} + \mathbf{F} - m_i \frac{\partial N_e \mathbf{U}}{\partial t} \right). \qquad (11)$$

Equation (11) shows, as indicated by terms in the parentheses, that the acceleration of the plasma and all non-collisional forces acting on it couple directly to the neutrals through neutral collisions.

Equations (8), (9), and (11) describe the behavior of the current, plasma motion, and neutral wind, all important quantities in solar wind-magnetosphere-ionosphere coupling, and will be used as the basic equations in our work.

2.2 1-D Steady State Incompressible three-fluid System

We base our analysis on a physical model similar to that described by Song and Vasyliūnas [2002] in which perturbations and changes (e.g. associated with reconnection) propagate from the magnetopause to the ionosphere along the magnetic field. As pointed out by Song and Vasyliūnas, the same model can be used for perturbations on the nightside associated with magnetotail reconnection. Instead of parallel propagation, we assume here, equivalently, a steady state in a horizontally uniform and vertically stratified (one-dimensional) ionosphere without vertical flow. These assumptions may be most applicable to regions, such as those near the poles, without any strong effects of Birkeland (field-aligned) currents and precipitating particles. An electric field, or equivalently a plasma flow velocity, is imposed at the magnetopause as a boundary condition; it is presumed to result from reconnection processes, which at the present stage of the model are not described in more detail.

Because we focus on electromagnetic coupling and collisions, we assume $\mathbf{F}_i = \mathbf{F}_e = 0$ and neglect compression. The incompressibility assumption combined with continuity and one-dimensionality implies that the velocity perturbation is perpendicular to the vertical direction and that the gradient term $\nabla \cdot \mathbf{K}_\zeta$ vanishes. In this regard, we do not consider global scale ionospheric dynamo effects in our discussion. Under these assumptions, it is clear from (11) that there is no nontrivial solution unless $\mathbf{F}_n$ is nonzero. As discussed by e.g. Schunk and Nagy [2000], the most important force in the neutral equation is the viscous force. Then the steady state equations become

$$0 = \mathbf{j} \times \mathbf{B} - N_e e (\mathbf{E} + \mathbf{U} \times \mathbf{B}) - N_e m_e (v_{en} - v_{in})(\mathbf{U} - \mathbf{u}_n) + \frac{m_e}{e} v_e \mathbf{j}, \tag{12}$$

$$0 = \mathbf{j} \times \mathbf{B} - N_e m_i v_{in} (\mathbf{U} - \mathbf{u}_n) + \frac{m_e}{e} (v_{en} - v_{in}) \mathbf{j}, \tag{13}$$

$$0 = \mathbf{j} \times \mathbf{B} - \nabla \bullet \eta \nabla \mathbf{u}_n \tag{14a}$$

where η is the (dynamic) viscosity coefficient of the neutral gas. In a Cartesian coordinate system with the background magnetic field in the downward (-Z) direction, (14a) becomes

$$0 = \mathbf{j} \times \mathbf{B} - \frac{d}{dz} \eta \frac{d\mathbf{u}_n}{dz}. \tag{14b}$$

The perturbation of the magnetic field by the current is assumed to be much smaller than the background field. With $\mathbf{U}$ and $\mathbf{u}_n$ horizontal and $\mathbf{B}$ along Z, $\mathbf{E}$ and $\mathbf{j}$ are also in the horizontal (X-Y) plane.

For the tenuous neutral gas at high altitudes, the dynamic viscosity coefficient depends on temperature and collision cross-section and is independent of density. It can be evaluated from a hard-sphere model [e.g., Schunk and Nagy, 2001] as

$$\eta = \frac{5\sqrt{\pi}}{16} \frac{\sqrt{m_n k T_n}}{\pi \sigma_n^2} \tag{15}$$

where T_n, σ_n, and k are the temperature, the diameter of the neutral particles, and the Boltzmann constant, respectively.

In this set of equations, the vertical spatial coordinate appears explicitly as an independent variable only in the viscous force. It is, however, implicit in the height-dependence of all quantities, such as densities and collision frequencies. Figure 1 shows the height-dependences which we use for our numerical solutions. The parameters are taken from the MSIS model [Hedin, 1987, 1988] and the International Reference Ionosphere (IRI) model [Bilitza, 2001], and the height-dependent gyro-frequencies, collision frequencies, plasma density, ion-to-neutral mass density ratio $\alpha = N_e m_i / N_n m_n$, and the viscosity η, at the northern magnetic pole during the winter solstice, are calculated based on Kelley [1989], Richmond [1995] and Schunk and Nagy [2000]. Near the pole, the background magnetic field is vertical, the Birkeland currents and the intensities of precipitating particles may be weak, and our 1-D approximation may be valid. The ion gyrofrequency decreases slightly with decreasing altitude, due to an increase of the mean ion mass. The viscosity coefficient does not change much. The plasma density shows peaks at 280 km (F-layer) and at 100 km (E-layer).

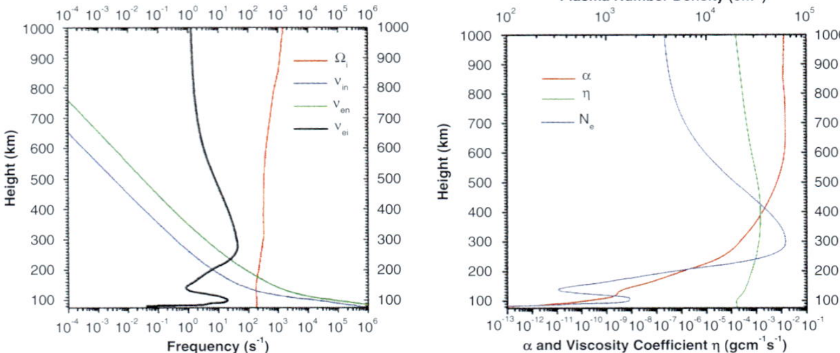

Figure 1. Plasma and neutral parameters for the winter northern pole as functions of height.

We assume that the electric field, **E**, at higher latitudes where we are modeling, results from the flow of the solar wind across open magnetic field lines. We impose the electric field in the +Y (dusk) direction at the upper boundary of the ionosphere to model the situation for a steady-state southward IMF. With the background magnetic field approximated as locally uniform, the electric field **E** is independent of the vertical coordinate Z, both because the conductivity along the magnetic field line is very high (as shown by e.g. Richmond [1995]) and by Faraday's law because the scale length is much shorter for vertical variations than for horizontal.

3. Steady State Solutions.

With the given dependence of the parameters in equations (12) to (14) on height, **j**, **U**, and **u**$_n$ can then be solved for as functions of height in a completely coupled manner. The three equations are combined to obtain a single equation for **u**$_n$ by eliminating **j** and **U**:

$$(D^2 + C)\frac{d}{dz}\eta\frac{d\mathbf{u}_n}{dz} = N_e e(C\mathbf{E}\times\mathbf{b} - B\mathbf{u}_n + D\mathbf{E} + D\mathbf{u}_n\times\mathbf{B}) \qquad (16)$$

where

$$C = \frac{\Omega_i}{v_{in}} + \frac{v_e}{\Omega_e} - \frac{\Omega_i(v_{en}-v_{in})^2}{\Omega_e^2 v_{in}}$$

$$D = 1 - 2\frac{\Omega_i(v_{en}-v_{in})}{\Omega_e v_{in}} \qquad (17)$$

$$\Omega_{i,e} = eB/m_{i,e}$$

The upper boundary is set at 1000 km where all collision effects are assumed to be negligible; then, from (13), the current is zero, hence, from (14), the viscous force is also zero, and from (12) the plasma flow is equal to the E×B drift. The lower boundary is set at 80 km where the neutrals are assumed to be at rest (in a frame of reference corotating with the Earth). The current is also zero at the lower boundary because the electron density becomes negligible. The boundary conditions thus are: at the upper boundary, $d\mathbf{u_n} / dZ = 0$; at the lower boundary, $\mathbf{u_n} = 0$. The equations imply that $\mathbf{U} = \mathbf{b} \times \mathbf{E} /B$ at the upper and $\mathbf{U} = 0$ at the lower boundary (from vanishing of $\mathbf{j}$ at both). Since significant height-dependence occurs only below 400 km, the location of the upper boundary of the ionosphere does not affect the results measurably when it is above 1000 km.

The standard differential equation solver in Mathematica is used to solve the equation. Figure 2 shows the numerical solutions of $\mathbf{j}$, $\mathbf{U}$, and $\mathbf{u_n}$ for viscosity coefficient equal to 1 (solid lines) and 10 (dashed lines) times the assumed nominal value given by (15). The greater viscosity is used for comparison, should the viscosity be much larger due to, for example, eddy motion in addition to inter-particle collisions.

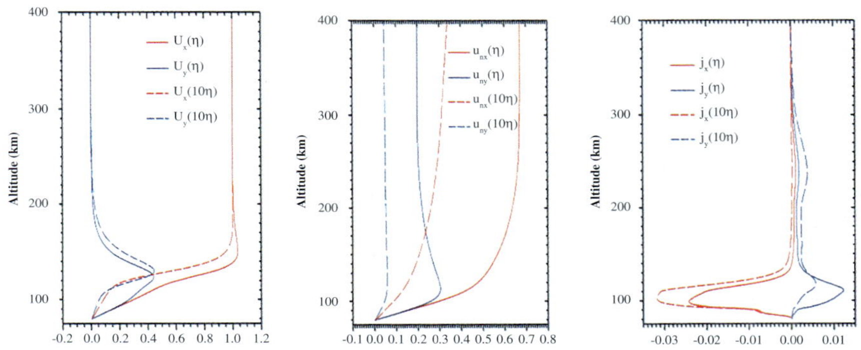

Figure 2. Plasma velocity, neutral wind velocity, and current distribution above the winter northern pole. Solid lines are based on hard-sphere collision model of the viscosity. Dashed lines are for 10 times the model viscosity.

The driver of this strongly coupled system is plasma flow above the ionosphere where collisions are negligible, given by the E×B drift of the electric field assumed at the upper boundary. At high altitudes where the neutral collisions are weak, the ions, which carry most of the plasma mass, convect with the electrons in the electric drift direction. At lower altitudes, the combination of the neutral collisions and electric field acceleration leads to a duskward motion, +Y direction, and reduced antisunward convection, -X direction. These features are well understood in the conventional picture. Our results provide a quantitative description with some new features. An interesting new feature is an enhanced antisunward flow, faster than the electric field drift, near 150~300 km for the low-viscosity case. This flow is related to an additional Hall current discussed below.

The neutrals are driven by the plasma flow through collisions which result in the anti-sunward neutral flow speed being an altitude-dependent fraction of the plasma velocity. At lower altitudes the neutral velocity is driven by the local plasma flow while at higher altitudes it is the result of momentum transfer from lower to higher altitudes by viscosity. The basic function of the viscosity is to smooth the neutral velocity profile. A higher viscosity is equivalent to a neutral population that is heavier (as indicated in equation (15)) and therefore moves more slowly (as can be seen by comparing the neutral velocity profiles for the two viscosity values.)

454

The Pedersen current, j_y, appears to have a two-layered structure with one peak in the F-layer and one in the E-layer. The major Hall current, j_x, is sunward, carried by antisunward electron drift, as expected in the conventional picture, and occurs in the E-layer. However, there is also a weak antisunward Hall current in the F-layer for the low- viscosity case, carried by ions that move in the antisunward direction faster than electrons, as mentioned above. It can be shown that the Lorentz force of this Hall current is needed to balance the viscous force of the duskward neutral flow.

4. Discussion

4.1. Physical Interpretation

It may be assumed, for the purposes of interpreting the results, that the electrons are frozen-in with the magnetic field and moving antisunward with the constant electric drift velocity in most regions in the system, because the electron gyrofrequency is much greater than the electron collision frequencies. (Note that at the lowest altitudes the electron gyrofrequency is 4~5 orders in magnitude greater than the ion gyrofrequency because of the dominance of heavy ions.) At the highest altitudes, the plasma flow follows the electric drift. Although the neutral wind is significantly different from the plasma flow, the neutral-collision frequencies are so small that the current is practically zero. In the F-layer, the plasma flow becomes slightly different from the electric drift but the collision frequencies remain low and the current increases but only slightly. A peak in the current profile occurs mainly because there is a peak in the plasma density. In the E-layer, the combination of the plasma density and the collision frequencies results in the maximum velocity differences, leading to larger currents. At the lowest altitudes, the current decreases because of decreasing electron density, forming a peak in the current profile just above the lower boundary. The difference in the currents for the two viscosity values is readily understood from the neutral-plasma velocity differences for the two cases.

4.2. Effects of Neutral Wind

Conventional M-I coupling models work in the neutral-wind frame of reference, using an ionospheric Ohm's law formulated in that frame which, as shown by Song et al. [2001], can be derived from the combination of (12) and (13). However, as we have shown in Figure 2, the neutral wind speed varies with height when the neutral wind effect is self-consistently included. No single neutral wind frame can be given for all heights, and thus the conventional formulation becomes inapplicable. Furthermore, as shown in (11), the neutral wind speed is also a function of time. In practice, most of M-I models discussed in the magnetospheric community discard equation (11) or its steady state form (14) altogether, so that the current can be solved by combining (12) and (13), and those discussed in the ionospheric community rarely include (8) and (9) as functions of height. Figure 3 compares the electric current distribution when the neutral wind (14) is not included with that self-consistently described by the three-fluid model.

When the neutral wind effects are neglected (dashed lines) the Hall current is mostly concentrated in the E-layer. The Pedersen current, on the other hand, is enhanced in the F-layer but less intense in the E-layer. The total current is greater than that without the neutral wind effect (as pointed out for M-I coupling at Jupiter by Huang and Hill [1989]). When comparing dashed lines in Figures 2 and 3, the current profiles without the neutral wind effect are similar to those with neutral wind effect but with a higher viscosity (note that the two figures use different scales). In fact, these two situations are physically equivalent: a higher viscosity is equivalent to a heavier neutral wind. When the neutral wind is extremely heavy, it cannot be driven, which is same as assuming that the neutral wind is not moving. The conventional theory is thus *not* the limit $\eta=0$ but, on the contrary, the limit $\eta=\infty$ and $\mathbf{u_n} = 0$ at all altitudes so that $\mathbf{j}\times\mathbf{B}$ in (14) becomes indeterminate.

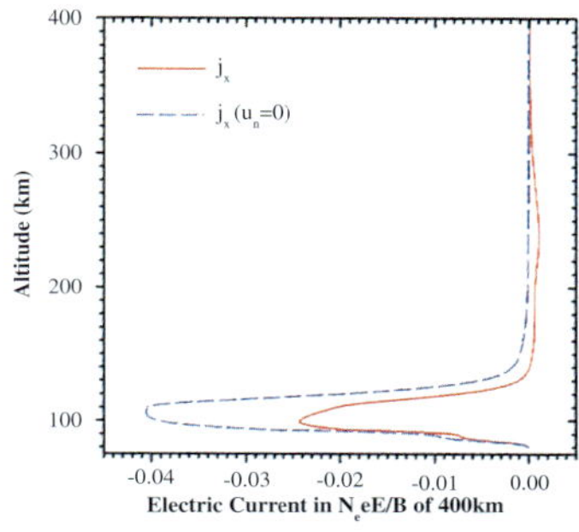
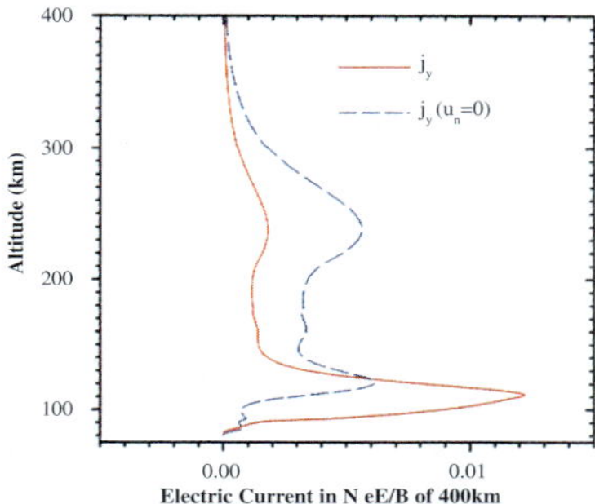

Figure 3. Comparison of the current with (solid lines) and without (dashed lines) neutral wind effect.

4.3 Comments on the Dynamics of M-I Coupling

One other limit in which the conventional theory, as represented by the dashed lines in Figure 3, applies is the case when the solar-wind driver has just turned on after a prolonged quiet period, for a time period short enough so the neutral wind has not yet been significantly accelerated. Since the solid lines represent a completely steady-state convection with a permanent driver, viewing each pair of lines as a time sequence gives a *very* rough idea of the dynamical development following a change in the solar wind.

Because of its higher neutral mass density and greater collision frequencies, the E-layer is expected to respond to solar wind/magnetospheric activities much more slowly than the F-layer. When the neutral wind effects are self-consistently included, the two ionospheric layers, E and F, are expected to behave differently. Most of the dynamical M-I coupling occurs in the F-layer, where the plasma and current are expected to change nearly instantaneously in response to an IMF reversal. A more detailed study of the dynamics requires calculating the time-dependent solutions of the equations.

5. Conclusions

We have developed a three-fluid formalism to describe the coupling from the solar wind to the magnetosphere and to the ionosphere and thermosphere, deriving a single set of equations which can be solved to calculate the electric current plasma velocity, neutral velocity and electromagnetic field as functions of space and time.

As the first step in understanding this system (to be followed, it is hoped, by analyses under more realistic conditions), we calculated the 1-D steady state solutions under the assumption of a duskward electric field and no vertical flow, to model the ionosphere-thermosphere between 70 km and 1000 km altitude, near the pole, for southward IMF situations. In order to have solutions, the viscous force must be included in the neutral momentum equation. The results show that

456

the plasma flow follows the electric drift down to about 200 km in height and then turns duskward and slows down. The neutrals are dragged by the plasma, gaining significant speed in the antisunward and duskward directions in high altitudes. Most of the current is concentrated in the E-layer, and a weak Pedersen current appears in the F-layer. This F-layer current may enhance when the magnetospheric electric field undergoes sudden changes and hence may play an essential role in the dynamics of the M-I coupling processes.

Acknowledgements

The work was supported by NSF under Award NSF-ATM0318643.

References

Akasofu, S.-I and S. Chapman, *Solar-Terrestrial Physics,* 901pp, Oxford University Press, London, 1972.

Bilitza, D., International Reference Ionosphere 2000, *Radio Science, 36,* 261, 2001.

Gombosi, T. I., *Gaskinetic Theory*, Cambridge Press, 1994.

Hedin, A. E., MSIS-86 Thermospheric Model, *J. Geophys. Res. 92,* 4649, 1987.

Hedin, A. E., High Altitude Atmospheric Modeling, *NASA Technical Memorandum 100707,* Scientific and Technical Information Office, Washington, D.C., 1988.

Huang, T. S., and T. W. Hill, Corotation lag of the Jovian atmosphere, ionosphere, and magnetosphere, *J. Geophys. Res. 94,* 3761, 1989.

Kelley, M. C., *The Earth's Ionosphere: Plasma Physics and Electrodynamics*, Academic Press, San Diego, 1989.

Richmond, A. D., Ionospheric Electrodynamics, in *Handbook of Atmospheric Electrodynamics,* Edited by H. Volland, CRC Press, London, 249, 1995.

Rossi, B., and S. Olbert, *Introduction to the Physics of Space*, McGraw-Hill, 1970.

Schunk, R. W., and A. F. Nagy, *Ionospheres*, Cambridge University Press, 2000.

Sen, H. K., and A. A. Wyller, On the generalization of the Appleton-Hartree magnetoionic formulas, *J. Geophys. Res., 65,* 3931, 1960.

Song, P., T. I. Gombosi, A. D. Ridley, Three-fluid Ohm's law, *J. Geophys. Res., 106,* 8149, 2001.

Song, P., and V. M. Vasyliūnas, Solar wind-magnetosphere-ionosphere coupling: Signal arrival time and perturbation relations, *J. Geophys. Res., 107*(A11), 10.1029/2002JA009364, 2002.

Vasyliūnas, V. M., Time scale for magnetic field changes after substorm onset: constraints from dimensional analysis, in *Physics of Space Plasmas (1995),* edited by T. Chang and J.R. Jasperse, MIT Center for Geo/Cosmo Plasma Physics, Cambridge, Massachusetts, 553, 1996.

Vasyliūnas, V. M., Electric field and plasma flow: What drives what, *Geophys. Res. Lett., 28,* 2177, 2001.

Vasyliūnas, V. M. and P. Song, Meaning of ionospheric Joule heating, *J. Geophys. Res.,* in press, 2004.

CHAPTER SEVEN

PRESENT AND FUTURE MUTLISCALE SPACE MISSIONS

3D PLASMA STRUCTURES OBSERVED BY THE FOUR CLUSTER SPACECRAFT

C. P. Escoubet and H. Laakso
ESA/ESTEC, SCI-SH, Keplerlaan 1, 2200 AG Noordwijk, The Netherlands

M. Goldstein
NASA/GSFC, Greenbelt, USA.

Abstract. After 3 years of operations, the Cluster mission is fulfilling successfully its scientific objectives. The mission, nominally for 2 years, has been extended by another 3 years, up through December 2005. The main goal of the Cluster mission is to study in three dimensions the small-scale plasma structures in the key plasma regions in the Earth's environment: solar wind and bow shock, magnetopause, polar cusps, magnetotail, and auroral zone. During the course of the mission, the relative distance between the four spacecraft is being varied from 100 and 18,000 km to study the scientific regions of interest at different scales. The inter-satellites distances achieved so far are 600, 2000, 100, 5000 km and recently 250 km. The change of the separation distances is a fairly complex process that lasts around 1.5 months and requires more than 40 individual spacecraft manoeuvres. The latest results, which include measurement of the bow shock thickness, computation of magnetic field curvature, waves geometry analysis and reconnection in the tail and at the magnetopause will be presented.. The access to data through the Cluster science data system, several public web servers, and the Cluster active archive development will be presented.

Keywords. Bow shock, magnetic gradients, waves, reconnection.

1. Introduction

Past multi-spacecraft missions have been investigating the role of small-scale structures in the Sun-Earth connection, in particular ISEE 1 and 2 [see e. g. *Russell*, 2000 for a review] and more recently Interball-1/Magion-4 [see e.g. *Zelenyi et al*, 2000 for a review]. Cluster is following on this work with four identical spacecraft to study these structures in three dimensions.

The Cluster mission was first proposed more than 20 years ago in response to an ESA call for proposals for the next series of scientific missions [*Haerendel et al.*, 1982]. At that time the proposal selected a mother and three daughter spacecraft that were replaced later on by four identical spacecraft for economic reasons. The assessment study was conducted in 1983 to prove the feasibility of the mission concept. Subsequently, the Phase-A study was conducted jointly with NASA during 1984-1985. At the end of 1985 the Cluster mission was presented to the scientific community and in February 1986, the Solar Terrestrial Science Pprogramme, combining both Cluster and Soho missions, was selected by ESA Science Programme Committee. After a joint ESA/NASA Announcement of Opportunity issued in March 1987, the 11 instruments making up the scientific payload were selected in March 1988. It took about 7 years for four spacecraft to be built and tested and made ready for launch. Unfortunately the

460

launch with the first Ariane 5 on 4 June 1996 was a failure as the rocket exploded 47 s after lift-off destroying the four spacecraft.

After the shock of seeing their work of 8 years annihilated in less that 1 min, the principal investigators and the project teams rolled up their sleeves and investigated how the mission objectives could be recovered. After ten science working team meetings and a few extraordinary ESA committees meeting, the Cluster scientists convinced the ESA Science Programme Committee (SPC) that it was essential for the European scientific community to rebuild the four spacecraft. This was agreed by the SPC in April 1997. Cluster II was born.

As the Cluster II spacecraft and instruments were essentially a rebuild of the original Cluster it took less than half the time to rebuild them (about 3 years). When the first Soyuz blasted off from Baikonur Cosmodrome on 16 July 2000, we knew that Cluster was well on the way to recovery from the previous launch setback. However, it was not until the second launch on 9 August 2000 and the proper injection of the second pair of spacecraft into orbit that we knew that the Cluster mission was back on track. In fact, the experimenters said that they knew they had a mission only after switching on their last instruments on the fourth spacecraft.

In a first section, the Cluster mission will be described with a focus on the orbit and the separation distances achieved so far and planned in the future. In a second part the instrumentation will be briefly described. Then a few examples of Cluster observations will be presented in a third part and finally the data distribution through the Cluster Science Data System and the plans for the Cluster active archive will be presented in the fourth section. The purpose of this paper is not to review what has been done in the past in the various magnetospheric missions but to present highlights of the Cluster results. The reader can find more references in the referenced papers.

2. Mission

The scientific objectives of the Cluster mission are to study the small-scale structures and the turbulence in the key regions of the magnetosphere. Such regions are: the solar wind, the bow shock, the magnetopause, the polar cusp and the magnetotail.

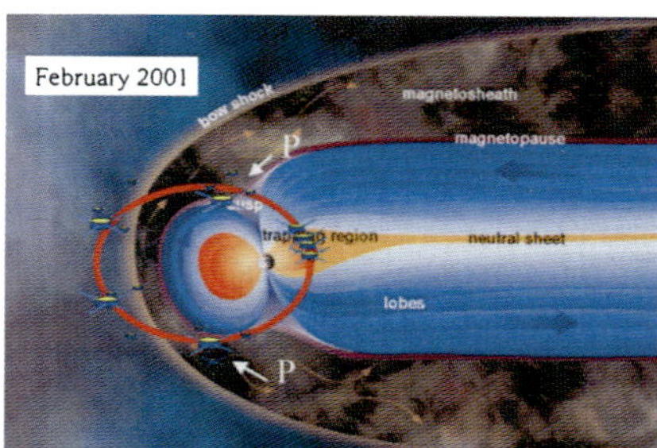

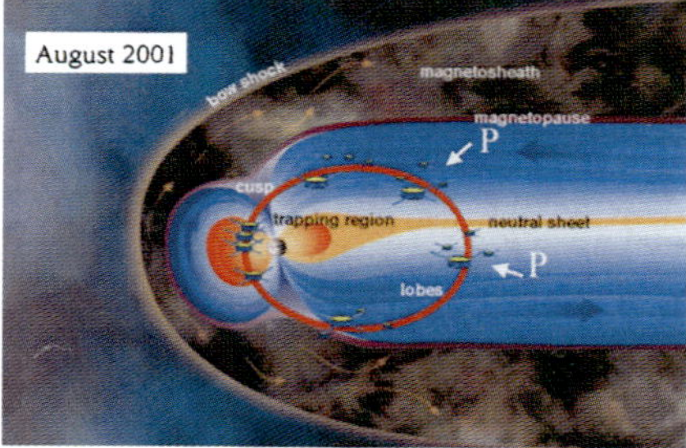

Figure 1. Regions of the magnetosphere crossed by the Cluster spacecraft. The left panel shows the orbit in February and the right panel in August. The perfect tetrahedron locations are marked with a "P".

In addition the temporal variations of structures observed in the auroral zone mid-altitude polar cusp, plasmasphere can be studied for the first time as the spacecraft are following each other as a "string of pearls" near perigee.

To perform these objectives, the Cluster spacecraft have been placed in a 4×19.6 R_E polar orbit (Figure 1). The spacecraft have slightly different orbits to form a perfect tetrahedron in key regions of space such as the Northern polar cusp, Southern polar cusp and plasmasheet (Figure 1).

During the first two years of the mission, the separation distance was changed approximately every 6 months (Table 1 and Figure 2). It was decided to start with small distances (down to 100 km) and then to increase it toward the end of the mission (up to 10000 km). All measurements at small distances have to be done first since after 10000 km, the remaining fuel will no longer allow to substantially change the separation distance. The Cluster mission has been extended an additional 3 years from January 2003 to December 2005 to cover more separation distances and spend more time in all key regions. After early 2003, the constellation manoeuvres are done only once a year to decrease the operational costs and to decrease the downtime of the instrument during the manoeuvres. This was possible with the innovative manoeuvre method used by the Flight Dynamics Team at the European Space Operations Centre (ESOC). This method allows to have a tetrahedron at the same time in the tail and in the Northern polar region, therefore covering both the tail and the Northern cusp 6 months later.

Table 1. Spacecraft separation distances

Year	Phase	Separation (km)
2001	Cusp	600
2001	Tail	2000
2002	Cusp	100
2002	Tail	4000
2003	Cusp	5000
2003	Tail	200
2004	Cusp	300
2004	Tail	1000
2005	Cusp	13000
2005	Tail	10000

Each constellation manoeuvre is done in many small step to minimize the fuel consumption and make sure that the target position of each spacecraft is reached precisely. Typically, it involves around 40 firings of the thrusters that last a few seconds or up to 2 hours (Table 2). After each thruster firing, a precise orbit determination is done which is then used to prepare the next manoeuvre. More details in the preparation and execution of the manoeuvres can be found in *Volpp et al.* [2004].

The small distances (200 km) in the magnetotail in Aug. 2003 were not in the initial planning of the Cluster mission, but were recommended afterwards by the International Space Science Institute substorm working group. The small scales in the tail are necessary to investigate the processes that produce geomagnetic substorms. There are two competing models: magnetic reconnection and current disruption. The existence of a small " diffusion " region where the plasma is rapidly accelerated is expected in the first model, while a disruption of cross-tail current is expected in the second model. Both phenomena have a scale size of approximately 500 km, which require a spacecraft separation distance of a few hundred kilometers to be studied.

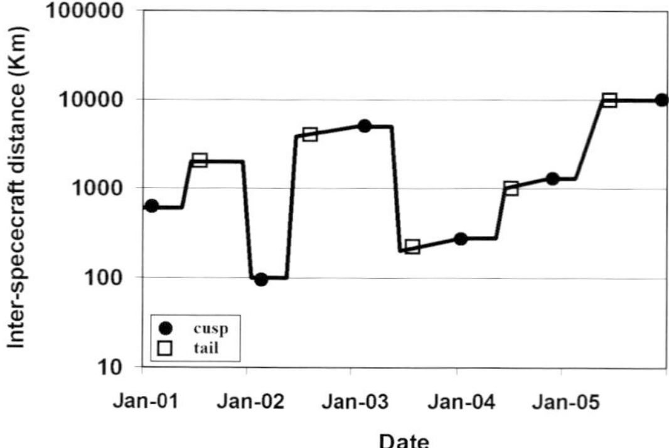

Fig. 2. Separation distances during the course of the mission

Table 2. Spacecraft constellation manoeuvres during summer 2003. The duration of each manoeuvre is given in minutes. A total of 43 manoeuvres were executed to move the spacecraft from a separation distance of 5000 km to 200 km.

Date	C1 (min)	C2	C3	C4
4/Jun/2003			34.4	
7/Jun/2003	99.0	125.0	111.0	32.0
9/Jun/2003	2.1	2.3	2.7	2.0
9/Jun/2003	0.1	0.2	0.4	0.1
14/Jun/2003	13.6	8.5	2.5	32.0
18/Jun/2003	5.0	1.0	0.1	6.5
21/Jun/2003	0.8	0.1	2.0	2.0
24/Jun/2003	0.5	1.5	0.5	0.4
29/Jun/2003	0.5	0.8	0.3	0.2
2/Jul/2003	2.2	0.1	2.5	
7/Jul/2003	0.1	0.1	0.1	0.1
12/Jul/2003	0.1		0.1	0.1

In addition, the mission data return has also been augmented by adding a second ground-station in Maspalomas (Spain). At the beginning of the mission, due to the very large amount of data produced by Cluster, the baseline data return was limited to 50% of the orbit. After a few months of operations, it was however realised that many highly bursty phenomena were missed due to their un-predictable behavior (sudden storm commencement, substorms, storms) and large scientific regions (e.g. magnetotail and North and South cusp) could not be fully observed. In February 2002, the ESA SPC agreed to extend both the data coverage and the mission. The 100% coverage started in June 2002. An example of the data return before and after data coverage extension is shown in Figure 3. In 2001, the data acquisition focused on the boundaries (Figure 3 left panel) while in 2003 the orbit is fully covered ((Figure 3 right panel). With the full coverage we clearly see that the field direction in the solar wind (central part around apogee in figure 3 right panel) is correlated with the rotation of the Sun such that the patterns are repeated every 28 days (this is the effect of the tilt of the magnetic pole of the Sun with respect to the Sun rotation axis). A few data gaps occur when constellations or attitude manoeuvres are performed.

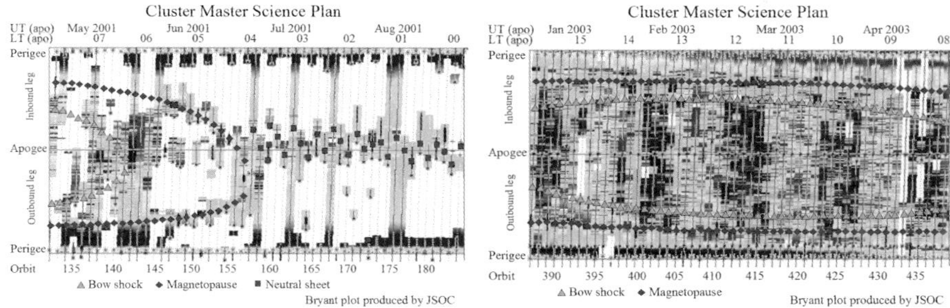

Figure 3. Master Science Plan during Summer 2001 (left panel) and Spring 2003 (right panel). Each vertical dashed line represents an orbit from perigee (bottom) to apogee (middle) and then to perigee (top). The Magnetic field direction (from FGM) in the GSE x-y plane is shown in grey scale (black=sunward, light grey=anti-sunward).

3. Instrumentation

Cluster spacecraft contains a complete suite of instruments to measure magnetic fields, electric field, electromagnetic waves, and particles (Table 3). In addition a potential control instrument keep the spacecraft potential close to a few Volts positive (typically 7 V) in very tenuous plasma. More details on the payload can be found in *Escoubet et al.* [2001].

Table 3: The 11 instruments on each of the four Cluster spacecraft. Instruments marked with a star are part of wave experiment consortium (WEC).

Instrument	Principal Investigator
ASPOC (Spacecraft potential control)	K. Torkar (IWF, A)
CIS (Ion composition, 0<E<40 keV)	H. Rème (CESR, F)
EDI (Plasma drift velocity)	G. Paschmann (MPE, D)
FGM (Magnetometer)	A. Balogh (IC, UK)
PEACE (Electrons, 0<E<30 keV)	A. Fazakerley (MSSL, UK)
RAPID (High energy electrons and ions, 20<Ee<400 keV, 10<Ei<1500 keV)	P. Daly (MPAe, D)
DWP * (Wave processor)	H. Alleyne (Sheffield, UK)
EFW * (Electric field and waves)	M. André (IRFU, S)
STAFF * (Magnetic and electric fluctuations)	N. Cornilleau (CETP, F)
WBD * (Electric field and wave forms)	D. Gurnett (IOWA, USA)
WHISPER * (Electron density and waves)	P. Décréau (LPCE, F)

464

4. Cluster results

Results from the first few months of operations were presented in previous papers [*Escoubet et al.*, 2001]. In this paper we will focus on specific phenomena that requires four spacecraft to be fully studied like the determination of the bow shock size, the electromagnetic waves characterization using the k-filtering technique, calculation of the electric current at the magnetopause, surface waves in the plasmasheet, magnetic curvature and reconnection in the tail. It is not the goal of this paper to describe each result in details but just to give a brief overview of the examples presented.

4.1 Size of the bow shock

One of the first scientific objectives of the Cluster mission is to characterize the boundaries of the magnetosphere using the four spacecraft located at different position along and across the boundary. The bow shock is one of the most important boundary of the magnetosphere environment since it is slowing down and thermalising the solar wind plasma that then interact directly with the magnetosphere. The physical processes at work are not yet well known and determining the characteristics of the shock are crucial for this understanding. *Bale et al.* [2003] have for the first time determined the size of the bow shock using the density measurement of the four spacecraft (figure 4).

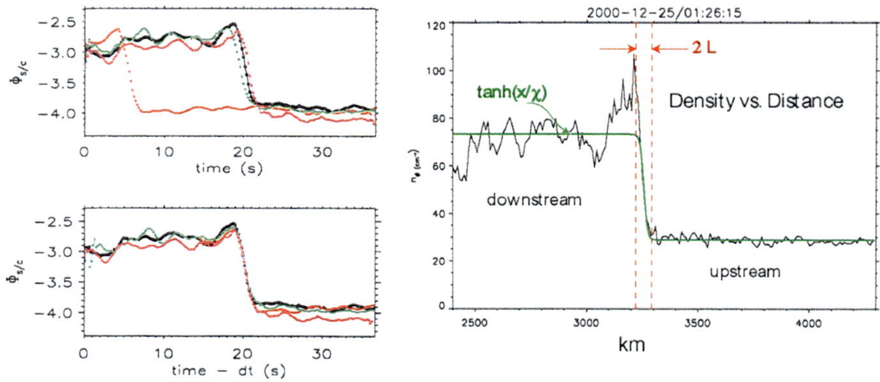

Figure 4. Top-left: Smoothed spacecraft potential measurements $-\Phi_{sc}$ from the four Cluster spacecraft (each as a different color: black: 1; red: 2; green: 3; purple: 4), at a collisionless shock on 25 Decembre 2000. Bottom-left-: the bow shock crossing have been synchronized with respect to spacecraft 1. The time shift between encounters on the 4 spacecraft is used to compute the shock normal vector and speed. Notice the coherence of the first cycle of the overshoot). Right: density transition from downstream to upstream. The green line is the hyperbolic tangent fit; red vertical lines show the density transition scale. Reprinted with permission from Bale, S. D., F. S. Mozer and T. S. Horbury, Density-Transition Scale at Quasiperpendicular Collisionless Shocks, *Physic Rev. Lett.*, 91, 26, DOI: 10.1103/PhysRevLett.91.265004, 2003. Copyright (2003) by the American Physical Society

For each of the 98 bow shock crossings studied, measurements of a spacecraft floating potential, on the four Cluster spacecraft, have been used as a proxy for electron plasma density. The time of the crossing by each spacecraft gives the shock speed and normal (see details on discontinuity analysis from multi-spacecraft in Dunlop and Woodward [1998]); the shock speed is used to convert the temporal measurement to a spatial one.

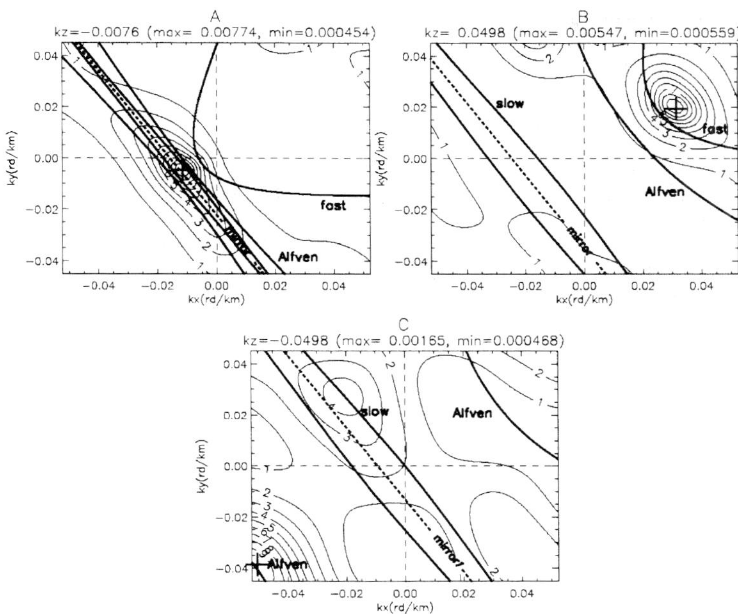

Figure 5: For f1 = 0.37 Hz (in the satellite frame), the three k_Z planes which contain the three maxima of magnetic energy (labelled by crosses) are displayed. The grey color isocontours are the experimental magnetic field energy distribution results, whereas the black lines are the theoretical dispersion relations. The given "max" and "min" values are the maximum and the minimum energy density [$nT^2/Hz(rd/km)^3$] for the given k_Z plane. For the same frequency the magnetic energy is distributed over four wave vectors which appear close to mirror, fast, Alfven, and slow modes. The peaks in (A, B, C) are ordered from the highest energy to the lowest one (from Sahraoui, et al., ULF wave identification in the magnetosheath: The k-filtering technique applied to Cluster II data, *J. Geophys. Res.*, 108(A9), 1335, doi:10.1029/2002JA009587, 2003. Copyriht 2003 American Geophysical Union. Reproduced by permission of American Geophysical Union).

On previous missions, the density transition scale has been rarely if ever investigated primarily because particle counting instruments traditionally have required one spacecraft spin to acquire a complete distribution function measurement and spacecraft spin rates are on the order of seconds. Of course one key assumption is that the bow shock does not change between the crossings of each spacecraft. This assumption could be considered valid if the time difference between each crossing is not too large and if the density profile of each crossing very similar. The new result of this study is that the thickness of the Earth's bow shock is proportional to the gyroradius of downstream ions, according to a large set of observations from the Cluster satellites. This dataset covers a broad distribution of thickness from a few km up to 400 km.

4.2 Characterization of magnetic fluctuations using the k-filtering technique

The magnetosheath is well known to be a place where a wide variety of electromagnetic waves are present. In the past, missions with one or two satellites had to make assumptions or select special events to characterise the main type of waves observed.

With four spacecraft, Cluster allow for the first time to identify the individual modes superimposed at a given frequency and to obtain for each of them the distribution in energy and the direction of propagation. The analysis method is the k-filtering, developed by *Pincon and Lefeuvre* [1991]. Assuming the wave-field is time stationary and homogeneous in space, it allows to obtain optimum estimates of the power spectra density, in the frequency wave vector space.

Figure 5 shows the k-filtering method applied to a magnetosheath interval on 18 February 2002 around 05:34 UT [from *Sarahoui et al.*, 2003], using the magnetic fluctuations measured by the STAFF instrument. The spacecraft were located at X_{GSE} = 5.6 , Y_{GSE}= 4.6 Re, Z_{GSE} = 8.4 R_E, and they were separated by about 100 km in a tetrahedron formation.

The plots show the experimental magnetic field energy distribution (isocontours) for a frequency of 0.37 Hz with the theoretical dispersions superimposed in colour. The panels A, B, and C show the maxima of the power at a 3 different k_Z direction. Preliminary analysis suggests that four modes, mirror, fast, Alfven and slow can be identified with a different propagation direction. The results presented above on the wave identification will have to be confirmed and completed in a more refined study, in particular, on the polarization associated with the different maxima. This very powerful method can also be applied to other region of the magnetosphere like the polar cusp.

4.3 The magnetic curvature

The magnetic field in the tail is elongated such that the magnetic curvature is small (radius of curvature is large) in the lobes and large (radius of curvature small) in the neutral sheet. The radius of curvature has been shown to play a key role in the motion of particles in the tail. This parameter can be computed for the first time using the measurement of the magnetic field at the four spacecraft position. An example of a neutral sheet crossing is shown in Figure 6 [from *Shen et al.*, 2003]. The spacecraft were first in the Northern lobes and crossed the neutral sheet around 09 UT. The radius of curvature (bottom panel in Figure 6) is below 2 R_E in the neutral sheet between 09:00 and 09:30 and then very large, around 10 R_E, in the lobes (before 08 or after 10). In addition Shen et al. [2003] have shown that the curvature radius is changing during the phase of the substorms: it is very small (less than 0.5 R_E) during the growth phase and expansion phase and larger during the recovery (above 0.8 R_E).

4.4 Magnetic reconnection in the magnetotail and in the cusp

Magnetic reconnection is a universal physical process that transfers magnetic energy into plasma energy, subsequently accelerating particles to very high energy. The effects of this process can be observed deep in the tail in the neutral sheet and in the external boundary of the magnetosphere, in the polar cusp. The main signatures of this process are the plasma jets, clearly visible in the ion flow, and the magnetic reconfiguration.

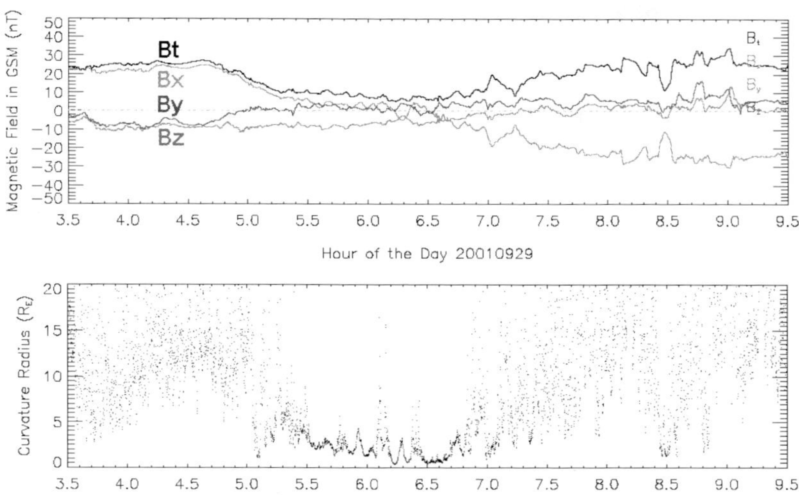

Figure 6: Magnetic field (top) and radius of curvature of the magnetic field (bottom) on 17 September 2001. B_X, B_Y, B_Z and B_T are shown in red, green, blue and black (from Shen, C., X. Li, M. Dunlop, Z. X. Liu, A. Balogh, D. N. Baker, M. Hapgood, and Xinyue Wang, Analyses on the geometrical structure of magnetic field in the current sheet based on Cluster measurements, *J. Geophys. Res.*, 108 (A5), doi: 10.1029/2002JA009612, 2003. Copyright 2003 American Geophysical Union. Reproduced by permission of American Geophysical Union).

The first example in Figure 7 shows the data collected by Cluster in the neutral sheet [from *Runov et al.*, 2003]. The top panels show B_Y as a function of B_X during two crossings of the neutral sheet of the four spacecraft.

During the first crossing (right panel) By changes from being negative at $B_X > 0$ (northern hemisphere) to positive at $B_X < 0$ (southern hemisphere). The other crossing (left panel) is characterized by positive B_Y at $B_X > 0$ and negative B_Y at $B_X < 0$. These results are consistent with the presence of magnetic disturbances due to Hall currents that are reversed on either side of the reconnection point [*Sonnerup*, 1979]. These currents are typically thought of being indicative of the decoupling of the ions from the magnetic field and the electron fluid. A sketch of the reconnected current sheet structure including the Hall magnetic currents (green) near the X-line is given in the bottom panel of Figure 7. The accelerated plasma is observed on each side of the reconnection point as indicated by the purple arrows.

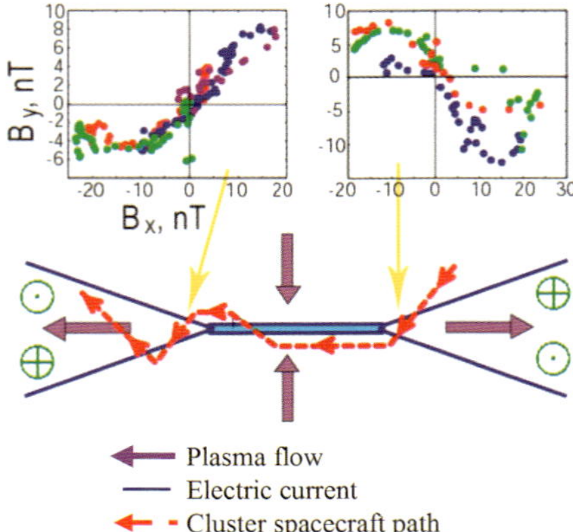

Figure 7: Two crossings of the neutral sheet on 1 October 2001. The top diagrams show the magnetic field measured on Cluster and the sketch below shows the reconnection event and the respective trajectory of Cluster (dashed red line) (from Runov et al., Current sheet structure near magnetic X-line observed by Cluster, *Geophys. Res. Lett.*, 30, 11, 1579, doi:10.1029/2002GL016730, 2003. Copyright 2003 American Geophysical Union. Reproduced by permission of American Geophysical Union).

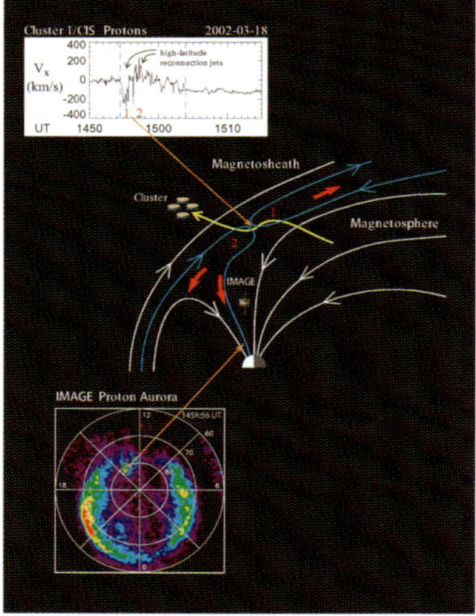

Figure 8: Cluster and IMAGE observations on 18 March 2003. Top: ion flow speed component V_x (GSE). Middle: sketch of lobes reconnection and Cluster trajectory. Bottom: Proton aurora from IMAGE spacecraft at 14:58 UT (from Phan, et al. Simultaneous Cluster and IMAGE observations of cusp reconnection and auroral proton spot for northward IMF, *Geophys. Res. Lett.*, 30, 10, doi:10.1029/2003GL016885, 2003. Copyright 2003. Reproduced by permission of American Geophysical Union).

Another example of reconnection observed in the polar cusp is shown in Figure 8 [from *Phan et al.*, 2003]. Cluster was crossing the poleward boundary of the cusp and observed first a tailward flow (1), followed by an Earthward flow (2). This flows are consistent with reconnection poleward of the cusp under Northward interplanetary magnetic field. At the same time the IMAGE spacecraft looking at the precipitation of protons in the ionosphere observed a bright spot, indicative of accelerated ions, located on the same field lines. This is a first direct evidence that protons aurora in the cusp are produced by reconnection on the magnetopause.

5. Cluster Data

5.1 Cluster Science Data System

The Cluster science data system has been set-up to distribute quicklook and processed data to all Cluster Principal and Co-Investigators, as well as to the scientific community. The Cluster community consists of 11 Principal Investigators and 259 Co-Investigators from 89 laboratories located in 24 countries. To distribute the data efficiently to all users, a system of nine data centres located in Austria, China, France, Germany, Hungary, Netherlands, Sweden, United-Kingdom and United-States and interconnected with each other has been set-up. Each data center stores the full database of processed and validated data from all instruments. Data from February 2001 until end of January 2004 are available through the web at http://sci2.estec.esa.nl/cluster/csds/csds.html.

A quicklook plot is also available between a few hours to a few days after data acquisition and includes time series plots and spectrograms from most of the instruments. This is very useful for scientists to pick-up interesting events and for the Cluster project to monitor the progress of the mission. The average download rate during the year 2003 was above 5 Gbytes/month (without including the US data centre).

5.2 Public access to other data sets

The PI teams are distributing other Cluster data sets; these include particle spectrograms, high resolution data, enhanced prime parameters data or summary plots. The web links are in table 4.

Table 4: Links to Cluster additional data sets

Instrument	Web address	Comment
ASPOC	http://saturn.iwf.oeaw.ac.at/acdc/acdc.html	Raw data
CIS	http://cis.cesr.fr:8000/CIS_sw_home-en.htm	6h/spectrograms,3SC
EDI	http://edi.sr.unh.edi	Prime parameters
EFW	http://www.cluster.irfu.se/efw/data/spinfit/index.html	E_X, E_Y at 4 s res.
FGM	http://www.sp.ph.ic.ac.uk/Cluster/	Prime parameters
PEACE	http://cluster2.space.swri.edu/	High resolution
RAPID	http://www.linmpi.mpg.de/english/projekte/cluster/rapid.html	High resolution
STAFF	http://www.cetp.ipsl.fr/CLUSTER/accueil/framepa.html	Spectrograms
WBD	http://www-pw.physics.uiowa.edu/plasma-wave/istp/cluster/	Spectrograms
WHISPER	http://www.whisper.cnrs-orleans.fr/	Spectrograms

5.3 Cluster active archive

470

The Cluster mission is successfully delivering summary and prime parameter data through the Cluster Science Data System. In February 2003, the ESA Science Programme Committee has agreed that the Cluster Project set-up a Cluster Active Archive (CAA) that will contain processed and validated high-resolution scientific data, as well as raw data, processing software, calibration data and documentation from all the Cluster instruments. The scientific rationale underpinning this proposal is as follows:

- Maximise the scientific return from the mission by making all Cluster data available to the worldwide scientific community.
- Ensure that the unique data set returned by the Cluster mission is preserved in a stable, long-term archive for scientific analysis beyond the end of the mission.
- Provide this archive as a contribution by ESA and the Cluster science community to the International Living With a Star programme.

The CAA will be a database of high-resolution data and other allied products that will be established and maintained under the overall control of ESA.

Two important aspects of the CAA are as follows. In view of the shortage of manpower in most of the institutes processing Cluster data, ESA is supporting manpower to be deployed in institutes where the relevant expertise exists, to assist in the preparation, validation, and documentation of the high-resolution data to be deposited in the archive.

In view of the urgency in starting the programme, and recognising that much of the in-house expertise might be lost when National Agency funding to the Cluster instrument teams is expected to be greatly reduced at the end of the Cluster mission, it was imperative to start these activities as soon as possible. The design phase was finished in 2003, and the development and implementation phase is taking place in 2004.

Processing and preparation of data to be archived is starting now, with data from all instruments entering the database at an average rate of two years of data per calendar year. The data from the year 2001 will be archived in 2004, then data from 2002-2003 in 2005, and 2004-2005 in 2006. The year 2007 is kept for reprocessing and finalization of the archive. Data from any individual experiment may be archived more or less rapidly, subject to the requirement that the archiving of all data should be completed at the conclusion of CAA phase. The archive will be accessible to all scientists. Once data are included in the archive, it will be public.

6. Conclusion

After 3 years of operations, the Cluster spacecraft have shown their full capability to make substantial advances in magnetospheric physics. For the first time plasma structures have been studied in three dimensions.

In this paper we have shown a few example of the Cluster capabilities, such as the measurement of the bow shock thickness, the computation of magnetic field curvature and the analysis of the waves geometry. Reconnection in the tail and at the magnetopause was observed in great details.
Next year Cluster will go for the first time to the large separation distances (10000 km) and should shed new lights on large scales phenomena such as the cusp geometry or the size of bursty bulk flow in the tail. The operations are funded up to end of 2005, however, if the

spacecraft and instrument are in good health a further extension may be proposed to the ESA Science Programme Committee at the end of 2004.

At the end of last year the Cluster mission has been complemented by a Chinese mission called Double Star. This mission consists of two spacecraft, one equatorial TC-1 (550 x 66,970 km, 28.5 deg. inclination) launched on 29 December 2003 and the second one polar TC-2 (700 x 39,000, 90 deg. inclination) to be launched in July 2004. The orbits of Double Star have been specially designed to maximize the conjunction with Cluster in the plasma sheet and in the polar cusp. Furthermore half of the Double Star payload is made of spare or duplicate of the Cluster instruments which should allow full comparison to be made.

The latest news and the access to Cluster data through the Cluster Science Data System can be found at: http://sci.esa.int/cluster/

Acknowledgements

The authors thank all PIs and their teams who provided the Cluster data, and the JSOC and ESOC team for their very efficient Cluster operations. In particular, the authors would like to acknowledge C. Perry and M. Hapgood for the Bryant plots, S. Bale (Berkeley, USA) for the bow shock analysis, F. Sarahoui (CETP, F) for the k-filtering application technique, A. Runov, R. Nakamura and W. Baumjohann (IWF, A) for the reconnection event, C. Shen (CSSAR, China) for the magnetic field curvature computation and T. Phan (SSL, USA) for the reconnection event observed by both Cluster and IMAGE.

References

Bale, S. D., F. S. Mozer and T. S. Horbury, Density-Transition Scale at Quasiperpendicular Collisionless Shocks, *Phys. Rev. Lett.*, *91*, DOI: 10.1103/PhysRevLett.91.265004, 2003.

Dunlop, M. W., and T. I. Woodward, Discontinuity analysis: Orientation and motion, *in Analysis Methods for Multispacecraft Data*, ed. by G. Paschmann and P. Daly, ISSI Sci. Rep. SR-001, *ESA publication*, 271-305, 1998.

Escoubet, C. P., M. Fehringer and M. Goldstein, The Cluster mission, *Ann. Geophys.*, *19*, 2001.

Haerendel, G., A. Roux, M. Blanc, G. Paschmann, D. Bryant, A. Korth, B. Hultqvist, Cluster, study in three dimensions of plasma turbulence and small-scale structure, *mission proposal submitted to ESA*, 1982.

Phan, T.; Frey, H. U.; Frey, S.; Peticolas, L.; Fuselier, S.; Carlson, C.; Rème, H.; Bosqued, J.-M.; Balogh, A.; Dunlop, M.; Kistler, L.; Mouikis, C.; Dandouras, I.; Sauvaud, J.-A.; Mende, S.; McFadden, J.; Parks, G.; Moebius, E.; Klecker, B.; Paschmann, G.; Fujimoto, M.; Petrinec, S.; Marcucci, M. F.; Korth, A.; Lundin, R. Simultaneous Cluster and IMAGE observations of cusp reconnection and auroral proton spot for northward IMF, *Geophys. Res. Lett.*, *30*, CiteId 1509, DOI:10.1029/2003GL016885, 2003

Pincon, J. L., and F. Lefeuvre, Local characterization of homogeneous turbulence in a space plasma from simultaneous measurements of field components at several points in space, *J. Geophys. Res.*, *96*, 1789–1802, 1991.

Runov A., R. Nakamura, W. Baumjohann, R. A. Treumann, T. L. Zhang, M. Volwerk, Z. Vo¨ro¨s, A. Balogh, K.-H. Glaßmeier, B. Klecker, H. Reme, and L. Kistler, Current sheet structure near magnetic X-line observed by Cluster, *Geophys. Res. Lett.*, *30*, CiteId 1579, DOI:10.1029/2002GL016730, 2003.

Russell, C. T., ISEE lessons learned for Cluster, in: *Proceedings of Cluster II workshop on Multiscale/Multipoint plasma measurements*, ESA SP 449, ed. by R. Harris, ESA publications, Noordwijk, 11–23, 2000.

Sahraoui, F., J. L. Pincon, G. Belmont, L. Rezeau, N. Cornilleau-Wehrlin, P. Robert, L. Mellul, J. M. Bosqued, A. Balogh, P. Canu, and G. Chanteur, ULF wave identification in the magnetosheath: The k-filtering technique applied to Cluster II data, *J. Geophys. Res.*, 108(A9), CiteId 1335, DOI:10.1029/2002JA009587, 2003.

Shen, C., X. Li, M. Dunlop, Z. X. Liu, A. Balogh, D. N. Baker, M. Hapgood, and Xinyue Wang, Analyses on the geometrical structure of magnetic field in the current sheet based on Cluster measurements, *J. Geophys. Res.*, 108(A5),CiteId 1168, DOI: 10.1029/2002JA009612, 2003

Sonnerup, B. U.O., Magnetic field reconnection, in vol. 3 of *Solar System Plasma Physics*, ed by C.F. Kennel, L.J. Lanzerotti and E.N. Parker (eds.), North Holland, 45-108, 1979.

Volpp, J., C. P. Escoubet, S. Foley, J. Godfrey, M. Hapgood, S. Pallaschke, Cluster constellation change manoeuvres – Management and Operations, SpaceOps 2004, Montreal, Canada, May, 2004.

Zelenyi, L., Zastenker, G., Dalin, P., Eiges, P., Nikolaeva, N., Safrankova, J., Nemecek, Z., Triska, P., Paularena, K., and Richardson, J., Variability and Structures in the solar wind – magnetosheath – magnetopause by multiscale multipoint observations, in: *Proceedings of Cluster II workshop on Multiscale/Multipoint plasma measurements*, ESA SP 449, ed. by R. Harris, ESA publications, Noordwijk, 29–38, 2000.

Published by Elsevier B.V.
Multiscale Coupling of Sun-Earth Processes
A.T.Y. Lui, Y. Kamide and G. Consolini, editors.

THE CENTRAL ROLE OF RECONNECTION IN SPACE PLASMA PHENOMENA TARGETED BY THE MAGNETOSPHERIC MULTISCALE MISSION

S.A. Curtis
*Magnetospheric Physics Branch, Laboratory of Extraterrestrial Physics,
NASA/Goddard Space Flight Center, Greenbelt, MD 20771*

P.E. Clark
*L3 Communications GSI, 3750 Centerview Drive, Chantilly, VA 20151 at
NASA/Goddard Space Flight Center, Greenbelt, MD 20771*

C.Y. Cheung
*Magnetospheric Physics Branch, Laboratory of Extraterrestrial Physics,
NASA/Goddard Space Flight Center, Greenbelt, MD 20771*

Abstract. The Magnetospheric Multiscale Mission, which consists of four identically instrumented spacecraft flying in close formation, will allow the first extensive characterization of the 3D structure and dynamic variations of the magnetosphere on scales down to electron inertial length. By targeting the magnetopause and magnetotail, two regions already known to provide observations crucial to understanding the reconnection process, the mission will determine the spatial and temporal scales of basic plasma processes.

Keywords. magnetospheric multiscale mission, multi-platform, formation, reconnection, space plasma processes, observations, temporal, spatial, 3D structure, dynamics, electron inertial length

1. Understanding of Magnetospheric Processes

The relatively low density, low temperature plasma system surrounding the Earth, termed the 'magnetosphere' by Gold (1959), and has been observed for decades. Its organization into macroscopic 'cells' characterized by different composition, temperature, and structure and separated by thin boundary layers has been noted relatively recently (Falthammar et al, 1978). By linking adjacent cells, the magnetic reconnection process occurring in key regions at these impermeable boundaries controls the flow of energy, mass, and momentum in the entire system.

Current understanding of processes of magnetospheric processes such as reconnection has evolved considerably over the last decade as small to moderate spatial scale measurements have been provided by magnetospheric missions, sequential observations from one platform mission such as POLAR and now simultaneous observations from four platforms of ESA Cluster. These observations have produced the current model of the magnetosphere (Figure 1) based on far greater understanding of individual particle dynamics and global plasma structure (STDT, 2000).

Earth's Magnetosphere

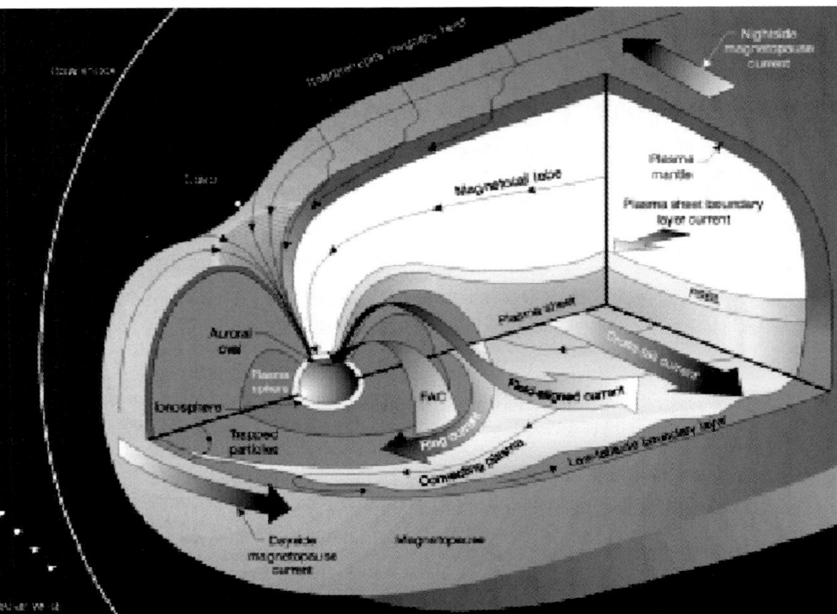

Figure 1: Current Model of the Magnetosphere based on more higher resolution spatial information provided over the last decade (STDT MMS Report, 2000).

Temporal scale variations at critical locations such as the magnetotail and magnetopause are still poorly understood. Combining observations of temporal and spatial variations will yield an understanding of the reconnection process itself and thus establish the link between local and global variations in the magnetosphere. By providing observations at critical locations (Figure 2) from four identically instrumented platforms for which separations can be accurately (±1%) varied from 10 kilometers to a few tens of thousands of kilometers, MMS should be able to differentiate between spatial structure and temporal evolution of fundamental magnetospheric processes, thus establishing the relationship between microscale, mesoscale, and global phenomena.

2. Current Understanding of Reconnection

2.1 Reconnection Models

Our present understanding of magnetic reconnection is illustrated from the perspective of a widely held model in Figures 3 and 4. Magnetic flux and plasma are generally, but not permanently, coupled to each other within a magnetic field. When the relationship breaks down, disrupting magnetic field lines, reconnection occurs, requiring transfer of flux rapidly across steep spatial gradients, a process particularly likely to occur when adjacent field lines are anti-parallel and plasma domains are oppositely polarized. Reconnection allows conversion of magnetic field

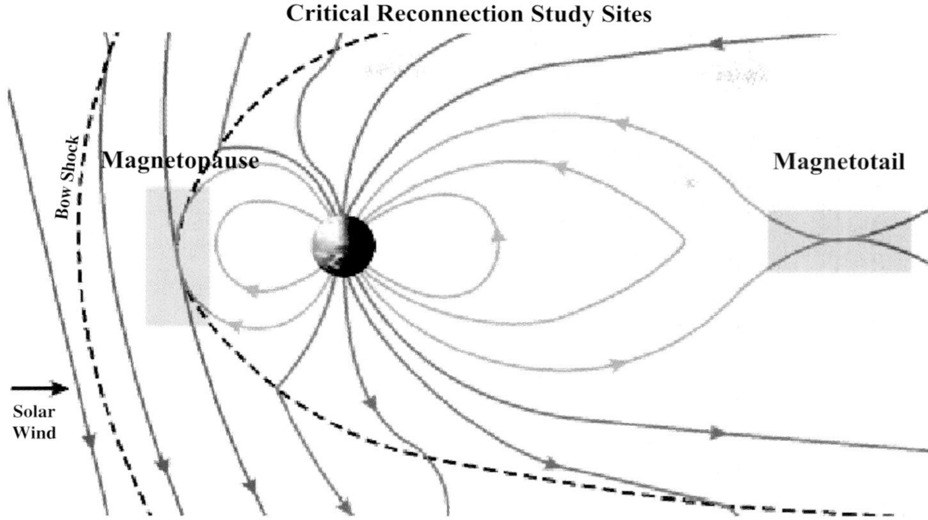

Figure 2: Critical sites to study reconnection (STDT MMS Report, 2000).

to kinetic energy of the plasma, and thus is one of the primary mechanisms for the transfer of energy into a form which allows the mobility of individual particles throughout the universe. The importance of reconnection is illustrated by the correlation between intensity of geomagnetic disturbances and southward magnetic direction of the solar wind, making it antiparallel to the Earth's magnetic field. These disturbances tend to occur in the dayside magnetopause and magnetotail, where solar wind interactions are most intense (Figure 3) (Birn et al, 2001; Dorelli and Birn, 2003).

The diffusion boundary between cells consists of two embedded regions of different physical parameters and sizes (Figure 4). The larger region, called the 'Hall Zone', exhibits typical dimensions of a few to about ten ion inertial lengths, which in many cases is roughly equivalent to the ion gyroradius. In typical space plasmas this dimension is a few hundreds of kilometers. The larger ion mass generates a region where the ions decouple from the electrons. The currents, which result when magnetized electrons respond to the EXB environment as a fluid, carry the asympototically coplanar magnetic fields out of the common plane and form a quadrupolar out-of-plane magnetic field structure that is roughly aligned with the reconnection line, also called the X-line.

Current studies of reconnection are designed to provide insight on 1) the conditions under which reconnection occurs, 2) whether some magnetic configurations are stable, 3) the nature of microscale processes that occur and their coupling to larger scale processes, and 4) parameters which control the rate of reconnection and the possibility of these processes occurring steadily as well as explosively.

2.2 Current Observations

Recent studies from the POLAR and Cluster missions illustrate the state-of-the-art in magnetospheric study.

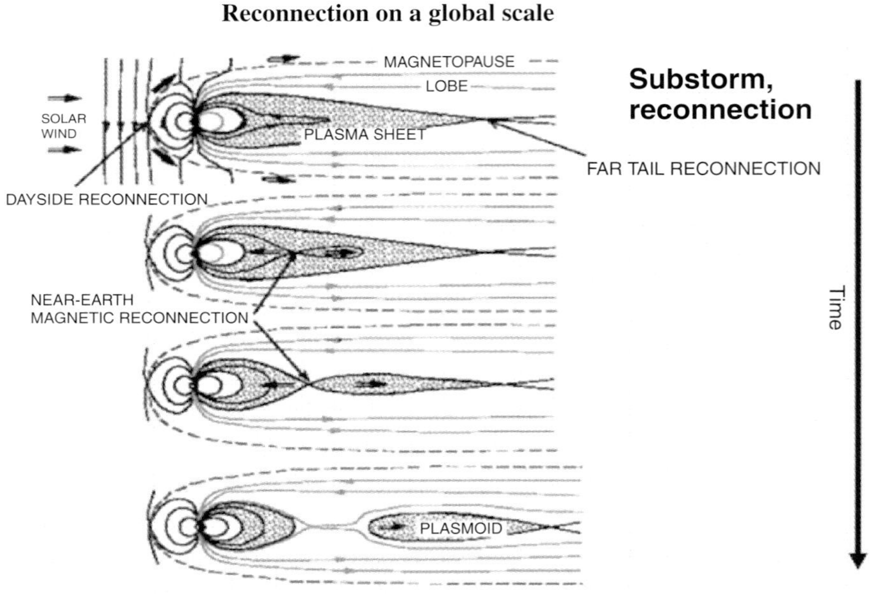

476

Reconnection on a global scale

Figure 3: Reconnection at the global scale showing primary activity sites and plasmoid formation from Birn et al, 2001.

The first experimental documentation of magnetic reconnection provided by Hydra, EFI, and MFE instruments on POLAR mission based on 2 million datapoints collected over 3.5 hours. Figure 5 illustrates the study of magnetopause reconnection by Drake and coworkers (2003), who simulated and observed solitary waves at the spacecraft crossing of the magnetopause, performed an epoch analysis of Polar magnetopause crossings, and made a comparison of measured versus predicted wave size. The differential motion of the spacecraft.relative to the reconnection site indicated that thesite was effectively traversed and sampled. This study confirmed that the process of collisionless magnetic connection works as theoretically foreseen and can be a quasi-stationary process.

Scudder and coworkers (2002) analyzed plots of electron, ion, and magnetic field parameters obtained on March 29, 1996 and observed an event involving reconnection (Figure 6). Observed electron pressure, $1/3\ T_rP_e$, was anti-correlated with concurrent magnetic pressure, $B^2/8\pi$. These results apparently demonstrate reconnection occurring outside the diffusion region (Scudder et al, 1999, 2002; Sonnerup et al, 1995), with dynamics that are inconsistent with gyrotropy, based on the implication that the electron inertial length is more than an order of magnitude smaller than the electron thermal gyroradius.

The four Cluster spacecraft were launched in the year 2000 into elliptical polar orbits, with dimensions of 19 000 by 119 000 km and with a 57 hour period. Separation distances have ranged from hundreds of kilometers (from 2000 to 2001 and from 2003 to 2004)) to thousands of kilometers (from 2001 to 2003) (Figure 7) and are designed to allow the exploration of mesoscale structures and processes.

Reconnection at the microscopic scale

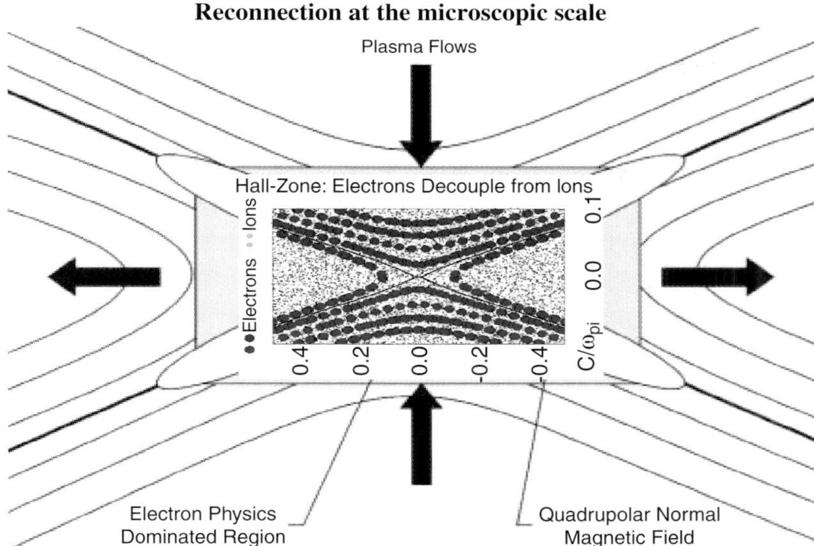

Figure 4: Reconnection at the microscopic scale showing Hall effect at the cell boundaries from STDT MMS Report, 2000 with insert from Ma and Bhattacharie, 1996.

Cluster provided a series of observations of the magnetopause boundary from a mesoscale perspective during major reconnection events in March, 2002 (Owen et al, 2001; Slavin et al, 2003) and March, 2003 (Khotyaintsev et al, 2003) (Figure 8), confirmed by POLAR visible imaging system. Observations such as these provide the first serious constraints on the propagation and point of origin of the event. Combined observations from several missions on August 27, 2001 gave evidence of a global magnetic storm suggesting reconnection in the midtail region (Figure 9) (Baker, 2003).

Reconnection in the magnetotail is believed to occur around 40 Earth radii, is thought to result in the release of plasmoids and flux ropes that propagate away from the reconnection point. Cluster, located at up to 20 Earth radii, is ideally located to study these phenomena. The four spacecraft are now being used to identify such flux ropes, and to determine their speed and direction of propagation very accurately (Escoubet et al, 2001).

3. Magnetospheric MultiScale (MMS) Concept

The Magnetospheric Multi Scale Mission will consist of four identical spacecraft equipped with a robust particle and fields payload with multiple plasma heads and the capability of 3D electric field detection. Spacecraft will have onboard accurate ranging capability which will operate in the ≤10 to ≤1000 kilometer range to provide knowledge of each spacecraft's position within a tetrahedral flying formation. Onboard Delta-V will be approximately 1 kilometer per second. The mission will have up to four distinct Orbital Phases (STDT) (Figures 10 and 11). The final orbit design will be determined from study team reports. Its scale of spatial determination ranges from ≤10 to ≤1000 kilometer, depending on orbital phase and processes

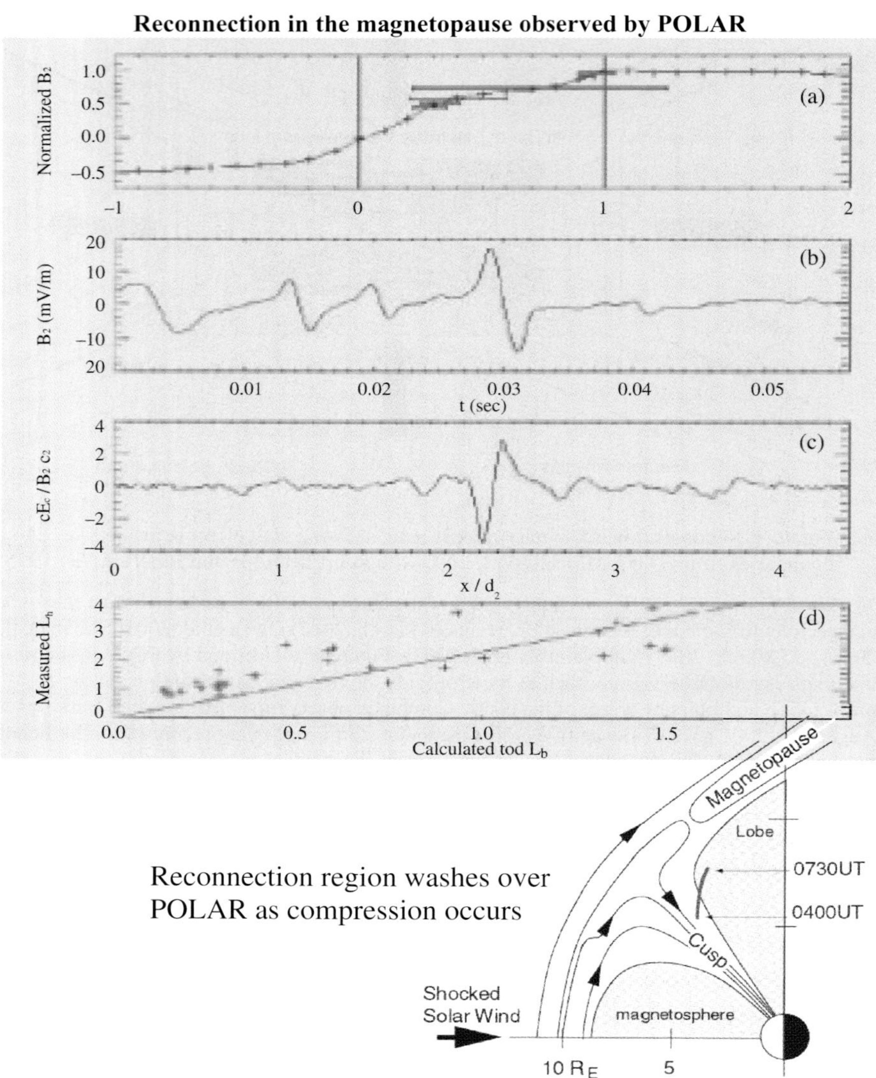

Figure 5: These figures from the POLAR data by Drake et al (2003) provide the first experimental documentation of magnetic reconnection. Left from top to bottom, epoch analysis of Polar magnetopause crossings, solitary waves at Polar magnetopause crossing and in simulation, and wave measured vs predicted size. In lower left, differential motion of spacecraft relative to reconnection site meant site was effectively traversed and sampled.

A Reconnection event observed by POLAR

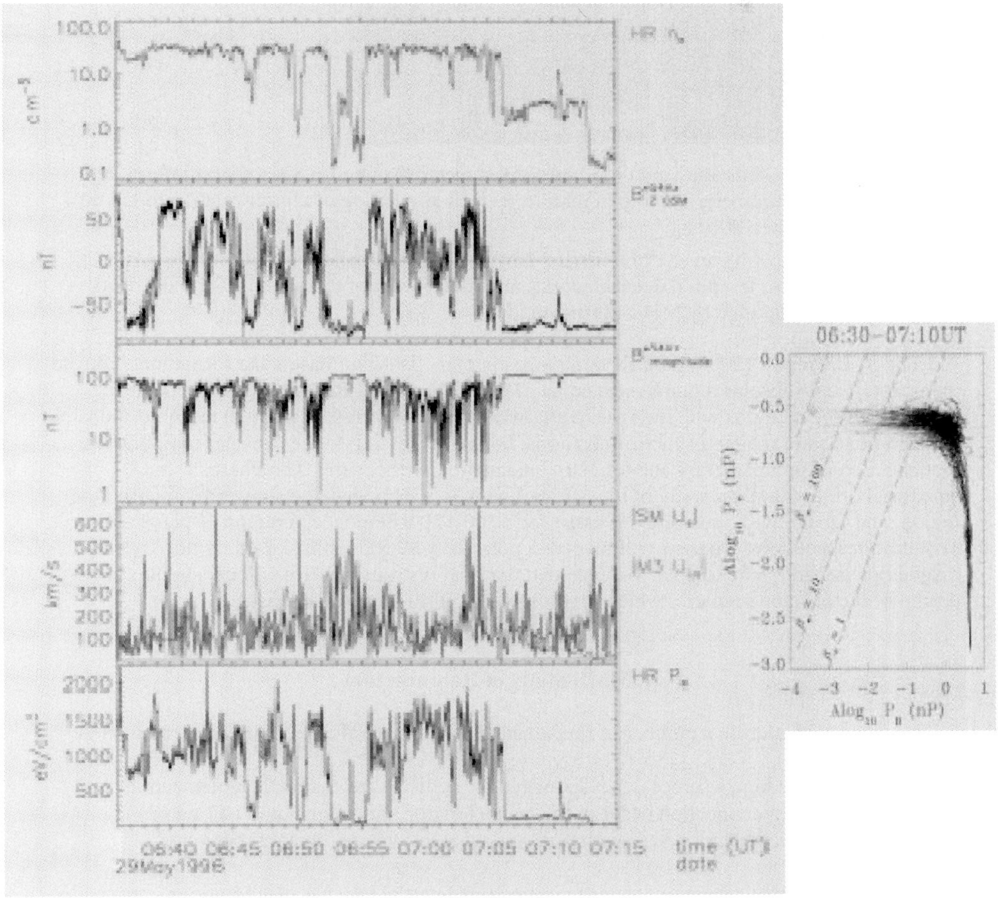

Figure 6 Plots of electron, ion, and magnetic field parameters obtained from Polar during the May 29, 1996 reconnection event. Insert on right shows anti-correlation of observed electron pressure, 1/3 TrPe, with concurrent magnetic pressure, B2/8π, consistent with magnetic reconnection outside the diffusion region (Scudder et al, 2002, 1999; Sonnerup et al, 1995).

480

Orbits and Configuration of Cluster Spacecraft within Magnetosphere

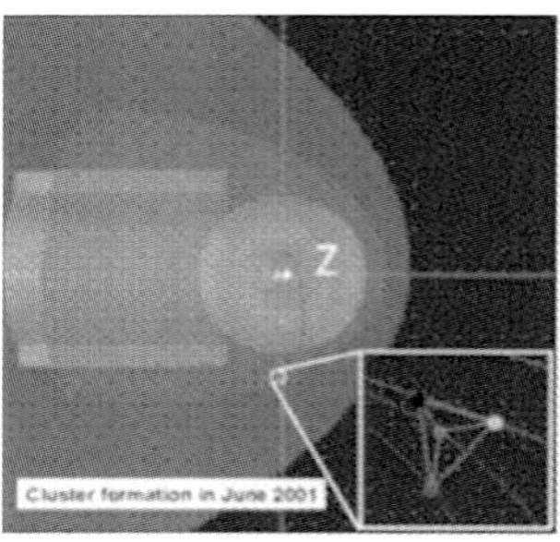

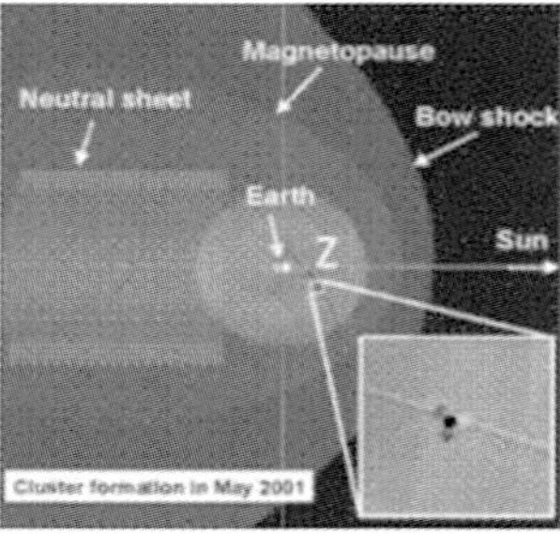

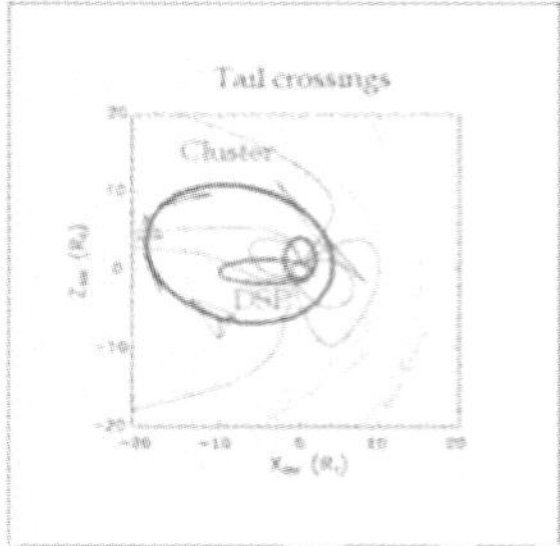

Figure 7: Cluster orbits showing thousands of kilometer spacing between spacecraft on left, and hundreds of kilometer spacing in middle, as well as position of cluster relative to tail on right (Escoubet et al, 2001).

being observed, allowing an electron inertial length of approximately 10 kilometers to be resolved. Its scale of temporal determinations are on the order of tens of milliseconds. Data acquisition will be guided by MHD plasma models.

The STDT report (2000) described the mission plan. During Phase 1 the formation will study the near tail and subsolar regions in an equatorial orbit with a 12 RE apogee equatorial orbit. Thus, the mission begins with the spacecraft dwelling in the current disruption region responsible for the initiation of magnetospheric substorms, before precessing to the subsolar region where reconnection of interplanetary and terrestrial magnetic fields occurs. The Phases 2 and 3 equatorial orbit will allow study of the Mid-tail with 30 RE apogee, the substorm reconnection region, and far-tail with lunar gravity assist for a 100-120 RE apogee, a region of plasmoid evolution respectively. Phase 4 will involve a polar 10 x 40 RE orbit, skimming the dayside magnetopause from pole to pole. We note that the final instrument team will decide the final design of the mission scenario, which may allow the STDT mission scenario.

4. MMS Study of Reconnection

MMS wiould address a number of fundamental science questions relating to reconnection:

1. What are the kinetic processes responsible for collisionless magnetic reconnection?
2. Where does reconnection occur in the magnetopause and magnetotail and why?
3. How does reconnection vary with time and what influences temporal behavior?

To date, answers to these questions have remained partial because of limitations imposed by dependence on measurements from single platforms. Nor, for the most part, will answers be provided by the multipoint data from Cluster, which will focus on different regions and have larger spacecraft separations appropriate to its main objective of investigating MHD turbulence.

4.1 Kinetic processes responsible for collisionless magnetic reconnection

Figure 3 illustrates the current understanding of diffusion at the boundary region. The inner diffusion region is dominated by electron physical processes, which generate an electric field aligned with the X-line with a component parallel to the magnetic field. The electric field can be

A Reconnection event observed by Cluster

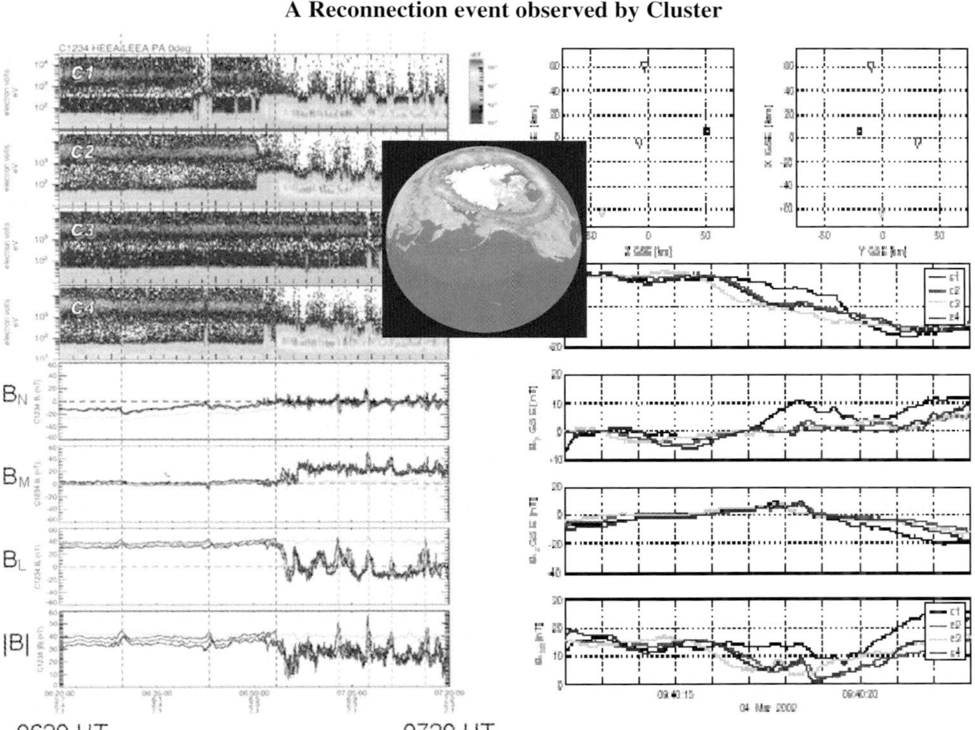

Figure 8: First time observations taken on either side of magnetopause boundary during reconnection using 4 Cluster spacecraft in March, 2002 (Owen, 2003, left) and March, 2003 (Khotyaintsev et al, 2003, right), with insert showing Auroral oval from POLAR visible imaging system during the 2002 event. Observations such as these provide serious constraints on the propagation and origin.

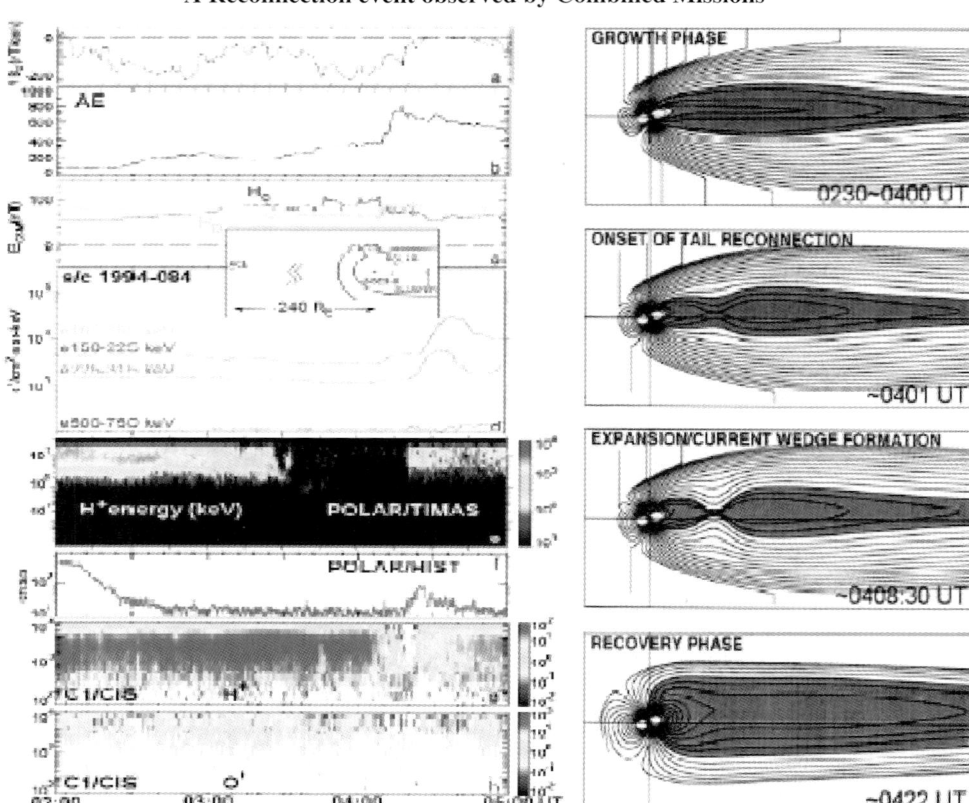

Figure 9: Combined mission observations from August 27, 2001 showing a global magnetic storm suggesting reconnection in the midtail region. From top to bottom, on the left, for ACE VBz, Auroral electrojet indices, GOES8 Magnetometer, LANL 50-500 keV energetic electron data, POLAR TIMAS H+ data and POLAR HIST electron flux, CLUSTER Spacecraft 1 H+ and O+ data. On right is model derived by Baker at al, 2003.

generated by bulk electron acceleration in the direction antiparallel to the main current flow, a process which appears to dominate in the case of very thin current sheets with thickness on the scale of the electron inertial length or in the presence of a guide magnetic field. This process is likely to occur at the magnetopause. Alternatively, the electric field can be generated by nongyrotropic electron pressure anisotropies. Scudder and coworkers (2002) demonstrated non-gyrotropic behavior of electrons has been demonstrated at and in the vicinity of the separator crossing. Non-gryotropic processes appear to dominate in the case of thick sheet thickness on the scale of ion inertial length. The rate at which the magnetotail current sheet thins may determine which mechanism operates and could be resolved and investigated using MMS 3D spatial structural and temporal variations in microscopic particle motions and wave-particle interactions.

4.2 Where and why reconnection occurs

Although the magnetopause is a key site for reconnection, observations have been limited to low and middle latitudes and the cusp. Models based on these observations predict either sustained reconnection along a line that runs across the magnetopause through the subsolar point that tilts according to the orientation and magnitude of the magnetosheath magnetic field or merging along merging lines in the northern and southern hemispheres where magnetosheath and terrestrial fields are antiparallel but not in the subsolar region. A dominant factor in determining where reconnection occurs is the orientation of the interplanetary magnetic field or magnetosheath field. As the orientation changes, the merging zones shift, moving from lower latitudes for southward to higher latitudes for northward fields. MMS data sequences include low and high inclination orbits and the undersampled high latitudes. Such observations should be able to provide far better constraints for reconnection in the magnetosphere.

MMS would directly sample the two regions in the magnetotail associated with reconnection, the near-Earth neutral line in the pre-midnight section and the distant neutral line downtail. The interplay of reconnection activity at these two lines determine how much energy is extracted from the solar wind and how it is distributed between the inner magnetosphere and the tail. Geotail has provided a great deal of information on activity in the tail, and has provided evidence that thin current sheets have localized instability leading to 'patchy' magnetic reconnection. Only spatially and temporally resolved observations, such as those MMS could provide, will determine the dynamic structure of this patchiness in the magnetotail, and its relationship to plasmoid/flux rope formation.

4.3 The temporal behavior of reconnection

The nature of temporal variations in the reconnection process is not well understood. Magnetopause observations indicate a variety of modes, including minutes long quasi-periodic merging modes, steplike discontinuities in the cusp, all interpreted as evidence for 'pulsed' reconnection. In the near-Earth tail, reconnection is thought to occur in bursty fashion, whereas in the far-tail it occurs in a slower, more steady fashion.

MMS would measure the local rate of reconnection in both locations, in three ways: first, by measuring the inflow velocity, or dimensions of the diffusion region; second, by measuring the magnetic field component normal to the magnetopause or magnetotail current sheet, a fraction of the total magnetic field; third, by measuring the tangential component of the electric field in the vicinity of the X-line. MMS multipoint measurements resolve temporal and spatial variations and thus allow determination of the boundary orientation and motion.

MMS Study of the Magnetosphere

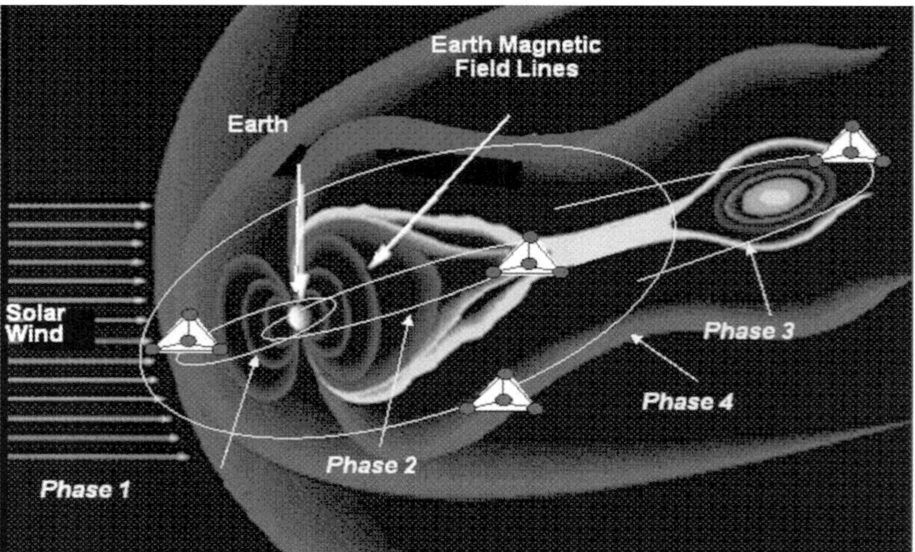

Figure 10: MMS will study reconnection in the magnetosphere in four distinct phases, each one focusing on a particular site crucial to reconnection processes.

5. Conclusions

The MMS mission would be capable of providing the first measurements down to the electronic inertial length of the 3D structure and dynamics of regions already known to provide observations crucial in understanding the reconnection process, including the subsolar magnetopause and magnetotail. In addition, the MMS instrument suite would determine the spatial and temporal scales of the basic plasma processes involved in reconnection.

Acknowledgements

Support for this work has been provided through NASA contract NAS5-99189, subcontract 0299189EER, with ITMI.

References

Baker, D.N., Telescopic and microscopic views of the magnetosphere: Multispacecraft observations, Space Sci Rev, 109 (1-4), 133-153, 2003.

Birn, J., Drake, J.F., Shay, M.A., Rogers BN, Denton RE, Hesse M, Kuznetsova M, Ma ZW, Bhattacharjee A, Otto A, Pritchett PL, Geospace environmental modeling (GEM) Magnetic Reconnection Challenge, JGR-Space, 106 (A3), 3715-3719, 2001.

Dorelli, J.C. and Birn, J., Whistler-mediated magnetic reconnection in large systems: Magnetic flux pileup and the formation of thin current sheets, JGR-Space 108 (A3), #1133, 2003.

Details of MMS Phases

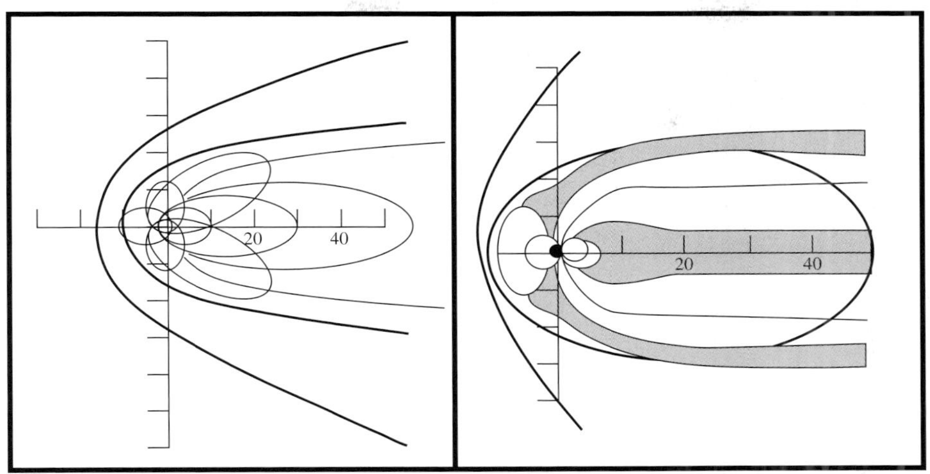

Figure 11: On the left equatorial orbital phases 1 through 3 are viewed from above the ecliptic. Phase 1 has 12 RE apogees precessing through the near tail and then subsolar regions, Phase 2 30 RE apogees precessing through the mid-tail region, and Phase 3 100-120 RE apogees precessing through the Far-tail region. On the right, 10 x 40 RE polar orbital phase 4 is viewed from the ecliptic plane.

Drake, J.F., Swisdak, M., Cattell, C., Shay, M.A., Rogers, B.N., Zeiler, A., Formation of electron holes and particle energization during magnetic reconnection, Science, 299 (5608), 873-877, 2003.

Escoubet, C.P., Fehringer, M., Goldstein, M., The Cluster mission – Introduction, ANNALES GEOPHYSICAE, 19 (10-12): 1197-1200 Sp. Iss. SI OCT-DEC 2001

Gold, T., Motions in the magnetosphere of the Earth, JGR, 64 (9), 1219-1224, 1959.

Falthammar, C.G., Akasofu, S.I., Alfven, H., Significance of magnetospheric research for progress in astrophysics, Nature, 275 (5677), 185-188, 1978.

Khotyaintsev, Y., Vaivads, A., Ogawa, Y., Popielawska, B., Andre, M., Buchert, S., Decreau, P., Lavraud, B., Reme, H., Magnetic reconnection in high-beta plasma: Cluster observations in the exterior cusp region, Ann Geophys, #12345, 2003.

Ma, Z.W. and Bhattacharjee, A., Fast impulsive reconnection and current sheet intensification due to electron pressure gradients in semi-collisional plasmas, GRL, 23 (13), 1673-1676, 1996.

Owen, C.J., Fazakerley, A.N., Carter, P.J., Coates, A.J., Krauklis, I.C., Scita, S., Taylor, M.G., Travnicek, P., Watson, G., Wilson, R.J., Balogh, A., Dunlop, M.W., Cluster PEACE observations o f electrons during magnetospheric flux transfer events, ANN GEOPHYS, 19 (10-12), 1509-1522, SI, 2001.

Scudder, J.D., Mozer, F.S., Maynard, N.C., Russell, C.T., Fingerprints of collisionless reconnection at the separator, I, Ambipolar-Hall signatures, JGR Space, 107 (A10), #1294, 2002.

Scudder, J.D., Puhl-Quinn, P.A., Mozer, F.S., Ogilvie, K.W., Russell, C.T., Generalized Walen tests through Alfven waves and rotational discontinuities using electron flow velocities, JGR-Space, 104 (A9), 19817-19833, 1999.

Slavin, J.A., Owen, C.J., Dunlop, M.W., Voralv, E., Moldwin, M.B., Sibeck, D.G., Tanskanen, E., Goldstein, M.L., Fazakerley, A., Balogh, A., Lucek, E., Richter, I., Reme, H., Bosqued, J.M., Cluster four spacecraft measurements of small traveling compression regions in the near-tail, GRL, 30 (23), #2208, 2003.

Sonnerup, B.U., Paschmann, I, Phan, T.D., Fluid aspects of reconnectin at the magnetopause: In situ observations, in Physics of the Magnetopause, 90, ed. P. Song, B. Sonnerup, and M. Thomsen, 167-180, 1995.

STDT, The Magnetospheric MultiScale Mission, Resolving fundamental processes in space physics, NASA/TM-2000-209883, 2000.

Multiscale Coupling of Sun-Earth Processes
A.T.Y. Lui, Y. Kamide and G. Consolini, editors.

MULTISCALE GEOSPACE PHYSICS IN CANADA

William Liu[1], Jonathan Burchill[2], Leroy Cogger[2],
Eric Donovan[2], Gordon James[3], David Kendall[1],
David Knudsen[2], Jianyong Lu[4], Ian Mann[4],
Réjean Michaud[1], Sandy Murphree[2], Robert Rankin[4],
John Samson[4], Emma Spanswick[2], George Sofko[5],
Trond Trondsen[2], and Andrew Yau[2]

[1] *Space Science Program, Canadian Space Agency, Saint-Hubert,
Québec, Canada*
[2] *Department of Physics and Astronomy, University of Calgary,
Calgary, Alberta, Canada*
[3] *Communications Research Centre, Ottawa, Canada*
[4] *Department of Physics, University of Alberta, Edmonton,
Alberta, Canada*
[5] *Department of Physics and Engineering Physics, University of
Saskatchewan, Saskatoon, Saskatchewan, Canada*

Abstract. Geospace physics research is entering a new era in Canada. Although this development has been a poorly kept secret among the informed observers, there has not been, to date, an attempt to summarize these changes in a single source and convey to the scientific world the vision and potential of the new "Northern perspective". In this paper we make a first attempt to fill this gap, without, however, claiming authoritativeness or completeness – such a claim would be defeated by the fast-paced development in Canada on many fronts. The new Canadian perspective is based on a keen awareness of the multiscale character of geospace, and on a realistic yet innovative outlook which integrates Canada's strengths into national efforts emphasizing the use of multi-instrument techniques to attack multiscale complexities in problems. We organize our discussion in four major sections: a description of the plan to enhance Canada's leadership position in ground-based geospace science, centered on the Canadian Geospace Monitoring program; a description of Canada's latest effort to achieve major breakthroughs in our understanding of geospace transition region dynamics, centered on the Canadian small satellite experiment ePOP; a description of Canada's participations in three major international geospace missions (THEMIS, SWARM, and AMISR); and, finally, a description of two potential new Canadian-led geospace missions, Ravens and Orbitals. We will integrate key scientific goals and strategies in our narrative about these missions, and use examples to illustrate the considerable potential of these Canadian efforts.

1. Introduction

Earth has much to teach the Universe. In the astrophysical context, geospace, defined here as the multiscale plasma system held together by the geomagnetic field, is unique and fundamentally important to science. While the uniqueness is easily appreciated in terms of amenability to in-situ measurements, the fundamental significance of geospace is not as widely known, or readily accepted, by the general physical science community. This importance is best understood through the concept of multiscale coupling, the topical phrase of the Kona conference, the proceedings of which this volume contains.

The multiscale nature of geospace relates to both Earth's vulnerability and resiliency. Earth's magnetic field lacks the strength to withstand the solar wind assault without adapting. By resiliency, the same field resorts to progressively strained geometrical and, occasionally, topological transformations to shed excessive mass and energy that accumulate during active times. Geometry, being a key variable in this context, can therefore assume a far richer variety than what is generally assumed for the Sun, stars, galaxies, and, for that matter, the entire cosmology. When geometry becomes as complicated as that characterizing the late-growth-phase magnetotail, or when topological changes occur on as short a timescale as the substorm expansive phase, scales become strongly coupled. Processes which under quieter conditions chug along in their own characteristic wavelengths become interwoven, and cause and effect begin to lose their linear clarity. Interpretation of data and consequently formulation of theory will not be correct, unless multiscale coupling is taken into account from the outset.

Geospace is also rich in diversity. Its plasma species range from collisionless (the solar wind and magnetosphere), to weakly collisional (upper ionosphere, geocorona, and radiation belts), to collision-dominated (the lower ionosphere). Coupling among these disparate plasma species adds to the complexity of geospace. These facts underscore the unique but still under-appreciated value of geospace plasma physics to science and are the ultimate reason why space agencies around the world are spending hundreds of millions of dollars on missions such as Cluster, THEMIS, and MMS, in order to untangle the convolved multiscale structure seen, for example, in the aurora. If geometry can also be used as a rhetoric device, geospace physicists are situated in a special plane, from which they study a natural plasma system from inside out, hence availing themselves some of the best angles to observe a number of universal problems (e.g., reconnection).

In this paper, we review the Canadian solar terrestrial science program, which is rapidly evolving in both appearance and contents. While we cannot be exhaustive in our discussion, the examples we choose will illustrate our outlook and strategy. In our vocabulary, multiscale is not a word or slogan. Rather, in this term is embedded a powerful outlook to a new way of science, consisting in a holistic, first-principle-based approach to understand a complex system from top down. Multiscale is antithetical to the serial thinking of "problems in their own right"; its new mantra might well be "think global and stay there." In this outlook, multiscale is a departure from the reductionist philosophy of divide-and-conquer. It demands an opposite impulse, to try to formulate and tackle a problem on as wide a scale hierarchy as possible, in as general a term as technologies would permit. This explains what is happening in the solar terrestrial science community today, with its armada of constellations, networks, assimilated data products, and computer models pushed relentlessly toward the reproduction of real-world events, instead of ethereal mathematical oddities awaiting a cosmological name. To us, these are the kernels of multiscale.

What is multiscale, if not also multilateral? Global in reach and manifestation, geospace science will not be complete, if any nation feeling its impact or exposed to its wonder is not also a contributor to its understanding. No one nation's resource, talent, or geographic location is all-encompassing to solve this grand challenge by its own. For this reason, geospace science has been among the most international in its tradition, with International Geophysical Year, International Magnetospheric Study, various initiatives undertaken by the Scientific Committee on Solar-Terrestrial Physics (SCOSTEP), International Solar-Terretrial Physics (ISTP) program, and now International Living With a Star (ILWS, http://ilws.gsfc.nasa.gov), which have drawn scientists of different backgrounds and cultures together. In our conception and implementation of a Canadian program, we are mindful of our place in the world, and how we can best contribute to a most challenging and enduring intellectual pursuit in modern science.

We describe the present and planned Canadian geospace science activities in four sections: the Canadian Geospace Monitoring (CGSM) program, the ePOP/CASSIOPE small satellite experiment, the Canadian participation in international constellation missions (THEMIS and potentially SWARM), and, looking forward, two new mission concepts currently under study (Ravens and Orbitals). While our goal is to give up-to-date information on the Canadian program, we will make a connection, in each case, with major geospace problems commanding our attention today.

2.   Canadian Geospace Monitoring

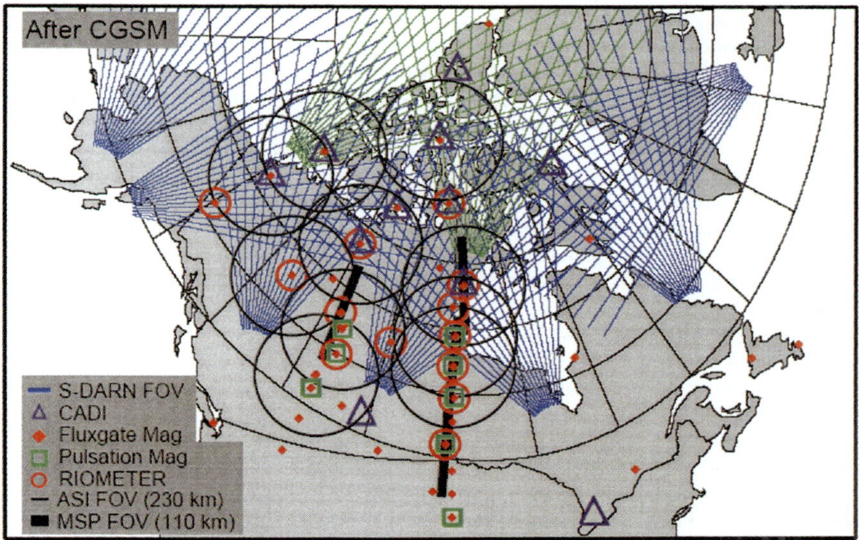

Figure 1. Ground-based coverage in the Canadian sector will undergo a major expansion in the coming years, anchored by the Canadian Geospace Monitoring array. Diamonds in the figure represent magnetometers (those with blue outlines are induction coil magnetometers). The circles represent the field of view of an All-Sky Imager. The rays represent the coverage region of SuperDARN radars. CADI, CANOPUS riometers and MPA, are not shown.

Canada's tradition and achievements in ground-based study of geospace science are well-known. The widespread use of CANOPUS data and the instrumental role Canada plays in the international SuperDARN program are but two examples. Since 2001, the Canadian geospace community has taken a series of steps to integrate the existing and planned ground-based observatories and facilities into a truly national program, known as Canadian Geospace Monitoring (CGSM). While CANOPUS supported by the CSA is a centerpiece of CGSM, many other non-CSA originated programs are being integrated under a sweeping vision. CGSM sets itself upon the following scientific objectives: 1. convection within and energy injection into the global magnetosphere; 2 trigger and control of magnetotail instabilities and flows; 3. generation, modulation, and structure of auroral particle acceleration; 4. energization, transport and loss of energetic magnetospheric particles and 5. injection, transport, and loss of low energy magnetospheric particles.

Programmatically, CGSM has 7 components: a) an expanded magnetometer array comprising of CANOPUS, Natural Resources Canada CANMOS, and 23 new magnetometers funded by Canada Foundation for Innovation (CFI); b) a continental-size imaging array comprised of 10 all-sky imagers (6 of which are funded by the CFI), 13 riometers, and 4 meridianal scanning photometers; c) the Canadian component of SuperDARN comprised of two existing HF radars (Saskatoon and Prince George) and potentially a new pair of PolarDARN; d) the Canadian Advanced Digital Ionosonde (CADI) array; e) the Herzberg Institute of Astrophysics solar radio telescope which has been monitoring solar F10.7 flux since 1947; f) the Facility for Data Assimilation and Modeling, and g) the Canadian Space Weather Forecast Service (www.spaceweather.ca). Figure 1 gives the projected distribution of ground-based instruments in the Canadian sector (including non-Canadian instruments), which will be fully operational in 2 years. (The map does not include the ground-based instruments to be deployed specifically to support THEMIS and other potential instruments to support the AMISR incoherent radar, part of which will be deployed in Resolute Bay, Canada). It is noted that the CSGM instruments will be supported by the HSi technology using very-small aperture terminals to transfer data in near real-time to science centers, as they are today under the CANOPUS and CANMOS arrays, but at a bandwidth ten times larger. Over time, CGSM data will become accessible through a CGSM data portal powered by distributed grid-computing technology, permitting a seamless integration of the rapidly developing Canadian multiscale models into the wealth of CGSM and allied data.

The upgraded infrastructure is to serve well-defined scientific needs. We discuss, through two examples, how an expanded CGSM data set can lead to new advances in science.

2.1 Magnetoseismology

The magnetosphere is filled with waves, many of which leave distinct signatures in the ionosphere and on the ground. Data assimilation techniques allow increasingly detailed characterization of these waves on global to meso-scales. The ground-based vantage has long been capitalized by Canadian scientists (e.g., CANOPUS and its predecessors) in the study of micropulsations and field-line resonances [cf., *Samson et al.*, 1971]. Of late, attempts have been made to use wave distributions on the ground to infer the structure of the magnetosphere, giving rise to the notion of magnetoseismology.

It is instructive to compare magnetoseismology with its more heralded counterpart, helioseismology. Both are founded on the general principle of using wave signatures on a surface to sound the interior of a difficult-to-reach object. In detail, however, the two are quite different, with magnetoseismology in general being the more difficult to handle computationally. First there is the difference in geometry. The solar geometry, in a first approximation, is spherical, with density structures treated as perturbations to this overall configuration. In the magnetosphere, the geometry is quite complex and sometimes unstable. This difference makes multiscale coupling a greater concern for magnetoseismology. The second difference is the wave modes. The dominant wave at the Sun is an acoustic-gravity mode, of both surface and body variety. In the magnetosphere, the dominant wave is magnetohydrodynamic, which is strongly anisotropic and closely coupled in an inhomogeneous medium. The third difference is perspective. Solar imaging has the classical astronomical view, whereas magnetospheric wave characterization can, at best, be done on a continental scale (Figure 1) and involves an 'inside-out' view. These differences render many techniques developed through helioseismology unfit for magnetoseismology; consequently, much of the theoretical foundation of magnetoseismology cannot be borrowed but must be built anew.

Ground-based instrument arrays like the one shown in Figure 1 are an essential tool to magnetoseismology, much as Michelson-Doppler Interferometry (MDI) is essential to helioseismology. The identification of the Alfvén continuum [Waters et al., 1995] and its subsequent application are an example of how ground-based wave observations can be used to infer magnetospheric density. However, recent results have cast doubt on whether the linear field-line resonance theory underpinning the existing magnetoseismic techniques is valid. Historically, field-line resonances have been studied in simplified Cartesian or dipolar geometry. Later more realistic geometries were used [Rankin et al., 2000]. Although it has been generally understood the FLR would behave differently under nondipolar and nonlinear conditions, the expectation has also been that the linear FLR theory is not too far-off, as a first approximation.

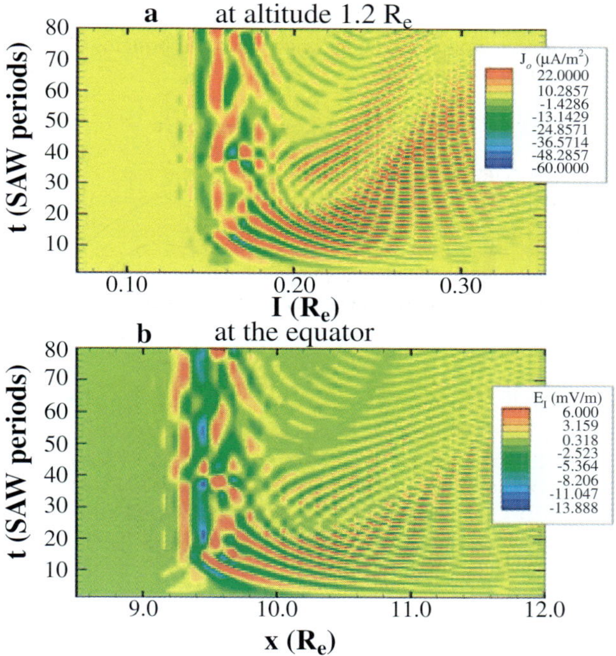

Figure 2. Nonlinear evolution of field-line resonances have shown a marked difference from linear predictions. The figure shows that a migration of the primary resonance from the location predicted by linear theory, in addition to multiscale features from nonlinear interactions.

The work of *Lu et al.* [2004] has called this assumption into question. The model is the most general theoretical treatment of FLR to date, featuring a nondipolar geometry and explicit account of ion bounce (which have a comparable timescale to the FLR period, making ion response to the wave nonadiabatic and leading to an effective nonlocal parallel conductivity). The initial objective of the authors was quite straightforward: Drive the model magnetosphere with a monochromatic wave and observe the formation of the corresponding FLR in the nonlinear regime. Although some departure from the linear theory was expected (e.g., the fringe features in Figure 2), what actually transpired went beyond this expectation. As shown in Figure 2, the localization of the wave does not occur anywhere near the location predicted by the linear theory, but migrates to a point where a nonlinear dispersion function minimizes. This result suggests that a linear FLR might be an unstable structure: Not only it compactifies with time, it also moves to a location that this steepening is least compromised by dispersion. Although the new FLR theory has yet to be fully validated, the results so far offer strong suggestions that much of what we take for granted in magnetoseismology needs reconsideration, and that quantitative magnetoseismology may be fundamentally nonlinear.

2.2 Multispective imaging

Imaging, for our purpose, is the technique of projecting information on a two-dimensional plane. One of the valuable aspects of ground-based observations is its two-dimensional information content. The value of the CGSM array consists in multiple layers of 2D information, each from an instrument group, overlaying each other and bringing different perspectives to give a fuller representation of the physics.

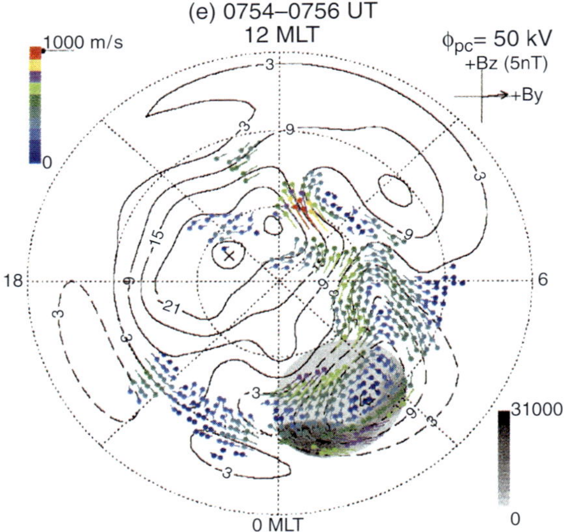

Figure 3. An overlay of NORSTAR ASI (b & w) observation on SuperDARN (vector) data in the field of view of Rankin Inlet ASI. The overlay technique can help resolve advective and propagatory motions in auroral forms.

Figure 3 is an example of the multiperspective imaging technique. Here a comparison is made between contemporaneous measurements over Rankin by the Northstar All-sky imager (ASI) and by SuperDARN. For years, auroral imagers, first on the ground and then in space, have captured the apparent motion of auroral forms on multiscales and in multiple wavelengths. There is an uncertainty in this information, as apparent motion is a combination of advection and wave propagation. To definitively associate auroral structures with waves of one kind or another, the latter needs to be distinguished from the apparent motion. Figure 3 is an overlay of two 2-D fields, the SuperDARN convection and temporally evolving auroral forms, giving one a tool to make this distinction.

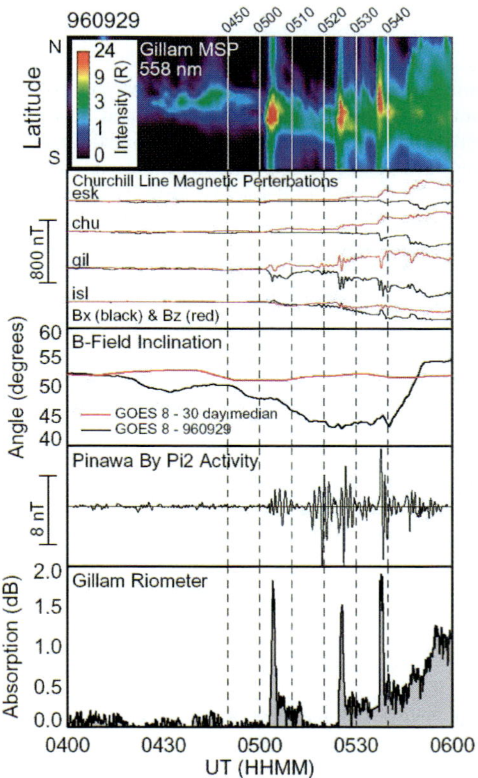

Figure 4. Use of multiple ground instruments for the study of substorms has been a proven technique. The figures shows how riometer absorption spikes (bottom panel) correlate with other signatures of substorms (meridianal scanning photometer, top panel; ground magnetometer, second from top; in-situ magnetic field, third from top, and Pi2 pulsations). The riometer data, relating to adiabatic electrons > 30 keV, can be used to trace the location and size of the substorm onset region.

Figure 4 is another example of multiperspective use of CGSM data, taken from *Spanswick et al.* [2004]. Aside from wave measurements and characterization in 2D (to enable magnetoseismological diagnostics), ground-based observations of particle precipitation can be used as a tracer of dynamical episodes such as substorms. Auroral morphology and intensity are direct functions of ion and electron motions. Many of the particles, however, move in an unknown wave environment and their mark in the atmospheric emissions is thus difficult to be related directly to magnetospheric geometry, without knowing how the particle motion is modified by the waves. Energetic electrons with energy in the 30-100 keV range provide an exception to the rule. Adiabatic even during the most disturbed period, they can be used to time and locate the substorm onset, without having to account for the effect of wave-particle interactions.

The example given in Figure 4 shows energetic electron precipitation (>30 keV) registered as riometer signals (bottom panel), on September 29, 1996, and how it corresponded with the 558 nm electron aurora seen by the CANOPUS Gillam Meridional Scanning Photometer (MSP) (top panel). The middle panels are magnetometer data, both ground-based and in-situ, to differentiate between pseudo-breakups (first two) and the actual onset (last spike). As suggested earlier, >30 keV electrons, their adiabaticity unaffected by temporal variations, are an excellent tracer of substorm onset. It can be shown that the riometer spikes in Figure 4 correspond in a very general way with conditions when magnetic field lines dipolarize suddenly. In its present form, a CANOPUS riometer works independently, and a true two-dimensional precipitation distribution is not yet possible. However, it can be envisaged that, an imaging riometer array deployed in the middle of the CGSM network, through the provision of a 2D riometer map every few seconds, can be used to not only diagnose the timing of substorm onsets, but also estimate its size and spatial characteristics. This capability would be a valuable addition to space missions such as THEMIS.

The above are but two examples of how we can exploit CGSM data through interoperable instruments and data-constrained computational modeling. Other noteworthy applications using CGSM (and its predecessors) data include the association of field-line resonances with different phases of the substorm [Samson et al., 1992a; Xu et al., 1993], constraint of the substorm onset location [Samson et al., 1992b], and the identification of the Alfvén continuum [Waters et al., 1995].

3. Geospace Transition Region Physics

Between approximately 300 km and 1 R above the surface of Earth, geospace undergoes a drastic transition from a tenuous plasma to a dense atmosphere; multiscale coupling is richly manifested in this transition. While the geometry in the region is simple, the physical forces and particle constitution are more complicated than the magnetosphere. Here, gravity, collision, and electromagnetic force can, at any given time, dominate. The extension of the atmosphere (geocorona), precipitating magnetospheric particles, and outflowing ionospheric particles all bring their own physics to the fray. The geospace transition region (GTR), though limited in size, boasts a large number of unsolved questions. In a laboratory setting, an interface between two plasmas of sharp dissimilarity often occurs in the form of a sheath. In GTR, this sheath, as a monolithic structure, is never observed. Put differently, the sheath is unstable, and the products of this instability is a wealth of small and meso-scale structures that mediate the interaction between the magnetosphere and ionosphere. Since plasma interfaces as the one found at Earth are common occurrences, particularly in planetary atmospheres, understanding multiscale coupling arising from the presence of competing forces and multiple species is an important problem which may have implications on the evolution of the atmosphere over the planetary timescale.

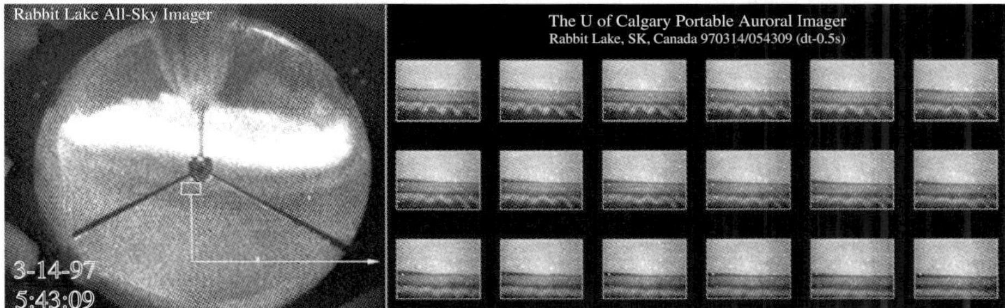

Figure 5. The Portable Auroral Imager developed by the University of Calgary reveals highly regular small-scale structures in the aurora with a spatial scale on the order of 100 m. The panel on the right shows in greater detail the area contained within the rectangle of the all-sky camera view on the left.

While the interaction between a planetary atmosphere and surrounding plasmas can be easily seen as an important problem, over the years the study of this problem has not matched the intensity of, say, the substorm research. The reasons for this are several, but today, interests in the GTR have grown substantially. It is well-known that auroral features do not exhibit a one-to-one scale correspondence to their assumed magnetospheric drivers. Figure 5 shows a series of auroral images taken by the University of Calgary Portable Auroral Imager with scale resolution down to 100 m [*Trondsen and Cogger*, 1997; *Trondsen and Cogger*, 2001]. The structures, some pulsating at the period of 1-2 s, do not have a ready magnetospheric explanation and are suspected to be a product of nonlinear multiscale processes occurring in the GTR. Therefore, the question of how the GTR processes large-scale magnetospheric inputs and how the resulting nonlinear wave-particle interaction produce small-scale auroral structures is a major unsolved problem.

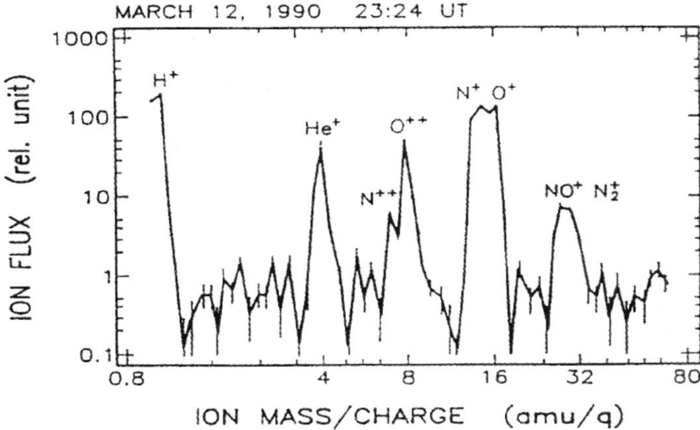

Figure 6. The Suprathermal Mass Spectrometer (SMS) on the Akebono satellite is the longest operating mass spectrometer in space. The example shows the mass outflow from the ionosphere during a geomagnetic storm on March 12, 1990.

496

Historically, GTR physics has been a central interest in the Canadian geospace research. Early examples include the confirmation of the polar wind by ISIS 2. Suprathermal Mass Spectrometer (SMS) onboard the Japanese satellite Akebono was another leap in terms of high-resolution mass spectroscopy in space (Figure 6). Still operating after 15 years, SMS has collected ion outflow data for more than a solar cycle. By virtue of its high sensitivity and longevity, SMS remains one of the best data set to characterize the ionosphere as a particle source to the magnetosphere, and how this contribution varies in accordance with geomagnetic activity the solar cycle [*Cully et al.*, 2003]. See Figure 7 for an example.

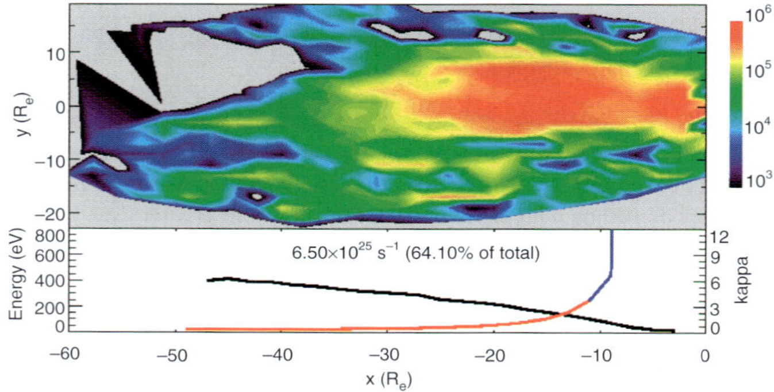

Figure 7. Tracing of SMS O+ outflow data from the exobase to the magnetosphere using the empirical Tsyganenko (magnetic field) and Weimar (electric field) models, for high *Kp* and southward IMF. Adapted from *Cully et al.*, [2003].

Higher-resolution measurements are needed to make out structures in GTR that correspond to small-scale auroral features seen in Figure 5. Building on the hemispherical electrostatic analyzer concept [*Whalen et al.*, 1994], Canadian scientists have developed two strands of the so-called imaging ion spectrometer (i.e., simultaneous formation of 2D or 3D particle distributions on the detector plane). The mass-resolving, 3D variety will be described in connection with our discussion of the ePOP mission. The Suprathermal Ion Imager (SII), on the other hand, puts the emphasis on speed and accuracy of distribution function reconstruction, through the use of a CCD detector partly borrowed from Viking and Freja Ultra-Violet Auroral Imager (UVAI) [*Knudsen et al.*, 2003]. The prototype of this instrument flew on the Canadian sounding rocket GEODESIC in February 2000 (with two subsequent launches onboard NASA'a CUSP2002 and Joule sounding rockets). Sampling at a rate of 10 ms per image, SII was able to resolve plasma structures in phase space on a scale of 10 m (sounding rockets) and 100 m (a would-be satellite platform). The latter scale corresponds to the auroral features shown in Figure 5. Figure 8 is a sample of GEODESIC ion distribution derived from the SII data [*Burchill et al.*, 2004].

Canada has also been a leader in active radio wave experiments in space. A recent example of this activity was the Oedipus-C (OC) sounding experiment in 1995. A double payload carrying a radio wave transmitter, an associated receiver and particle detectors was launched. On the flight upleg, the transmitter and the receiver were connected by a conducting tether that joined the two rockets. The resulting data permitted a first test of the dispersion relation for electromagnetic surface waves propagating along a conducting tether in a magnetoplasma. The experiment established that a conducting wire in a plasma is an efficient guide of RF energy. On the flight down leg, the tether connecting the rockets was severed, and two-point plane-wave propagation experiments were carried out with the transmitter and receiver. The principal fringes were attributed to the Faraday rotation of the plane of linear electric-field polarization resulting from the sum of two cold-plasma wave modes. The agreement between observations and cold-plasma theory suggests that future bistatic propagation facilities can exploit the Faraday effect to measure density, density gradient and parameters dependent upon wave polarization (see Figure 9).

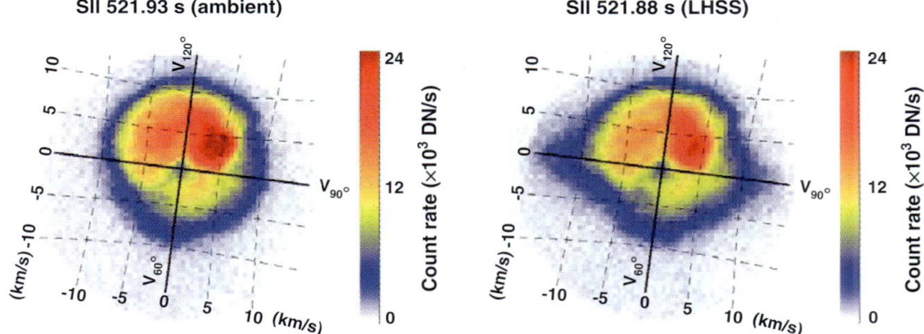

Figure 8. The Suprathermal Ion Imager flown on the Canadian sounding rocket GEODESIC proved both the feasibility and scientific value of using CCD detectors for particle imaging. The example shows the 2D distribution function of oxygen ions observed in the auroral zone during the rocket flight. At 10 ms time resolution, the movies of the distribution function allows a direct view on how wave-particle interactions take place in phase space. Adapted from Burchill et al., [2004].

The OC experiment allowed a detailed study of the so-called sounder-accelerated electrons (SAE). As the OC separation between payload ends increased from 153 to 537 m on the up leg, enhanced electron fluxes were detected at energies up to 20 keV, at the receiver end of the tether. The physical cause of SAE remains a theoretical mystery. The experiment proved that an active antenna can serve as an artificial injector of moderate-intensity auroral fluxes, but without the disruptive beam-plasma discharge experienced by dense localized beams from electron guns.

Investigations of the radiation characteristics of electric dipoles for both emission and reception of various electromagnetic magnetoplasma modes were carried out in the two-point free propagation experiment. The investigation of quasi-electrostatic whistler-mode waves propagating near the lower oblique resonance cone led to the discovery that the effective length of a dipole receiving such waves can be many times its physical length. In contrast, the radiation and reception of electromagnetic propagation can be accounted for using the cold-plasma dispersion relation and the short-dipole theory for the OC dipoles. OC antenna metrology has thus provided both guidance and warnings about the use of dipoles for observing wave magnitudes of spontaneous emissions around the electron characteristic frequencies.

Canada's two traditional streams of GTR physics research are finding an intersection in the Enhanced Polar Outflow Probe (ePOP) mission to be flown in 2007. A sophisticated experiment comprised of eight instruments, ePOP proposes to combine high-resolution imaging particle detectors and wave instruments covering frequencies from 10 Hz to several GHz, in order to probe simultaneously the two aspects of wave-particle interactions. The ePOP payload will fly on a CSA satellite known as CASSIOPE, along with a technology demonstration payload (Cascade) developed by the Canadian space company MDA. The decision to combine two payloads was not entirely a calculation based on economy; the Cascade technology, nicknamed Fedex in the Sky, calls for the use of Ka band for same-day gigabyte data delivery to and from remote locations.

With a downlink bandwidth >30 Mbps (compare to the typical S-band downlink rate of 1 Mbps), Cascade will allow ePOP to telemeter high-resolution data like that shown in Figure 8 to the ground, significantly increasing the expected science throughput.

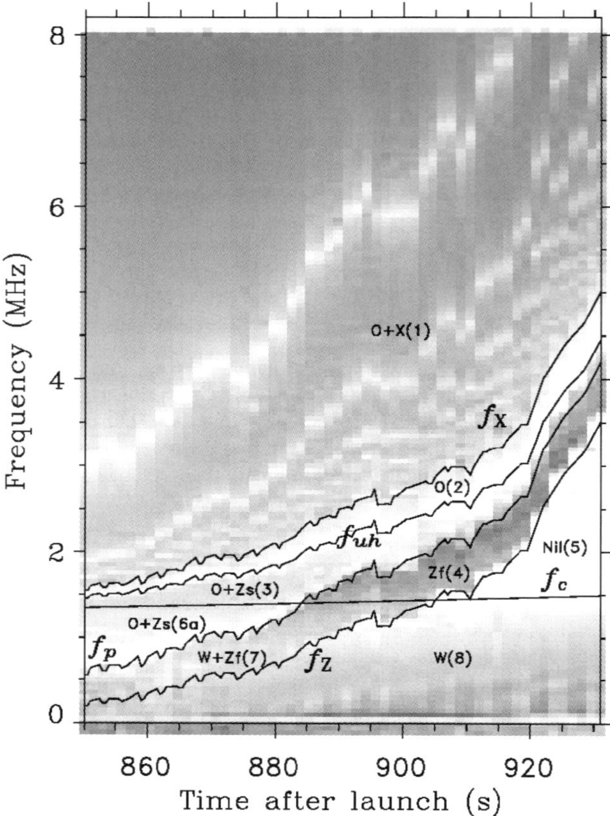

Figure 9. Survey showing strength of two-point propagation observed during the last 80 s of the OEDIPUS-C flight, in various regions of Clemmow-Mullaly-Allis (CMA) diagram. The regions are defined by the characteristic cold-plasma frequencies. Inside these boundaries are indicated the cold-plasma modes that propagate: O, X, Zs(slow Z), Zf(fast Z) and W(whistler), in each case followed by the CMA region number in parentheses. The white discontinuous streaks in CMA1 are Faraday-rotation fringes (adapted from *Horita and James*, [2004]).

The planned orbit of ePOP is an inclined elliptical orbit, with apogee at 1,500 km and perigee at 300 km. The initial inclination is approximately 80 degrees, and the orbital precession over the 1 year mission life (extendable) will allow the sampling of a wide range of latitudes. The eight ePOP science instruments feature an eclectic mix of in-situ and remote-sensing measurements, with reaches from global (>1000 km in horizontal size in the ionosphere), to meso (10-1000 km), and to small (<10 km) scales. The eight ePOP instruments are: an ion rapid imaging spectrometer (IRM, University of Calgary), a suprathermal electron imager (SEI, University of Calgary), a fast-auroral imager (FAI, University of Calgary), a radio receiver instrument (RRI, Communications Research Centre), a GPS receiver package (GPS, University of New Brunswick), a three-axis fluxgate magnetometer (MGF, Magnetometrics, Ottawa), a neutral mass spectrometer (NMS, Institute of Space and Astronautical Science, Japan), and a Coherent EM radiation tomography experiment (CER, Naval Research Laboratory, US).

We briefly describe the multiscale science envisaged under the ePOP experiment.

Small-Scale Science. Previously we discussed the importance of interface dynamics between the dense atmosphere and comparatively tenuous magnetosphere, and the apparently unstable nature of this interface which resulted in a rich variety of small-scale structures detected, for example by DE-2 and Freja. ePOP will represent the next level in our measurement of nonlinear waves permeating the GTR, and the wave-particle interactions which account for much of the particle dynamics, including ion outflow. A key concern is how the initial acceleration of ions to the escape speed is effected. Because of ePOP's orbit and, more importantly, the high-resolution and high-bandwidth data optimized for the interplay between waves and particles in phase space, we expect new understanding to emerge.

At the heart of ePOP's charged particle experiment are the Imaging and Rapid-scanning Mass spectrometer (IRM) and Suprathermal Electron Imager (SEI). IRM is an exemplar of the 3D strand of the original hemispherical electrostatic analyzer. SEI is the same SII discussed earlier, but biased to collect electrons rather than ions. Both instruments are capable of forming two-dimensional particle distribution functions in 10 ms, which, for the typical spacecraft speed, translates into a spatial resolution of less than 100 m. In comparison, the gyro-diameter of a 10-eV oxygen ion is about 80 m; therefore, IRM will be able to provide the first ever inside look of heavy ions' response to waves on kinematic scales. Because of its lighter mass, electrons are mostly fluid-like in their behavior, but it is known that electron inertial length, $\lambda_e = c / \omega_{pe}$, is a key scale parameter in the formation of parallel electric field in an Alfvén wave [*Wei et al.,* 1993]. The electron inertial length is 150 m for a $n = 1000$ cm^{-3} plasma.

The ePOP radio receiver instrument (RRI) can operate at frequencies between the ULF and HF bands, inclusive. It will be commanded into modes for characterizing ion and plasma waves. For example, the oxygen gyrofrequency in the GTR is of the order of 0.1 kHz, and the lower-hybrid resonance frequency for an O+ dominated topside plasma is about 10 kHz. The RRI thus has the capacity to sample several types of plasma waves known to interact strongly with ions and electrons.

Meso-scale Science. Meso scale, the gray area between large and small, has a subjective element in its definition. In this article, we take the range of meso scale to be between the kinematic scale of heavy ions (100 m) and overall size of the system (1000 km, at GTR altitude). Within this range there is a central division, approximately 1-100 km, where some of the most important meso-scale processes take place. The exact numbers notwithstanding, meso-scale is the pathway through which energy first generated through large-scale MHD coupling is funneled and ultimately dissipated in small-scale interactions (whether it is wave-particle interactions or classical collisions in the atmosphere). Many of the known drivers of GTR dynamics are unmistakably meso-scale in nature. For example, the field-line resonance, with its latitudinal width between 10 and 100 km, is a quintessential meso scale process, resulting, on the one hand, from the concentration of large-scale waves into a narrow shell of resonant oscillation of field lines, and leading, on the other hand, to further nonlinear kinetic modification of incident Alfvén waves on the electron inertial scale [*Wei et al.,* 1993]. From the observational point of view, the lateral width of discrete auroral arcs often preceding substorm onsets has the same order of magnitude as the FLR width, suggesting a possible link between the two phenomena.

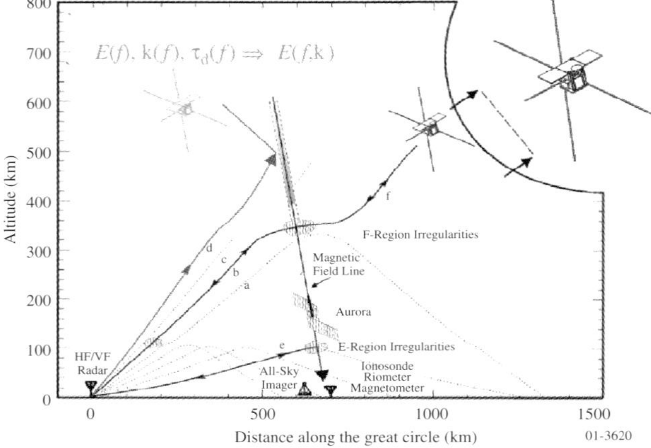

Figure 10. The HF (transionospheric) part of the ePOP RRI experiment. Radio waves from ground transmitters undergo severe deflections by meso-scale plasma structures. A space-based radio receiver observes the integrated change. Theory and inversion techniques can help determine the underlying structures.

The ePOP instruments with primarily a meso-scale focus are the fast auroral imager (FAI), the HF band of the RRI, and the CER beacon. FAI will image in the visible and infrared range. The FAI-SV camera will zoom in on the 630 nm line, with a cadence of 1 minute; the FAI-SI camera will sample the NIR band between 650 and 900 nm, at 10 images per second. This combination gives the experimenter both a quantitative reference (with respect to the well studied 630 nm red line) and a fast sampler of morphology at an adjacent band. The low altitude of the ePOP satellite allows FAI to have a spatial resolution of 4.8 km at apogee, making it possible to see how meso-scale processes (at least those with auroral signatures) feed and influence small-scale wave-particle interactions. Second, the RRI-HF will conduct the classical transionosphere propagation study, in conjunction with SuperDARN and CADI transmitters on the ground (Figure 10). The overflying ePOP will collect HF signals bent by underlying plasma irregularities, and through inversion, give a crude map of electron density irregularities of the size 100 km or less. The CER experiment is largely the RRI-HF in reverse, with the transmitter (beacon) on the spacecraft, and receivers on the ground. The CER frequencies are in or close to the GHz range, and the rays are not as drastically bent as the HF rays shown in Figure 9. This gives ePOP a differential capability important for calibration and quantitative inversion.

Large-scale Science. The ePOP radio experiment has a large-scale component in the GPS occultation instrument GAP (4 patch plus 1 helical antennas). By intercepting GPS signals refracted by the limb ionosphere, the GAP experiment can provide large-scale (1000's km) information on how the total electron content responds to magnetospheric perturbations. We discuss the neutral mass and velocity spectrometer (NMS) under the large-scale category, although the relevant physics are meso and small-scale in origin. It has been known for years that ion drag can be a significant driver of themospheric wind, able to accelerate neutrals to a speed comparable to ionospheric convection. It is logical to ask whether this interaction applies to field-aligned motion to the same extent; that is, whether the outflowing ions can drag atmospheric neutrals into escaping orbits. This speculation will be tested for the first time through ePOP NMS. If the positive case is established, the enhanced neutral loss could have significant implications on Earth's atmospheric evolution over the planetary timescale. Even if the negative were established, the ePOP NMS experiment could provide a useful comparative benchmark and important data to extrapolate to planets with a weaker magnetic field or lower escape speed (e.g., Mars), which might be more susceptible to ion-neutral interactions leading to atmospheric loss.

ePOP/CASSIOPE as a single low-Earth orbit satellite has its intrinsic limitations in multiscale observations. However, through the above discussion, we showed that a thoughtful mixing of instruments can go a long way toward overcoming the limitation.

3. International Collaborations

Canada's resource base is not enough to sustain an experimental space physics program that is entirely national in character. Besides this factor of necessity, an international outlook has helped Canada broaden its perspective and enlarge its opportunities. Therefore, international collaboration remains a dominant theme in the Canadian STS program. Here we discuss a few near-term opportunities.

4.1 THEMIS

The NASA Middle-Class Explorer THEMIS (Time History of Events and Macroscale Interations during Substorms) is our latest attempt, and best hope as well, to put the substorm controversy to rest. With five satellites aligning in the midnight sector every four days, the constellation is expected to encounter more than 100 substorms over the 2-year mission time. Each spacecraft, carrying an electric field instrument, a search coil and a fluxgate magnetometer, an electrostatic analyzer, and a solid-state particle telescope, will capture a range of substorm signatures as they run past the aligned satellites. By carefully comparing the timing of distinct substorm events, THEMIS will establish the causal relationship that has been at the heart of substorm controversy for decades.

For all its sweeping vision and originality, the THEMIS alignment strategy exposes itself to a potential ambiguity, as magnetospheric processes are not required to be colinear, i.e., if the cause and effect do not occur on the same line, the timing would be subject to an uncontrolled error. Context definition through ground-based observations is therefore a critical mission objective, so that the level of uncertainty can be reduced or eliminated. Four Canadian scientists have been invited to become THEMIS co-I's to lend Canada's ground-based observations and experiences in making and interpreting such observations. The orbital phasing strategy of THEMIS does not only call for the spacecraft to align every four days but the aligned spacecraft must be in magnetic-longitude conjunction with the CANOPUS Churchill line. In addition to making the entire CGSM capability available to THEMIS, the Canadian contribution will also involve the deployment of additional instruments. THEMIS mission has a specific requirement to resolve the onset timing to within 1 second, which is beyond the capability of multi-wavelength ASIs in the CGSM array. THEMIS will acquire 20 high-cadence white-light ASI and position them in Canada and Alaska (see map in Figure 11). Canadian scientists will deploy and operate the 16 ASIs in Canada, and have been participating in the design and testing of the THEMIS GBO since Phase A. Also shown in Figure 11 is the mapped field of view of THEMIS ASIs in the magnetospheric equatorial plane (through the T89 field model), which clearly shows the context-defining power of the THEMIS ASIs. Coupling this with the wave characterization by the CGSM

502

magnetometers (plus a number of new THEMIS magnetometers to fill the CGSM gap), the ground-based THEMIS component will be an indispensable component in the overall THEMIS data.

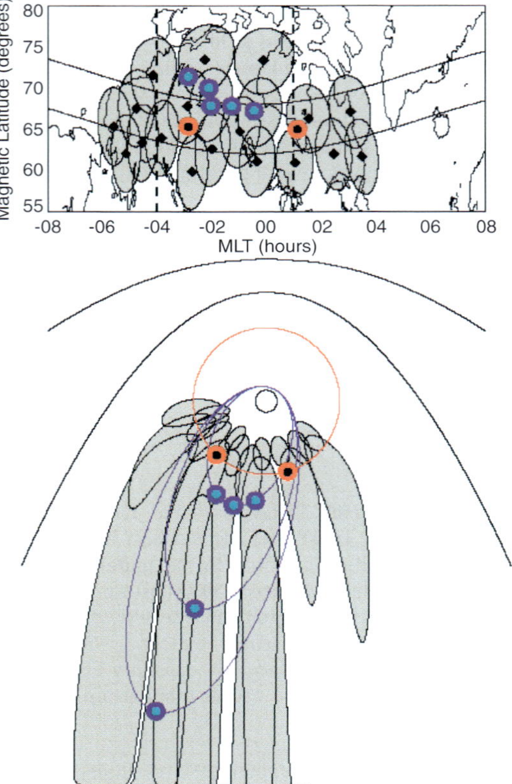

Figure 11. THEMIS (blue) and GOES 8 and 10 (red), in the context of the THEMIS ASIs to be deployed in Canada.

4.2. SWARM

SWARM is an Earth Explorer mission recently selected for development by the Earth Observation Program, European Space Agency. The primary objective of SWARM as an Earth Observation mission is short and intermediate-term geomagnetic field variations. There has been much publicity around the recent reports that the dipolar magnetic field strength is undergoing an accelerating decrease and the Northern Magnetic Pole has drifted several hundred kilometers over the past decades [*Newitt et al.,* 2001]. Whether these changes are due to long-term trends in the geodynamo (which reverses direction once every 200,000 years or so) or fluctuations associated with core-crust interactions (e.g., geomagnetic jerks) is an open question. SWARM, by focusing on the short- and intermediate-term changes, would provide more definitive information to help answer this question.

SWARM must distinguish magnetic fields of the terrestrial origin from those due to ionospheric and magnetospheric currents. The SWARM plan calls for a 3-spacecraft constellation flying in two altitudes at 450 and 500 km, respectively. Each spacecraft would carry a vector fluxgate and scalar magnetometer (the latter for in-flight calibration), an electric field instrument, an accelerometer, and a GPS receiver, the satellites will survey the global geomagnetic field for four years. As an "electromagnetometric" mission, the requirements on measurement accuracy and resolution exceed the current state of the art by an order of magnitude, making the detection of previously unseen small-scale or subtle features possible.

Canada would contribute the electric field instrument onboard SWARM, based on the imaging ion spectrometer techniques discussed earlier. Although the limit in telemetry would not allow the full power of an imaging particle detector to be utilized (i.e., full 2D ion distributions would be telemetered at one image every few minutes, rather than the 10 ms sampling capability), the electric field (through the proxy of ion drift) data would be the first from not one but three 3-axis stabilized satellites. This in turn would provide precise information on multiscale structures hitherto unseen in field-aligned currents, Poynting flux, field-aligned electric field, and nonlinear evolution of Alfvén waves.

4.3. AMISR

The Advanced Modular Incoherent Scatter Radar (AMISR) is a major $44 M upper atmospheric research facility funded by the US National Science Foundation in August 2003. Featuring a novel modular design, AMISR offers the scientists two advantages: relocatability to different locations around the globe to facilitate comparative studies, and remote steering and control to allow scientists to zoom in on interesting auroral and ionospheric features in real time.

Two of the three gigantic AMISR "faces" (each 32 m2 in size, with 128 modular panels) will be built at Resolute Bay, Canada, where a host of Canadian and international ground instruments already operate, making it one of the best locations to study the underexplored polar cap region. Many polar cap phenomena are direct signatures of solar wind-magnetosphere interaction, and proxies of magnetospheric topology (theta auroral, auroral patches, and traveling vortices); the debate on convection topology under northward IMF conditions is seen by many as a litmus test of competing theories of dayside reconnection. However, the absence of an incoherent radar facility in the polar cap has limited the study of how these large- and meso-scale structures cascade to small scales. AMISR will fill this gap by providing data on how ionospheric instabilities can be excited by external forcing and how the resulting wave-particle interactions may control polar wind dynamics, among other potential effects.

Canada's involvements in AMISR is still developing. The poleward extension of the HF radar network, appropriately named PolarDARN, is one such possibility. The complementarity between PolarDARN and AMISR can be appreciated by the following consideration. AMISR is a powerful instrument to probe ionospheric waves and instabilities on the meso and small scales; however, its limited field of view does not allow it to see the context in which these processes occur. For example, not knowing whether the convection is four-cell or two-cell limits one's ability to connect ionospheric observations to magnetospheric reconnection and convection. PolarDARN is an ideal complement in giving the large-scale information, much like the complementary value of THEMIS GBO to the overall mission. Other potential Canadian support instruments include Canadian Advanced Digital Ionosonde (CADI) and optical instrumentation (including potentially THEMIS-type imagers) placed in strategic locations to allow stereoscopic imaging of auroral features underlying AMISR observations.

5. Future Plan

While the business analogy is not necessarily apt, one can view space science as a competition of ideas and products; in the space science business so defined, a great many ideas compete for a very scarce resource, flight opportunity. While experiments in space are not an end-all or an end in itself, they are a necessary condition for discovery and knowledge. A recipe for success, which may be teasingly called Newton's fourth law, can be phrased as follows: action is equal to proaction squared. The Canadian community is taking proactive steps to shape its future. We discuss some examples.

5.1 Advanced instrument development

Canada's reputation in space was earned in large part through our historical emphasis on instrument development. Even during the difficult period of the mid 90's when the Canadian community underwent a major structural dislocation and funding crisis, this activity was kept up (leading to the SII and TPA/IRM). Today, as our program recovers, the need to anticipate and, to a degree, set the international trend for the future has become an important consideration.

In recently years, the Canadian Space Agency has started a series of Concept Studies, Advanced Studies, and Small Payload test flights, with a view toward replenishing our instrument bank account. The Suprathermal Ion Imager was one of the success stories. As we look forward, we see some exciting opportunities lurking over the horizon.

A derivative benefit of ePOP is low-cost, low-mass GPS occultation instruments built with commercial parts – so far the device built with the Novatel circuit boards have withstood some tough irradiation tests. The GPS occultation technique is well-known, but the cost of a space-qualified GPS at \$2 M per unit has been an inhibiting factor to populating the near-Earth space with such devices for tomographic studies. On the VLF side, Canadian scientists are looking into the possibility of using LEO satellites to emit powerful radio waves into the magnetosphere to allow both fundamental research on plasma wave theory using active techniques, as well as remediation of a "contaminated" radiation belt.

Canada's traditional strength in imaging geospace is being renewed. The IMAGE mission has shown the power and value of having simultaneous global images at different wavelengths. Opportunities in magnetospheric imaging are not exhausted, however. There are ways to improve instrument sensitivity to look farther into the plasma sheet, to use field-widened techniques to view a larger portion of space, to improve detector sensitivity and hence data quality, and to explore new imager complement to yield quantitative information on particle dynamics in the magnetosphere.

As one of the countries most affected by space weather, Canada has a keen interest in energetic particles in geospace. As part of a proposed Canadian radiation belt mission (to be discussed below), Canadian scientists are studying a new design of energetic proton telescope able to capture protons accelerated internally in the magnetosphere and of the SEP origin in one single spectrum. On the applied side, a Canadian company is investigating the transition of a highly sensitive radiation dosimeter flown as an experiment on ISS to monitor astronauts' radiation exposure to a minimum-resource (a fraction of 1 W, 100's g) device suitable for flight on nano-satellite platforms, laying the groundwork for a global above-the-atmosphere radiation monitoring network.

The ePOP neutral particle spectrometer has a fairly simple design suitable mainly for the deduction of bulk properties. In the future, as our interest in neutral-ion interaction becomes more nuanced, better instrument resolution is needed, in order to see the exchange of energy and momentum on a kinetic level. To meet this anticipated requirement, a next-generation neutral particle analyzer, similar to the SMS in some ways, is being studied in Canada.

5.2 ORBITALS

Outer Radiation Belt Injection, Transport, Acceleration, Loss Satellite (ORTIBALS) has recently been funded by the Canadian Space Agency for a Concept Study. Responding to the growing international interest in the origin and dynamics of MeV particles in the radiation belt, ORBITALS is a natural offshoot of the recent interest and contribution of Canadian scientists in this field [Rostoker et al., 1998; Summers and Ma, 2000; Liu et al., 1999; Mathie and Mann, 2000]. ORBITALS is concerned with the following problems: 1. How are relativistic electrons accelerated, transported and subsequently lost in the inner magnetosphere? 2. What controls the dynamics of the ring current? 3. How does the colder plasmasphere interact with the more energetic radiation belt populations, (e.g., the generation of plasma hiss at the plasmapause)?

ORBITALS proposes to carry a magnetometer package (search coil and fluxgate, University of Alberta), an electric field instrument (University of Minnesota), a low-energy ion analyzer (SII, University of Calgary), an energetic particle detector (Aerospace Corporation), and a solid-state particle telescope (University of Alberta). A potential Canadian contribution to restore the original vision of a three-petaled radiation belt storm probes, ORBITALS could be part of an internationalized effort to fly a coordinated multisatellite mission in the inner magnetosphere for the first time.

5.3 RAVENS

RAVENS is another Concept Study funded by the Canadian Space Agency. Deriving its name from the Norse legend in which two ravens, Hugin and Munin, fly around Earth as the informants for the Norse god Odin, Ravens is a idea whose simplicity makes its realization that much overdue. By proposing to place two microsatellites in a contra-posed (one at apogee when the other is at perigee) polar orbit around Earth, RAVENS aims to achieve 24/7 coverage of the northern auroral ionosphere for the first time, and in so doing, allows the complete imaging of a geomagnetic storm from start to finish.

The strawman instrument package consists of a FUV spectrograph imager flown on IMAGE (a contribution from the University of Liége, Belgium) and a two-camera UV imager (University of Calgary) – one imaging in the 140-160 nm band affected by O_2 absorption and another in the 170-200 nm range, where the absorption is weak. This differential technique can be used to derive quantitative, as well as morphological information, on auroras, such the energy of precipitating particles. While a novel mission concept in its own right, the tremendous synergy of RAVENS with upcoming ILWS missions such as MMS and THEMIS has been widely noted.

6. Conclusion: Toward A Canadian Geospace Superconstellation

With Cluster, and soon THEMIS, MMS, and SWARM, satellite constellations have become commonplace in our terminology and a method of choice for experimental scientists. On a working level, all constellations are not the same. There are the classical vector (tetrahedral) constellations (Cluster and MMS), phase constellations (THEMIS), survey constellations (SWARM), and "wolfpack" constellations (MagCon). If one relaxes the definition, the occultation techniques discussed in connection with the ePOP GPS experiment might be said to have formed a parasite constellation (ePOP the pathogen and its various GPS satellite hosts). However, despite the many shapes and forms of constellations, a niche has not been fully developed – use constellation techniques to study the coupled magnetosphere-ionosphere system. While ISTP was clearly a pioneering attempt to tackle the problem, the underlying satellites were not constellation-based. For lack of a better term, we call this missing link the superconstellation.

Luck has it that Canada is poised to participate in nine satellites, ePOP/CASSIOPE, THEMIS and SWARM. Given the tough financial situations facing the solar-terrestrial community, Canadian scientists are extremely fortunate that all three missions have been confirmed for full implementation, and all will fly in the same time window (2007-2008). Luck has it further that the orbits and scientific objectives of these satellites are highly complementary and allow us to conceive a superconstellation for coordinated observations of the coupled magnetosphere-ionosphere system.

Imagine a conjunction of the nine satellites near the midnight meridian during a magnetospheric substorm. THEMIS will capture the sequence of energy transport and intensity of response in equatorial plane, with quantitative information on the flow, particle distribution, and fields between 10 and 30 R. With GOES 8 and 10, the distance range extends to geostationary orbit, and with THEMIS GBO, the in-situ measurements are set in a global context. SWARM, on the other hand, will provide a detailed picture of the electrodynamical response of the ionosphere to the magnetospheric perturbations, including multiscale structures of the Poynting flux and field-aligned current just above the ionosphere (which, incidentally, are more difficult to measure on a single satellite). ePOP, finally, will shuttle between THEMIS and SWARM, characterizing in-situ the plasma wave environment and how the resulting wave-particle interactions affect the physical coupling between the ionosphere and magnetosphere and the repopulation of the plasmasphere, ring current, and plasma sheet. Although coordinated satellite missions are not a novel concept (the latest being the ISTP fleet), a superconstellation based on constellations as building blocks is, however, a new reality. It is therefore almost a certainty that the THEMIS-ePOP-SWARM superconstellation will enable a qualitative leap in our ability to characterize the quantitative response and feedback of one geospace region on another, because of superior resolution, larger field of view, and orbital complementarity. Specifically, we anticipate progress in the following questions: How do magnetospheric substorms intensify the field-aligned current system and create multiscale current filaments/Poynting flux bursts? How does the ionosphere respond to the magnetosphere quantitative throughout the scale hierarchy? How does an excited ionosphere, as a source of plasma waves and outflowing ions, back-react on the magnetosphere?

Our optimism is further buttressed by the much improved ground observatories in the northern hemisphere, particularly in Canada. The expanded CGSM, THEMIS GBO array, AMISR, and the likely expansion of the SuperDARN network both equatorward (StormDARN) and poleward (PolarDARN) will give the ground-based scientists an unprecedented tool, and associated with it, power to frame multiscale coupling questions to influence the future of space missions. Canada's experience and perspective in this respect are unique, and the presence of a geospace superconstellation will give us the opportunity to experiment new ways to integrate ground observations with the upcoming space missions.

Although superconstellation as a concept is a significant step on the ladder of space observation techniques and holds much promise, we are cognizant of the increased demand this concept places on the accuracy of our theory and computational models. If the primary goal of a superconstellation is to characterize quantitatively the response and interactions between various subsystems of geospace, it entails the use of computational models as an interpretative tool to an extent that has not yet become the norm. Our previous discussion of quantitative magnetoseismology demonstrates the point. To relate information across scales and to relate data between spatially distant points of observation, first-principle models are imperative. While it is the nature of our science that from time to time, we may miss a turn or stumble on the path to knowledge, the general principle remains that empirical knowledge, as a circuitous pun, is unprincipled knowledge. Over the next decade, we expect in many ways an unprecedented geospace dataset to emerge from Canada; it is crucial that the theory and modeling community take maximum advantage of the situation. We close with a commentary on some general theoretical challenges we face and how we may use the emerging dataset toward their solution.

1. The mesoscale is the most critical and least understood in the multiscale family. The deficiencies of MHD models need to remedied on this scale, as we incorporate nonideal nonlocal effects into our formalism. Advance in this field will very much depend on our success to make coordinated optical, radio, and magnetic observations utilizing the multiscale potential of ground-based stations;
2. Data assimilation, as a corollary, easy access by the geospace community to assimilated data products with increasing sophistication and self-consistency, will become a necessary condition, either as simulation driver or checks in knowledge validation, or even tweakers and nudgers in an auto-correcting algorithm;
3. Superconstellations of satellites will present new challenges to our theory. In essence, to achieve quantitative knowledge in a massively underdetermined system is a test of our ingenuity. It would be highly beneficial that theorists be included in the planning of future superconstellations, so that the development of computer models can be optimized in the context of real needs of real missions and, in so doing, increase the likelihood of closure of problems;
4. The geospace transition region will rise in importance as a front-edge theoretical problem, especially as we explore the electrodynamical connections of geospace to the mesosphere and the terrestrial electric circuit further down. The question whether or not solar variations mediated through the magnetosphere have any major impact on Earth's weather and climate will command more attention;
5. Knowledge of multiscale coupling gained through the study of geospace will make inroads in helping solar physicists and astrophysicists understand how plasma processes shape the objects of their interest;
6. The controversy on substorms will continue to rage, defying aging and other evidence of time.

Our last prognostication, given in good fun, is nonetheless a truism in science. We relish the prospect that more unknowns will be unearthed, the more we know, a prospect that will keep us employed, in more than one sense of the word.

References

Burchill, J. K., D. J. Knudsen, B. J. J. Bock, R. F. Pfaff, D. D. Wallis, J. H. Clemmons, S. R. Bounds, and H. Stenbaek-Nielsen, Core ion interactions with BB ELF, lower hybrid, and Alfvén waves in the high-latitude topside ionosphere, J. Geophys. Res., in press, 2004.

508

Cully, C. M., E. F. Donovan, A. W. Yau, and H. J. Opgenoorth, Supply of thermal ionospheric ions to the central plasma sheet, J. Geophys. Res., 108, doi:10.1029/2002JA009457, 2003.

Horita, R., and H. G. James, Two-point studies of fast Z-mode waves from dipoles in the ionosphere, in press, Radio Sci., 2004.

James, H. G., Electromagnetic whistler mode radiation from a dipole in the magnetosphere, Radio Sci., 38, 1009, doi:10.1029/2002RS002609, 2003.

Knudsen, D. J., J. K. Burchill, K. Berg, T. Cameron, G. A. Enno, C. G. Marellus, E. P. King, and I.menyWevers, A low-energy charged particle distribution imager with a compact sensor for space applications, Rev. Sci. Instr., 74, 202, 2003.

Liu, W. W., G. Rostoker, and D. N. Baker, Magnetic pumping acceleration of relativistic electrons by large-amplitude ULF pulsations, J. Geophys. Res., 104, 17391, 1999.

Lu, J. Y. R. Rankin, R. Marchand, V. T. Tihonchuk, and J. Wanliss, Finite elements modeling of nonlinear dispersive field line resonances: trapped shear Alfvén waves inside field aligned density structures, J. Geophys. Res., 108, 1394, doi:10.1029/2003JA010035, 2003.

Mathie, R. A., and I. R. Mann, On the solar wind control of Pc5 ULF pulsation power at mid-latitude: Implications for MeV electron acceleration in the outer radiation belt, J. Geophys. Res., 106, 29783, 2001.

Newitt, L. et al., Recent acceleration of the North Magnetic Pole linked to magnetic jerk, EOS Trans., 83, 381, American Geophysical Union, 2002.

Rankin, R., F. Fenrich, and V. T. Tihonchuk, Shear Alfvén waves on stretched magnetic field lines near midnight in Earth's magnetosphere, Geophys. Res. Lett., 27, 3265, 2000.

Samson, J. C., J. A. Jacobs, and G. Rostoker, Latitude-dependent characteristics of long-period geomagnetic micropulsations, J. Geophys. Res., 1971.

Samson, J. C., L. R. Lyons, P. T. Newell, F. Creutzberg, and B. Xu, Proton aurora and substorm intensifications, Geophys. Res. Lett., 19, 2167, 1992a.

Samson, J. C., D. D. Wallis, T. J. Hughes, F. Creutzberg, J. M. Ruohoniemi, and R. A. Greenwald, Substorm intensification and field line resonances in the nightside magnetosphere, J. Geophys. Res., 97, 8495, 1992b.

Spanswick, E., E. Donovan, W. Liu, D. Wallis, A. Aasnes, T. Hiebert, B. Jackel, and M. Henderson, Substorm associated spikes in high-energy particle precipitation, Physics and Modeling of the Inner Magnetosphere, AGU Book Series, in press, 2004.

Summers, D., and C.-Y. Ma, A model for generating relativistic electrons in the Earth's inner magnetosphere, J. Geophys. Res., 105, 2625, 2000.

Rostoker, G., S. Skone, and D. N. Baker, On the origin of relativistic electrons in the magnetosphere, Geophys. Res. Lett., 25, 3701, 1998.

Trondsen, T. S., and L. L. Cogger, High-resolution television observations of black aurora, J. Geophys. Res., 102, 363-378, 1997.

Trondsen, T. S., and L. L. Cogger Fine-Scale Optical Observations of Aurora, Phys. Chem. Earth (C), 26, No. 1-3, pp. 179-188, 2001.

Xu, B. L., J. C. Samson, W. W. Liu, F. Creutzburg, and T. J. Hughes, Observations of modulated auroral arcs by resonant Alfven waves, J. Geophys. Res., 98, 11531, 1993.

Wei, C.Q., J. C. Samson, R. Rankin, and P. Frycz, Electron inertial effects on geomagnetic field line resonances, J. Geophys. Res., 99, 11265, 1994.

Whalen, B. A., et al., The Freja F3C cold plasma analyzer, Space Sci. Rev., p. 541, 1994.

Yau, A. W., and M. Andre, Source processes in the high-latitude ionosphere, Space Sci. Rev., 80, 1-25, 1997.

Multiscale Coupling of Sun-Earth Processes

A.T.Y. Lui, Y. Kamide and G. Consolini, editors.

A CHINESE-EUROPEAN MULTISCALE MISSION: THE DOUBLE STAR PROGRAM

Z. X. Liu[1], P. Escoubet[2], J. B. Cao[1]

[1]*Center for Space Science and Applied Research, Beijing, 100080, China*
[2]*Space Science Department ESA/ESTEC, 2200 AG Noordwijk, The Netherlands*

Abstract. The Double Star Program (DSP) is a two-satellite space mission designed to (1) investigate physical processes of magnetospheric space storms, (2) establish physical models of magnetospheric space storms, and (3) develop dynamic models and prediction techniques of the near-Earth space environment. It is the first joint endeavor between the Chinese National Space Administration and the European Space Agency. The first satellite, launched in December 30, 2003, has an eccentric equatorial orbit and the second satellite will be launched to an eccentric polar orbit in July 2004. These two satellites will make measurements of the inner magnetosphere, dayside magnetopause, magnetosheath, bow shock, and near-Earth magnetotail. The orbits are synchronized with the four-satellite Cluster mission in local time sampling of magnetospheric regions, making this combination a six-point observing constellation in geospace.

Keywords. Magnetic storm, substorm, magnetotail, multiscale space mission.

1. Introduction

The modern space era with artificial satellites sent for direct measurements of the near-Earth space began in the early 1960's. In spite of the past four decades of satellite measurements of the Earth's magnetosphere, it is generally agreed that the Earth's magnetosphere is still sparsely sampled simply due to its enormous volume. This fact naturally poses a hindrance to achieve comprehensive understanding of many magnetospheric phenomena. Compounding to this hindrance is the increasing evidence that many challenging magnetospheric problems are associated with physical processes involving multiple spatial and/or temporal scales. There is tight coupling between microphysical phenomena with large-scale ones. Consequently, many magnetospheric investigations and space missions to date emphasize multi-point measurements. Achievement of multi-point measurements in space often calls for strenuous efforts and immense resources, which can be more efficiently and economically accomplished through international collaboration. Case in point is the Double Star Program (DSP) (*Liu et al.*, 2000), which is the first joint endeavor between the Chinese National Space Administration (CNSA) and the European Space Agency (ESA). This DSP involves two satellites and is designed to dovetail with the four-satellite Cluster mission (*Escoubet et al.*, 1997), complementing the Cluster mission with *in situ* measurements of the inner plasma sheet and equatorial dayside magnetopause as well as *in situ* measurements and remote sensing of the inner magnetosphere.

In this paper, we provide a brief overview of the DSP mission. A public brochure describing this program is available from the Center for Space Science and Applied Research. Its scientific objectives are outlined in Section 2. This is followed by a description of the spacecraft and their orbits in Section 3. The scientific payloads on these two satellites are then given in Section 4.

510

2. Scientific Objectives

As mentioned in the previous section, the DSP mission consists of two satellites, one at an eccentric equatorial orbit with an apogee of about 13.4 R_E and the other at a polar orbit with an apogee of about 6.0 RE. More details of these satellites and their orbits are given in the next section. With these two satellites, the DSP can detect the temporal-spatial variations of fields and particles in the near-Earth equatorial and polar active regions.

The main scientific objectives of DSP are as follows:
(1) To determine the multiple temporal-spatial scale driving processes of magnetospheric storms.
(2) To investigate the multiple temporal-spatial scale triggering processes of storms and substorms, physical model and prediction method of magnetospheric storms. In particular, the coordinated measurements of the equatorial satellite of DSP and Cluster on the night side magnetotail will possibly determine the spatial region of substorm onset and to some extent contribute to the evaluation of several candidate models of substorm trigger.
(3) To examine the dynamical coupling processes of the magnetosphere, ionosphere and thermosphere.

The equatorial satellite of DSP (TC-1 in China) with onboard particle and field instruments will detect the physical processes of magnetic storms and magnetospheric substorms in the near-Earth magnetotail as well as the energy transfer from the solar wind to the magnetosphere via the dayside magnetopause. The polar satellite of DSP, capable of making remote sensing observations as well, will detect energy transfer from the solar wind and near-Earth magnetotail to the polar ionosphere and upper atmosphere, including ionized particle transfer from the ionosphere to the magnetosphere.

The DSP mission is also designed to make coordinated observations with Cluster constellation by synchronizing the local time sector sampled by these two missions. The following coordinated exploration data of DSP and Cluster on the dayside, geotail and polar region can be used to investigate the scientific objectives above:

(1) Solar wind – magnetosheath – dayside magnetopause （see Figure 1）

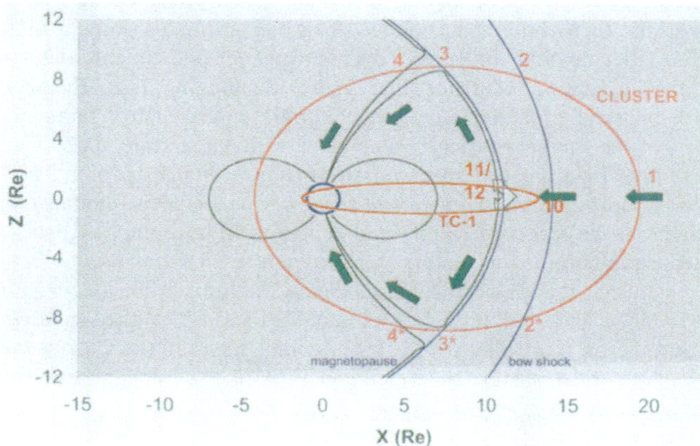

Figure 1. The schematic of space configuration of DSP and Cluster missions on the dayside

(2) Near-earth magnetotail plasma sheet – ring current region – radiation belt – plasmasphere (see Figure 2)

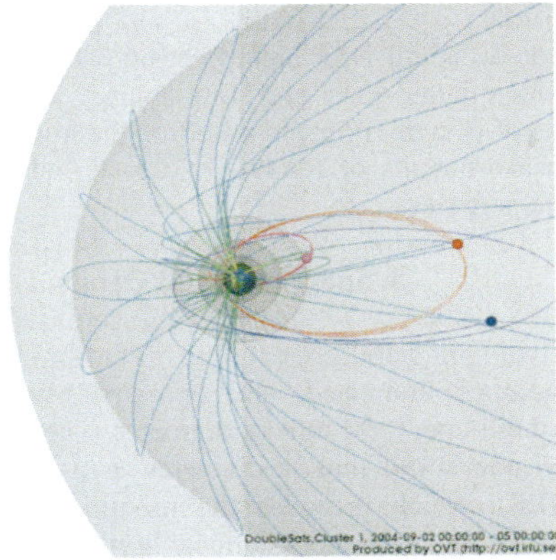

Figure 2. The schematic of space configuration of DSP and Cluster missions on the nightside (Cluster: blue and DSP: red)

(3) Cusp region, auroral acceleration region, and high latitude boundary layer region （see Figure 3）.

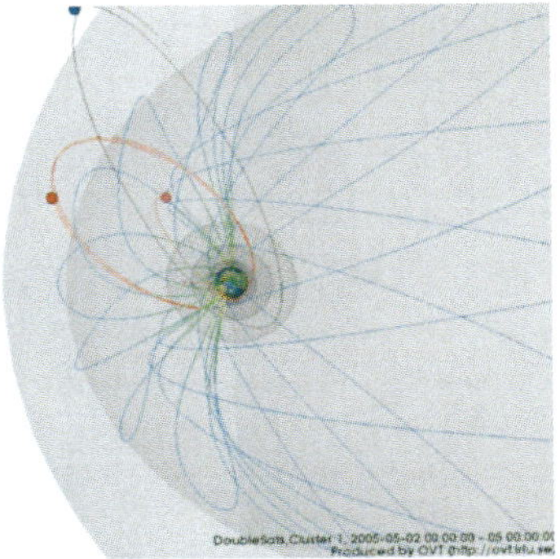

Figure 3. The schematic of space configuration of DSP and Cluster missions in the high latitude region (Cluster: blue and DSP: red).

Accordingly, the measurements of high-altitude polar cusp by Cluster will be complemented by DSP with measurements of mid- to low-altitude polar cusp at some occasions and with incoming solar wind parameters at other occasions. The measurements of dayside magnetopause by DSP-E (TC-1) will be complemented by Cluster with measurements at the dayside solar wind which is close to the bowshock. On the nightside, the measurements of mid-tail disturbances by Cluster will be complemented by DSP with measurements of the inner plasma sheet and inner magnetosphere. The combination of DSP and Cluster forms a six-point observing constellation in the near-Earth geospace environment.

512

In addition, the DSP is also coordinated with the ground-based observation program called "Meridian Chain Projection", which is a key scientific program to be implemented by the Chinese government.

The scientific payloads on board DSP include basic plasma measurements such as magnetic field, wave analyzer, low energy ions and low energy electrons, ensuring the successful coordinated measurements of two missions. In addition, the high energy particle measurements of DSP overlap with the measurements of RAPID of Cluster, allowing also possible coordinated measurements of energetic particles.

3. Spacecraft and Orbits

The DSP consists of two spin-stabilized satellites. The spin rate is planned to be 15±1.5 rpm. They are designed, developed, launched, and operated by the CNSA. Each spacecraft is cylindrical with 980 mm high and 1800 mm in diameter. The curved solar-array panels are 740 mm high. The central cylinder 1194 mm in diameter is inside the spacecraft body. The Main Equipment Platform (MEP), which provides the mounting area for most of the spacecraft subsystems, is fabricated as an aluminum-skinned honeycomb panel reinforced by an outer aluminum ring. There are six aluminum honeycomb braced panels symmetrically arranged on the upper surface of MEP internal to the central cylinder. Figure 4 show the DSP-1 satellite that was being tested.

Figure 4. DSP-1 Satellite in test

The first satellite, refereed to as DSP-1 (TC-1 in china), has an equatorial orbit. It was launched on December 30, 2003 by a Chinese Long March 2C rocket. The mass of scientific payloads is about 28.5 kg. The average power and maximum power of scientific payloads are 27.0 W and 30.7 W respectively. The DSP-1 satellite has achieved an orbit with an apogee geocentric distance of about 13.4 R_E, a perigee geocentric distance of about 6934 km, an inclination of about 28.2°, an orbital period of about 27.4 hrs, and a spin rate of about 15 rpm.

The second satellite, referred to as DSP-2, will be launched in July 2004 with a Chinese Long March 2C rocket also. The mass of scientific payloads is about 28.76 kg. The average power and maximum power of scientific payloads are 26.9 W and 30.6 W respectively. The plan is for it to be launched to a 700 × 39,000 km polar orbit, with 90° inclination, an orbital period of about 7.3 hrs, and a spin rate of 15±1.5 rpm.

Data reception will be undertaken jointly by Miyun Station of Beijing, Sheshan station of Shanghai, and Villafranca station of the European Space Agency.

4. Scientific Payloads

There are eight scientific instruments onboard the equatorial satellite DSP-1. They are Active Spacecraft Potential Control (ASPOC) (*Riedler et al.*, 1997), FluxGate Magnetometer (FGM) (*Balogh et al.*, 1997), Hot Ion Analyzer (HIA) (*Reme et al.*, 1997), Heavy Ion Detector (HID) (Liu et al., 2000), High Energy Electron Detector (HEED) (*Liu et al.*, 2000), High Energy Proton Detector (HEPD) (*Liu et al.*, 2000), Low Frequency Electromagnetic Wave Detector (which is comprised of Spatio-Temporal Analysis of Field Fluctuations (STAFF) and Digital Wave Processor (DWP)) (*Liu et al.*, 2000), and Plasma Electron and Current Experiment (PEACE) (*Johnstone et al.*, 2000). Further specification of these instruments is given in Table 1.

Table 1. The scientific payloads for DSP-1

Instrument (Acronym)	Range of Measurements	PI (Institution)
Active Spacecraft Potential Control (ASPOC)	Max. current: 50 – 80 μA	K. Torkar (IWF, Graz, Austria)
Fluxgate Magnetometer (FGM)	–65536 nT to 65536 nT	C. Carr (Imperial College, UK)
Plasma Electron and Current Experiment (PEACE)	1 eV – 1 keV; 30 eV – 26 keV	A. Fazakerley (MSSL, Dorking, UK)
Hot Ion Analyzer (HIA)	30 eV – 40 keV	H. Reme (CESR, Toulouse, France)
Spatio-Temporal Analysis of Field Fluctuations (STAFF) and Digital Wave Processor (DWP)	8 Hz – 4 kHz	N. Cornilleau (CETP, Velizy, France) and H. Alleyne (Sheffield Univ., UK)
High Energy Electron Detector (HEED)	200 keV – 10 MeV	W. Zhang (Engineering) and J. B. Cao (Science)(CSSAR, China)
High Energy Proton Detector (HEPD)	3 – 400 MeV	J. Liang (Engineering) and J. B. Cao (Science)(CSSAR, China)
Heavy Ion Detector (HID)	10 MeV (He) – 8 GeV (Fe)	Y. Zhai (Engineering) and J. B. Cao (Science) (CSSAR, China)

There are also eight scientific instruments for the polar satellite DSP-2 (see Table 2). Five of these instruments are identical to that for DSP-1. The three other instruments in the payloads of DSP-2 are the Neutral Atom Imager (NUADU) in place of the Active Spacecraft Potential Control instrument, LEID (Low Energy Ion Detector) in the place of HIA, and LFEW (Low Frequency Electromagnetic Wave Detector) in the place of STAFF.

514

Table 2. The scientific payloads for DSP-2

Instrument (Acronym)	Range of Measurements	PI (Institution)
Neutral Atom Imager (NUADU)	0.1 keV – 140 keV.	S. McKenna-Lawlor (Ireland University, Ireland).
Fluxgate Magnetometer (FGM)	–65536 nT to 65536 nT	T. Zhang (IWF, Graz, Austria)
Plasma Electron and Current Experiment (PEACE)	1 eV – 1 keV; 30 eV – 26 keV	A. Fazakerley (MSSL, Dorking, UK)
Low Energy Ion Detector (LEID)	60 eV – 40 keV	Q. Ren (Engineering) and J. B. Cao (Science)(CSSAR, China)
Low Frequency Electromagnetic Wave Detector(LFEW)	8 Hz – 10 kHz	Z. Wang (Engineering) and J. B. Cao (Science)(CSSAR, China)
High Energy Electron Detector (HEED)	200 keV – 10 MeV	W. Zhang (Engineering) and J. B. Cao (Science)(CSSAR, China)
High Energy Proton Detector (HEPD)	3 – 400 MeV	J. Liang (Engineering) and J. B. Cao (Science)(CSSAR, China)
Heavy Ion Detector (HID)	10 MeV (He) – 8 GeV (Fe)	Y. Zhai (Engineering) and J. B. Cao (Science) (CSSAR, China)

3. Future Outlook

The Chinese-European Double Star Program will play an important role in the development of space physics research and space technology. This space mission complements the Cluster mission and is coordinated with other ground-based observations as well. It is anticipated that scientific investigations based on these multi-point observations will bring significant progress in the understanding of geospace phenomena for the 23-cycle solar activities.

Acknowledgements

The authors are grateful to all staff affiliated with the Double Star Program for their dedicated and extraneous efforts, contributing immensely to the success of this space mission.

References

Balogh, A. et al., The Cluster magnetic field investigation, *Space Science Review*, 79, 1-2, 65-92, 1997.

Escoubet et al., Cluster-ccienec and mission overview, *Space Science Review*, 79, 11-32, 1997.

Johnstone, A. D. et al., PEACE: A plasma electron and current experiment, *Space Science Review*, 79, 351-398, 1997.

Liu et al., Report on the phase a study of Geospace Double Star Project (DSP), CSSAR, Beijing, China, 2000.

Reme et al., The Cluster Ion Spectrometry experiment, *Space Science Review*, 79, 303-350, 1997

Riedler, W. et al., Active spacecraft potential control, *Space Science Review*, 79, 271-302, 1997.

Author Index